P9-AFE-934

"*The Great Christ Comet* is a stunning book. Colin R. Nicholl develops a convincing case for what exactly the Star of Bethlehem was. The book reads like a detective novel, and while it is full of evidence, information, and argumentation, it is accessible and enjoyable to read. This work is now the definitive treatment of the subject. I highly recommend it."

J. P. Moreland, Distinguished Professor of Philosophy, Biola University; author, *The Soul: How We Know It's Real and Why It Matters*

"I am simply in awe of this book. *The Great Christ Comet* is an absolutely astonishing triumph of interdisciplinary scholarship so rarely seen and so tremendously illuminating as to merit bright comparison with the very celestial phenomenon it describes. Both lead us to the Manger and to the Great Poet within, whose syllables are the Moon and Sun and Stars."

Eric Metaxas, *New York Times* bestselling author of *Miracles* and *Bonhoeffer*

"The most comprehensive interdisciplinary synthesis of biblical and astronomical data yet produced. It is a remarkable feat that a biblical scholar has been able to master the scientific data at such a level of erudition. No discussion of the historicity of the Star of Bethlehem can afford to ignore this book."

Simon Gathercole, Senior Lecturer in New Testament, University of Cambridge; author, *Where Is Boasting?* and *The Preexistent Son*

"In this erudite, engrossing, and compelling book, Colin R. Nicholl painstakingly develops a new solution for the enduring mystery of the Star of Bethlehem, bringing together the biblical story and ancient descriptions of the sky with modern understandings of astronomy. Nicholl's argument—that the celestial visitor was actually a phenomenal comet that passed perilously close by Earth in 6 BC—is certain to be discussed and debated for years to come."

Duncan Steel, Visiting Astronomer, Armagh Observatory; Visiting Professor, University of Buckingham; author, *Eclipse* and *Marking Time*

"This is an amazing study. It reads like an absorbing detective story. Nicholl starts with a detailed reading of Matthew's account of the visit of the Magi. He makes the case, based on ancient and modern astronomy, that the star of Bethlehem was a great comet whose behavior in the sky would have been interpreted by ancient astrologers as announcing the birth of a Jewish Messiah. The depth and breadth of learning that Nicholl displays is prodigious and persuasive, and all future studies will have to take its proposals most seriously."

Gordon Wenham, Adjunct Professor of Old Testament, Trinity College, Bristol

"This is an outstanding book, quite breathtaking in the range of its scholarship, yet a page-turner in terms of its accessibility. Colin R. Nicholl is eminently followable, using detective skills to assess the biblical, historical, and astronomical evidence that lead him to conclude that the 'star' of Bethlehem was a comet. A real tour de force that I recommend unreservedly to a broad readership."

John C. Lennox, Professor of Mathematics, University of Oxford

"Colin R. Nicholl brilliantly tackles a subject that has been debated for centuries. *The Great Christ Comet* is a captivating book on the Star of Bethlehem. You will not be able to put this book down!"

Louie Giglio, Pastor, Passion City Church, Atlanta, Georgia; Founder, Passion Conferences

"Readers of this book will learn a lot of astronomy, history, and theology. Nicholl has produced a remarkable and fascinating book that combines the best of recent scientific scholarship with the best biblical scholarship. *The Great Christ Comet* is a model of the integration of science and Scripture, and presents a tightly reasoned and highly plausible argument that the Star was a comet. A terrific read!"

Donald A. Hagner, George Eldon Ladd Professor Emeritus of New Testament, Fuller Theological Seminary; author, *Matthew* (Word Biblical Commentary)

"Fascinating reading. Clearly the author has not only done his homework but has meticulously mined both quarries, theological and astronomical."

Paul L. Maier, Russell H. Siebert Professor of Ancient History, Western Michigan University

"Nicholl breaks important new ground in the quest for the historical Star of Bethlehem. Not only does he develop a formidable case for identifying the Star as a great comet; he also proposes a fresh explanation as to what it may have done to so impress the Magi. Nicholl has a clear understanding of the relevant areas of modern astronomy, and especially of the nature, evolution, and orbital dynamics of comets as currently understood. This work will be of great interest to astronomers, theologians, historians of science, and the general public, and will hopefully stimulate important new lines of scientific enquiry."

Mark E. Bailey MBE, Director, Armagh Observatory; coauthor, *The Origin of Comets*

"Colin R. Nicholl's magnum opus, which interprets Matthew's Nativity 'star' as a spectacular comet, is fascinating and illuminating. He supports his thesis by appealing to Babylonian, classical, and patristic texts as well as modern astronomical research on comets. His comprehensive mastery of the data enables him to present a detailed scenario of the Magi's initial sighting, subsequent observations, journey, and visit to the house in Bethlehem to view the newborn Christ child."

Edwin M. Yamauchi, Professor Emeritus of History, Miami University

"This is the only book I know of by a biblical scholar on the Star of Bethlehem. It is rooted in a detailed analysis of the biblical text and offers a comprehensive scientific explanation for the Star of Bethlehem. Nicholl makes a compelling case that the Star was a comet, supporting this conclusion with a mass of evidence from a variety of sources. I strongly recommend his work on one of the most fascinating biblical mysteries."

Sir Colin Humphreys, Professor and Director of Research, Department of Materials Science and Metallurgy, University of Cambridge; author, *The Miracles of Exodus*

"This rigorous and compelling book sets a new standard for the study of the Star of Bethlehem. No prior investigation of this mystery has brought the disciplines of biblical studies and astronomy together in such a clear, thoroughly researched, and decisive way. Nicholl lets us observe the skies with the Magi and walk with them all the way to the baby Jesus in Bethlehem. This richly illustrated and pleasantly accessible work is a must-read for everyone even vaguely interested in the Magi's Star. I enthusiastically recommend this eye-opening book!"

John J. Hartmann, former Assistant Lecturer of Greek, University of Cambridge; Pastor, New Reformation Church, St. Louis, Missouri

"Colin R. Nicholl offers an impressive case for understanding the Magi's star as a comet. He has produced a readable and beautifully illustrated introduction to relevant fields of astronomy, and has laid out pertinent historical data with proportion, care, and integrity. Based on detailed biblical study and current astronomical knowledge, Nicholl develops a fascinating reconstruction of the unprecedented events relating to the Star and the Magi."

John Nolland, Tutor in New Testament, Trinity College, Bristol; Visiting Professor, University of Bristol; author, *The Gospel of Matthew* (The New International Greek Testament Commentary)

"*The Great Christ Comet* is a significant new contribution to the long-running debate over the nature of the Star of Bethlehem. One of the book's many strengths is its critique of earlier, widely discussed hypotheses proposed to explain the Star. The book also explains the relevant astronomy very clearly at a level the general reader should have no trouble following. The case Nicholl makes for the Star being a great comet is certainly worthy of serious consideration."

Martin Gaskell, Department of Astronomy, University of California at Santa Cruz

"It is a real pleasure to commend *The Great Christ Comet* to everyone who has ever wondered what could possibly account for the appearance of the Star of Bethlehem. Few have expended as much earnest research, or written as clearly, on the astronomical basis for this special event as has Colin R. Nicholl. When you're reading this book, the pages turn rapidly—similar to the way the pages fly when you're engrossed in a mystery novel. All readers will be richly rewarded!"

Walter C. Kaiser, Jr., Colman M. Mockler Distinguished Professor Emeritus of Old Testament and President Emeritus, Gordon-Conwell Theological Seminary

The Great Christ Comet

Revealing the True Star of Bethlehem

COLIN R. NICHOLL

WHEATON, ILLINOIS

The Great Christ Comet: Revealing the True Star of Bethlehem

Published by Crossway
1300 Crescent Street
Wheaton, Illinois 60187

Cover design: Josh Dennis

Cover image: Sirscha Nicholl

First printing 2015

Printed in China

Single words or short phrases have been quoted from the American Standard Version (ASV), Common English Bible (CEB), Holman Christian Standard Bible (HCSB), International Standard Version (ISV), King James Version (KJV), NET Bible (NET), New American Standard Bible (NASB), New English Bible (NEB), Revised English Bible (REB), and Revised Version (RV).

In the Bible translations quoted, "Yahweh" is regularly substituted for "the LORD."

Italics in Bible translations have been added by the author.

Translations of Biblical passages, phrases, or words by the author are marked as such.

Hardcover ISBN: 978-1-4335-4213-8
ePub ISBN: 978-1-4335-4216-9
PDF ISBN: 978-1-4335-4214-5
Mobipocket ISBN: 978-1-4335-4215-2

Library of Congress Cataloging-in-Publication Data

Nicholl, Colin R., author.
The great Christ comet : revealing the true Star of Bethlehem / Colin R. Nicholl.
pages cm
Includes bibliographical references and index.
ISBN 978-1-4335-4213-8 (hardcover)
ISBN 978-1-4335-4216-9 (ePub)
ISBN 978-1-4335-4214-5 (PDF)
ISBN 978-1-4335-4215-2 (Mobipocket)
1. Comets. 2. Star of Bethlehem. 3. Astronomy in the Bible. 4. Bible. Matthew. I. Title.
QB724.N53 2015
226.2'085236—dc23 2014024971

Crossway is a publishing ministry of Good News Publishers.

OGP 25 24 23 22 21 20 19 18 17 16 15
15 14 13 12 11 10 9 8 7 6 5 4 3 2 1

To my parents, Drew and Florence Nicholl

Contents

Illustrations

Tables

Foreword

Since observing Comet Kohoutek in my senior year of high school, I have had a passion for comets. I have observed more than 300 comets and have written six volumes documenting recorded comets from the first millennium BC all the way through to modern times (*Cometography: A Catalog of Comets*, 6 vols. [Cambridge: Cambridge University Press, 1999-]).

I have also been interested in the Star of Bethlehem for many years and have been of the strong belief that it was a comet rather than something else like a planetary conjunction or a nova. I was pleasantly surprised when Colin Nicholl approached me on the subject in 2011, and I quickly became fascinated by his fresh approach to the topic and the new information he was contributing to the centuries-old debate about the identity of the Star. Colin was being aided by David Asher and Mark Bailey of the Armagh Observatory. I too was eager to assist him in his exciting research project. What Colin was doing with respect to the Star of Bethlehem reminded me of what historian John T. Ramsay and astronomer A. Lewis Licht had done with reference to Caesar's Comet of 44 BC.

Colin had developed a new case for the Star of Bethlehem being a comet and was proposing that Revelation 12:1–5, a Biblical text previously unharvested in discussions about the Star, revealed what the Magi had seen in the eastern sky that prompted them to travel westward to Judea. Having deduced that only a comet could do what this passage described, Colin had developed a timeline of the comet's appearances and movement across the sky. From this I recognized that a close approach to Earth was necessary, as well as an unusual Earth-comet-Sun geometry.

Over the following months Colin and I dialogued as he continued to study cometary astronomy. During that time Colin came to the sensible conclusion that the Christ Comet was a long-period comet, and he figured out that the comet must have orbited in a retrograde direction. As for me, I was feverishly working on the sixth volume of *Cometography* and struggling to find time to complete my orbital calculations based on Colin's data. Meanwhile, Colin took it upon himself to work out the orbit within the set parameters. He then presented me with his considered orbit and I found that it was in accord with what I had been concluding based on my own preliminary investigations.

Using my own research into the probable brightness of Halley's Comet in ancient times and the likely magnitude required for its discovery at each apparition, I was able to calculate brightness parameters for the Christ Comet based on Colin's orbit. This revealed that the Star became a spectacular object visible in broad daylight for a period of time. It was brightest at the time when it heliacally rose in the eastern sky in the aftermath of its perihelion. The comet's maximum brightness was reminiscent of what was observed for the Great September Comet of 1882 and Ikeya-Seki in 1965, the brightest comets in the last two centuries. 30 to 40 days after the comet switched to the western sky, the Christ Comet would have been a striking object in the southern evening sky, being visible from about 6 p.m. until it set upright in the west around midnight. The profile of the comet that emerges is impressive—it would have been the largest comet to come within Earth's orbit in recorded history and, as Colin claims, the greatest comet in history.

I was and remain excited about and fascinated by Colin's work on the Star of Bethlehem, and I very much enjoyed working on the project with him. I read through the majority of the manuscript and liked everything that I read. The research is solid and everything is explained very well. In every respect this volume is a remarkable achievement. I regard it as the most important book ever published on the Star of Bethlehem and enthusiastically commend it.

Gary W. Kronk

Preface

The Star of Bethlehem is one of the greatest mysteries in the Bible, in history, and in astronomy. What was the Star, and what precisely did it do that so deeply impressed the Magi? The modern academic quest to identify the Star can be traced back to Johannes Kepler in the seventeenth century. Regrettably, after four centuries of scholarly discussion, we are still far from a solution.

Part of the reason for the lack of any significant advance is that Biblical scholars, intimidated by the astronomical aspects of the task or preferring to think of the Magi's Star as miraculous or mythical, have given the debate a wide berth. If they mention the topic at all, their comments tend to be brief, superficial, and inaccurate.

As a Biblical scholar with a high regard for the historical credentials of the Scriptural text, I have felt constrained to leave my comfort zone and wade into the field of astronomy in order to follow up Biblical leads concerning the Star. Doing so has been stretching and uncomfortable at times, but has also been incredibly rewarding and invigorating. In this book I present the fruits of my research—what I believe is a decisive breakthrough in the quest for the historical Star of Bethlehem.

Any progress that I have made is in no small part due to help from others. In particular, I wish to express my profound gratitude to three astronomers who aided me greatly in my work from a very early stage.

First, a word of special thanks to David Asher, Research Fellow at the Armagh Observatory in Northern Ireland. David is one of the top astronomers in his field, respected the world over. How often I have coveted his brain! Mercifully, David graciously and sacrificially shared his with me, simply because he believes in the academic enterprise and was fascinated by my research. He made many complicated calculations for me, spent two whole days in conversation with me in Armagh, and wrote countless painstaking, long emails answering my questions.

Mark E. Bailey, MBE, Director of the Armagh Observatory, played an important part in guiding me in astronomical matters. Mark is always brimming with creative, fresh, and penetrating ideas, questions, and insights. He read over an early draft of part of the book and offered valuable feedback, and he took the time to answer my questions and keep me informed of pertinent developments in astronomy.

Gary W. Kronk, highly esteemed author of the authoritative six-volume *Cometography* (Cambridge University Press), offered extensive assistance, support, and encouragement. Gary read several drafts of many chapters of the book, made calculations for me, answered countless questions, and graciously agreed to write the foreword.

I shared my discoveries and research openly with each of these three scholars, and their engagements with me influenced my thinking at many points. As much as they helped me, however, they bear no responsibility for any failings that remain. Nor, of course, should it be assumed that they concur with all that I have written.

Two scholars assisted me by reading through an entire draft of the book and offering incisive criticism and counsel: John J. Hartmann, Pastor of New Reformation Church in St. Louis; and Paul L. Maier, Russell H. Siebert Professor of Ancient History at Western Michigan University.

Other academics were also extremely helpful, even when they did not know the details of the project I was working on. Andreas Kammerer, a highly esteemed German amateur astronomer; David W. Pankenier, Professor of Chinese at Lehigh University in Bethlehem, Pennsylvania; and Josefina Rodríguez Arribas, researcher at the Warburg Institute of the University of London, each graciously responded to a number of questions by email about comets, Imperial Chinese history, and medieval Jewish astrology, respectively. Professor Rodríguez Arribas also read through a few pages I wrote to check that my understanding of Jewish medieval astrology was correct. I am grateful to Peter Jenniskens, senior research scientist at the Carl Sagan Center of the SETI Institute and at NASA's Ames Research Center, for doing orbital calculations for me and answering my questions about meteors and meteor showers.

Thanks are due the following scholars for their assistance: Peter V. Bias, Professor of Business and Economics at Florida Southern College and author of *Meteors and Meteor Showers*; Paul C. Hewett of Corpus Christi College, Cambridge, and, until recently, Director of the Institute of Astronomy of the University of Cambridge; Chandra Wickramasinghe, Director of the Buckingham Centre for Astrobiology, University of Buckingham; Sir Colin J. Humphreys, former Goldsmiths' Professor of Materials Science at the University of Cambridge; and Roberta Olson, Curator of Drawings at New-York Historical Society. Thanks also to Roger MacFarlane and Paul Mills for graciously giving me prepublication access to relevant parts of their forthcoming work (the first of its kind) *Hipparchus' Commentary on the Phaenomena of Aratus and Eudoxus*.

Tyndale House Library (and, in particular, Ian Wilson) in Cambridge, England, provided an invaluable service by sending scanned versions of journal articles in Biblical studies that I was otherwise unable to obtain. Atlantic Productions generously sent me a complimentary DVD copy of the documentary "Star of Bethlehem: Behind the Myth" that they produced.

I am grateful to the makers of the following planetarium software programs: Project Pluto's Guide, Simulation Curriculum Corporation's Starry Night® Pro, and United Soft Media's Redshift. Calculations regarding comet brightness and size are based on Guide 9.0. Star and planet brightness values and the apparent positions of a comet in the sky are largely derived from Starry Night® Pro. Determinations of the location of planets and comets in space are based on Redshift 7. Thanks are due Seth Meyers of Simulation Curriculum Corporation for permission to make use of images from Starry Night® Pro.

Images of the planets in the illustrations are courtesy of NASA.

I am grateful to the many organizations and individuals who granted permission for the use of their images. I am also thankful to a number of artists and designers for guidance regarding illustrations and images: Ross Wilson, a leading Northern Irish artist; Christoph Kaiser of Christoph Kaiser LLC, a designer and architect; and Josh Dennis, Senior Vice President, Creative Department, at Crossway.

Special thanks must go to my literary agent, Mark Sweeney, and the whole team at Mark Sweeney & Associates for all their labors on behalf of this book.

I would also like to express my gratitude to Justin Taylor and his colleagues at Crossway for all their work on this project. As an act of kindness, Justin read the first four chapters at an early stage and offered valuable feedback, guidance, and encouragement. Little did he or I know that Crossway would end up being the publisher! Thanks also to Bill Deckard for his stellar editorial work.

How grateful I am to my daughters Gabriella and Evangelia. They were amazingly patient with Mommy and Daddy and readily made sacrifices so that this book could be finished. The girls also made sure that, no matter how stressed Daddy was, he regularly unwound. Gabriella was my astronomy buddy, learning with me and accompanying me on trips to see the Northern Lights, comets, planets, and meteor showers. Evangelia frequently thought she had spotted the Star of Bethlehem in the sky from the backseat of our car. Who knows? One day she might actually see it!

I am thankful for the backing of my family. In particular, I would like to express my gratitude to my brother Roy for his friendship, kindness, wise counsel, and enthusiastic support. In addition, my father-in-law engaged me in many conversations regarding the Star of Bethlehem and helped kindle in me an interest in the modern debate.

No one has done more to assist me in the writing of this book than my sacrificial, patient, and meticulous wife, Sirscha. She has strongly supported and encouraged me every step of the way and has done everything she could to free me up to do my research and writing. Moreover, I am particularly grateful to her for expending so much time and artistic energy to produce a beautiful set of illustrations for the book.

This book is dedicated to my father and mother, Drew and Florence Nicholl. I count it a great privilege to have them as my parents. Their love, faithfulness, prayers, and example have been a firm bedrock in my life. From my earliest years they have consistently got behind me as I have pursued my passion to study the Bible. Anything good that issues from my life may be credited to God and to them.

Colin Nicholl
North Coast, Northern Ireland
SDG

1

"Star of Wonder"

Introducing the Bethlehem Star

The Bethlehem Star is, without doubt, the most famous and celebrated astronomical entity in history. No other celestial object captures the attention of the world like it, particularly at Christmastime. This enormous fascination is found among Christians and non-Christians, young and old, and most peoples of the world.

THE STAR IN THE MODERN WORLD

Every December, planetariums and television channels put on special shows to discuss the Star, keenly aware that few, if any, astronomical issues or Biblical mysteries hold the spell over the popular imagination that the Magi's celestial phenomenon does. Astronomers, even some who scarcely believe in God, give public lectures on it.

Each Christmas the Star features prominently in our celebrations—it is often found as the crowning glory of Christmas trees, embossed on Christmas cards, or perched loftily over the sets of nativity plays. Many of our Christmas carols mention the Star, and a good number are focused on it. Among the many is "We Three Kings of Orient Are." In it we actually join one of the Magi in addressing the Star:

> Star of wonder, star of light,
> Star with royal beauty bright,
> Westward leading, still proceeding,
> Guide us to thy perfect light.[1]

This love for the Star is apparent all across the globe, but in some places the Magi's Star enjoys special distinction.

Understandably, in few locations is the Star more celebrated than in the town of Bethlehem in the West Bank. There pilgrims can stay at the Bethlehem Star Hotel, do some shopping at the annual Christmas Market on Star Street, have a coffee at Stars & Bucks Café, and visit Manger Square, where images of the Bethlehem Star abound, none more striking than the massive illuminated comet set up on a pole. On Christmas Eve the annual procession to the Church of the Nativity culminates at a grotto (cave), where a large silver 14-pointed star marks the spot where, it is claimed, Jesus was born.

Bethlehem, Pennsylvania, known as

[1] "We Three Kings," by John Henry Hopkins (1792–1868).

"Christmas City, USA," relishes in the stellar associations of its name. The most prominent manifestation of this is a spectacular 90-foot-high, 8-rayed star, illuminated by 250 bulbs, on South Mountain, that stunningly beams out over the city and is visible 20 miles away.[2]

Similarly, Palmer Lake in Colorado boasts a 500-foot-tall, 5-pointed, incandescent Star of Bethlehem on the side of Sundance Mountain.[3]

Perhaps no people celebrates the Star of Bethlehem more than the Poles. To them Christmas is known as "Little Star." Festivities formally commence around sunset on Christmas Eve, when the first star is spotted and is called "the Star of Bethlehem." The Poles enjoy a "Star Supper," during which a "heavenly Star" cookie might be served. Then, according to the tradition practiced in many parts of Poland, the "Star Man," the Santa-like gift giver in Poland who represents "the Little Star" itself,[4] appears, bearing presents from "Star Land."[5] He is accompanied by the "Star Boys," carollers dressed up as the Magi or other characters from the Christmas Story, and who carry a Star lantern.[6]

In parts of western Alaska, Orthodox believers with a Ukrainian heritage practice the yuletide tradition of "Starring"—carollers spin brightly decorated 8-pointed stars as they go from house to house, singing and giving gifts to children.[7]

And we should not neglect to mention the people of Mexico, who remember the Bethlehem Star when they create (and then smash!) vibrantly colored, 7-pointed Christmas piñatas, and when they decorate their houses with poinsettias.

The pervasive influence of the Star of Bethlehem can be detected in fields as diverse as horticulture and space astronomy. Many star-shaped flowers have been named after it, including one kind of orchid, two types of lilies, and five species of the perennial *Ornithogalum*. Moreover, the spacecraft Giotto, sent to investigate Halley's Comet, was named after the artist who painted a magnificent fresco on the ceiling of the Arena Chapel in Padua, Italy, in which he portrayed the Bethlehem Star as a comet (see fig. 6.1).

POPULAR PORTRAYALS OF THE STAR

As famous and beloved as the Bethlehem Star is in the modern world, conceptions of it are surprisingly varied. Anyone leafing through carol books, collections of religious-themed Christmas cards, or nativity storybooks for children, or watching nativity plays or cinematic portrayals of the birth of Jesus will be exposed to a wide range of ideas concerning the Star. There is a consensus that the Star was an objective phenomenon, was beautiful and bright, and accompanied the Magi as they traveled westward to Judea; but beyond that, there is little agreement.

With respect to Christmas songs, while some suggest that the Star was a new astronomical entity and one even claims that it was a tailed comet, most are content to leave the question of the Star's identity mysterious. The object is sometimes presented as so bright that its light bleaches out that of the other stars, and indeed as visible not only during the hours of darkness but also during the day-

[2] Wikipedia, s.v. "Bethlehem, Pennsylvania," http://en.wikipedia.org/wiki/Bethlehem,_Pennsylvania#Christmas_star (last modified April 1, 2013).
[3] See "Star Light, Star Bright," Palmer Lake Historical Society, http://palmerdividehistory.org/startale.html (last modified May 7, 2011).
[4] Wikipedia, s.v. "Christmas in Poland," http://en.wikipedia.org/wiki/Christmas_in_Poland (last modified February 20, 2013).
[5] Maria Hubert von Staufer, "Christmas in Poland," http://www.christmasarchives.com/wpoland.html (last modified October 25, 2010).
[6] Ibid.; Barbara Rolek, "Polish Christmas Traditions," http://easteuropeanfood.about.com/od/christmaseve/a/Polishxmas.htm (last modified May 3, 2013).
[7] Rebecca Luczycki, "Starring in the Night," *Alaska* magazine, http://www.alaskamagazine.com/article/77/09/starring_in_the_night (accessed May 3, 2014).

time. As to color, some express a preference for its being silver.

Christmas cards, children's storybooks, and nativity play sets portray the Star as intensely bright, often rivaling the full Moon, as stunningly golden or silver, and as very beautiful. Sometimes it is presented as a curved-tailed comet, but most of the time it is an extremely bright, multi-rayed star with a particularly long downward ray. The Star is frequently depicted hanging over the manger, in which cases the downward ray is pointing down to Jesus below, with or without the shepherds and/or Magi in attendance. Of course, the Star is also included in scenes of the Magi traveling from their homeland toward Judea by camel.

Over the last couple of decades, a number of popular computer-generated imagery (CGI) films have portrayed the Star. Two of them present it as a conjunction, or alignment, of planets and stars.

The 2006 multi-million-dollar-budget movie *The Nativity Story*[8] suggests that the Star consisted of the planets Jupiter and Venus becoming perfectly aligned with the star Regulus in the constellation Leo. As the Magi leave Jerusalem for Bethlehem, the three bright spots suddenly merge (at great speed, it must be said!), with the result that a long downward beam of light with an intensity approaching that of the Sun shines through a gap in the clouds into the cave where Mary is delivering Jesus.

Similarly, according to the British four-part TV drama *The Nativity*, first broadcast on BBC television in 2010,[9] the Star consisted of Jupiter, Saturn, and Regulus in perfect conjunction. When the Magi enter Bethlehem just as the child is emerging from Mary's womb, Jupiter, Saturn, and Regulus are merging immediately above her, causing a brightness like that of the full Moon to shine forth.

However, other CGI films have sought to portray the Star in more traditional terms. The 1999 Universal Studios made-for-TV movie *Mary, Mother of Jesus*[10] shows it as a very large four-rayed, cross-shaped new star hanging over Bethlehem. This star is brighter than Venus, although not as intensely bright as the full Moon. Moreover, the popular 2013 live-action History Channel miniseries *The Bible*[11] portrays the Star as an incredibly bright star with eight rays.

It is clear, then, that in the modern world conceptions of the Star vary greatly.

THE ONGOING MYSTERY OF THE STAR

The diversity of contemporary portrayals of the Star is simply a reflection of the scholarly debate concerning it. As bright as the Star evidently was, for almost two millennia its identity has been enshrouded in a mysterious darkness.

The phenomenon witnessed by the Magi was the subject of speculation in the first millennium, and fascination regarding it endures to the present. The modern debate began with Johannes Kepler in the early seventeenth century. Even now, in the twenty-first century, the number of theories offered to explain the Star seems to grow each year. Any interested party will have to sort through countless hypotheses—was it a planet like Jupiter, a star, a conjunction of planets, a nova or supernova, a comet, a miraculous phenomenon, or something else? It has justifiably been described by one respected astronomer as "the greatest of all

[8] Mike Rich, *The Nativity Story*, directed by Catherine Hardwicke, produced by Wyck Godfrey, cinematic release December 1, 2006 (Los Angeles: New Line Cinema, 2007), DVD.
[9] Tony Jordan, *The Nativity*, television miniseries, directed by Coky Giedroyc, produced by Ruth Kenley-Letts, aired on the BBC, December 2010 (Ampthill, England: Red Planet Pictures, 2011), DVD.
[10] Albert Ross, *Mary, Mother of Jesus*, television film, directed by Kevin Connor, produced by Eunice Kennedy Shriver, aired on NBC TV, November 14, 1999 (Universal City, CA: Universal Studios, 2010), DVD.
[11] Roma Downey and Mark Burnett, *The Bible*, television miniseries, produced by Roma Downey and Mark Burnett, aired on the History Channel, March 3–31, 2013 (Beverly Hills, CA: Lightworkers Media, 2013), DVD.

detective stories"[12] and "perhaps the greatest of all astronomical mysteries."[13]

Newspapers and magazines publicize any new hypothesis as though it might just be the critical breakthrough that enables us to identify the Star, no matter how profoundly and obviously flawed the theory may be. Whole books and DVDs are devoted to the task of identifying the Star. Some websites that promote particular theories get hundreds of thousands or even millions of hits. Internet discussion groups are weighed down with countless armchair experts sharing their hunches with the world.

The multiplicity of views and extravagant claims made by some, coupled with the lack of any real progress in the debate, has naturally engendered cynicism among many regarding the whole task. Indeed people would be forgiven for doubting that any definitive explanation of the Star will ever be presented, and for approaching any new proposed solution to the age-old problem with a healthy degree of skepticism. The whole debate about the Star of Bethlehem, after all, has become disconcertingly speculative. All too often, scholars have put forward astronomical explanations of the Star that are only superficially rooted in Matthew's account and, upon close inspection, fail to take seriously key aspects of the narrative.

MOVING BEYOND FUTILE SPECULATION

Part of the problem is that academic interest in the Star of Bethlehem has been largely confined to the astronomical community, with relatively few contributions from the theological community. This has resulted in a scholarly debate that is heavy on astronomy but light on theology and history. Astronomer Mark Kidger has humbly admitted that "An astronomer may not be the best qualified person to take on such a task [the quest for the historical Star], particularly in areas where even the greatest experts have profound and fundamental differences, and where many of the agreed-upon facts are based on the penetrating and complicated interpretation of subtle clues."[14]

The history of literature on the Star bears out the truth of Kidger's admission: the contributions of astronomers untrained in Biblical studies are all too often flawed in their treatment of the source material. They do not take sufficient account of critically important matters such as genre, subgenre, grammar, and historical background. Even the best astronomical essays on the topic have a tendency to discount cavalierly aspects of the Matthean account.

It should surely go without saying that any quest for the historical Star must be built firmly on the foundation of a rigorous analysis of Matthew 1:18–2:18. Only when this text has been mastered and the profile of the Star fully laid out can one realistically hope to deduce the precise astronomical phenomenon in view. What the whole debate concerning the Star calls for, then, is input from Biblical scholars.

Of course, just as astronomers contributing to the debate face the problem of being out of their comfort zone and area of specialization when they examine the pertinent Biblical and early Christian texts, so also Biblical scholars find themselves in alien territory when they move into the astronomical aspects of the task. What is needed, then, is interdisciplinary work and cooperation between the astronomical and theological communities.[15]

By training, I am a Biblical scholar. I have

[12] Mark R. Kidger, *The Star of Bethlehem: An Astronomer's View* (Princeton, NJ: Princeton University Press, 1999), xi.
[13] Ibid., ix.
[14] Ibid., ix–x.
[15] One positive recent development on this front was the multidisciplinary colloquium on the Star of Bethlehem at the University of Groningen on October 23–24, 2014; see http://www.astro.rug.nl/~khan/bethlehem (accessed July 5, 2014). A collection of papers presented at the colloquium is due to be published by Brill.

therefore been forced to spend the last few years trying to come to grips with relevant fields of astronomy so that I could develop the implications of the Biblical data in the astronomical arena. Equally importantly, respected members of the international astronomical community, professional and amateur, have graciously and indeed sacrificially given of their time and energy to assist me on the celestial end of things.

In this book I offer what I am convinced is the solution to the age-old mystery of the Star of Bethlehem. What I propose is rooted in a careful consideration of the relevant Biblical material and is, I believe, able to explain everything said about the Star in a natural and compelling way and in harmony with current astronomical knowledge.

When it comes to claims of major advances in the understanding of long-debated Biblical mysteries, many people are naturally very skeptical. We have all seen too many television documentaries on mysteries of the Bible. A grandiose claim is made at the start of the program, and we wait patiently—or, more often, impatiently—through commercials and a long, drawn-out build-up for the narrator finally to unveil the supposedly great discovery. When the program does eventually get there, almost without exception we end up rolling our eyes and regretting that we have just wasted an hour of our lives. The most common, but by no means the only, problem is that key details of the Biblical text have been twisted or ignored in order to accommodate the featured hypothesis.

Over against this tendency, my treatment of the Star of Bethlehem mystery will be driven by the Biblical data and will not play fast and loose with it.

Surprisingly, many have dismissed any possibility that a straightforward reading of Matthew's text is compatible with a natural astronomical phenomenon. R. A. Oriti has written, "Those who believe in a literal [interpretation] of the Bible may choose to believe that the Star literally moved and stood over the young child. Such an interpretation must rule out any astronomical explanation."[16] Even Mark Kidger has claimed that, if the Scriptural narrative is interpreted "literally," any scientific explanation is impossible.[17]

Certainly, if we were to judge the accuracy of Matthew's text according to how well it matches the various proposals regarding the Bethlehem Star offered in recent centuries, we might be tempted to concur with these negative assessments. The respected New Testament scholar Raymond Brown made the following statement: "*Really no one, including the astronomers, takes everything in the Matthean account as literal history.* Matt[hew] says that the magi saw the star (not planets, not a comet) of the King of the Jews at its rising (or in the East), and that it went before them from Jerusalem to Bethlehem and came to rest over where the child was. In recent literature I have not found an astronomical proposal that fits that literally."[18]

Of course, we should not excuse a low view of the accuracy of Matthew's narrative about the Magi and the Star on the ground that no astronomical hypothesis has yet succeeded in explaining all of its details.

As we shall see in the following chapter, Matthew's Gospel should be classified as an ancient Greco-Roman biography with a definite interest in historical accuracy. If Matthew, our major source regarding the Star, cannot be trusted in the little information that he gives us about it, then the quest for the historical Star is doomed to failure. Everyone in the debate should accept that the

[16] R. A. Oriti, "The Star of Bethlehem," *The Griffith Observer* 39.12 (December 1975): 9.
[17] Kidger, *Star of Bethlehem*, viii.
[18] Raymond Brown, *The Birth of the Messiah: A Commentary on the Infancy Narratives in the Gospels of Matthew and Luke*, 2nd ed. (New York: Doubleday, 1993), 612 (italics his). He goes on to state that Matthew's nativity account might be deemed essentially or partly historically accurate if we allow for poetic license and the use of symbolism.

preferred hypothesis is the one that matches Matthew's account most closely, and the ideal hypothesis is the one that fits it perfectly. Hence the more any hypothesis is in tension with the data in Matthew's Gospel, the more it should be regarded as inferior. Only when we begin with Matthew's account, interpret it straightforwardly and sympathetically, and resist the temptation to veer away from it in order to accommodate a pet theory can we rescue the discussion from the mire of endless speculation.

MATTHEW'S ACCOUNT OF THE MAGI AND THE STAR

We must now turn our attention to the Gospel of Matthew's account of the Nativity and, in particular, the Magi and the Star. We find this in Matthew 1:18–2:23. According to Matthew, an extraordinary astronomical phenomenon caused some magi from the east to make a long journey west to Judea in order to welcome the newborn King of the Jews. Remarkably, when they arrived at their final destination, Bethlehem of Judea, the Star pointed out to them the house where the baby Messiah and his virgin mother were staying:

> Now the birth of Jesus Christ took place in this way. When his mother Mary had been betrothed to Joseph, before they came together she was found to be with child from the Holy Spirit. And her husband Joseph, being a just man and unwilling to put her to shame, resolved to divorce her quietly. But as he considered these things, behold, an angel of the Lord appeared to him in a dream, saying, "Joseph, son of David, do not fear to take Mary as your wife, for that which is conceived in her is from the Holy Spirit. She will bear a son, and you shall call his name Jesus, for he will save his people from their sins." All this took place to fulfill what the Lord had spoken by the prophet:
>
> "Behold, the virgin shall be with
> child[19] and bear a son,
> and they shall call his name
> Immanuel"
>
> (which means, God with us). When Joseph woke from sleep, he did as the angel of the Lord commanded him: he took his wife, but knew her not until she had given birth to a son. And he called his name Jesus.
>
> Now after Jesus was born in Bethlehem of Judea in the days of Herod the king, behold, magi[20] from the east came to Jerusalem, saying, "Where is he who has been born king of the Jews? For we saw his star at its rising[21] and have come to worship him." When Herod the king heard this, he was troubled, and all Jerusalem with him; and assembling all the chief priests and scribes of the people, he inquired of them where the Christ was to be born. They told him, "In Bethlehem of Judea, for so it is written by the prophet:
>
> 'And you, O Bethlehem, in the land
> of Judah,
> are by no means least among
> the rulers of Judah;
> for from you shall come a ruler
> who will shepherd my people
> Israel.'"
>
> Then Herod summoned the Magi secretly and ascertained from them

[19] My translation.
[20] I prefer "magi" (ESV footnote) to the rendering "wise men" (ESV text) and therefore will use that term (and "the Magi") throughout.
[21] My translation.

what time the star had appeared. And he sent them to Bethlehem, saying, "Go and search diligently for the child, and when you have found him, bring me word, that I too may come and worship him." After listening to the king, they went on their way. And behold, the star that they had seen at its rising went before them until it came and stood[22] over the place where the child was. When they saw the star, they rejoiced exceedingly with great joy. And going into the house they saw the child with Mary his mother, and they fell down and worshiped him. Then, opening their treasures, they offered him gifts, gold and frankincense and myrrh. And being warned in a dream not to return to Herod, they departed to their own country by another way.

Now when they had departed, behold, an angel of the Lord appeared to Joseph in a dream and said, "Rise, take the child and his mother, and flee to Egypt, and remain there until I tell you, for Herod is about to search for the child, to destroy him." And he rose and took the child and his mother by night and departed to Egypt and remained there until the death of Herod. This was to fulfill what the Lord had spoken by the prophet, "Out of Egypt I called my son."

Then Herod, when he saw that he had been tricked[23] by the Magi, became furious, and he sent and killed all the male children in Bethlehem and in all that region who were in their second year or under,[24] according to the time that he had ascertained from the Magi. Then was fulfilled what was spoken by the prophet Jeremiah:

> "A voice was heard in Ramah,
> weeping and loud lamentation,
> Rachel weeping for her children;
> she refused to be comforted,
> because they are no more."

But when Herod died, behold, an angel of the Lord appeared in a dream to Joseph in Egypt, saying, "Rise, take the child and his mother and go to the land of Israel, for those who sought the child's life are dead." And he rose and took the child and his mother and went to the land of Israel. But when he heard that Archelaus was reigning over Judea in place of his father Herod, he was afraid to go there, and being warned in a dream he withdrew to the district of Galilee. And he went and lived in a city called Nazareth, so that what was spoken by the prophets might be fulfilled, that he would be called a Nazarene.

So many questions fill the mind of the reader. What was this "star"? What was so striking about it? What convinced the Magi to make their trek westward? How could the star appear in the east, guide the Magi to Judea in the west, and then lead them southward from Jerusalem to Bethlehem, and finally pinpoint the very place in Bethlehem where the Virgin Mary and her holy son were? There can scarcely be any doubt that the Star, if it really did exist, was the most extraordinary astronomical phenomenon ever seen by humans.[25] Moreover, if it occurred and coincided with the birth of Jesus, then it would constitute a

[22]In deference to the preposition that follows ("over" or "above"), I have translated the phrase "came and *stood*" (cf. NASB) rather than "came to rest" (cf. ESV). See Josephus, *J.W.* 6.5.3 (§289) for the expression "stood over" in an astronomical context.
[23]In this context, "double-crossed" would probably be a preferable translation to "tricked."
[24]Matthew's "two years old or under" actually means "one year old or under," since the Jews reckoned that a baby in the first year of life was already "one year old" and in the second year of life was "two years old." (See, for example, Robert H. Gundry, *Matthew: A Commentary on His Handbook for a Mixed Church under Persecution*, 2nd ed. [Grand Rapids, MI: Eerdmans, 1994], 35). "In their second year or under" is a preferable, dynamic-equivalent translation of the Greek.
[25]Cf. David W. Hughes, *The Star of Bethlehem Mystery* (London: J. M. Dent, 1979), 13.

dramatic corroboration of Jesus's claim to be the Messiah.

OVERVIEW OF THE BOOK

In this book we shall first consider Matthew's Gospel and its account of the Nativity, in particular the story of the Magi's visit to Bethlehem (chapters 2–3). Based on this, we will draw up a suitable set of facts that candidates for the role of the Star of Bethlehem must be able to explain satisfactorily. Thereafter we shall consider the main hypotheses that have been put forward to explain the Star (chapter 4) and make our own case that one particular astronomical phenomenon lies behind the story of the Magi's Star (chapters 5–6). Following that, we shall attempt to extrapolate a more precise picture of the celestial phenomenon witnessed by the Magi in their homeland (chapter 7) and on what basis they came to the conclusion that it signified the birth of the King of the Jews (chapter 8). Next, our focus shall turn to the task of building a profile of the particular celestial entity known to us as the Bethlehem Star, based on the Biblical text, and, in light of this, draw on modern astronomical knowledge to discover as much as possible about it (chapters 9–10). Finally, after considering the entity in relation to its astronomical counterparts (chapter 11), we shall conclude our study by telling the story of the main participants in the story of the Magi and the Star from the point at which the Massacre of the Innocents occurred (chapter 12).

2

"We Beheld (It Is No Fable)"

The Testimony of Matthew's Gospel

Before we examine closely the well-known account of the Star of Bethlehem (in the following chapter), it is important to introduce the source in which the story is found, the Gospel of Matthew.

Matthew's Gospel is a beautifully written and structured theological narrative about Jesus's life and ministry—particularly his words and deeds—and his death and resurrection. It is this Gospel that has given us the most familiar versions of the Lord's Prayer and the Beatitudes ("Blessed are . . . , for they shall be . . .").

AUTHORSHIP

Although the author of Matthew's Gospel is technically anonymous, early church tradition (most notably the early second-century AD church father Papias) unanimously attributed it to Jesus's disciple, the former tax collector, Matthew. The attribution seems reasonably secure: if Matthew did not pen the Gospel, it is difficult to imagine how it came to be associated with him. He was, after all, it would seem, a relatively obscure member of the early Christian movement.

DATE

Matthew wrote his Gospel around the year AD 70, most likely shortly before or after that year. Certainly Matthew was noticeably interested in including sayings of Jesus that foretold the destruction of Jerusalem (which occurred at the hands of the Romans in AD 70) (Matt. 22:7; 24:15). That he did not write much before AD 70 is suggested by Matthew's heavy use of Mark's Gospel, which is probably dated to the mid-60s AD, during Nero's persecution of Christians in Rome. Accordingly, it seems that Matthew penned his Gospel about three or four decades after Jesus's ministry in Galilee and Judea and about seven or eight decades after Jesus's birth.

SOURCES

The disciple made extensive use of sources, not just Mark's Gospel (which, according to the early church fathers, consists of traditions proclaimed by the apostle Peter), but other materials too, some of which were also used by Luke and others of which were not. With respect to chapters 1–2, Matthew evidently was drawing largely upon stories preserved by the family of Jesus, most naturally Mary, and subsequently safeguarded by the apostles.

JEWISH NATURE

The Gospel is strongly Jewish in nature, emphasizing that Jesus was the son of David, the Messiah who fulfilled the Hebrew Scriptures and was engaged in ministry to the "lost sheep" of Israel (Matt. 10:6; 15:24). Matthew wrote as a Christian Jew strongly critical of his fellow Jews who had rejected Jesus as their Messiah.

KEY EMPHASES

THE GENTILES

At the same time, Matthew's Gospel emphasizes that the invitation to enjoy the salvation secured by the Messiah has been extended to the Gentiles. The Gospel climaxes with Jesus sending the disciples to preach the good news to all the nations (Matt. 28:18–20). It also records Jesus's healing of a Roman centurion's servant (8:5–13) and the exorcism-at-a-distance of a Gentile woman's demonized daughter (15:21–28). Matthew's mention of Jesus's high praise of this soldier and woman is striking. The Gospel's inclusion of the story of the Magi's journey to see the baby Messiah fits with this editorial emphasis.

FULFILLMENT OF THE HEBREW SCRIPTURES

Matthew emphasizes that Jesus is the fulfillment of the Hebrew Scriptures. Not only does his Gospel associate Jesus with Abraham and David by means of the genealogy (Matt. 1:1–17), but it also portrays Jesus as reenacting the exodus (2:15), and presents him in terms of the temple (12:6), Israel in the wilderness (4:1–11), Moses (2:20; 5:21–22), David (12:3–5), Solomon (12:42), and Jonah (12:40–41). More importantly for our purposes, Matthew commonly claims that particular Old Testament texts have been fulfilled. The Gospel often uses one somewhat formulaic phrase: "This took place to fulfill what was spoken. . . ." We see this in Matthew 1:22–23, which declares that Mary's conception of the Messiah through the intervention of the Holy Spirit occurred in fulfillment of Isaiah 7:14:

> All this took place to fulfill what the Lord had spoken by the prophet:
>
> "Behold, the virgin shall be with
> child and bear a son,
> and they shall call his name
> Immanuel"
>
> (which means, God with us).

But the Gospel also introduces Old Testament quotations in other ways. For example, when the Magi visit Jerusalem, seeking the King of the Jews, Herod the Great inquires of the chief priests and scribes concerning the location of the Messiah's birth. Their answer becomes the means by which Matthew makes the point that Jesus's birth in Bethlehem was a direct fulfillment of an oracle by the prophet Micah, in Micah 5:2.

GENRE

One key concern for those engaged in the quest for the historical Star is the genre of Matthew and the attendant matter of the Gospel's historical reliability.

The Gospels are biographies of Jesus. In that the Gospels are written about a single individual, this may seem to be stating the obvious. And so it was, until the early twentieth century. At that time many Bible scholars abandoned this approach in favor of the view that the Gospels were more like popular folk literature based on oral traditions, and so had little historical value as regards the life of the historical Jesus. However, the works of David Aune[1] and Richard Burridge[2] in the last three decades of the twentieth century caused most New Testament scholars to return to the view that the Gospels are biographies—not modern biographies, but ancient Greco-Roman biographies or *bioi*, like Plutarch's *Lives*.[3] We may speak of the Gospels as historical biographies with theological agenda.

Burridge pointed out that, with respect to their opening, size, narrow central focus, and essential chronological structure; coverage of ancestry, great deeds, virtues, and death; and respectful tone, emphasis on the final years, continuous prose narrative, combination of different subgenres, use of different sources, display of the subject's character, apologetic and polemical nature, and goal of preserving the memory of the subject, the Gospels are all strongly reminiscent of ancient *bioi*.[4] The similarity is strongest between the Gospels and the biographies from the same general period, like those of Plutarch (late first century AD) and Suetonius (early second century AD).[5]

Ancient Greco-Roman biographies, like ancient Greco-Roman histories[6] and indeed modern biographies and histories, could vary in their historical reliability.[7] However, "Biographies were normally essentially historical works."[8] While ancient biographies tended to be one-sided in their assessment of their subjects, they were "still firmly rooted in historical fact rather than literary fiction. Thus, while the Evangelists clearly had an important theological agenda, the very fact that they chose to adapt Greco-Roman biographical conventions to tell the story of Jesus indicates that they were centrally concerned to communicate what they thought really happened."[9]

Biographies written about subjects who lived in the recent past, relative to the time of writing (e.g., those by Tacitus and Suetonius), especially those penned in the early empire, were generally marked by a greater concern for factual accuracy.[10] In such cases, biographers were expected to reject implausibilities and to seek to write what was true. The Gospels, penned within fifty years or so of Jesus's ministry, fall into this category.

That the Gospel writers were determined to produce an accurate account of Jesus's life is especially clear in the opening of one of them, the Gospel of Luke (Luke 1:1–4). This prologue is very much in the mold of Thucydides, Polybius, and Josephus. Luke clearly claims to be writing history conforming to the highest standards of Greco-Roman historiography: he depended on eyewitnesses, personally investigated everything from the beginning, and strove for historical accuracy. The very nature of early Christianity, founded on the historical claim that Jesus rose from

[1] For example, David E. Aune, "The Gospel as Hellenistic Biography," *Mosaic* 20 (1987): 1–10; idem, *The New Testament in Its Literary Environment* (Cambridge: James Clarke, 1988). Cf. Charles H. Talbert, *What Is a Gospel? The Genre of the Canonical Gospels* (Philadelphia: Fortress, 1977).

[2] Richard A. Burridge, *What Are the Gospels? A Comparison with Graeco-Roman Biography*, 2nd ed. (Grand Rapids, MI: Eerdmans, 2004).

[3] Note the statement of Graham Stanton, *Jesus and Gospel* (Cambridge: Cambridge University Press, 2004), 192: "The gospels are now widely considered to be a sub-set of the broad ancient literary genre of biographies." Also Craig S. Keener, *The Gospel of Matthew: A Socio-Rhetorical Commentary* (Grand Rapids, MI: Eerdmans, 2009), 24; idem, *The Historical Jesus of the Gospels* (Grand Rapids, MI: Eerdmans, 2009), 78–79; Richard A. Burridge, "About People, by People, for People," in *The Gospels for All Christians: Rethinking the Gospel Audiences*, ed. Richard Bauckham (Grand Rapids, MI: Eerdmans, 1998), 120–121.

[4] Burridge, *What Are the Gospels?*, 105–232. Burridge notes that "The genre of *bios* is flexible and diverse, with variation in the pattern of features from one *bios* to another. The gospels also diverge from the pattern in some aspects, but not to any greater degree than other *bioi*; in other words, they have at least as much in common with Graeco-Roman *bioi*, as the *bioi* have with each other. Therefore, the gospels must belong to the genre of *bioi*" (250).

[5] Keener, *Historical Jesus of the Gospels*, 78–81.

[6] Ancient biographies and ancient histories were different in genre but they could be very similar in practice—it is striking that Luke–Acts is a two-volume work, the first volume of which is biographical and the second of which is historical (see ibid., 81).

[7] See ibid., 84.

[8] Ibid., 80.

[9] David E. Aune, "Greco-Roman Biography," in *Greco-Roman Literature and the New Testament* (Atlanta: Scholars Press, 1988), 125. As Keener (*Historical Jesus of the Gospels*, 81) observes, where a biographer deliberately falsified events, he was departing from the conventions of ancient biographical writing.

[10] Keener, *Historical Jesus of the Gospels*, 83–84.

the dead, explains why the Gospel writers were so committed to restricting themselves to traditions that they were convinced were historically accurate.

Most Biblical scholars today would agree that the Gospel writers believed that what they were writing was historically accurate and worthy of acceptance, and that the first readers of these literary works would have approached them with the expectation that they were describing what had actually taken place in history.[11]

The production of literary Gospels was obviously intended to ensure that the testimony of the eyewitnesses was not lost to the Christian movement or susceptible to contamination in the aftermath of the disciples' deaths.

That the Gospels are Greco-Roman biographies that present the testimony of the authorized eyewitnesses who preserved the Jesus tradition has recently been powerfully argued by Richard Bauckham in *Jesus and the Eyewitnesses: The Gospels as Eyewitness Testimony*. He writes, "The kind of historiography they are is testimony," which, he goes on to say, is "a form of human utterance that . . . asks to be trusted."[12] The natural question to ask is whether the traditions recorded by the Gospel writers merit this trust. In other words, were the writers of the Gospels correct in their claim that what they wrote was historically accurate?

HISTORICAL RELIABILITY

Contrary to what many have claimed, we have every reason to believe that the stories about and sayings of Jesus that we find in the Gospels were stably transmitted and not embellished or corrupted over the decades. The key factor is that, up until the time when the Gospels were written, the traditions were preserved and guarded by a circle of eyewitnesses,[13] chief among whom were the apostles.[14] That the traditions were passed on faithfully, without contamination or innovation, is strongly suggested by the fact that the Jesus tradition preserved in the Gospels is not what we would have expected had it been shaped in the early decades of the Christian movement. For example, it is remarkable that in the Gospels there is no saying attributed to Jesus regarding circumcision, and that the main title employed by Jesus for himself is "Son of Man," which was not popular among the first Christians. Writing in AD 50–54, Paul cites from the Jesus tradition in letters to his churches (1 Thess. 4:15–17a; 1 Corinthians 7, 11–12, 15), revealing that by that time he had in his possession a written collection of Jesus material that he wholeheartedly trusted. That the stories about and teachings of Jesus were quickly committed to writing is hardly surprising in view of the fact that Jesus's followers esteemed him as Messiah and Lord and would obviously therefore have been eager to conserve and safeguard what they knew and remembered about him.

Bauckham draws on psychology of memory research to assess whether the Gospel traditions were the kind that would tend to be accurately preserved in the memories of the eyewitnesses.[15] He points out that the stories were related to unusual, indeed often unique, events, and that they were vivid and extraordinarily consequential for the eyewitnesses, and that they would have been profoundly emotional for them and frequently rehearsed by them, beginning very shortly after the events.[16] On this basis, he concludes that "the

[11] Cf. Martin Hengel, *Acts and the History of Earliest Christianity* (London: SCM, 1979), 3–68.
[12] Richard Bauckham, *Jesus and the Eyewitnesses: The Gospels as Eyewitness Testimony* (Grand Rapids, MI: Eerdmans, 2007), 5.
[13] Ibid., 93–182, 305–357.
[14] Ibid., 93–113.
[15] Ibid., 319–357, building on Samuel Byrskog, *Story as History—History as Story* (Leiden: Brill, 2002).
[16] Bauckham, *Jesus and the Eyewitnesses*, 331–335, based on W. F. Brewer, "What Is Recollective Memory?," in *Remembering Our Past: Studies in Autobiographical Memory*, ed. D. C. Rubin (Cambridge: Cambridge University Press, 1996), 35–57; and Gillian Cohen, *Memory in the Real World* (Hillsdale, NJ: Erlbaum, 1989), 118–125.

memories of eyewitnesses of the history of Jesus score highly by the criteria for likely reliability that have been established by the psychological study of recollective memory."[17]

AN APPROPRIATE APPROACH TO THE GOSPEL OF MATTHEW

What, then, should be our attitude to the testimony of the Gospels? Bauckham points out that "Trusting testimony is indispensable to historiography. This trust need not be blind faith. In the 'critical realist' historian's reception and use of testimony there is a dialectic of trust and critical assessment. . . . For most purposes, testimony is all we have."[18] The reader is therefore put in the position of having to judge whether to trust or distrust the testimony offered by the eyewitnesses. The correct way to read the Gospels as Greco-Roman biographies is therefore to approach them not with a radically suspicious mind-set that assumes that every story or saying is unreliable unless proven otherwise, but rather to approach them with a sensitivity to their historiographical claim to be testimonies, with an appropriate level of trust in the credibility of the witnesses. The judgment of whether particular testimonies are to be trusted or not must be based on "internal consistency and coherence, and consistency and coherence with whatever other relevant historical evidence we have and whatever else we know about the historical context."[19]

Many stumble over the extraordinary nature of the events described in the Gospels. Many stories in the Gospels are rejected as unhistorical by critics because they have exceptional elements in them. However, as Bauckham writes, "We must beware of a historical methodology that prejudices inquiry against exceptionality in history and is biased toward the leveling down of the extraordinary to the ordinary."[20] As for the charge that the fact that those making such extraordinary claims were biased undermines their credibility, in truth their bias should engender confidence in their testimony:

> The testimony of involved participants is especially valuable in the case of exceptional events. . . . The degree of commitment to their testimony such witnesses usually have should not in itself arouse our suspicions; in more ordinary cases we usually take such commitment as a reason for taking especially seriously what a witness has to say. It is by no means irrational to take the risk of crediting the testimony of involved and committed participants to the extraordinary and the exceptional in history.[21]

All in all, we should therefore approach the Gospel narratives, including Matthew, with due sympathy and respect, aware of their

[17] Bauckham, *Jesus and the Eyewitnesses*, 346.

[18] Ibid., 490.

[19] Ibid., 506. To change the legal analogy, the Gospels should be regarded as innocent until proven guilty rather than vice versa. See particularly Joachim Jeremias, *New Testament Theology*, trans. John Bowden (New York: Scribner, 1971), 37; R. T. France, "The Authenticity of the Sayings of Jesus," in *History, Criticism, and Faith*, ed. Colin Brown (Downers Grove, IL: InterVarsity Press, 1976), 107; I. H. Marshall, *I Believe in the Historical Jesus* (Grand Rapids, MI: Eerdmans, 1977), 199–200; W. G. Kümmel, *Heilsgeschehen und Geschichte*, vol. 2 (Marburg: N. G. Elwert Verlag, 1978), 187–190; Robert H. Stein, "The 'Criteria' for Authenticity," in *Gospel Perspectives* 1, ed. R. T. France and David Wenham (Sheffield: JSOT Press, 1980), 225–253; S. C. Goetz and C. L. Blomberg, "The Burden of Proof," *Journal for the Study of the New Testament* 11 (1981): 39–63; Donald A. Hagner, "Interpreting the Gospels: The Landscape and the Quest," *Journal of the Evangelical Theological Society* 24 (1981): 31–32; R. T. France, *Jesus and the Old Testament* (Grand Rapids, MI: Baker, 1982), chapter 1; Craig Blomberg, *The Historical Reliability of the Gospels* (Downers Grove, IL: InterVarsity Press, 1987), 240–243; Ben Witherington, *The Christology of Jesus* (Minneapolis: Augsburg Fortress, 1990), chapter 1; and Keener, *Gospel of Matthew*, 29, who observes that

> The burden of proof thus rests with New Testament scholars who betray an unduly skeptical bias toward the Gospel accounts . . . ; such scholars must imply that disciples who considered Jesus Lord were far more careless with his words in the earliest generations of Christianity than first- and second-generation students of most other ancient teachers were. . . . Especially given how much of Jesus' teaching was disseminated in public during his lifetime, the sort of "radical amnesia" this skepticism requires of Jesus' first followers . . . is certainly not typical of schools of other early sages.

[20] Bauckham, *Jesus and the Eyewitnesses*, 506–508.

[21] Ibid., 507.

theological agenda but not disregarding their implicit claim to be historically trustworthy. The testimony presented is simultaneously both theological and historical. This indeed is one major aspect of their magnificence—theology and history do not vie against but rather complement each other.[22] For example, Matthew wrote to demonstrate that Jesus was the messianic King promised by the Prophets, and to unveil the nature of the kingdom that he inaugurated during his ministry. Matthew accomplished this not by freely mixing the unhistorical with the historical, but by basing his narrative on historically reliable records and reports.

[22] See Blomberg, *Historical Reliability*, 73–152, on the reliability of Matthew, Mark, and Luke, the three so-called Synoptic Gospels.

3

"They Looked Up and Saw a Star"

The Story of the Star

We turn now to devote our attention to Matthew's account of the Magi and the Star, which is found in Matthew 2:1–18.

THE HISTORICAL RELIABILITY OF MATTHEW'S ACCOUNT OF THE STAR

MATTHEW'S BELIEF

Can we trust Matthew's narrative concerning the Star, which purports to document an event that occurred some three decades before Jesus began his ministry? Clearly, Matthew believed that his source for this episode was reliable, and he was convinced that the account was historically accurate.[1] The very fact that he includes the episode and suggests that what happened fulfilled the Scriptures demonstrates this.

HISTORICAL PLAUSIBILITY

But was Matthew right to judge that the story was historically accurate? There are a number of elements in the story that are striking for their historical plausibility.

For one thing, we know from Josephus that Herod the Great in his final years was extraordinarily cruel and capable of the most terrible atrocities.[2] Therefore the Massacre of the Innocents recorded by Matthew in this passage fits perfectly into the framework of the historical period.

Second, what the Magi did in undertaking a long journey westward to greet a king is not implausible, but, as we shall see, is very similar to what other magi did about seven decades later, in the time of Nero.

Third, most devout Jews and Christians despised astrologers and would not normally have been inclined to trust their testimony. Therefore you would not have expected someone fabricating a nativity narrative to choose astrologers as among the first to welcome the newborn Messiah into the world. A fabricator would most likely have stayed away from any elements that seemed theologically suspect and risked offending the intended readership.

[1] Alfred Plummer, *An Exegetical Commentary on the Gospel according to S. Matthew* (London: Elliot Stock, 1909), 11; W. C. Allen, *The Gospel according to St. Matthew*, 2nd ed., International Critical Commentary (New York: Scribner, 1907), 14; John Nolland, *The Gospel of Matthew*, New International Greek Testament Commentary (Grand Rapids, MI: Eerdmans, 2005), 106–107; and Gregory W. Dawes, "Why Historicity Still Matters: Raymond Brown and the Infancy Narratives," *Pacifica* 19 (2006): 163, who also adds the following comment: "the evangelists' theological message not only takes the accuracy of these reports for granted; it actually *requires* that they be accurate" (164), since the theological message is built on the foundation of the historical claim (175).
[2] See Josephus, *Ant.* 17.

Fourth, the fact that the Star is said to have first appeared at least a year before the Massacre of the Innocents, and that Herod determined the age of the infants to be killed based on this information, speaks strongly for historicity, since it is difficult to explain otherwise.[3]

"The main outline of the story" is, as W. C. Allen put it, "noteworthy for its historical probability."[4]

In addition, key features of the Matthean account are corroborated by another first-century writer, Luke, who makes much of his credentials as a historian (Luke 1:1–4). In particular, Luke authenticates Jesus's birth to Mary in Bethlehem (2:1–7) and the unusual circumstances surrounding his conception (2:26–38). Luke may also quietly attest to the historicity of the Star (1:78–79).

Moreover, if the story of the Star is not rooted in history, then in what is it rooted? No plausible alternative explanation of the story's origin has ever been offered.[5]

Furthermore, recent psychology of memory research supports the claim of Matthew 2 to be considered historical. A natural source for much of Matthew 2 is the family of Jesus, in particular Mary. Joseph was probably deceased by the time Jesus began his ministry, whereas Mary lived to witness Jesus's crucifixion and evidently for some considerable time afterwards, looked after by John (John 19:25–27). It would have been surprising if the early Christians, including Matthew, did not inquire of her concerning the circumstances of Jesus's conception and birth. Therefore, when we read the account of the Magi and the Star in Matthew 2, we are almost certainly coming into close contact with the precious memories of the historical Mary. Consequently, as we read the story, we are put in the position of having to respond to her indirect testimony about the extraordinary events that surrounded the birth of Jesus. According to recent studies in the psychology of memory to which we have already referred,[6] the Magi's visit was for Mary the kind of event that tends to be remembered accurately by eyewitnesses. It was a very memorable and vivid unique occasion, an important and deeply emotional moment, relating to the birth of her special eldest son, and it would undoubtedly have been something she would have frequently rehearsed mentally and orally, beginning immediately afterwards and continuing on until her death. Luke appears to confirm that Mary did indeed frequently mentally rehearse unusual events relating to Jesus: speaking of what happened in the wake of Passover when Jesus was twelve years old, Luke says that Mary "treasured up all these things in her heart" (Luke 2:51).

The Magi's story could simply have been mediated to Matthew through Mary, but it is also possible that (and would hardly be surprising if) some of the early Christians tracked down and interviewed the Magi themselves (or other members of their traveling party) about what had happened, and that Matthew made use of this material as well. Needless to say, for the Magi (and their fellow travelers) this was a once-in-a-lifetime, profoundly emotional, and indeed life-changing event which they would have discussed often, beginning as soon as they left the holy family. In addition, written astronomical records would have reinforced the memories of the Magi. There is every reason to believe that, four or five decades later, any of the Magi (or their fellow travelers) who remained alive

[3] See George M. Soares Prabhu, *The Formula Quotations in the Infancy Narratives of Matthew: An Enquiry into the Tradition History of Mt 1–2* (Rome: Pontifical Biblical Institute, 1976), 298; also R. T. France, "Herod and the Children of Bethlehem," *Novum Testamentum* 21 (1979): 113.

[4] Allen, *Gospel according to St. Matthew*, 14–15.

[5] Plummer, *Matthew*, 11–12.

[6] W. F. Brewer, "What Is Recollective Memory?," in *Remembering Our Past: Studies in Autobiographical Memory*, ed. D. C. Rubin (Cambridge: Cambridge University Press, 1996), 35–57; Alan D. Baddeley, *Human Memory: Theory and Practice*, rev. ed. (Hove, England: Psychology Press, 1997), 213–222; Gillian Cohen, *Memory in the Real World* (Hillsdale, NJ: Erlbaum, 1989), 118–125.

would have accurately remembered what had happened.

Finally, if what Matthew records concerning the Star is found to be in accord with astronomical knowledge, that would constitute further powerful evidence in favor of the account's historical credentials. Indeed, since scarcely any episode in the Gospel of Matthew is more commonly rejected as unhistorical than the story of the Magi's visit to baby Jesus, authenticating the historicity of the Star would be an important validation of the historical reliability of the Gospel as a whole.

COUNTERING OBJECTIONS TO HISTORICITY

Those who have been unwilling to accept the historical basis of the Star narrative have resorted to rather feeble arguments to justify their cynicism. For example, J. N. M. Wijngaards questions whether Herod would have convoked the Sanhedrin, highlights the redundancy of the Star leading the Magi to Bethlehem (since they already would have known the way),[7] suggests that Herod would have sent a spy along, and wonders why an episode such as the slaughter of the infants went unmentioned by other sources.[8]

However, when Herod was in urgent need of religious instruction, he would, of course, have arranged for a gathering of the top theologians in the land (Matt. 2:4: "all the chief priests and scribes of the people"). Knowing that within Judaism and indeed even within individual movements like Pharisaism there was a variety of interpretive traditions, Herod would have been eager to be exposed to the full range of exegetical opinion on the question of the Messiah's birthplace. Too much was at stake for Herod to restrict his theological counsel to one or two religious leaders.

Moreover, the appearance of the Star on the final stage of the Magi's journey was not entirely devoid of purpose, since the Star did pinpoint the precise location within Bethlehem where the messianic child was. To the extent that the Star's role in guiding the Magi from Jerusalem to Bethlehem was redundant, since the Magi presumably had been informed as to where Bethlehem was, this actually speaks for the authenticity of the story rather than against it. Historically accurate narrative is full of redundancies, whereas fiction prefers to avoid them. At the same time, historically, the Star's presence when the Magi were traveling from Jerusalem to Bethlehem may have had various functions: to confirm the Magi in their mission, to heighten their sense of anticipation, and to engender in them the feeling that they were being ushered into the presence of the Messiah.

The claim that Herod would have sent a spy along with the Magi fails to take due account of the dynamics of the story as told by Matthew: The Magi were obviously extremely naive, gullibly believing that Herod really did want to make a personal journey to worship the Messiah. They had willingly become his agents and promised to report back to him.[9] There was, quite simply, no reason for Herod to doubt that they would return to the palace in due course, and hence there was no need to send a spy along.

As regards the lack of explicit references to the massacre elsewhere, this is hardly surprising—the incident, while horrific, was confined to a small area around Bethlehem and may have involved the deaths of no more than 20–40 infants. No historian records every event, as a reading of Tacitus, Suetonius, and Cassius Dio on any given period makes clear. Josephus naturally focused his account on the many events that seemed to him of greater

[7] Cf. Paul Gaechter, *Das Matthäus-Evangelium* (Innsbruck: Tyrolia, 1963), 290.
[8] J. N. M. Wijngaards, "The Episode of the Magi and Christian *Kerygma*," *Indian Journal of Theology* 16 (1967): 32–33.
[9] Cf. Warren Carter, "Matthew 1–2 and Roman Political Power," in *New Perspectives on the Nativity*, ed. Jeremy Corley (Edinburgh: T. & T. Clark, 2009), 86.

immediate political significance to Herod's reign and his succession.[10]

However, strikingly, an implicit reference to the incident may be found in the pseudepigraphal *Assumption of Moses* (6:4), which very probably dates to early in the first century AD (AD 6–30). It reports in prophetic form what Herod did during his reign: "And he will cut off their chief men with the sword, and will destroy (them) in secret places, so that no one may know where their bodies are. He will slay the old and the young, and he shall not spare. Then the fear of him shall be bitter unto them in their land. And he shall execute judgments on them as the Egyptians executed upon them, during thirty and four years, and he shall punish them."[11] Historically, we know of no other event during Herod's reign that would explain the peculiar reference to his merciless slaying of the young than the Massacre of the Innocents.[12] C. E. B. Cranfield rightly wonders if this is remarkably early evidence of Herod's atrocity from an independent source.[13]

Furthermore, the incident is mentioned in the *Protevangelium of James* (22:1: "When Herod perceived that he had been outwitted by the Magi, he was enraged, and sent murderers, instructing them to slay the children in their second year or under"[14]), which dates to around AD 150.

In conclusion, it would seem that the case against the historicity of the Star narrative is contrived and weak. The evidence is strongly in favor of the account's historical authenticity.

MATTHEW 1–2: THE NATIVITY AND THE STAR

We must now consider the relevant section of Matthew's Gospel.

Immediately after setting out Jesus's genealogy (Matt. 1:1–17), Matthew relates the story of the Nativity (vv. 18–25) and the visit of the Magi to Jesus (2:1–12). The focus in 1:18–2:12 is clearly on Jesus's birth, as highlighted by the introductory summary in 1:18a: "Now the birth of Jesus Christ took place in this way." The emphasis on the birth of Jesus is apparent in 1:21, 23, 25; 2:1, 2, and 4–6. The Star seen by the Magi in the east was evidently interpreted by them as heralding his birth (note 2:2: "Where is he who has been born king of the Jews? For we saw his star at its rising . . ."). Matthew claims that the prophets Isaiah (Isa. 7:14) and Micah (5:2) had prophesied about the circumstances of the Messiah's birth, and that these prophecies were fulfilled in connection with the birth of Jesus.

Before turning to Matthew 2:1–18, we must consider briefly 1:18–25.

MATTHEW 1:18–25: A BRIEF OVERVIEW

Joseph the Descendant of David

Matthew's focus in this brief account of the nativity story is on Joseph as Jesus's legal father. That Joseph has this role is emphasized in verse 25b, where we read that he "called [the child's] name Jesus." In the words of Hare, "Joseph's naming of Mary's baby constituted in this instance an acknowledgment that, by God's will and act, the boy is authentically his son."[15] Because of this, it is Jesus's

[10] F. W. Farrar, *The Herods* (London: James Nisbet, 1899), 154, points out that it is hardly surprising that the episode is not mentioned by Josephus, since it was only one of many terrible atrocities during Herod's reign.

[11] Translation from R. H. Charles, *The Assumption of Moses* (London: A. & C. Black, 1897), 22. Johannes Tromp, *The Assumption of Moses: A Critical Edition with Commentary* (Leiden: Brill, 1993), 201, highlights the correspondences between what is described here and Josephus's account of the career of Herod the Great.

[12] The non-Christian Ambrosius Theodosius Macrobius, around the turn of the fifth century, mentioned the Massacre of the Innocents in his work *Saturnalia* (2.4.11), although it is of questionable reliability.

[13] C. E. B. Cranfield, *On Romans, and Other New Testament Essays* (Edinburgh: T. & T. Clark, 1998), 159; cf. Ethelbert Stauffer, *Jesus and His Story* (New York: Knopf, 1960), 38–42.

[14] My translation of the Greek text in Emile de Strycker, *La forme la plus ancienne du Protevangile de Jacques* (Brussels: Société des Bollandistes, 1961), 174.

[15] Douglas Hare, *Matthew*, Interpretation (Louisville, KY: Westminster John Knox, 1993), 12.

family tree through Joseph that is set out in verses 1–17. This is important because Joseph is in the line of David.

Joseph the Legal Father of Jesus

According to Matthew, Mary was already betrothed to Joseph when she conceived by the Holy Spirit. Naturally, when Joseph learned of Mary's conception, he, knowing for certain that he was not the father of the child in Mary's womb, assumed that Mary had been sexually promiscuous. However, Matthew informs us that, although some men might in these circumstances have acted ruthlessly and rashly, publicly shaming their betrothed, Joseph was "a just man and [was] unwilling to put her to shame" and therefore "resolved to divorce her quietly" (1:19). Of course, such an action, even though well-intentioned, would have derailed the work of God, for Joseph was the Messiah's God-ordained legal father from the line of David. For this reason, according to Matthew's account, God intervened to ensure that his plan remained on track: an angel appeared to Joseph in a dream, assuring him that the child in Mary's womb was not the product of immorality, but of a miraculous conception by the Holy Spirit, and calling on him not to be afraid to carry on with his original plan to marry her. The angel explained that Mary would "bear a son" and that he, Joseph, would call "his name Jesus" (v. 21). Joseph responded to the divine instruction with obedience, taking Mary as his wife, avoiding sexual relations with her until after she had given birth, and naming the child Jesus.

Fulfillment of Isaiah 7:14

Of particular importance to Matthew is that what transpired—the virginal conception and birth and Joseph's formal acceptance of Jesus as his son—fulfilled a key prophecy in the Hebrew Scriptures, Isaiah 7:14. In this prophecy Isaiah declared to the eighth-century BC Davidic king Ahaz, "Therefore the Lord himself will give you a sign. Behold, the virgin shall be with child[16] and bear a son, and shall call his name Immanuel." Matthew clearly regarded this prophecy as in some sense awaiting fulfillment until the virgin birth of Jesus the Messiah. This interpretation of Isaiah's prophecy may well have set Matthew apart from most contemporary non-Christian Jewish interpreters, who did not believe that the Messiah would be born without a biological father.[17]

MATTHEW 2:1–12: A DETAILED LOOK

We shall now walk through Matthew 2:1–12 in a bid to understand the story of the Magi more precisely and to glean a more detailed profile of the Star.

The visit of the Magi took place in the wake of the birth of Jesus. Verse 1 states, "Now after Jesus was born in Bethlehem of Judea in the days of Herod the king, behold, magi from the east came to Jerusalem." The sequence of events makes good sense, since the Magi have come to Jerusalem on account of witnessing a celestial sign that they have interpreted as announcing the birth of the King of the Jews.

Jesus's Birth: The Place

The town where Jesus was born is identified here for the first time as Bethlehem. This prepares for verses 4–6, where the chief priests inform Herod that the Messiah was, according to the prophet Micah, to be born in Bethlehem. The town had strong Davidic connections. As the book of Ruth makes plain, Bethlehem was where Boaz and Ruth, David's great-grandparents, met and married and bore David's grandfather Obed. Moreover, it was in Bethlehem that Obed fathered Jesse and that Jesse fathered David.

Matthew highlights that Bethlehem was "of Judea." It is possible that "of Judea" is

[16] My translation.
[17] However, as we shall see, it is possible that the Septuagint does reflect a messianic interpretation of Isaiah's oracle.

intended to distinguish the Bethlehem just south of Jerusalem from the Bethlehem in Galilee,[18] or that it is merely a stereotypical phrase. However, it is more likely that it anticipates verses 5–6, which quote Micah's prophecy to the effect that the Messiah would be born in the territory of the tribe of Judah, ultimately recalling Genesis 49:9–10 ("Judah is a lion's cub. . . . The scepter shall not depart from Judah, nor the ruler's staff from between his feet, until tribute comes to him; and to him shall be the obedience of the peoples").

Jesus's Birth: The Time

The time when Jesus was born is defined as being "in the days of Herod the king" (Matt. 2:1), referring to Herod the Great. Although there is some debate, it is generally believed that Herod died in 4 BC. Coins minted under the reign of Herod's sons date to 4 BC, indicating that this was indeed the year of Herod the Great's death. Moreover, Josephus, *Ant.* 17.6.4 (§§164–167), informs us that there was a lunar eclipse less than a month before the Passover that presaged Herod's death, and the only plausible candidate is the partial lunar eclipse on the night of March 12–13 in 4 BC (during which year Passover fell on April 11). Herod, then, passed away at the end of March or beginning of April in 4 BC. In order to accommodate the events of Matthew 2:12–18, it seems best to conclude that Jesus's birth took place in either 6 BC or 5 BC. A date in either of these years would be compatible with Luke 3:23's statement that Jesus began his ministry (in the late 20s AD) when he was "about thirty years of age," which would readily accommodate an age of up to 34.

It must be understood that the BC/AD (BCE/CE) system with which we are so familiar was built on the work of Dionysius Exiguus, a sixth-century Scythian monk. Unfortunately, in his dating of Jesus's nativity he made a couple of miscalculations which mean that, contrary to what one might have expected, the birth of Jesus does not correspond to the year 0 (zero). First, the monk did not include a year 0, so that one moves from 1 BC straight to AD 1. Second, he failed to take into consideration the four years when Caesar Augustus reigned under his original name Octavian. Consequently, even before considering the Biblical evidence, the BC/AD transition is off target by 5 years.

The Magi's Entry into Jerusalem

The use of "behold" in association with the entry of the Magi into Jerusalem (Matt. 2:1) highlights that it was a memorable marvel.

The Procession of Magi in AD 66. As noted at the beginning of this chapter, the journey of the Magi in Matthew's account was similar to a journey by another group of magi some seven decades later—the procession of Tiridates and the magi to Nero in Italy in AD 66. Tiridates, the king of Armenia, his royal court, a train of servants, three Parthian royal sons, and 3,000 Parthian horsemen made "a quasi-triumphal procession through the whole country west from the Euphrates" to Italy (Cassius Dio 63.1–2).[19] The presence of magi is mentioned by Pliny the Elder, *Natural History* 30.6.16–17. The procession lasted nine months in total, and the traveling party was greeted with fanfare and acclamations as it passed through various towns and cities (Cassius Dio 63.2). Upon meeting Nero, Tiridates "knelt down upon the ground, and with arms crossed called him master and did obeisance" (63.2).[20] Later on, at a public celebration at the Forum, Tiridates repeated his gesture of obeisance to Nero, as the emperor

[18] So, for example, Donald A. Carson, "Matthew," in *Expositor's Bible Commentary*, rev. ed., ed. Tremper Longman III and David E. Garland, vol. 9 (Grand Rapids, MI: Zondervan, 2010), 109; and Ulrich Luz, *Matthew 1–7: A Continental Commentary*, trans. Wilhelm C. Linss (Minneapolis: Augsburg Fortress, 1989), 107.
[19] *Dio's Rome, Volume 5*, trans. Herbert Baldwin Foster (New York: Pafraets, 1906), 59.
[20] Ibid., 60.

was seated on his curule chair on the Rostra, wearing triumphal dress (Suetonius, *Nero* 13; Cassius Dio 63.2). Then Tiridates formally addressed Nero: "Master, I am the descendant of Arsaces, brother of the princes Vologaesus and Pacorus, and your slave. And I have come to you, my deity, to worship you as I do Mithra. The destiny you spin for me shall be mine; for you are my Fortune and my Fate" (63.4–5).[21] We would not expect the procession of the Magi at the time of Jesus's birth to be anywhere near as large or dramatic as that of Tiridates in AD 66. At the same time, there may well have been more than three Magi in the visiting party. Regardless of how many Magi there were, the extraordinary sight of them entering Jerusalem must have lived long in the memories of all who witnessed it.

What Were "Magi"? Matthew refers to the visitors as "magi." As Brown puts it, "the term 'magi' refers to those engaged in occult arts and covers a wide range of astronomers, fortune tellers, priestly augurers, and magicians of varying plausibility."[22] The Greek version of Daniel by Theodotion uses the word to translate Hebrew and Aramaic terms for "enchanters" (NIV, ESV), "conjurers" (NASB), or "astrologers" (KJV; NET) (Dan. 1:20; 2:2, 10, 27; 4:7; 5:7, 11, 15). In Matthew 2, where the focus is so strongly on a star's behavior, it is clear that the "magi" in view are first and foremost professional astronomers and astrologers.[23] Their lives were dedicated to gleaning from the heavens insights concerning human affairs. It was evidently as they were making their normal astronomical observations that they saw the Star.

There can be no serious doubt that the Magi were Gentiles.[24]

The fact that the visitors are described as "magi" does not necessitate that the celestial phenomenon they witnessed was astrological in nature. It only suggests that they were in the habit of examining the heavens for astrological information. What precisely they saw and its interpretation cannot be prejudged on the basis of the identity of the beholding eyes. Certainly, as Plummer highlighted, "There is not one word in the narrative to indicate that the Magi did wrong in drawing inferences from what they saw in the heavens, or that their knowledge of the birth of the Messiah was obtained from evil spirits or by the practice of any black art."[25] Astrology is condemned in the Hebrew Scriptures (Jer. 10:1–2; cf. Deut. 18:9–14; Isa. 47:13). If we assume, as would seem wise, that Matthew adhered to the established Scriptural and official Jewish stance toward astrology,[26] then he could hardly have believed that the sign witnessed by the Magi was fundamentally astrological in nature. It may have had meaning to astrologers, but its correct interpretation probably did not require astrological knowledge. This, of course, begs the question,

[21] Ibid., 61.

[22] Raymond Brown, *The Birth of the Messiah: A Commentary on the Infancy Narratives in the Gospels of Matthew and Luke*, 2nd ed. (New York: Doubleday, 1993), 167.

[23] As Hare (*Matthew*, 13) points out, the scholarly consensus favors the view that *magoi* here designates astrologers or astronomers, not magicians or Zoroastrian Persian priests (the original meaning of the term). On the term, see W. Bauer, W. F. Arndt, F. W. Gingrich, and F. W. Danker, *A Greek-English Lexicon of the New Testament and Other Early Christian Literature*, 3rd ed. (Chicago: University of Chicago Press, 2000), 608, and especially G. Delling, "*magos*," in *Theological Dictionary of the New Testament*, ed. Gerhard Kittel and Gerhard Friedrich, 10 vols. (Grand Rapids, MI: Eerdmans, 1964–1976), 4:356–359.

[24] David W. Hughes, *The Star of Bethlehem Mystery* (London: J. M. Dent, 1979), 52, suggests that they may have been Jews who had lapsed into astrology, but this is an unnecessary and implausible theory.

[25] Plummer, *Matthew*, 15.

[26] Craig Evans, *Matthew*, New Cambridge Bible Commentary (Cambridge: Cambridge University Press, 2012), 51; Edwin M. Yamauchi, "The Episode of the Magi," in *Chronos, Kairos, Christos*, ed. Jerry Vardaman and E. M. Yamauchi (Winona Lake, IN: Eisenbrauns, 1989), 28; Craig S. Keener, *The Gospel of Matthew: A Socio-Rhetorical Commentary* (Grand Rapids, MI: Eerdmans, 2009), 84. The second-century BC *Sibylline Oracles* 3:221–230 praises the Hebrew nation for abstaining from astrology and Chaldean divination. Yamauchi highlights that the New Testament (see Acts 8:9–24 and 13:6–11), the Apostolic Fathers, the Apologists, and second-century Christians unanimously held to a negative view of astrology ("Episode of the Magi," 27–28). At the same time, it seems that astrology had made significant inroads into the thinking and practice of Jews by the turn of the ages (see Keener, *Gospel of Matthew*, 101).

What could they have seen that so impacted them but did not require astrological training to understand?

The East. The Magi hailed from "the east" (Matt. 2:1). The country from which the Magi came is much debated, with scholars suggesting Arabia, Persia/Parthia, and Babylon.

Arabia? Arabia has often been proposed,[27] largely due to the fact that it was sometimes said to be in the east (Gen. 10:30; Judg. 6:3; Job 1:3; Isa. 11:14; Ezek. 25:4, 10) and because of the reference in Isaiah 60:6 (cf. Ps. 72:10, 15) to people from the Arabian Peninsula (Midian, Ephah, and Sheba) in the eschatological era making a pilgrimage to Jerusalem, during which they bring "gold and frankincense." However, while Matthew may well have Isaiah 60:6 in view as he writes, he is hardly claiming that this prophecy was completely fulfilled by the Magi who visited Jesus at his birth; consequently, it is unwise to read Isaiah's mention of Arabia into Matthew 2:1. At most, Matthew is presenting the Magi's coming as an anticipation of what would take place in the future. Moreover, as Keener points out, "In most accounts Magi hail from Persia or Babylon (e.g., Cic. *De Leg.* 2.10.26; Philo, *Spec.* 3.100; *Prob.* 74; Dio Chrys. *Or.* 36; Lucian, *Runaways* 8; Diog Laert. 8.1.3; Char. *Chaer.* 5.9.4; Philost. *V.A.* 1.24)."[28]

Persia? A number of scholars have favored Persia as the place from which the Magi set out on their journey.[29] Certainly Persia is located in the east and was associated with astrology and astronomy, although we know next to nothing about astronomy/astrology in Persia in that period.[30]

Babylon? The other possibility is Babylon.[31] Babylon is to the east of Judea and was described as being so by ancients (for example, *Assumption of Moses* 3:13, where "the land of the east" is Babylon). Moreover, the fact that the Magi seem to have been familiar with Jewish prophecies and traditions is more readily explicable if they hailed from Babylonia, since there was a significant population of Diaspora Jews there. In addition, as the book of Daniel makes abundantly clear, magi had long been associated with Babylon (Dan. 2:2, 10). Babylon was an important Parthian city and a renowned international center of astronomy.[32] We have many Babylonian astronomical almanacs from the third century BC to the first century AD. They reflect an excellent knowledge of the celestial realm. Also in favor of Babylon as the home of the Magi is the fact that, as Eric Bishop points out, the formidable trio of Origen, Jerome, and Augustine all believed this.[33] The opinions of Origen and Jerome are especially weighty, since these men spent a lot of time in Palestine and Jerome even lived in Bethlehem for more than thirty years.[34]

I suggest, therefore, that the Magi may well have come from Babylon.

Babylon at the Turn of the Era. What can we say about Babylon at the turn of the era?

First, the city was still very much inhabited. As T. Boiy points out, "All miscellaneous sources concerning Babylon after 60 BC together clearly prove that there still

[27] E.g., Justin, Tertullian, Epiphanius, and the *Dialogue of Athanasius and Zacchaeus*.
[28] Keener, *Gospel of Matthew*, 99.
[29] So, for example, Clement of Alexandria, Chrysostom, Cyril of Jerusalem, Cosmas Indicopleustes, the *Arabic Gospel of the Infancy*, and early iconographic tradition; Craig Blomberg, *Matthew*, New American Commentary (Nashville: Broadman, 1992), 61–62.
[30] David Pingree, "Astronomy and Astrology in India and Iran," *Isis* 54 (1963): 240–241.
[31] So Celsus; Origen; Jerome; Augustine of Hippo; also Allen, *Gospel according to St. Matthew*, 11; Edwin M. Yamauchi, *Persia and the Bible* (Grand Rapids, MI: Baker, 1990), 481; Carson, "Matthew," 111; Donald A. Hagner, *Matthew*, 2 vols., Word Biblical Commentary (Dallas: Word, 1993–1995), 1:27.
[32] See Simo Parpola, "The Magi and the Star: Babylonian Astronomy Dates Jesus' Birth," in *The First Christmas: The Story of Jesus' Birth in History and Tradition*, ed. Sara Murphy (Washington, DC: Biblical Archaeology Society, 2009), 14.
[33] Eric F. F. Bishop, "Some Reflections on Justin Martyr and the Nativity Narratives," *Evangelical Quarterly* 39 (1967): 33.
[34] Ibid.

was habitation in Babylon until at least the third century AD."[35]

Second, it had a sizable Jewish population. Josephus, *Ant.* 15.2.2 (§14), referring to Parthian king Phraates IV's decision to let Hyrcanus II (an old ally of Herod the Great) establish his residence in Babylon around 40 BC, comments that it was "a place where many Jews lived."[36] Josephus also mentions that at times Jews in Babylon suffered persecution, leading some to emigrate to Seleucia (*Ant.* 18.9.8 [§373]). Philo (*Legatio ad Gaium* 31, line 216) states that, in the days of Emperor Caligula (AD 37–41), "Babylon and many other satrapies were inhabited by Jews."[37] Furthermore, the Babylonian Talmud makes frequent references to Jewish rabbis visiting the city's extensive Jewish community as late as the third and fourth century AD (e.g., *Megillah* 22a; *Ta'anit* 28b; *Baba Batra* 22a and *Eruvin* 63a; *Berakot* 31a, 57b).[38]

Third, the city remained socioeconomically vibrant. It withstood a siege in the middle of the first century BC, and its theatre was repaired in the second century AD.[39] Moreover, an important trade station was located there until at least AD 25 and the city continued to exercise key municipal roles within the general region.[40]

Fourth, with respect to religion, Babylon contained the famous Esagil, the temple of Bel Marduk, together with its priests,[41] and it maintained the ancient system of rituals and offerings and festivals, including the very ancient New Year Festival.[42]

Astronomy in Babylon. Babylon continued to maintain its longstanding and strong commitment to astronomy.

Astronomy. Of all the sciences, none was more closely identified with Babylon or Mesopotamia than astronomy.[43] We have astronomical tablets from the city from as late as AD 74/75.[44] The temple of Bel-Marduk presumably remained the hub of observational and theoretical astronomical activity and record-keeping that it had been in previous centuries.[45] This link to the temple was the main reason why astronomy persisted in Babylon for so long, even after it ceased to have political importance.[46]

Astrology. The astronomy done in Babylon was largely in the service of astrology. As Boiy highlights, the most important means of divination in the Greco-Roman era was astrology, and it necessitated astronomical know-how and indeed astronomical records.[47]

Teukros. The one famous Babylonian astronomer from the turn of the ages about whom we know is Teukros (Teucros/Teucer),[48]

[35] T. Boiy, *Late Achaemenid and Hellenistic Babylon* (Leiden: Brill, 2005), 191. Strangely, however, there was a widespread tendency among Greco-Roman authors to state that Babylon was little more than a desert with the temple of Bel in it: Strabo 16.1.5; Pausanias 8.33.3; Pliny the Elder, *Natural History* 6.121–122; Martianus Capella 6.701; Diodorus 2.9.9; cf. Cassius Dio 68.30.1.
[36] My translation.
[37] My translation.
[38] Boiy, *Babylon*, 189.
[39] Ibid., 192.
[40] Ibid., 191–192.
[41] Ibid., 191.
[42] Ibid., 320.
[43] Cf. ibid., 297.
[44] Ibid., 187–188, 191.
[45] Ibid., 297–303.
[46] Francesca Rochberg, *Babylonian Horoscopes* (Philadelphia: American Philosophical Society, 1998), 12.
[47] Boiy, *Babylon*, 302.
[48] Porphyrius in the third century AD and later authors called Teukros "the Babylonian." A number of scholars have struggled to accept that he was from Mesopotamian Babylon and postulated that he was from a little-known fortress town near Cairo called Babylon. However, as Otto Neugebauer, *The Exact Sciences in Antiquity* (Mineola, NY: Dover, 1969), 189, highlighted, there is no evidence to substantiate this view (see also Franz Boll, *Sphaera* [Leipzig: Teubner, 1903; Hildesheim: Georg Olms, 1967], 158–159; Wolfgang Hübner, "Teukros im Spätmittelalter," *International Journal of the Classical Tradition* 1.2 [1994–1995]: 45). The mere fact that his writing betrays a strong Greco-Egyptian influence should not be regarded as inconsistent with Mesopotamian Babylon. As Boll (*Sphaera*, 158–159) pointed out, there was much interplay between the Greek, Egyptian, and Chaldean worlds, including in the realm of astronomy, around the time of Teukros. Teukros was probably—as most, it seems, now accept—from the astronomical capital of the ancient world: Babylon on the River Euphrates.

who did much to promote Babylonian astronomy and astrology across the ancient world.

At the same time, much concerning the history of Babylonian astronomy remains a mystery.[49] However, it is evident that Babylonian astronomy greatly influenced Greek astronomers such as Hipparchus (second century BC).[50]

The Magi's Journey. If the Magi had come due west from Babylon to Jerusalem, they would have covered about 550 miles.

Time. The long distance raises the question of how long it took the Magi to get to Jerusalem. Kidger proposes that the Magi would have taken 2 weeks to prepare for their journey from Babylon to Jerusalem and that their camel caravan then traveled at approximately 2 miles per hour, or 16 miles per day, with the result that they arrived in Bethlehem about 2 months after they had seen the star in the east.[51] Colin Humphreys proposed that the journey lasted 1–2 months.[52] There is no reason to believe that it took the Magi long to depart, and every reason to think that they would have left quickly—after all, they interpreted the celestial sign to mean that the King of the Jews had just been born in Judea, and they were eager to worship him. We can probably safely presume that they left in the immediate aftermath of the completion of the sign in the east. Assuming they traveled by camel caravan, an arrival time in Bethlehem approximately 28–37 days after they had set out would have been normal by ancient standards (an average speed of 15–20 miles per day).

Destination: Jerusalem. The Magi made their way from their eastern homeland "to Jerusalem." Jerusalem was the capital city of Judea, and it had been the royal city of the Davidic dynasty from the turn of the first millennium BC until the dynasty's eventual demise early in the sixth century BC.

It seems that the Magi had discerned from the celestial phenomenon they had witnessed in their homeland that the long-awaited Messiah had just been born. It was, of course, most natural for them to presuppose that this nativity of the King of the Jews had transpired in Judea and, in particular, in its capital, Jerusalem. The Magi might theoretically have assumed that the new king would be a member of Herod's own household. However, the fact that Herod's first and only encounter with the Magi occurs in Matthew 2:7–9a, after he had summoned them into his presence, suggests that they did not immediately make their way to the palace but simply made enquiries around Jerusalem.

Was the Star present during their journey from Babylon to Jerusalem? The text is not explicit regarding what the Star did after its time in the eastern sky (v. 2) but before its appearance during the final, Jerusalem-to-Bethlehem stage of the Magi's journey (v. 9). Did the Star disappear during that period or did it remain visible?

Some scholars argue that the Star was unobservable for the duration of the Magi's westward journey.[53] This interpretation could be regarded as favored by: (a) the fact that the Magi, rather than going straight to Bethlehem, went to Jerusalem and there enquired as to where the newborn messianic King was (vv. 1–9a); (b) verse 9's reference to the Star of Bethlehem as "the star that they had seen at its rising"; and (c) the Magi's ecstatic joy upon seeing the Star at the climax of their journey (v. 10).

However, in response to (a), the Magi may

[49] As Otto Neugebauer, *A History of Ancient Mathematical Astronomy*, 3 vols. (Berlin: Springer, 1975), 555, concluded.
[50] Boiy, *Babylon*, 308–309.
[51] Mark R. Kidger, *The Star of Bethlehem: An Astronomer's View* (Princeton, NJ: Princeton University Press, 1999), 29–30.
[52] Colin J. Humphreys, "The Star of Bethlehem, a Comet in 5 B.C., and the Date of the Christ's Birth," *Tyndale Bulletin* 43 (1992): 34 and 48. See also idem, "The Star of Bethlehem—A Comet in 5 B.C.—And the Date of the Birth of Christ," *Quarterly Journal of the Royal Astronomical Society* 32 (1991): 389–407; idem, "The Star of Bethlehem," *Science and Christian Belief* 5 (1995): 83–101.
[53] E.g., Kenneth D. Boa, "The Star of Bethlehem" (ThM thesis, Dallas Theological Seminary, 1972), 34 (http://www.kenboa.org/downloads/pdf/TheStarofBethlehem.pdf, accessed March 12, 2013).

have visited Judea's capital city because the Star's guidance was general rather than specific, because they wrongly concluded that the Star had pinpointed Jerusalem or completed its mission, or because clouds prevented them from seeing it in the period immediately prior to their arrival in Jerusalem.

In response to (b), the reference to the Star during the final stage of the Magi's journey as the one that they had seen earlier, at its rising, need not imply its absence in the intervening period, but may merely underscore the natal significance of the Star. It was, after all—as we shall highlight below—what the Star did in connection with its rising that disclosed to the Magi that the King of the Jews had been born. When the Star appears in order to usher the Magi to Bethlehem and ultimately to the Messiah himself, it is very natural to recall that this same Star had by its behavior in the eastern sky prompted the Magi to make their remarkable pilgrimage to Judea in search of the newborn Messiah.

In response to (c), as we shall see, the Magi's joy upon seeing the Star in verse 10 probably does not relate to the appearance of the Star in the sky as they left Jerusalem, but rather to the standing up of the Star over the house in Bethlehem where Jesus was, which was the focus of verse 9b.[54]

In favor of the view that the Star did not disappear from sight for the duration of the Magi's journey to Judea is that the Star that they saw on the last phase of their trip, from Jerusalem to Bethlehem, is explicitly identified as the very same one they had previously observed, back in their eastern homeland (v. 9). Had there been any notable period of absence between the two stages of the Star's appearance, there could have been no certainty that the same identical Star was in view. This is particularly the case because the Star at the climax of the Magi's journey is in a completely different part of the sky—the south—whereas at the time of its rising it was in the east. The only logical conclusion to draw is that the Magi had been tracking the Star's movements in the intervening period.

That raises the question of where the Star was while the Magi were traveling from their homeland to Jerusalem. Did it quickly move to the west to guide them toward Judea? A westward-leading Star might have played some part in the Magi's decision to travel 550 miles to worship the Messiah, and the western route that they seem to have taken. Certainly by the end of their journey it had migrated to the southern sky, but that does not mean that it was not, in the days leading up to their arrival in Jerusalem, being perceived to be guiding them westward. Any star that appears at a reasonable altitude in the southern sky in the first half of the night, as evidently did the Star of Bethlehem at the climax of its appearance, will naturally proceed over the course of the following hours to set in the western sky. Accordingly, as the Star advanced toward and then over the horizon each night, it may well have seemed to the Magi to be going on ahead of them to Jerusalem. Indeed it is very possible that, the night before the Magi arrived in Jerusalem, the Star set in the direction of Judea's capital city, prompting them to conclude that it was their destination. If the Magi had come to regard the Star as their very own celestial guide over the preceding weeks, their perception that it was, on the final stage of their journey, ushering them to Bethlehem and then pinpointing the Messiah's location within Bethlehem would make especial sense.

While therefore the text does not make explicit what the Star did between its appearance in the eastern sky and its appearance when the Magi departed from Jerusalem toward

[54] Even if one were to assume that the Magi's joy marked the appearance of the Star as they embarked on their journey from Jerusalem to Bethlehem, that would hardly require that they had not seen it for the duration of their journey. The Eastern astrologers might conceivably have assumed that the Star, having led them to Judea, had done its job and would therefore no longer play a role in their pilgrimage.

Bethlehem, it is probable that the Star was present throughout this time and that it was perceived by them to be a westward guide.

The Wonder in the Eastern Sky. Matthew 2:2 records that when the Magi arrived in Jerusalem, they asked, "Where is he who has been born king of the Jews? For we saw his star at its rising and have come to worship him." This is the only statement attributed to the Magi in the narrative. From it we get some idea of what the Magi saw in the heavens that motivated them to travel westward, and how they interpreted it. That they mention that they *saw* the Star's rising indicates that it was an observed phenomenon rather than something that they merely calculated, and this may well imply that what happened at that time was, as far as they were concerned, unexpected and marvelous.

Announcing a birth. The Magi believed that the astronomical wonder they had seen indicated that a special birth had taken place. In light of this, and considering the likelihood that they left promptly and traveled some 28–37 days from their homeland to Jerusalem, we conclude that they would have fully expected to find a recently born baby boy.

King of the Jews. Somehow the Magi had deduced that the one whose birth had been announced in the heavens was in particular the King of the Jews.

The Messiah's Star. The Eastern Magi justified their question regarding the location of the recently born messianic baby by explaining that they had seen "his star" (Matt. 2:2). The Magi clearly regarded one particular celestial body as closely identified with the Messiah; it was "his star." How they made this determination is not stated. Perhaps there was something about the celestial phenomenon itself that was suggestive of a Jewish personage, but the key was probably Jewish tradition, more specifically the Hebrew Scriptures. In particular, most scholars have correctly detected here an allusion to a messianic oracle of the prophet Balaam, recorded in Numbers 24:17. In this prophetic word the seer from Mesopotamia (Deut. 23:4) foresaw that in the distant future "a star shall come [Hebrew; the Greek Septuagint (LXX) reads "a star shall rise"] out of Jacob, and a scepter shall rise out of Israel." (We will explore this prophecy in detail in chapter 8.)

Matthew's account makes it clear that the Magi had seen something extraordinary back in their homeland, something that engendered in them a certainty that the Jewish Messiah had been born. At the same time, the revelation they had was limited, for they did not know at this time where precisely within Judea the baby king was. Indeed they may have assumed that he would be in Jerusalem.

The Star's rising. The astronomical wonder that had convinced the Magi that the Messiah had recently been born in Judea occurred "at [the Star's] rising" (Matt. 2:2).[55]

The Greek word *anatolē* could theoretically refer either to a celestial entity's nightly rise above the horizon *or* to its first rising above the eastern horizon just prior to

[55] Most modern scholars believe that the use of the singular phrase *en tē anatolē* should be rendered "at its rising." However, a minority favor "in the east," claiming that the preposition with the article can on rare occasions be used of compass directions, as in Hermas, *Vis* 1.4.1, 3. Furthermore, they point out that the singular form of *anatolē* can be used of "the east," as in Rev. 21:13 and Hermas, *Vis* 1.4.1, 3, and as is common in Josephus (Friedrich Blass, Albert Debrunner, and R. W. Funk, *A Greek Grammar of the New Testament and Other Early Christian Literature* [Chicago: University of Chicago Press, 1961], §141.2). They try to explain the change from the plural (*anatolōn*) in Matt. 2:1 to the singular (*anatolē*) in v. 2 as merely stylistic. Luz, *Matthew 1–7*, 128n1, states that it is awkward to assign different senses to *anatolē* in vv. 1 and 2 (although he does opt for the meaning "rising" rather than "east" here). However, the case for "at its rising" in v. 2 is stronger: (a) This meaning is more likely in an astronomical context. (b) The preposition with the article is only rarely used of compass directions (see Blass et al., *Greek Grammar*, §253.5). (c) The employment of the singular in v. 2 (*en tē anatolē*), in contrast to the plural form in v. 1 (*apo anatolōn*, "in the east"), suggests that the sense is different. Of course, whichever way we translate it, the phenomenon was probably seen in the east, since generally heliacal risings occur in the east (on "heliacal" risings, see note 57, and the accompanying main text). In the final analysis, "in the east" is possible, but "at its rising" is to be preferred.

dawn.[56] Once every year, fixed stars (other than those in close proximity to the poles), after disappearing from the night sky for a period, reappear low in the eastern sky just before the Sun rises at dawn. When the Sun is still some distance under the horizon (how far depends on the entity's brightness, location, and other factors), the object makes a brief appearance before its light is overwhelmed by the light of the sunrise. This is called a heliacal rising, and in ancient times it was considered the most important regular event in the career of a fixed star or constellation.[57] In the case of a fixed star, the heliacal rising occurs annually on one particular day. In the case of a constellation, it obviously stretches out over a longer period, the precise duration depending on the size of the constellation and the brightness of its stars. With respect to "wandering stars" like planets, heliacal risings do not occur at one particular set time each year, but according to a different, more complicated schedule, depending on the planet's orbital course. As regards comets, generally they are capable of heliacally rising in the eastern sky only whenever they are in the inner solar system[58] (whether their return is every four years or every million years) and when their orbit takes them into conjunction with the Sun and then to its east side, from Earth's perspective. Because comets are not point sources but rather extended objects, technically they heliacally rise over a period of time, like constellations. In ancient Babylonian and Chinese comet reports, however, the focus was on the coma or "head" (that is, the sunward part of the comet) and hence a report of a comet's heliacal rising would most naturally refer to its coma. Of course, the date of a heliacal rising was strongly dependent on the weather.[59]

It seems clear that *anatolē* in Matthew 2:2 is referring to a heliacal rising rather than a standard nightly rising.[60] The fact that the Magi even mention the context of the Star's display ("at its rising") strongly favors the view that this is significant information that played a key part in their assessment of the astronomical phenomenon. It is unclear what that significance could be if it were merely the Star's nightly rising. Moreover, since the Magi (and Matthew, the narrator) employ the possessive article—"at *its* rising"—it is evident that a distinctive occasion in the Star's career is in mind. One should also observe that ordinary, nightly risings of stars, signs, and constellations over the horizon had no astro-

[56] For the nontechnical usage of *anatolē* referring to a heliacal rising, see the first-century BC Greek astronomer and mathematician Geminos's *Introduction to the Phaenomena* 13.3–5. Note especially *Introduction aux phénomènes*, ed. and trans. Germaine Aujac (Paris: Les Belles Lettres, 1975), 68n1, which points out that, although Geminos insisted on using distinct terms for heliacal and daily risings and criticized others for employing them interchangeably, he actually did this himself on occasion, as did other astronomical writers, for example Autolykos (see James Evans and J. Lennart Berggren, *Geminos's* Introduction to the Phenomena*: A Translation and Study of a Hellenistic Survey of Astronomy* [Princeton, NJ: Princeton University Press, 2006], 70–71). Other instances of *anatolē* being used in the same nontechnical way as Matthew to refer to a heliacal rising include Homer, *Odyssey* 12:4; Plato, *Politicus* 269a; Euripides, *Phoenissae* 504; Testament of Levi 18:3; and *Papyri Graecae Magicae* 13:1027. The term for a heliacal rising preferred by Ptolemy was *epitolē*.

[57] Cf. Clive L. N. Ruggles, *Ancient Astronomy: An Encyclopedia of Cosmologies and Myth* (Santa Barbara, CA: ABC-CLIO, 2005), 398. For a contemporary Greco-Roman explanation of heliacal risings, see Geminos's *Introduction to the Phenomena* 13.3, 5 and 9–10 (Evans and Berggren, *Geminos's* Introduction, 200–201). For the importance of heliacal risings in Babylonian astronomy, see Rochberg, *Babylonian Horoscopes*, 6, 124. With respect to the Moon or a comet, a heliacal rising may also occur in the western evening sky, as the entity moves away from the Sun after being in conjunction with it. See the discussion in Courtney Roberts, *The Star of the Magi: The Mystery That Heralded the Coming of Christ* (Franklin, NJ: Career Press, 2007), 120–121.

[58] By "inner solar system" I am referring to the area from the asteroid belt to the Sun. Some distinguished comets, like Hale-Bopp, are capable of heliacally rising also when in the "outer solar system" (i.e., the region from Neptune to Jupiter).

[59] The Babylonian Diaries contain many records of the weather and reveal that, surprisingly often, astronomical observations were impossible due to clouds, rain, mist, and fog, and that frequently over a number of nights in a row observations of the stars were rendered impossible (Noel M. Swerdlow, *The Babylonian Theory of the Planets* [Princeton, NJ: Princeton University Press, 1998], 17–18).

[60] Among the many who have appreciated that a heliacal rising is in view here are: A. H. McNeile, *The Gospel according to St. Matthew: Greek Text with Introduction, Notes, and Indices* (London: Macmillan, 1915), 15; W. D. Davies and Dale C. Allison, *A Critical and Exegetical Commentary on the Gospel according to Saint Matthew*, 3 vols., International Critical Commentary (Edinburgh: T. & T. Clark, 1988–1997), 1:235–236; Kidger, *Star of Bethlehem*, 27; J. Neville Birdsall, in Owen Gingerich, "Review Symposium: The Star of Bethlehem," *Journal of Biblical Literature* 33 (2002): 391, 393, 394; Tim Hegedus, *Early Christianity and Ancient Astrology* (New York: Peter Lang, 2007), 202; and Richard Coates, "A Linguist's Angle on the Star of Bethlehem," *Astronomy and Geophysics* 49.5 (October 2008): 28.

logical importance, whereas heliacal risings could be perceived to have great astrological significance. Ancient astrologers developed horoscopes based on the zodiacal sign that had been heliacally rising at the point when a client was born.[61] Of course, low over the eastern horizon shortly before dawn, a celestial body is seen in the context of a heliacally rising sign and constellation. Particularly if the Star was within the zodiac, it was natural for the Magi to consider the possibility that it might be communicating something of natal significance against the backdrop of the heliacally rising constellation.

In addition, when the Magi mention the Star's rising, they may well be alluding to Balaam's oracle about a star-scepter. Prophesying about the coming of the Messiah, the Mesopotamian seer declared that the messianic star-scepter would "rise." By this he seems to have been referring to a heliacal rising in the eastern sky (cf. LXX Zech. 6:12; Luke 1:78; 2 Pet. 1:19; Rev. 2:26–28; 22:16). The Magi's apparent allusion to Numbers 24:17 would favor the conclusion that *anatolē* in Matthew 2:2 has in view a heliacal rising.

In light of all of this, it is likely that when the Magi mentioned the "rising" of the messianic star, they were thinking not of a daily rising but rather of a heliacal rising, that is, the predawn emergence of the Star over the eastern horizon after having been obscured by the Sun.

However, it is unclear whether the phrase "at [the time of] its rising" is being used narrowly or generally. If narrowly, the Magi would be referring only to what they saw during the short window (or, in the case of extended objects, windows) of time when the celestial entity appeared over the eastern horizon before the dawning Sun's rays bleached the sight. If the phrase is being used generally, the Magi would presumably be referring to the period when the Star was in the eastern sky, beginning with its heliacal rising.[62] By and large, after a celestial body has heliacally risen over the eastern horizon, it continues to separate itself from the Sun and so appears earlier and in a darker sky, becoming easier to see. The remarkable conclusion reached by the Magi—that the celestial wonder was announcing the birth of the divine Messiah—may favor the view that the Magi were speaking generally of the Star's whole time in the eastern sky.

We suggest therefore that the Magi were convinced that the Messiah's birth was taking place when they saw the Star at or around the time of its heliacal rising. Evidently, the Magi perceived significance in the Star's location within the constellations, its form, and/or its behavior, and/or in the timing of the heavenly wonder.

Babylonian astrology and births. Babylonian astrologers in the last centuries of the first millennium BC were convinced that a person's fate was encoded in the heavens at the point of birth.[63] Accordingly, they created horoscopes for the date (occasionally even the part of the day or night or the hour) of the subject's birth,[64] based on their astronomical almanacs, diaries, and other collections of data.[65] Twenty-eight Babylonian tablets con-

[61] Brown, *Birth of the Messiah*, 173. The "ecliptic" is the apparent path of the Sun through the sky. The "zodiac" in astronomy refers to the band of sky around the ecliptic through which the Sun, Moon, and planets appear to traverse. The zodiacal constellations are the 13 constellations through which the ecliptic passes. On the ecliptic and the zodiacal constellations, see fig. 7.12. The vast majority of constellations fall outside the zodiac. Zodiacal signs—fixed 30-degree geometric zones of the ecliptical band—should not be confused with zodiacal constellations, which are star groupings of unequal sizes.

[62] Matthew employs *en* plus the dative in a general time reference in, for example, 3:1; 11:25; 12:1 and 14:1. In Luke 14:14 Jesus states that eschatological rewards will be dispensed "at the resurrection" (*en tē anastasei*), and in 20:33 the Sadducees ask whose wife the woman who was married to the seven brothers will be "at the resurrection" (*en tē anastasei*). In each case the temporal *en*-phrase is used generally of the period that begins with the resurrection. "At its rising" in Matt. 2:2 could be regarded as synecdoche (a figure of speech in which a part is used for the whole, or vice versa) for the entire period during which the Star was in the eastern sky; cf. Matt. 5:45, where the Sun's rising represents its shining during its entire daily course through the sky.

[63] Rochberg, *Babylonian Horoscopes*, 15.

[64] Ibid., 1–2, 33–39.

[65] Ibid., 33.

taining horoscopes have survived, covering 410 to 69 BC.[66] Typically, they specify the positions of the Moon, Sun, Jupiter, Venus, Mercury, Saturn, and Mars within the zodiacal signs, in that order, noting where a planet was not visible due to conjunction with the Sun.[67] They also detail other astronomical data relevant to the month or year of the birth, particularly lunar phenomena and eclipses.[68]

In essence, Babylonian astrologers could detail two kinds of omens to mark a nativity: one (horoscopic) type related to the zodiacal sign in which the subject was born; the other (non-horoscopic) type related to some astronomical phenomenon that occurred on the birthdate (e.g., Jupiter's heliacal rising or a solar eclipse) which was perceived to disclose the subject's destiny.[69]

It is clear from Matthew that what impressed the Magi and prompted them to travel to Jerusalem related to what a Star did at or around the time of its heliacal rising. Obviously, therefore, what they saw occurred in the period shortly before dawn, around the time when certain other stars and constellations were heliacally rising. To that extent, what they saw was reminiscent of a horoscopic omen. However, in this case the focus was not on the place of the planets within the zodiacal signs but rather on the behavior of the one particular "star" that they associated with the Jewish Messiah. What the star did that was so meaningful to the Magi, it did in connection with its own heliacal rising, against the backdrop of the constellations. So the celestial sign was fundamentally non-horoscopic in nature, concerning an astronomical event that coincided with the Messiah's birth.

Normally, the relevance and significance of an astronomical natal omen was determined retrospectively, decades after the birthdate. However, in the case of Jesus's birth omen, the Magi perceived the meaning and significance of the celestial phenomenon as it happened. Based on it, they were able to deduce that the Messiah had been born in Judea.

The Magi's Response to the Wonder. The Magi were deeply impacted by the behavior of the Star in the eastern sky.

Pilgrimage. The astrologers set out on their long journey to Jerusalem, inspired by what the Star had done in connection with its heliacal rising.

Intention: worship. The Magi were intent on worshiping the messianic child. The Greek verb *proskuneō* often means "to express by attitude and possibly by position [namely, prostration] one's allegiance to and regard for deity."[70] It may also be used in nonreligious contexts of simple kneeling before one of higher rank, and is used thus in Matthew 18:26. However, Matthew almost always used the verb of an act of worship. When he is speaking of what people do in Jesus's presence, it always refers to worship.[71] This is in keeping with Matthew's high Christology and his strong emphasis on Jesus as the Son of God.[72] W. D. Davies and Dale Allison argue strongly that the verb implies worship here in Matthew 2:2, based on the fact that the verb "come" (*erchomai*) followed by "worship" (*proskuneō*) in the Greek translation of the Hebrew Scriptures (LXX) usually refers to a cultic act, and because Jews regarded full prostration (*proskynesis*) as appropriate only when directed to

[66] Ibid., 3.
[67] Ibid., 7.
[68] Ibid., 11, 39–45.
[69] Ibid., 14.
[70] Johannes P. Louw and Eugene A. Nida, *Greek-English Lexicon of the New Testament: Based on Semantic Domains* (New York: United Bible Societies, 1988), §53.56.
[71] See, for example, Matt. 4:9–10; 14:33; 28:9, 17; so H. Greeven, "*proskuneō*," in *Theological Dictionary of the New Testament*, ed. Gerhard Kittel and Gerhard Friedrich, 10 vols. (Grand Rapids, MI: Eerdmans, 1964–1976), 6:763–764; Davies and Allison, *Matthew*, 1:236–237; John P. Meier, *Matthew*, New Testament Message (Dublin: Veritas, 1980), 11.
[72] David L. Turner, *Matthew*, Baker Exegetical Commentary on the New Testament (Grand Rapids, MI: Baker Academic, 2008), 81.

God (cf. Matt. 4:9–10; Acts 10:25–26; Rev. 19:10; 22:8–9; Philo, *Legatio ad Gaium* 31, line 116; idem, *De decalogo* 64).[73] The Magi, then, seem to have come with the intention not just of paying homage to the newborn King of the Jews, but of worshiping him as a deity. This obviously forces us to ask the question, What convinced the Magi that the recently born ruler was worthy of worship? This is difficult to explain unless the celestial sign that they saw directly or indirectly convinced them that the King of the Jews was both human and divine.

Herod and the People of Jerusalem

The Magi clearly expected the people of Jerusalem, or at least some of them, to know that the Messiah had been born, and where he was now. However, in this and in their trust of Herod later in the story, the Magi were mistaken: "When Herod the king heard this, he was troubled, and all Jerusalem with him; and assembling all the chief priests and scribes of the people, he inquired of them where the Christ was to be born" (Matt. 2:3–4).

Herod the Great. The repetition of Herod's title "the king" here stands in sharp tension with the Magi's claim in verse 2 that the King of the Jews has just been born.

Historical backdrop. Herod had been endowed with the title King of the Jews by the Roman Senate.[74] As the reigning king in his final years, Herod was utterly obsessed with securing his dynasty by choosing from his sons a worthy successor or successors.

Herod had ten wives and many sons. His most important children as regards the succession were Alexander and Aristobulus (sons of Mariamne I); Antipater (son of Doris); Archelaus and Antipas (sons of Malthrace); Philip I (son of Mariamne II); and Philip II (son of Cleopatra of Jerusalem).

When Alexander and Aristobulus returned from Rome to Judea in 17 BC, Herod let his favorable sentiment toward them be known. However, Herod's sister Salome spread the rumor that these sons were conspiring against him. As a result, Herod turned his favor upon his eldest son and the child of his first marriage, Antipater, appointing him the sole heir. As for Alexander and Aristobulus, he decided to make charges against them before the Roman emperor. However, Herod in due course changed his mind about Alexander and Aristobulus and was reconciled with them. Thereafter, in 12 BC he incorporated them back into his will, so that each of them would be a ruler over a part of the territory. Unfortunately for Herod, this ideal state of affairs did not last for long. In 7 BC new rumors began to circulate to the effect that Alexander and Aristobulus were plotting to assassinate Herod; whether these rumors had a basis in fact or were merely manufactured by Antipater we do not know. Certainly Herod believed them, and he responded with fury and had the two siblings tried and executed. The king decided that Antipater should be the sole king, although now he specified that Philip I would be Antipater's successor. Then, in 6 BC, Philip I was removed from the will, because Herod suspected that his mother was guilty of conspiracy against him, leaving Antipater as the sole specified heir.[75]

It is around this time, in 6 or 5 BC, when Herod was acutely paranoid and focused on the succession, and perhaps thinking that he had finally sorted out the whole messy business, that the Magi entered Jerusalem asking where the newborn King of the Jews was and

[73] Davies and Allison, *Matthew*, 1:236–237, 248. They also argue that Matthew, in his redaction, tends to use the word only with reference to God (237).

[74] E.g., Josephus, *Ant.* 16.10.2 §311.

[75] See Josephus, *Ant.* 15–17; Harold W. Hoehner, *Herod Antipas* (Cambridge: Cambridge University Press, 1972), 269–276; Peter Richardson, *Herod: King of the Jews and Friend of the Romans* (Columbia: University of South Carolina Press, 1996), 33–36; Jerry Knoblet, *Herod the Great* (Lanham, MD: University Press of America, 2005), 133–138.

declaring that they had seen his star in the eastern sky.

In fact, Herod's paranoia and dynastic woes continued until his death. Early in 5 BC the king discovered that Antipater, before departing for Rome in 6 BC, had been conspiring to poison him. When Antipater returned to Judea late in 5 BC, Herod had him imprisoned and reported the crime to the emperor. He then named his youngest son, Antipas, sole heir. However, just prior to his death in the spring of 4 BC, Herod had yet another change of heart and divided up his kingdom between Antipas, Archelaus, and Philip II. Five days before Herod died, Antipater, his firstborn son, was executed.

Herod's Response to the Magi. Herod was "troubled" by the announcement of the Magi concerning the birth of the newborn King of the Jews. The historical context helps us make sense of this. As we have seen, during the last four years of Herod's life he was extremely paranoid, and with some justification. He had already killed two of his sons and 300 military officers supposedly conspiring with them in 7 BC and, within a few years, would have cause to have another son tried for conspiracy and executed. Consequently, Herod was unlikely to take kindly to any threat to his dynasty.

In addition, Richardson suggests that Herod may have been strongly hostile to messianic movements generally.[76]

At the same time, Herod clearly believed that the one who had just been born was the actual Messiah. Later, Matthew tells us of how Herod assembled the chief priests and scribes to determine where the Messiah had been born, based on the Hebrew Scriptures, and passed this information on to the Magi, anticipating that they would find the newborn King of the Jews there (Matt. 2:4–8). So convinced was Herod that the Messiah had been born that he slaughtered every baby boy in the region of Bethlehem in their first or second year of life in a desperate attempt to assassinate him (v. 16). His fear therefore probably reflects his belief that the prophesied Messiah would pose a formidable threat to his dynasty.[77]

Herod was part Jewish and has been generously described by one biographer as a man of "piety" who adhered to "simple and uncluttered" Judaism.[78] Certainly he was, at the very time when the Magi visited, overseeing the reconstruction and beautification of the Jerusalem temple. And yet Herod, though persuaded by the Magi that the Messiah had been born, did not rejoice, but recoiled with horror, because this momentous event did not accord with his succession plans.

The People of Jerusalem. Surprisingly, "all of Jerusalem" was also "troubled" (v. 3) by the Magi's announcement. Although some scholars have argued that the city's religious leaders are in view here,[79] that is too narrow a reading of the phrase. The more natural interpretation is that it refers to the general population of the city. However, we might well wonder why the people of Jerusalem responded so negatively to the Magi's proclamation. It can hardly be that the Jerusalemites preferred Herod to the Messiah. More likely the people in Jerusalem were troubled because they liked the status quo and were certain that Herod would respond with brutality to any serious threat to his dynasty. They may also have been afraid that Judea could degenerate into civil war. While some degree of fear

[76] Richardson, *Herod*, 295, who qualifies his statement by pointing out that Herod took a more tolerant approach to the Essene movement.

[77] Notably, when speaking to the Magi, Herod claimed to hold to a high Christology ("that I too may come and worship him"), evidently because his Eastern visitors did (v. 8).

[78] Richardson, *Herod*, 295.

[79] Blomberg, *Matthew*, 63; R. A. Horsley, *The Liberation of Christmas: The Infancy Narratives in Social Context* (New York: Crossroad, 1989), 49–52; and Turner, *Matthew*, 81.

might be expected, the lack of any positive rejoicing at the news that the long-awaited Messiah has finally been born is disturbing and, within the context of Matthew's Gospel, anticipates the city's rejection of Jesus at his trial (Matt. 27:15–26).

Had Herod and the Jerusalemites Seen the Star? The response of Herod and the people of Jerusalem has sometimes been taken to indicate that they had not seen the Star themselves.[80] However, this is most unlikely. It would be very surprising if the people of Judea would have accepted as a celestial sign of the Messiah's birth any phenomenon capable of being observed only by pagan Gentiles in Babylon and not at all by the Messiah's own people in Judea. Moreover, the strength of the reaction of the king and people to the arrival of the Magi's entourage and their query makes better sense if they had seen for themselves and been deeply impressed by the Star but had not perceived its momentous messianic significance.[81] Had they not seen the Star for themselves, they would hardly have been so shaken by the Magi's enquiry. What was new to the people of Jerusalem was not that there was a Star or even that the Star had done something unusual in connection with its heliacal rising, but rather that the Star had categorically signaled that the Messiah had recently been born. Exposed to that startling and evidently compelling interpretive key by some of the world's most respected astronomers and astrologers, who were so certain of their interpretation that they had just traveled hundreds of miles to welcome the newborn Messiah, suddenly Herod and the Jerusalemites became disturbed concerning the Star.

Of course, it is possible that not everything the Star did was detected by those in Jerusalem, whether because of inclement weather, a lack of dedicated observation, or an inopportune time of occurrence.

Herod's Meeting with the Jewish Teachers

Herod's Ignorance. It is clear that Herod did not know where the Messiah was to be born. Apparently Micah 5:2, with its disclosure of the location of the Messiah's birth, was not widely known or, at any rate, not widely understood. The Magi, Herod, and the population in Jerusalem as a whole were, it would seem, unaware that this verse held the key to identifying the place of the Messiah's birth.

The king therefore assembled "all the chief priests and scribes of the people" (Matt. 2:4), which may perhaps mean that he summoned the whole Sanhedrin[82] or simply that he gathered a sizable group of respected Bible scholars (in the Gospels, the Sanhedrin is normally designated "the chief priests, the scribes, and the elders," but the elders are not mentioned here).

The Teachers' Response. Herod presented the religious experts with his simple question: Where was the Messiah to be born? This half-Jewish king of Judea was clearly intent on assassinating the Messiah while he was still a baby. Completely devoid of any fear of God, he was prepared to use the revelation God had given concerning his plan of salvation in the Hebrew Scriptures—to thwart the divine plan! The hard-heartedness and audacity of this man who had made the Second Temple one of the most glorious structures in the ancient world are mind-boggling. So self-deluded is this king of Judea that he actually imagines that he can take on God and win!

According to Matthew 2:5–6, the chief priests and scribes "told [Herod], 'In Bethlehem of Judea, for so it is written by the prophet:

[80] So, for example, Kenneth Boa and William Proctor, *The Return of the Star of Bethlehem: Comet, Stellar Explosion, or Signal from Above?* (New York: Doubleday, 1980), 24, 38.
[81] Cf. Kidger, *Star of Bethlehem*, 29.
[82] Brown, *Birth of the Messiah*, 175.

"And you, O Bethlehem, in the land of
Judah,
are by no means least among the
rulers of Judah;
for from you shall come a ruler
who will shepherd my people
Israel.'"

The response of the Jewish religious leadership to Herod's question reveals a lot. Their answer reflects a high view of the Hebrew Scriptures. They regard Micah's oracle as the word of God channeled through a prophetic agent ("it is written by the prophet"), and they interpret it in a literal and straightforward manner to refer to the Davidic Messiah. The chief priests and scribes manifestly do have a basic grasp of God's plan of salvation through the Messiah. It is striking that Matthew is content to let these Jewish leaders introduce Micah 5:2 into the narrative concerning the birth of Jesus. Matthew does not explicitly state that the religious leadership was aware of the report of the Magi from the east. However, word concerning the Magi had spread like wildfire through the city, so that "all Jerusalem" heard it, and it is hard to justify excluding the Jewish religious leaders from this, particularly because Jerusalem was so oriented around the temple. Accordingly, when they answered Herod's question concerning the birthplace of the Messiah by appealing to Micah's prophecy, they were effectively testifying that, if what the Magi had seen was indeed the Messiah's natal sign, the Messiah was at that very moment a newborn baby in Bethlehem.

Remarkably, however, the Jewish religious leaders, despite having a knowledge of the Word of God considerably greater than that of the Gentile Magi, made no effort to travel the five or six miles south to Bethlehem to see if indeed the Messiah had been born in fulfillment of the Prophets. They evidently despised the report, and perhaps those who brought it, and so they remained in Jerusalem. They were content with the status quo and did not crave the promised salvation of God.

Micah's Oracle. The quotation from Micah 5:2 is significant:

But you, O Bethlehem Ephrathah,
who are too little to be among the
clans of Judah,
from you shall come forth for me
one who is to be ruler in Israel,
whose origin[83] is from of old,
from ancient days.
Therefore he shall give them up until
the time
when she who is in labor has given
birth.

Micah, a contemporary of Isaiah, here declares that the messianic King, the one through whom Yahweh will supremely fulfill his covenant promises to David, will be born in Bethlehem, the very same town in which King David was born. The apparent implication of Micah's prophecy concerning Bethlehem was that the Messiah would have strong Davidic ancestry.[84]

Micah's reference to "she who is in labor" giving birth seems to recall Isaiah 7:14, where Isaiah foretold that "the virgin shall be with child and bear a son, and shall call his name Immanuel." If so, Micah has evidently interpreted Isaiah's prophecy as referring to the Messiah's birth.[85] Notably, Matthew quotes Isaiah 7:14 in Matthew 1:23, claiming that it was fulfilled when Jesus was born to the Virgin Mary.

Matthew omits the latter part of Micah's

[83] My translation.
[84] Incidentally, Luke 2:4–5 informs us that Jesus's legal father, Joseph, was forced to go with his betrothed to Bethlehem for a census because "he was of the house and lineage of David."
[85] At the same time, Micah seems to portray Israel/Zion as being in labor and giving birth to the Messiah.

prophecy of the Messiah's birth: "whose origin is from of old, from ancient days. Therefore he shall give them up until the time when she who is in labor has given birth" (Mic. 5:3). However, as Davies and Allison suggest, the readers are probably supposed to fill this in for themselves.[86]

By replacing Micah's "Ephrathah" with "in the land of Judah," Matthew highlights that the Messiah had to be a member of the tribe of Judah, in accordance with Genesis 49:9–10, where Jacob prophesied that "Judah is a lion's cub. . . . The scepter shall not depart from Judah, nor the ruler's staff from between his feet, until tribute comes to him; and to him shall be the obedience of the peoples."

Matthew also edits Micah's "ruler in Israel" to say "ruler who will shepherd my people Israel" (Matt. 2:6). The introduction of shepherd imagery seems intended to recall God's promise to David in 2 Samuel 5:2 ("You shall be shepherd of my people Israel, and you shall be prince over Israel") and is thus very much in tune with the thrust of Micah's prophecy.[87]

Matthew's most striking change to the quotation from Micah is to transform "are too little to be among the clans of Judah" into "are by no means least among the rulers of Judah" (Matt. 2:6). In Micah, "small" highlights that the city in the era before fulfillment is of little importance, implying that its status will change fundamentally after the Messiah is born there. Matthew's "by no means" is consistent with this.[88] There has been a remarkable change in Bethlehem's significance as a result of the Messiah's birth there. From the eighth century BC, when Micah wrote, until Jesus's birth at the end of the first century BC, Bethlehem had been an insignificant town, with its sole claim to fame being that King David had been born there. However, now it was guaranteed to be esteemed and famous forever, because the Messiah himself had been born there.

Herod's Meeting with the Magi

Matthew 2:7–9a relates the story of Herod's meeting with the Magi: "Then Herod summoned the Magi secretly and ascertained from them what time the star had appeared. And he sent them to Bethlehem, saying, 'Go and search diligently for the child, and when you have found him, bring me word, that I too may come and worship him.' After listening to the king, they went on their way." This meeting between Herod and the Magi (see fig. 3.1) was apparently the first and only one of its kind.

Secrecy. The fact that Herod summoned the Magi "secretly" demands explanation. Why did Herod not want the people of Jerusalem to know about his meeting with the Magi? Most likely he had at least two main motives for his secrecy. First, he would not have wanted to lend credence to the Magi's announcement that the Messiah had been born.[89] Second and more important, Herod did not want to endanger his dastardly secret plan to kill the Messiah.[90] The Magi were ignorant of Herod's scheming ways, but the people of Jerusalem and Judea were not. Herod would have realized that some in Judea would have been enthusiastic about the announcement of the Magi concerning the arrival of the Messiah on the earthly scene. Were word to get out that Herod had told the Magi to report back to him regarding the precise location of the Messiah so that he could go and worship the

[86] Davies and Allison, *Matthew*, 1:244.
[87] See Carson, "Matthew," 115.
[88] R. T. France, *The Gospel of Matthew*, New International Commentary on the New Testament (Grand Rapids, MI: Eerdmans, 2007), 73.
[89] Cf. Stanley Hauerwas, *Matthew*, Brazos Theological Commentary on the Bible (Grand Rapids, MI: Brazos, 2006), 40.
[90] Most scholars appreciate that Herod had already concocted his plan to assassinate the newborn Messiah (so, for example, Luz, *Matthew 1–7*, 136; Carson, "Matthew," 115).

FIG. 3.1 A watercolor, "Les rois mages chez Hérode" ("The Magi in the House of Herod"), by James Tissot. Image credit: Brooklyn Museum, New York.

newborn King of the Jews, some pro-Messiah Judeans would undoubtedly have alerted the Magi to the king's obviously malign agenda.

Herod's Agenda. Herod treated the Magi's report with deadly earnestness. He wanted to know from them two critical pieces of information: the maximum age of the Messiah, and where exactly within Bethlehem he resided. Armed with these facts, he would be in an excellent position to strike down the infant King.

Discovering the Messiah's age. Herod acquired from the Magi accurate information concerning the precise time at which the Star had appeared (Matt. 2:7). Whether they knew this off the top of their heads or had written records of this is not stated. However, since record-keeping was very important to Babylonian astronomers, and since one would expect them to bring with them copies of pertinent records to show interested parties, we should probably envision the Magi consulting these in order to answer Herod's question.

Before the Magi arrived in Jerusalem, Herod may well have known about the Star's behavior in recent weeks or months, but he did not know when the Star had first appeared. How much of the Star's history he was aware of since that first appearance, we simply cannot know for sure.

In the wake of the Magi's arrival in Jerusalem but prior to Herod's meeting with them, the king only knew that the Magi were claiming that recent astronomical events had revealed that the Messiah had just been born.

At the covert discussion, Herod presumably got a more complete account of the Star. Most importantly, he discovered the particular day and month that the star had appeared in the sky—then, as nowadays, only professional astronomers could have been expected to have such information. Herod evidently figured that, if he knew this, he would know the maximum age of the Messiah.

Since Herod was providing the Magi with the key piece of information that they needed to complete their mission to worship the Messiah—the place of the Messiah's birth—and was even effectively commissioning them to find the newborn Messiah, the Magi had every reason to think that they could trust him. No doubt they attributed his curiosity to a spirit of joyful awe and wonder.

Discovering the Messiah's location. The half-Jewish king requested that the Magi, after finding the Messiah, bring back word to Jerusalem, claiming that he himself wanted to go and worship him. He obviously judged that these foreign magi had no inkling of his true agenda: to discover where precisely within Bethlehem the Messiah was located, so that he could assassinate him.

Evidently, Herod had a Plan A and a Plan B: Plan A was targeted assassination of the messianic baby, and Plan B was mass infanticide in Bethlehem. He clearly preferred the "cleaner" Plan A, which would entail only a short wait until the gullible Magi returned to his palace with detailed information regarding the precise whereabouts of the messianic child. However, the evil king had an atrocious backup plan that could be implemented if, for any reason, the favored plan failed.

The Star's First Appearance. Herod "ascertained the exact time"[91] (*akriboō ton chronon*) that the star had "appeared" (*phainomai*) (Matt. 2:7). The verb Matthew chose to describe the star's behavior has been interpreted by some to be synonymous with "rising" (vv. 2, 9) and hence to refer to "the time when the star came up over the horizon, the year, the month, and the day."[92] However, there is no reason to assume that "appeared" in verse 7 is equated with "rising" in verses 2 and 9. Indeed if the two are equated, the fact that Herod based his massacre of baby boys in their second year or under on the date of the first appearing of the Star would mean that the Magi had taken at least twelve lunar months to get to Jerusalem, which is completely implausible, or that the wonder in the eastern sky associated with the Star's heliacal rising lasted a ridiculously long time. It is much more natural to believe that the first appearance preceded the rising by many months. That is, after being visible in the sky for a long time, the Star was in conjunction with the Sun and then rose in the eastern sky in advance of the Sun.

Herod was obviously of the opinion that the child might have been born at some stage of the Star's apparition prior to its heliacal rising. The king, who was apparently influenced by astrology (see, for example, Josephus, *Ant.* 17.6.4 [§167]) and undoubtedly would have been knowledgeable about astronomy, may have wondered if the first appearance of the Star itself had coincided with the birth of the Messiah. The Magi clearly were convinced that what they had seen the Star do in connection with its heliacal rising marked the occasion of the birth, but Herod was eager to allow for an alternative interpretation.

Herod's Commissioning of the Magi. The magnitude of the task facing the Magi at Bethlehem is highlighted in what Herod says to them: "Go and search diligently for the child, and when you have found him . . ." (v. 8). Herod has told them that the Messiah's birthplace is Bethlehem, and the short (5–6 miles) journey to Bethlehem would not pose any problems. However, Herod does not

[91] My translation.
[92] Brown, *Birth of the Messiah*, 175.

know where precisely within Bethlehem the child is. The city of Bethlehem (and environs) was large enough to present a formidable challenge to the Magi as they sought to locate the newborn Messiah. The Magi would have had no choice but to move from door to door, asking for information. Moreover, how would they know which baby was the Messiah? Would the people of Bethlehem be as clueless as the people of Jerusalem? If the Bethlehemites did know, would they cooperate with the foreigners in their quest? Since the holy family was evidently keeping information out of the public domain, would they make themselves known to the Magi? The Magi faced an incredibly daunting task on the last phase of their journey to worship the Messiah.

The Star's Ushering of the Magi to Bethlehem and Jesus

In Matthew 2:9b–10 we read, "And behold, the star that they had seen at its rising went before them until it came and stood over the place where the child was. When they saw the star, they rejoiced exceedingly with great joy."

Traveling around Sunset. From what Matthew writes, it would seem that, on that day, the Star was not visible prior to the Magi's meeting with Herod but became visible during their journey from Jerusalem to Bethlehem. Therefore it is likely that they traveled in the evening, when the stars were emerging around the time of the Sun's setting.[93] The text is not explicit regarding when precisely during their journey to Bethlehem the Star appeared, but the implication seems to be that it accompanied them for the majority of their short trip. Therefore they probably saw the Star near the start of their trek southward. Travel by night in Judea was uncommon, but a relatively short journey by camel caravan in the evening would have been safe. Certainly the Magi would not have wished to delay the fulfillment of their urgent mission. Accordingly, the Magi's meeting with Herod most likely occurred in the late afternoon.[94]

Going ahead of the Magi to Bethlehem. The Star went before the Magi until it pinpointed the very house where baby Jesus was. Strictly speaking, of course, the Star's guidance was not needed to get the Magi to Bethlehem (Herod presumably informed the Magi where the town was), but only to direct them within Bethlehem. However, it seems that the Star ushered them to the town of David before pinpointing the particular house where the Messiah and his mother were.[95] The Star's going "before" or "ahead of" them (ESV; NIV) need not entail movement within the backdrop of the fixed stars and constellations. Most celestial entities (other than meteors, and comets very close to Earth), of course, have a natural daily westward course through the heavens.

Descending. Having reached Bethlehem within a couple of hours, the Magi would have seen the Star in the now-dark sky. At this point the Star is described as "coming" ("came"; v. 9). Since the Star had just led them to Bethlehem in the south and hence was at its highest point in the sky (its "culmination" at the meridian [the great imaginary circle that

[93] As we shall see, when the Magi saw the Star, it was in the southern sky (the direction of Bethlehem from Jerusalem) and hence at its culmination (the highest point of a celestial body's nightly path across the sky). That rules out the possibility that the Star had been below the horizon before the Magi saw it. If skies were clear, in the hours before it was seen by the Magi the Star must have been in the dome of the sky but below the threshold for easy daytime visibility. Accordingly, assuming clear skies, the Star must have appeared around sunset.

[94] The secret summoning of the Magi has been interpreted by some scholars (e.g., Robert H. Gundry, *Matthew: A Commentary on His Handbook for a Mixed Church under Persecution*, 2nd ed. [Grand Rapids, MI: Eerdmans, 1994], 30; Luz, *Matthew 1–7*, 137) as disclosing that the meeting occurred under cover of darkness. However, secretive behavior is not restricted to the deep darkness of night.

[95] The main Jerusalem-to-Hebron (north-south) road, the Way to Ephrath, was the obvious choice for travelers heading from Jerusalem to Bethlehem. When the Star guided the Magi to Bethlehem, it must have been moving toward the south-southwest.

passes through the zenith, the celestial poles, and the horizon's north and south]) and was now straight in front of them, its "coming" must refer to the Star's drop in altitude as it moved on its course toward the western horizon. The Star was preparing to point out for the Magi the place where the messianic baby was located.

Standing over the House. The text is rather clear in its description: from the perspective of the Magi, the Star eventually came to "stand"[96] over the place where the messianic child was. Many scholars, often in a bid to rescue the account from sheer implausibility, have insisted that the place over which the Star stood was the town of Bethlehem as a whole. Carson, for example, comments, "The Greek text does not imply that the star pointed out the house where Jesus was . . . ; it may simply have hovered over Bethlehem as the Magi approached it. They would then have found the exact house through discrete inquiry. . . ."[97] Hagner claims that verse 9 renders "difficult" any attempted explanation of the Star in terms of an astronomical phenomenon.[98]

Appreciating that the verse is the chief stumbling block for many Bethlehem Star hypotheses, Hughes maintains that verse 9 was not intended to be taken at face value or literally, as though the Star guided them as they went from Jerusalem to Bethlehem and then pinpointed the particular dwelling where Mary and Jesus were. According to Hughes, if Matthew's description is in accord with reality to any extent, the Star's leading could only have been general.[99] However, it is surely preferable to revise or abandon one's hypothesis rather than resort to special pleading in order to escape the natural force of the single most important description of the Star's behavior in Matthew's account.

The Magi did not need help finding Bethlehem. As verse 8 ("Go and search diligently for the child, and when you have found him . . .") makes abundantly clear, the challenge was in locating the infant within Bethlehem. R. T. France put it well:

> They already knew from Herod that Bethlehem (a mere five or six miles from Jerusalem) was their destination, so that they did not need the star to tell them that; their extravagantly expressed joy . . . is hard to explain unless the star somehow indicated the actual house rather than just the village as a whole. It seems, then, that the star's movement gave them the final supernatural direction they needed to the specific house "where the child was."[100]

That the Star stood over the individual building where the infant was is clearly implied in verse 11: "And going into the house they saw the child with Mary his mother." The "house" here most naturally picks up on verse 9b's "place where the child was."

The suggestion that no astronomical entity could pinpoint a house[101] is inaccurate. It could, depending on what it looks like, where the observers are located, and where the house is relative to the visible horizon. If a house is located on the visible horizon, an astronomical entity that is notably bright and large can seem to onlookers to be standing over it.

[96] Although the verb *histēmi* can mean "stop" when the subject has previously been moving, here, where it is followed by the preposition *epanō* ("over"), the most natural meaning is "stood over," without any necessary nuance of cessation of movement (see "came and stood over," KJV, ASV, NASB; contra ESV, RSV, NRSV, NIV, NET). Note that Josephus, *J. W.* 6.5.3 (§289) refers to a sword-like star that "stood over the city" of Rome, using the same verb with the preposition *huper* ("over").
[97] Carson, "Matthew," 115.
[98] Hagner, *Matthew*, 1:30, who goes on to say that if the Star was an astronomical phenomenon, v. 9 would have to be regarded as "romantic myth" or a theological touch.
[99] Hughes, *Star of Bethlehem Mystery*, 20–21.
[100] France, *Gospel of Matthew*, 74; cf. Gundry, *Matthew*, 31.
[101] So, for example, Hughes, *Star of Bethlehem Mystery*, 22.

In having the Star go before the Magi to the very house where the infant Messiah was, God was, Matthew implies, intervening to confirm them in their sacred mission and enable them to complete it. The star's presence at this point in the Magi's journey makes for beautiful symmetry. Just as the Star had marked the start of their mission, so it also marked its conclusion. More than this, the Star itself was now pointing out the precise whereabouts on the earth of the child whose birth it had earlier announced in the heavens.

Three Phases of the Star's Ushering. It is important to appreciate that there is a fundamental continuity between the Star's guidance of the Magi to Bethlehem and its standing over the house: "The star . . . went before them until, having come,[102] it stood over the place where the child was" (v. 9). This indicates that the Star guided them first to Bethlehem and then, after coming (that is, down in altitude, toward the horizon), to the very house where baby Jesus was. The Star, then, did three things that night: (a) It seemed to travel toward Bethlehem ahead of the Magi. (b) After the Star had led them to Bethlehem, the Star then entered a phase of descending in altitude. (c) As the Star descended to just above the visible horizon that night, it "stood" over the place where the Messiah was. From what Matthew writes, we can deduce the following: having left their homeland shortly after the completion of the sign in the eastern sky, and having traveled some 28–37 days (the length of a camel caravan trip from Babylon to Jerusalem) to Judea, the Magi saw the Star appear in the south-southeastern sky in the evening and then, along with the rest of the stars, over a couple of hours move to the south-southwest (the direction of Bethlehem from Jerusalem). From that point it descended toward the horizon and finally "stood."

The Magi's Joy. Verse 10 is somewhat ambiguous and could be interpreted in one of two ways: as revealing the response of the Magi to the appearance of the Star as they traveled from Jerusalem to Bethlehem,[103] or as indicating their reaction to the standing of the Star over one particular house.[104] The former interpretation fits with the "behold" of verse 9 but interrupts the narrative flow and would make verse 10 parenthetical. It would probably imply that the Magi had assumed that the Star, having guided them to Jerusalem, would not reappear. The latter interpretation fits the sequence of the story better—verse 9 has just climaxed with the Star standing over the particular location where the baby Messiah is, and verse 11 reports that they entered the house. I suggest that verse 10 is referring to the Magi's great joy at seeing the Star standing over the house at the culmination of their long trek to worship the messianic baby. The Magi certainly felt a sense of wonder at seeing the Star as they journeyed from Jerusalem to Bethlehem—that is conveyed by "behold" in verse 9. But they were even more astonished and overjoyed when they realized that the Star was pinpointing an exact location, which they interpreted to be where the Messiah was. Evidently they perceived the celestial marvel to be a divine confirmation and vindication of their journey. The Star had, as it were, intervened to ensure that their pilgrimage came to a successful conclusion.

Astonishingly, that night the Star had appeared to be making the same journey as the Magi. It had led them from Jerusalem to Bethlehem and right to the house where the messianic baby was staying. Indeed the Star evidently looked like it was about to enter

[102] My translation. It is regrettable that some modern English translations treat the Greek participle "coming" as redundant (e.g., NIV; NET).
[103] E.g., Leon Morris, *The Gospel according to Matthew* (Downers Grove, IL: InterVarsity Press, 1992), 41.
[104] E.g., France, *Gospel of Matthew*, 74.

the house. In the wake of this spectacular phenomenon, there could be no doubt that the newborn baby in this house was the one whose birth had been proclaimed previously in the eastern sky (cf. v. 2).

The Magi Meet the Messiah

Matthew 2:11 recounts what happened in the aftermath of the Magi's observation of the Star standing over the place where the infant Messiah was: "And going into the house they saw the child with Mary his mother, and they fell down and worshiped him. Then, opening their treasures, they offered him gifts, gold and frankincense and myrrh."

A House. Having been guided by the Star to the very place where the messianic child was located, the Magi made their way to the building and went inside. We discover in this verse that Jesus was in a "house." Wherever Jesus was born, whether in a stable, a cave, or the part of a house normally used by animals, he was certainly in a house by the time the Magi arrived on the scene. Presumably, as soon as the census was over and the population of Bethlehem had returned to normal levels, Mary, Joseph, and Jesus moved into living quarters more suitable for human habitation and remained there until they eventually fled from Bethlehem to Egypt (v. 14).

The Child and His Mother. It is striking that the Magi see inside the house "the child with Mary his mother" (v. 11). The celestial phenomenon that the Magi had seen in the eastern sky back in their homeland had been divinely orchestrated to get them to see this very sight: the recently born holy child with his mother. Now, finally, the Magi could feel joy and relief at having fulfilled their divinely appointed mission to welcome the Messiah to the earth.

The absence of Joseph, Jesus's legal father, from the description is notable. It reinforces the impression that the focus of the Magi was on the Virgin Mary and her holy child. As Luz points out, the description here recalls 1:18–25, where Matthew narrates that Jesus was born to a virgin in fulfillment of Isaiah 7:14.[105] It is fitting that this quiet allusion to Jesus's divine parentage is followed by the statement that the Magi fell down prostrate to worship him.

Falling down to Worship. Falling down on one's face before another is a powerful gesture of acknowledgement of higher status. It is particularly appropriate when one is expressing one's submission to and worship and fear of the divine. In this connection, Davies and Allison rightly comment that in Jewish circles prostration was regarded as something that should be directed only toward God.[106] The religious significance of the word used here (*piptō*) is present elsewhere in Matthew (4:9; 17:6; 26:39). Here, where it is accompanied by worship (cf. 4:9; 1 Cor. 14:25; Rev. 4:10; 7:11; 22:8), the religious nature of the prostration is clear. The heartfelt worship of Jesus offered by these Gentile astrologers from the east can be explained only if the celestial sign they had seen had been perceived by them directly and/or indirectly to disclose the divine nature of the newborn King.

Gifts. The Magi then offered Jesus their gifts of gold, frankincense, and myrrh. The choice of these particular gifts naturally brings to mind Isaiah 60:6, where Yahweh spoke inspiringly of the eschatological restoration, when Jerusalem and its temple would be the center of the earth and the focus of the nations' pilgrimages: "Herds of camels will cover your land, young camels of Midian and Ephah. And all from Sheba will come, bearing gold and incense and proclaiming the praise of Yahweh" (NIV). It is impossible to

[105] Luz, *Matthew 1–7*, 137.
[106] Davies and Allison, *Matthew*, 1:248.

believe that Matthew did not think of this text when he mentioned the gifts brought by the Magi. Indeed it is difficult to think that the Magi's choice of gifts was not influenced by Isaiah's eschatological oracle. Certainly the Magi had some knowledge of Jewish messianic expectations. In bringing their gifts, however, the Magi could hardly have been imagining that they were completely fulfilling Isaiah 60:6. They would presumably have felt that they were anticipating the fulfillment of this prophecy, looking forward to the period when the Messiah would reign over the whole earth.

Gold, according to Haggai 2:8, belongs to Yahweh, the Sovereign over all. Gold was used extensively in the building of the tabernacle and its vessels and furniture, most notably the ark of the covenant (Exodus 25), and the high priest's accoutrements (his crown, breastplate, and ephod) were made of it (Exodus 39). It was imported by Solomon for lavish use in the construction of the temple of God, most notably the Most Holy Place and the royal palace (1 Kings 6 and 10). Gold, then, was especially appropriate as a gift to God and to a king.

Like gold, frankincense and myrrh were expensive, luxury items. Frankincense is a yellowy white gum tapped from the frankincense tree, genus *Boswellia*, in the southern Arabian Peninsula or the Horn of Africa, which gives off a sweet, balsamic odor when burned or heated. Pure frankincense was widely used in the ancient Near East for cultic purposes.[107] In Israelite religion it was placed alongside the showbread in the tabernacle and then temple (Lev. 24:7), accompanied the cereal offerings (Lev. 2:1–2, 14–16; 6:14–18), and was the key ingredient in the sacred incense that was placed before the ark in the Most Holy Place (Ex. 30:34–36). The offering of frankincense to the messianic infant probably implied that the Magi were acknowledging his priestly and/or divine status.

Myrrh is a fragrant yellowy-brown gum resin tapped from the *Commiphora myrrha* tree, which is also found in southern Arabia and the Horn of Africa.[108] It was employed as an ingredient in perfumes (Est. 2:12; Ps. 45:8; Song 1:13) and in the manufacture of holy anointing oil (Ex. 30:23). It was also a painkiller (Mark 15:23) and a burial spice (John 19:39). In view of the fact that the gold and frankincense gifts were selected under the influence of Isaiah 60, it seems likely that the Magi's bringing of myrrh reflected a messianic interpretation of Isaiah 53's Suffering Servant Song, and that the Magi believed that the newborn Messiah would in due course suffer for the transgressions of his people.

The gifts of the Magi therefore most likely imply that, by the time they arrived in Judea, they believed that the one whom they were going to see was royal and divine, and yet was destined to be killed and buried. The Magi's offerings seem to have reflected a surprisingly enlightened grasp of Jewish messianic ideas in the Hebrew Scriptures, particularly in Isaiah. This is consistent with their having been aided in the interpretation of the Star by a Jew (or Jews) in Babylon educated in the Hebrew Scriptures.

The Magi may have secured the gold, frankincense, and myrrh in their homeland or purchased them en route to Judea.[109] In view of their mission, they would not have wished to arrive in Judea without having gifts fit for the newborn King.[110]

[107] See W. W. Müller, "Frankincense," in *The Anchor Bible Dictionary*, ed. D. N. Freedman, 6 vols. (New York: Doubleday, 1992), 2:854.

[108] Davies and Allison, *Matthew*, 1:249; Victor Matthews, "Perfumes and Spices," in Freedman, *Anchor Bible Dictionary*, 5:226–227.

[109] Nabataea, through which travelers taking a reasonably direct route from Babylon to Jerusalem would have passed, was a major hub for international trade in gold and especially frankincense and myrrh (Diodorus Siculus 19.94.5; Strabo 17.1.13).

[110] Frankincense and myrrh were the costliest spices in the Near East (Müller, "Frankincense," 2:854).

The Thwarting of Herod's Scheme

According to Matthew 2:12, "And being warned in a dream not to return to Herod, they [the Magi] departed to their own country by another way."

It is clear from this verse that the Magi had been completely fooled by Herod. They were evidently fully intent on reporting back to him concerning their successful mission to find the holy child. They were oblivious to the fact that the evil tyrant would seek to use that information to execute the one they had just journeyed 28–37 days to worship. We may presume that the Magi kept their agreement with Herod to themselves or, at any rate, that anyone to whom they passed this information shared their conviction that Herod had good motives.

According to Matthew, God had previously communicated with the Magi by means of celestial phenomena, which they interpreted through the lens of Jewish and particularly Biblical traditions, and by means of the Prophetic Word conveyed to them from the religious leaders through Herod. Now, God spoke to them by a dream. Whereas the prior revelations were directed toward getting the Magi to meet the newborn messianic King, this new revelation was intended to protect the child from the evil ruler's vile scheming. Because of the divine warning, the Magi avoided Jerusalem on their way back home.[111]

MATTHEW 2:13–18: A KEY PROBLEM

In Matthew 2:13–18 we are told of how Herod, upon discovering that the Magi had reneged on their commitment to return and tell him where the messianic child was, flew into a rage and "sent and killed all the male children in Bethlehem and in all that region who were in their second year or under,[112] according to the time that he had ascertained from the Magi" (v. 16).

Why Did Herod Order the Massacre of Boys in Their Second Year?

Many scholars have made much of the fact that Herod chose to eliminate all infants up to "two years old," maintaining that this reveals that the Magi arrived on the Judean scene a very long time after Jesus's birth. For example, Hauerwas and Turner argue that this is a clue indicating that the Magi did not arrive on the scene until about two years after the birth of Jesus.[113] However, that is implausible. We have already seen that the Magi believed that the celestial sign they had seen in the eastern sky—and that had been the catalyst for their heading westward—marked the occasion of the birth of the Messiah. Moreover, the fact that Mary and Joseph were still in Bethlehem when the Magi arrived suggests that the interval between Jesus's birth and the Magi's arrival could not have been long at all. In view of Luke 2:4–6's statement that Joseph and Mary went to Bethlehem on a temporary basis to register for the census, it is difficult to believe that they intended to remain in Bethlehem for long after they had offered their purification offerings and dedicated their infant on the fortieth day after the birth, in accordance with the Torah.[114]

Why, then, did Herod eliminate infants in their second year of life in the region of Bethlehem, when he was perfectly well aware that the Magi believed that Jesus had been born within the last couple of months?

Could the answer be found in the length of time it took for Herod to become certain that the Magi were not returning to see him? This

[111] Did the Magi inform Joseph of their warning dream, exposing Herod's scheme and putting Joseph on high alert in the brief period leading up to his own dream?

[112] See chapter 1, note 24.

[113] Hauerwas, *Matthew*, 41; and Turner, *Matthew*, 78.

[114] For a successful defense of the historical credibility of Luke's account of the Presentation of Jesus at the temple (Luke 2:22–24), see Richard Bauckham, "Luke's Infancy Narrative as Oral History in Scriptural Form," in *The Gospels: History and Christology: The Search of Joseph Ratzinger–Benedict XVI*, ed. Bernardo Estrada, Ermenegildo Manicardi, and Armand Puig i Tàrrech, vol. 1 (Vatican City: Libreria Editrice Vaticana, 2013), 399–417.

is most unlikely—it is difficult to imagine that this period was more than five or six days.

A key may be found in Herod's gruesome nature. If the tyrant was to ensure that this baby was eliminated from the scene, he could not afford to take any chances. He needed to allow for a wide margin of error.[115] The fact that he killed infants not only in Bethlehem proper but throughout the surrounding districts confirms that Herod was indeed seeking to spread his net sufficiently widely to be sure to catch his prey. The brutality entailed in this episode is very much in character with what we know of Herod from *Josephus*: not only had he had three of his own sons and one of his wives executed (*Ant.* 16.11.7 [§392]; 17.7 [§§182–187]), but he also even sought to ensure mourning at his passing by having one member of every noble Jewish family throughout the nation killed (*Ant.* 17.6.6 [§§180–181]).

Matthew suggests that Herod was fully cognizant that he was taking on God's Messiah. Aware of the special nature of the newborn infant, Herod may have been concerned that the child might appear older than he actually was.

However, the reason given by Matthew for Herod's decision to slaughter infants in their first and second year is that he had carefully ascertained from the Magi when the Star had first appeared. The implication is, of course, that the Star first appeared a considerable time before it rose heliacally on the eastern horizon. Since Jews typically reckoned years inclusively, the star must have appeared more than one year (twelve or thirteen lunar months, depending on whether there was an intercalary [or leap] month in Judea during the course of that year[116]) before the point at which Herod gave the order. In electing to slaughter Bethlehemite infants up to one year old based on the date of the Star's first appearance, Herod was covering the possibility that the Magi might have made a mistake when they concluded that what the Star did in connection with its heliacal rising coincided with the Messiah's birth. The Judean king was allowing that the Messiah might have been born at an earlier stage of the Star's apparition, either at the time of the Star's first appearance[117] or during the period between that event and the Star's "rising." Herod wanted to make absolutely sure that he successfully executed the Messiah.

The Failure of Herod's Assassination Attempt

As brutal, sweeping, and grievous as Herod's atrocity was, the massacre of the male infants in Bethlehem and the surrounding area did not succeed in its objective—the elimination of the Messiah. The reason for this is given in Matthew 2:13–15a: "Now when [the Magi] had departed, behold, an angel of the Lord appeared to Joseph in a dream and said, 'Rise, take the child and his mother, and flee to Egypt, and remain there until I tell you, for Herod is about to search for the child, to destroy him.' And he rose and took the child and his mother by night and departed to Egypt and remained there until the death of Herod."

The episode calls to mind Psalm 2, with its mysterious prophecy that the kings of the earth would conspire against Yahweh and his Messiah, but would do so in vain, because Yahweh had infallibly decreed that his Son, the Messiah, would reign over all nations.

CONCLUSION

In this chapter we have sought to unpack Matthew's account of the birth of Jesus and particularly his account of the journey of

[115] Davies and Allison, *Matthew*, 1:245.

[116] Those following a lunar calendar of twelve months (with years totaling about 354 days) must add a leap month every few years to get it back into sync with the 365-day solar year.

[117] Note Davies and Allison, *Matthew*, 1:244; Davies and Allison suggest that the assumption being made by Herod is that the Star's first appearance occurred at the time of the birth of the child.

the Magi to worship the baby Messiah. In the course of doing so we have developed a clearer understanding of the Star that played such an important role in the story.

In light of our detailed study of Matthew 1–2, we are now in a position to set out some data that a compelling hypothesis regarding the identity of the Bethlehem Star must be able to accommodate.

1. Matthew's Gospel is a theological biography with a definite interest in history. Matthew believed that the account of the Magi and the Star was historically reliable. Further, the narrative has a strong claim to be considered historically accurate.

2. The Magi were professional astronomers and astrologers, their astronomical observations being in the service of astrology. They would have done horoscopes for clients and so were particularly concerned with the heliacal risings of the zodiacal signs and the place of the planets within the zodiac. They took note of any unusual celestial phenomena. Such astronomical data was generally used to identify the fortunes of clients based on the state of the heavens on the occasion of a birth.

3. The Magi were probably from Babylon, an important center of astronomy and astrology and a city with a sizeable Jewish population. It was about 550 miles from Jerusalem.

4. The Magi saw a "star." That means that, whatever phenomenon explains the mystery of the Star of Bethlehem, it must have been counted a star by at least some of the ancients. The ancients could consider a number of different phenomena "stars," including fixed stars, planets, comets, and meteors.

5. The Star was an objective rather than a subjective phenomenon. This is strongly suggested by: (a) the fact that it was seen by a group of professional astronomers/astrologers (Matt. 2:1); (b) the phrase "at its rising" (*en tē anatolē*) in verse 2; (c) the distress felt by Herod and the people of Jerusalem (v. 3); (d) Herod's ascertaining from the Magi the precise time of the Star's appearance (v. 7); and (e) the guidance that the very same entity gave to the Magi when they were on the final stage of their journey, leading them to the very house where the Messiah was in Bethlehem (vv. 9b, 11). The Star was observable by the Magi in their eastern homeland and presumably across the whole Near East, including Judea, and beyond, subject to favorable atmospheric conditions.

6. The Star first appeared more than one luni-solar year, that is, twelve or thirteen lunar months (depending on whether there was an intercalary month in Judea during the relevant year), before Herod gave the order to slaughter the infants of Bethlehem. The Magi took careful note of the date of the Star's first appearance.

7. Some months after the Star's first appearance, it rose heliacally (note the word "rising" [*anatolē*] in Matt. 2:2, 9). That is, after becoming invisible for a time because it was close to the Sun, the Star was seen rising above the eastern horizon just in advance of the rising Sun.

8. The astronomical wonder relating to the rising of the Star was clearly extraordinary, indeed momentous, in the judgment of the Magi. This is demonstrated by the fact that the Magi responded to the celestial sight by undertaking a long journey—in fact, a pilgrimage—to Judea. One can safely presume that the Magi were not in the habit of making such urgent and challenging journeys.

9. The celestial marvel in connection with the rising of the Star was interpreted by the Magi as signifying a special birth, with the Star itself representing the important person being born. That the Magi perceived that a special birth had taken place is clear from the fact that the question they ask the people in Jerusalem relates to the location of one just born (v. 2a). That they understood the Star to represent the person born is demonstrated by their statement that they "saw *his* star" at its rising (v. 2b).

10. What the Star did at or around the time of its heliacal rising persuaded the Magi that the King of the Jews, the Messiah, was born and could be found in Judea. This is demonstrated in their traveling to Judea (v. 1) and their question to the people in Jerusalem, "Where is he who has been born king of the Jews?" (v. 2). The Magi were evidently aided in their interpretation of the meaning and significance of the astronomical scene by prophetic traditions in the Hebrew Bible, in particular Balaam's oracle concerning a rising star-scepter in Numbers 24:17 and messianic and eschatological prophecies in the book of Isaiah. (We shall consider these prophecies in greater detail in chapter 8.)

11. The heavenly sign seen in the east communicated directly and/or indirectly to the Magi that the one born was divine in nature. This is evident in Matthew 2:2b, where the Magi assert that they had "come to worship" the newborn baby (cf. v. 8). It is also clear in verse 11a, where we read that the Magi, upon going into the house and seeing "the child with Mary his mother," "fell down and worshiped him." Since the celestial phenomenon indicated a birth, the divine one was also human.

12. Matthew's favorable treatment of the Star and of the Magi's pilgrimage strongly suggests that what the Star did to announce the Messiah's birth was something that demanded more than merely a pagan astrological interpretation.

13. The Magi responded to the celestial phenomenon relating to the Star's rising by making a pilgrimage westward to Judea with the intention of worshiping the Messiah. They probably departed very shortly after the conclusion of the spectacle in the eastern sky (i.e., within a few days) and traveled westward by camel caravan 28–37 days to Jerusalem.

14. After the Star's appearance in the eastern sky, the celestial entity more than likely remained observable to the Magi as they traveled westward from their homeland to Jerusalem. It seems to have shifted quickly from the eastern morning sky to the western evening sky, and thereafter migrated to the southern evening sky. Each night, it may have seemed by the location of its setting in the west to be guiding the Magi toward Judea.

15. The appearance of the Star was perceived by Herod to be a serious threat to his royal dynasty. According to verse 3, he was "troubled" by the report of the Magi. We discover in verses 4–18 that he immediately determined to eradicate the threat by executing the one whose birth the Star had announced. He had a Plan A and a Plan B. Plan A was the targeted assassination of the baby Messiah based on the assumption that the Magi, after visiting Bethlehem, would return to Jerusalem to inform him where the messianic baby was located. Failing that, Plan B was the wholesale massacre of the babies of Bethlehem and the surrounding area, based on what the Magi had reported to him about the date of the Star's first appearance (which Herod regarded as revealing the Messiah's maximum age, if not his actual birthday) and what the Jewish teachers had told him regarding the prophesied place of the Messiah's birth.

16. In the wake of the Magi's arrival in Jerusalem, the people of the city came to regard the Star's appearance as a threat to the status quo, a provocation to Herod, and a herald of bloody regime change. They, too, were therefore "troubled." The Magi's messianic interpretation of the celestial event was evidently alien to them. This may have been because the Jerusalemites missed key parts of the astronomical phenomenon due to cloudy conditions, or simple oversight, or, more likely, because they had disregarded the astronomical phenomenon or interpreted it differently.

17. As the Magi set out on their way southward from Jerusalem to Bethlehem, probably in the evening, around sunset, they observed the Star appearing to go ahead of them. The Star was up in the sky in front of them and seemed to move forward, leading

them (*proagō*; v. 9). It was probably culminating around this time (i.e., it was at the meridian and hence basically due south). This celestial guidance of the Magi to Bethlehem was not, strictly speaking, necessary, because, after their meeting with Herod, they would undoubtedly have known where Bethlehem was. Probably, however, the presence of the Star at this point functioned to confirm the Magi in their mission and to endue in them a sense that they were being ushered into the Messiah's presence, thereby heightening their sense of anticipation.

18. The Star, having gone before the Magi on their way to Bethlehem, proceeded to descend and then stand over one particular house on Bethlehem's skyline. Verse 8 highlights that finding the infant Messiah would have been an enormous challenge. The sight of the Star standing over the structure therefore filled the Magi with extraordinary joy—the Star that had prompted them to make this journey to worship the Messiah was now pinpointing the very place where he was located.

19. Upon entering the house, the Magi were satisfied, when they saw "the child with Mary his mother" (v. 11a), that this was what the celestial sign they had observed in the east had commissioned them to go see. This is demonstrated by their worship of Jesus and the giving of their gifts of gold, frankincense, and myrrh (v. 11).

20. What the Star did in connection with its heliacal rising to convince the Magi that the Messiah was born coincided historically with the birth of Jesus in Bethlehem.

21. The Star's rising and its role in pinpointing the location of the baby Messiah occurred in the year 6 or 5 BC.

22. Matthew does not inform us what the Star did after it had guided the Magi to baby Jesus.

In light of these points, the question naturally arises: can any hypothesis regarding the identity of this Star accommodate all of this data? We shall now examine the major proposals put forward to identify the Star of Bethlehem.

4

"What Star Is This?"

Evaluating the Major Hypotheses

In this chapter we shall review and evaluate most of the major hypotheses proposed to explain the Star of Bethlehem. In particular we shall consider the views that the Star was: the triple conjunction of Jupiter and Saturn in Pisces in 7 BC; Jupiter, with the focus on two lunar "occultations" (obscurings) of Jupiter in Aries in 6 BC; a nova or supernova; two meteors; an ordinary star; Jupiter, with the focus on its movements in 3–2 BC; or a combination of some of the above. Then we shall evaluate the position that the Star was a supernatural phenomenon and the proposal that the Bethlehem Star has no foundation in fact.[1]

TRIPLE CONJUNCTION OF JUPITER AND SATURN IN PISCES IN 7 BC

The hypothesis that the Star of Bethlehem was the triple conjunction of Jupiter and Saturn in 7 BC is perhaps the single most popular view. Two notable proponents are David Hughes and Simo Parpola.[2]

In order to understand the theory, it is important to realize that Jupiter and Saturn are farther away from the Sun than the other naked-eye planets and take longer to complete their revolutions—Jupiter takes 12 years to orbit the Sun, and Saturn takes 30 years. Within the dome of the sky they seem to move at a snail's pace compared to the inner planets—Mars, Venus, and Mercury. Occasionally Jupiter and Saturn appear, from the perspective of Earth, to come together in the sky. When this happens, the planets move toward each other until they are in the same place in the sky, and then move apart. More rarely the two planets come together three times within a year or so, a phenomenon known as a triple conjunction. This occurs only when Earth, Jupiter, and Saturn are at particular points in their orbits, so that Earth

[1] If you download planetarium software, you will be able to discover what the sky looked like at any particular moment in history, even thousands of years ago. Those unfamiliar with astronomy are often taken aback by this fact. But when we remember that all the celestial bodies, including Earth, the Moon, the stars, and the planets, operate by well-understood set laws and regularities, it makes sense.

[2] David W. Hughes, *The Star of Bethlehem Mystery* (London: J. M. Dent, 1979); and Simo Parpola, "The Magi and the Star: Babylonian Astronomy Dates Jesus' Birth," in *The First Christmas: The Story of Jesus' Birth in History and Tradition*, ed. Sara Murphy (Washington, DC: Biblical Archaeology Society, 2009), 13–24; cf. A. Strobel, "Weltenjahr, große Konjunktion und Messiasstern, Ein themageschichtlicher Überblick," *Aufstieg und Niedergang der Römischen Welt* 2.20.2 (1987): 988–1187. Jeanne K. Hanson, *The Star of Bethlehem: The History, Mystery, and Beauty of the Christmas Star* (New York: Hearst, 1994), 51–55, also takes this view.

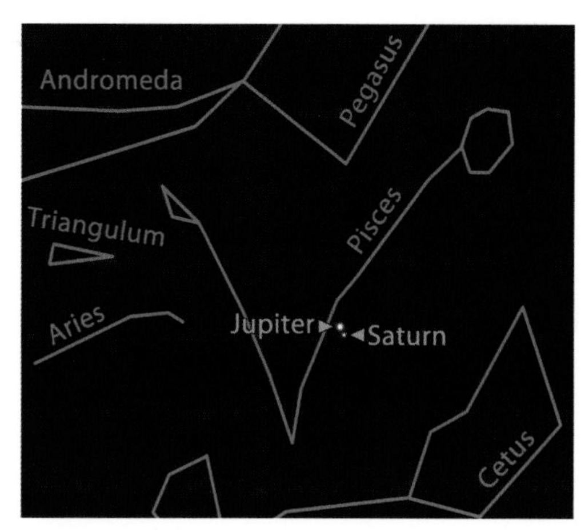

FIG. 4.1 The Jupiter-Saturn conjunction on May 29, 7 BC. This was the first of three Jupiter-Saturn conjunctions in 7 BC. Image credit: Sirscha Nicholl.

overtakes Jupiter and Saturn at the very time when Jupiter overtakes Saturn.[3]

In the year 7 BC there was a triple conjunction of Jupiter and Saturn between May and December: the planets came together once in May/June (fig. 4.1), once in September/October, and once in November/December. On May 27–31 and September 26–October 5 they came to within 1 degree, while on December 1–8 they were separated by just under 1 degree and 4 arcminutes.[4] This triple conjunction in 7 BC occurred in the zodiacal constellation of Pisces.

Normal Jupiter-Saturn conjunctions occur every two decades or so, while triple conjunctions take place on average every 139 years.[5] Triple conjunctions of Saturn and Jupiter had occurred in the constellation Leo in 821/820 BC, in Taurus in 563/562 BC, in Virgo in 523/522 BC, and in Cancer in 146/145 BC. The previous triple conjunctions in Pisces had taken place in 980/979 BC and 861/860 BC.

Hughes claims that, in ancient astrological thinking, a conjunction of these two great planets in Pisces would have been regarded by both Jews and Babylonians as an obvious sign marking the Jewish Messiah's birth.[6] Most proponents of this theory suggest that the Magi regarded Jupiter as the planet of the ruler of the world and Saturn as the planet representing Palestine or the westland generally.[7] Many also claim that the constellation/sign Pisces had a special association with the Hebrews and the last days.

According to Ethelbert Stauffer and Parpola, the conjunction meant that the latter-day king would, during this year, be born in Palestine.[8]

Parpola claims that the triple conjunction would have had astrological and political significance for Babylonian astrologers. Jupiter was the chief Babylonian god Bel-Marduk's "star,"[9] and Saturn was the "star" of his terrestrial representative, the king.[10] As for Pisces, it was the last of the twelve zodiacal signs (Aries being the first) and hence was liable to be interpreted as speaking of "the end of the old world order."[11] At the same time, Pisces was associated with the god of

[3] See Mark R. Kidger, *The Star of Bethlehem: An Astronomer's View* (Princeton, NJ: Princeton University Press, 1999), 202; and especially Parpola, "Magi and the Star," 21–23.
[4] Calculations were done on planetarium software, specifically Starry Night® Pro 6.4.3, Simulation Curriculum Corporation, 11900 Wayzata Blvd, Suite 126, Minnetonka, MN 55305, http://astronomy.starrynight.com. For more on the conjunctions, see Hughes, *Star of Bethlehem Mystery*, 139; U. Holzmeister, "La stella dei Magi," *Civiltà Cattolica* 93 (1942): 12–15. "Degree" and "arcminute" are astronomical measurements. A circle is 360 degrees. From the horizon to directly above your head (the zenith) is 90 degrees. There are 60 arcminutes in 1 degree.
[5] Hughes, *Star of Bethlehem Mystery*, 131.
[6] Ibid., 213–214.
[7] Ulrich Luz, *Matthew 1–7: A Continental Commentary*, trans. Wilhelm C. Linss (Minneapolis: Augsburg Fortress, 1989), 132, comments that the triple conjunction hypothesis is a reasonable candidate for the Star of Bethlehem because Jupiter could be regarded as the Star of royalty, and Saturn as the Star of Sabbath and the Jews. He refers in a footnote (132n25) to Albinus Tibullus 1.3.18; Tacitus, *Hist.* 5.4; Sextus Julius Frontinus, *Strategemata* 2.1.17, ed. Gotthold Gundermann (Leipzig: Teubner, 1888); and Cassius Dio 37.17–18.
[8] Ethelbert Stauffer, *Jesus and His Story* (New York: Knopf, 1960), 36–37; Parpola, "Magi and the Star," 20.
[9] Parpola (ibid., 18) wrongly claims that Jupiter is the "brightest planet." That distinction belongs to Venus. At the time of the second conjunction of Jupiter and Saturn in 7 BC, Venus was almost two magnitudes brighter than Jupiter.
[10] Ibid.
[11] Ibid.

creation and wisdom and hence could also be regarded as portending the nativity of a divinely appointed royal savior.[12] In Parpola's opinion, the Magi were most deeply impacted by the second conjunction, on Tishri 22, 7 BC, when the two planets were "shining at their brightest" (with reference to their performance during each of the three conjunctions) and constituted "a brilliant and suggestive sight."[13] To the Magi it seemed very significant that at that time Marduk's "star" was "directly above"[14] the king's "star," appearing to "embrace" it. Since "Tishri was known as the month of Amurru [the West],"[15] they concluded that the newborn was to be found in Syria-Palestine. At that point, therefore, they began to head west. Parpola asserts that Mars "joined the conjunction" during its final stage and that this was perceived as a signal of the terrestrial location—since Mars was the "star" of Amurru (the West), the royal figure had to be in Syria-Palestine.[16]

As regards the day when the Magi perceived that the Messiah was born, Hughes maintains that it was probably September 15, 7 BC, because that was, according to him, when the acronychal rising of Jupiter and Saturn took place (the "acronychal" rising of a planet refers to the planet's rising in the east as the Sun is setting in the west.)[17]

With respect to the standing of the Star over the place where Jesus was, Parpola ascribes this to the apparent pause of Jupiter in the sky on November 7 of 7 BC, between the second and third conjunctions.[18]

As popular as this triple conjunction theory is, it has serious flaws.

First, during the triple conjunction of 7 BC, the two planets never came sufficiently close together to appear as a single entity. Hughes admits that a 1-degree gap—twice the Moon's diameter, or the width of a little finger at the end of an extended arm—is not particularly close as far as conjunctions go, and that no ancient would have regarded the two planets as a single star.[19] By contrast, he concedes, the three conjunctions of 146/145 BC all saw Jupiter and Saturn come closer to each other than one quarter of one degree (equivalent to the Moon's radius).[20] The separation between the planets during the 7 BC triple conjunction constitutes a significant problem for those who identify it as the Bethlehem Star, because Matthew is clear that the Star was a single entity.[21]

Moreover, we might well ask why the Magi would have responded so dramatically to what was a rather unspectacular phenomenon.

As it happens, we have four copies of an astronomical almanac for the year 7/6 BC from Babylon.[22] Babylonian astronomical almanacs were written in the year prior to the year for which their predictions of astronomical events were made.[23] Strikingly, the almanac entry for 7/6 BC does not take particular note of the conjunctions of Jupiter and Saturn. With regard to the general time of the first conjunction, it simply mentions the stationary points of Jupiter and then Saturn in Pisces, without specifying that there would

[12] Ibid.
[13] Ibid., 19.
[14] It seems to me, however, that during the second conjunction this description is not particularly appropriate.
[15] Ibid.
[16] Ibid., 18.
[17] Hughes, *Star of Bethlehem Mystery*, 220. It is, however, worth noting that the acronychal risings of Jupiter and Saturn probably did not occur on the same day or as late as September 15 in 7 BC.
[18] Parpola, "Magi and the Star," 23.
[19] Hughes, *Star of Bethlehem Mystery*, 124, 139, 152.
[20] Kidger, *Star of Bethlehem*, 206.
[21] Cf. Patrick Moore, *The Star of Bethlehem* (Bath, England: Canopus, 2001), 45; Luz, *Matthew 1–7*, 132.
[22] A. J. Sachs and C. B. F. Walker, "Kepler's View of the Star of Bethlehem and the Babylonian Almanac for 7/6 B.C.," *Iraq* 46 (1984): 47, insist that the tablets all hail from Babylon or possibly Borsippa, which is about 11 miles from Babylon.
[23] Ibid., 46.

be a conjunction. With respect to the general time of the second conjunction, only Jupiter's acronychal rising is mentioned, not its stationary point, not the behavior of Saturn, and not the conjunction. Concerning the general time of the third conjunction, the stationary points of Jupiter and Saturn in Pisces are mentioned, but not the conjunction.[24] Sachs and Walker are therefore, strictly speaking, correct to point out that the 7/6 BC entry does not refer to the conjunctions.[25] That does not mean that the Babylonians assigned no significance to the triple conjunction, but it does suggest that they did not regard it as being of momentous significance, at least not in and of itself or in advance.[26]

Second, both the Magi and the narrator in Matthew 2 refer to the celestial entity as a "star," but contemporary evidence that a conjunction of planets could be considered "a star" is lacking.[27]

Third and related to this, as Brown points out, the fact that the Magi reported that they had seen the "rising" of the Messiah's star is difficult to reconcile with the idea that the Star was two planets.[28] In 7 BC, Jupiter and Saturn rose heliacally on different days, when they were more than 5 degrees apart, many weeks before the first of the three conjunctions. The word "rising" is, of course, more naturally used of individual astronomical entities—a star, a comet, a nova/supernova, or a planet—or constellations.

Fourth, the idea that magi at the turn of the ages would have interpreted a simple triple conjunction of Jupiter and Saturn in Pisces to signify the birth of a royal figure is very questionable.[29] As we noted above, Parpola suggests that Jupiter is Marduk's star and Saturn is the king's. According to him, Pisces, as the last zodiacal constellation, introduced the idea of the end of the old age and the commencement of the new age, more particularly the birth of a Savior and King elected by God.[30] However, that is an unsubstantiated and tenuous interpretation—there is no foundation for the claim that, to the ancient Near Easterners, Pisces could have conveyed the idea that the end of the age was in view or that a new divinely appointed savior had been born.

Similarly, while there is plenty of evidence to suggest that, in the ancient world, the planet Jupiter was viewed as the planet of the chief God, and that Saturn was sometimes thought to have a special association with the Jews,[31] the idea that the constellation Pisces had a particular association with Palestine or the Jewish people is lacking support in contemporary literature.

[24] For a translation of the relevant sections, see ibid., 43–45.

[25] Ibid., 45–46. The almanac mentions the behavior of Jupiter and Saturn in the midst of references to the locations of other astronomical entities like Venus, Mars, Mercury, and Sirius.

[26] Parpola, "Magi and the Star," 17–18, claims that the fact that the manuscript recording the triple conjunction of 7 BC exists in four copies is extraordinary and highlights the conjunction's rarity and importance. However, Sachs and Walker make the point that it was not uncommon for an almanac to be preserved in more than one copy and they comment that, in the case of the 7–6 BC almanac, it is unclear whether the copies are each original texts with independent astronomical calculations or whether two or more of them were copied from one of the others or from a common original (Sachs and Walker, "Kepler's View," 47).

[27] See Franz Boll, "Der Stern der Weisen," *Zeitschrift für die neutestamentliche Wissenschaft und die Kunde des Urchristentums* 18 (1917/1918): 40–43.

[28] Cf. Raymond Brown, *The Birth of the Messiah: A Commentary on the Infancy Narratives in the Gospels of Matthew and Luke*, 2nd ed. (New York: Doubleday, 1993), 173.

[29] W. D. Davies and Dale C. Allison, *A Critical and Exegetical Commentary on the Gospel according to Saint Matthew*, 3 vols., International Critical Commentary (Edinburgh: T. & T. Clark, 1988–1997), 1:233–234, however, point to evidence that some ancient astrologers—namely, Pseudo-Callisthenes 1:12; and Firmicus Maternus, *Math.* 6:1; 8:31—regarded particular conjunctions as hailing a royal birth.

[30] Parpola, "Magi and the Star," 18.

[31] That Saturn could be regarded by Gentiles in this period as the planet of the Jews may be suggested by the anti-Semitic Roman historian Tacitus (*Hist.* 5.4). This association of Saturn and the Jews is present also in the writings of Augustine (*On the Harmony of the Gospels* 1.21–22). Shlomo Sela, *Abraham Ibn Ezra and the Rise of Medieval Hebrew Science* (Leiden: Brill, 2003), 152–153, points out that it is uncertain when precisely the idea was first accepted by the Jews, but that the Babylonian Talmud did refer to Saturn as *Shabtay* (Sabbath). Sela suggests that Ibn Ezra was the first Jewish intellectual to develop the Saturn-Jews association in a macro-astrological scheme (153). Strikingly, Amos 5:25–26 (see also Acts 7:41–43) declared that the Israelites worshiped Saturn (named Sikkuth and Kiyyun in Amos) as their God in the eighth century BC and possibly even during the wilderness years.

Astrological geography identified signs/constellations with regions and/or ethnic groups, but different ancient writers assigned different regions/peoples to different signs/constellations. To the best of our knowledge, no writer ever subsumed Israel/Palestine under Pisces.

Moreover, many advocates of the hypothesis that the triple conjunction in Pisces in 7 BC was the Bethlehem Star appeal for support to the commentary on the book of Daniel (*The Wells of Salvation*, AD 1497) by the Jewish scholar Don Isaac Abarbanel (1437–1508). They claim that Abarbanel's conviction that a Jupiter-Saturn conjunction in Pisces would mark the coming of the Messiah was shared by the Biblical Magi some 1,500 years beforehand.[32] However, the idea that a Jupiter-Saturn conjunction in Pisces would mark the birth of the Messiah belongs to the Medieval period and not to the first millennium BC. Indeed, even according to the principles of Jewish Medieval macro-astrology espoused by these thinkers, the triple conjunction of Jupiter and Saturn in 7 BC was an unimportant one.[33]

Fifth, the triple conjunction view cannot explain the behavior of the Star at the climax of the Magi's journey, as described in Matthew 2:9b. The Star was observed to stand over a single house in Bethlehem, pinpointing it as the place where Jesus was. To attribute such behavior to a planet is most unnatural. The apparent pause of a planet in the sky before it reverses direction[34] is not visually perceptible within the short window of time available—a single night (Matt. 2:10). Of course, the Magi may have known of this pause in advance by their own mathematical calculations, but it should be noted that they would also have been aware that such a pause was simply the prelude to an apparent reversal of direction. Quite why any magus would have taken this particular pause as disclosing the Messiah's location is unclear. Moreover, it is not easy to come up with a plausible explanation of how advance knowledge of the pause in Jupiter's motion might have enabled the Magi to pinpoint one particular building in Bethlehem.

Sixth, the hypothesis that the Star was actually a triple conjunction in 7 BC struggles to come up with a birth date for the Messiah. Hughes suggests that the key moment within the period of the triple conjunction must have been the acronychal rising (the planets' rising in the east as the Sun was setting in the west) on September 15.[35] However, why would the acronychal rising of two distinct planets have been regarded as an astrological indication of a birth? Moreover, as we have already noted, the "rising" of the Star in Matthew 2:2 probably refers to a *heliacal* rising.

Seventh, the triple conjunction theory cannot account for the Magi's clear conviction that a ruler who was worthy of worship had been born, nor can it account for their decision to make a pilgrimage to Judea to find him.[36] It is hard to believe that the Magi headed to Jerusalem simply because of Mars's close approach to Jupiter and Saturn and/or Tishri being known as the month of Amurru. Even if they did think that Mars and Tishri were suggestive of the west country, Syria would have been an equally or more plausible destination. Moreover, since Parpola believes that the Magi headed westward in September/

[32] So, for example, Hughes, *Star of Bethlehem Mystery*, 90, 211; Jean-Pierre Isbouts, *Young Jesus: Restoring the "Lost Years" of a Social Activist and Religious Dissident* (New York: Sterling, 2008), 58–59; cf. Kidger, *Star of Bethlehem*, 206.
[33] The medieval Jewish thinker Ibn Ezra believed that it was a "Great" Conjunction in the sign Leo (205 BC) that had importance with respect to Jesus's birth, while Bar Hiyya associated the following "Great" Conjunction in Virgo in AD 34 with the emergence of Christianity (Sela, *Abraham Ibn Ezra*, 293–294; Josefina Rodríguez Arribas, "The Terminology of Historical Astrology according to Abraham Bar Hiyya and Abraham Ibn Ezra," *Aleph: Historical Studies in Science and Judaism* 11 [2011]: 22, 29).
[34] Parpola, "Magi and the Star," 19.
[35] Hughes, *Star of Bethlehem Mystery*, 147–148.
[36] Cf. Konradin Ferrari-D'Occhieppo, "The Star of the Magi and Babylonian Astronomy," in *Chronos, Kairos, Christos*, ed. Jerry Vardaman and E. M. Yamauchi (Winona Lake, IN: Eisenbrauns, 1989), 46.

October,[37] long before Mars's close approach to Jupiter and Saturn,[38] Mars could have played no role in the Magi's decision regarding direction. Further, needless to say, the proposal that the direction of the Magi's travel was determined by a perceived association between a month and a region is tenuous at best.

Eighth, the timing of the triple conjunction, in 7 BC, is too early to be a realistic candidate for the Star of Bethlehem. Jesus's birth is properly dated to 6 or 5 BC.

Ninth, the fact that Herod executed all infants "in their second year or under," based on when the Magi had said that the Star first appeared, cannot be plausibly explained by this hypothesis.

Consequently, we judge that the popular triple conjunction hypothesis is deeply flawed and should be ruled off the table. The Magi would certainly have looked up to see the triple conjunction in Pisces—it was the most notable predictable astronomical event in 7 BC—and may well have regarded it to be of some astrological significance, but it was certainly not the Star of Bethlehem.

OCCULTATIONS OF JUPITER IN ARIES IN 6 BC

Michael R. Molnar has made a case for the Star of Bethlehem being Jupiter, with the focus on two lunar occultations of Jupiter (that is, two occasions when the Moon obscured Jupiter) in Aries in 6 BC, the first being on March 20 and the second on April 17.[39]

As regards the March 20 event, shortly before the Sun set in Judea, Jupiter was occulted by the Moon while in Aries. The phenomenon ended a little more than 30 minutes later when Jupiter was low over the horizon in the west. Jupiter can sometimes be visible during the daytime to someone who knows exactly where to look, but not when it is on the far side of the Sun from the perspective of Earth and close to the Sun in the sky, as it was on March 20. Therefore the planet could not have been seen at that time. In addition, the Moon was too close to the Sun to be visible at all on that day. Nevertheless, Molnar states that the ancient astronomers were well able to calculate mathematically the courses of both the Moon and Jupiter and so would have known that they were in conjunction and perhaps even that the Moon was occulting Jupiter.[40]

The second occultation, on April 17, 6 BC, occurred shortly after noon in Judea and Babylon. The Moon had completed one revolution and was again in Aries and stationed over Jupiter. The occultation lasted for about 50 minutes for those in Babylon (about 75 minutes for those in Judea). Jupiter was invisible in the hours during which the occultation occurred.[41] In addition, the Moon, a 28½-day waning crescent, was not bright enough to be seen during the daytime. Significantly, according to Molnar, April 17 was the date of Jupiter's heliacal rising (its brief reappearance in the eastern sky just in advance of sunrise).[42] Consequently, a horo-

[37] Parpola, "Magi and the Star," 18–19, suggests that the Magi left Babylon on Tishri 22 (coinciding with the second conjunction) and then, a few paragraphs later, he contradicts himself by suggesting that they departed in "early Tishri." It is all very confusing because Parpola claims that early Tishri corresponds to October, when earlier he stated that Tishri 22 = October 6. Moreover, if the Magi left in early Tishri, then the second conjunction had not taken place and hence they had (according to the hypothesis proposed by Parpola) no basis for going in a westward direction. Probably Parpola meant to write "early October" rather than "early Tishri."

[38] In truth, Mars did not arrive with Jupiter and Saturn until mid-February of 6 BC, well after the final conjunction in early December of 7 BC. Mars was still a full zodiacal sign away from Pisces in the second week of December. It reached the sign of Pisces only in mid-January.

[39] Michael R. Molnar, *The Star of Bethlehem: The Legacy of the Magi* (New Brunswick, NJ: Rutgers University Press, 1999), 86–96.

[40] Ibid., 86.

[41] As Molnar concedes (ibid.).

[42] Ibid., 89. However, Jupiter's heliacal rising probably took place a week or so after this date—see *Planetary, Lunar, and Stellar Visibility* software (version 3.1.0; November 20, 2006), developed by Rainer Lange of alcyone software and Noel M. Swerdlow of the University of Chicago and available at http://www.alcyone.de. It is also interesting to note that the Babylonian almanac for 7/6 BC, which covers the period up to April 19, 6 BC, does not mention Jupiter's heliacal rising, which suggests that it did not occur when Molnar thinks. Molnar's theory is heavily dependent on the idea that the second occultation coincided with Jupiter's heliacal rising, and so the implications of the planet's heliacal rising falling on a day other than April 17, 6 BC, are significant.

scope drawn up for April 17, 6 BC would, in Molnar's opinion, have been deemed important by those operating out of a Greek astrological framework. Indeed he reckons that this second occultation would have been regarded as a sign marking the birth of a royal figure in Judea.[43] It was this later occultation that stimulated the Magi to travel in search of the Messiah. They arrived in Judea in mid-December of 6 BC.

Molnar reasons that, because the Magi's estimation of the significance of the Star was so strongly rooted in astrological principles, the people of Jerusalem would not have shared their assessment of the occultations.[44]

He envisions Jupiter performing the role of the Star in the aftermath of the initial sign.[45] As regards the key verse, Matthew 2:9b, Molnar suggests that the Star's going before the Magi is simply astrological parlance for a celestial entity (in this case, Jupiter) moving in the same direction as the heavens generally, while the Star's standing over the place where the child was located refers to Jupiter's apparent stationary status in mid-December of 6 BC, immediately prior to its commencing retrograde motion relative to the fixed stars.[46]

Molnar makes much of some Roman coins which portray a ram looking at a star—he believes that they indicate that Aries is where astrologers would have looked for celestial indications that the Messiah had been born.[47] These coins were minted in Syria. Molnar speculates, however, that the coins might have commemorated the Roman annexation of Judea in AD 6.[48]

However, we find Molnar's hypothesis unpersuasive.

First, it beggars belief that an invisible sign spurred the Magi to undertake a major journey westward to Judea in search for a newborn King. Surely a great, indeed divine, king warranted a more impressive sign than this. Indeed the Magi explicitly claim that they had "seen" the star of the King of the Jews back in their homeland (v. 2). By Molnar's own acknowledgement, neither occultation of Jupiter in 6 BC was visible in Babylonia. Neither the very new Moon nor Jupiter would have been visible at all on March 20. Moreover, Jupiter and the waning crescent Moon would have been visibly drowned out by the noonday Sun on April 17.

Second, there is no conceivable way that an occultation, particularly an invisible one, could be regarded as a "star."[49]

Third, a lunar occultation of Jupiter in Aries was not particularly rare. As Molnar concedes, another one followed on April 4, AD 54.[50] Why, then, would magi have made the long journey westward to worship the Messiah in 6 BC but not before or afterwards?

Fourth, the fact that Matthew is so favorable in his treatment of the Magi and the Star strongly suggests that his estimation of the Star's significance could not have been entirely dependent on astrological presuppositions.

Fifth, Molnar's theory cannot explain the Magi's conviction that the newborn infant was worthy of worship as a deity.

Sixth, it is difficult to see how Jupiter managed to guide the Magi to Bethlehem and then to a particular location within the town. The claim of Molnar that the standing still is due to Jupiter's becoming stationary[51] fails to convince: a change in Jupiter's motion would not have been detectable in the short window

[43] Molnar, *Star of Bethlehem*, 89, 96–97.
[44] Ibid., 86.
[45] Ibid., 87–96.
[46] Ibid., 90–92, 95–96.
[47] Ibid., 5.
[48] Ibid.
[49] As Hegedus, *Early Christianity and Ancient Astrology*, 202–203, points out.
[50] Molnar, *Star of Bethlehem*, 102.
[51] Ibid., 92, 96.

of time when the Magi were in Bethlehem searching for the messianic child. Even if the Magi were relying on advance calculations, these would not have been adequate to enable them to pinpoint a particular town, never mind a house within it. Moreover, identifying the Star's "standing over" with Jupiter's becoming stationary on December 19, 6 BC, creates an unrealistically long journey time for the Magi—some eight months.

Seventh, as Parpola observes, the significance of the astronomical phenomenon described by Molnar is unlikely to have been interpreted in the way suggested by him, for in the ancient Near East a lunar occultation of Jupiter was a bad omen, signifying disaster for a nation or kingdom or death to a king.[52]

Eighth, Molnar's hypothesis cannot explain why Herod ordered that infants a year old be slaughtered "according to the time that he had ascertained from the Magi" (Matt. 2:16).

Ninth, with respect to what territory the imagery on Molnar's coins had in view, Syria is a much more plausible candidate than Judea. After all, the coins were minted there. Molnar's proposed connection to Judea, namely that the coins might have been commemorating the Roman annexation of Judea in AD 6, is strained. Moreover, Aries was very closely associated with Syria and a number of other territories. Vettius Valens of Antioch claimed that Aries controlled Syria and the neighboring territories (*Anthology* 1.2), while Manilius, *Astronomica* 4.744–54, regarded Aries as representing Syria and northern Egypt.[53] The connection between Aries and Judea was apparently much less widespread. Ancient astrological writers were divided concerning where Judea fit in the scheme of astrological geography. Claudius Ptolemy, *Tetrabiblos* 2.3, did include Judea/Palestine among the lands of Aries, but he also included Syria, Idumea, Gaul, Britain, and Germany. In the first centuries AD, Judea could be subsumed not only under Aries but also under Gemini, Scorpius, or Aquarius.[54] Moreover, one ancient text (BM [British Museum] 47494) suggests that the astrologers of Babylon associated the signs of Gemini, Libra, and Aquarius with Amurru (the West).[55]

In conclusion, we find the hypothesis that lunar occultations of Jupiter in Aries explain the mystery of the Star seen by the Magi to be extremely problematic.[56]

NOVA OR SUPERNOVA HYPOTHESIS

A nova is a massive nuclear explosion (or eruption) of a still-white-hot, old, dying star called a white dwarf. To put it in simple terms, many stars exist in pairs (or "binary systems"), consisting of a relatively cool star and a white dwarf. In some cases the white dwarf may draw gases such as hydrogen and helium from the surface of its companion star (fig. 4.2). These gases, which accumulate in a layer around the white dwarf's surface, are subject to intense compression and heating. In due course there is a great nuclear reaction on the white dwarf that blasts the gases from

[52] Parpola, "Magi and the Star," 60n1. Note especially one particular ancient Near Eastern omen: "When the Moon occults Jupiter, that year a king will die (or) an eclipse of the Moon and Sun will take place. A great king will die. When Jupiter enters the midst of the Moon, there will be want in Aḫarrû. The King of Elam will be slain with the sword: in Subart[u] . . (?) will revolt. When Jupiter enters the midst of the Moon, the market of the land will be low. When Jupiter goes out from behind the Moon, there will be hostility in the land." Translation by R. Campbell Thompson, *The Reports of the Magicians and Astrologers of Nineveh and Babylon in the British Museum*, vol. 2 (London: Luzac & Co., 1900), lxvii, no. 192.

[53] For English versions of these works, see Mark Riley's translation of Vettius Valens, *Anthologies*, at http://www.csus.edu/indiv/r/rileymt/Vettius%20Valens%20entire.pdf (p. 3 for the geographical associations of Aries) (last modified January 5, 2011); and Manilius, *Astronomica*, trans. G. P. Goold (Cambridge, MA: Harvard University Press, 1989) (pp. xci–xcii on astrological geography).

[54] Frederick H. Cramer, *Astrology in Roman Law and Politics* (Philadelphia: American Philosophical Society, 1954), 23; see James M. Scott, *Geography in Early Judaism and Christianity: The Book of Jubilees* (Cambridge: Cambridge University Press, 2005), 74–75 table 3.

[55] See Francesca Rochberg, *Babylonian Horoscopes* (Philadelphia: American Philosophical Society, 1998), 109.

[56] It is surprising that the October 2014 multi-disciplinary colloquium on the Star of Bethlehem at the University of Groningen had as its primary purpose the examination of Molnar's theory; see http://www.astro.rug.nl/~khan/bethlehem/scientific-rationale.php (accessed July 5, 2014).

its surface. This causes a very bright explosion of light that brightens the star by some 5–15 magnitudes (100–1,000,000 times) over the space of a few days or weeks (to put that into perspective, the Sun is about 14 magnitudes or 400,000 times brighter than the full Moon). The brightness fades over the following 1–3 months. Afterwards, the binary system is restored to its former state.[57] On rare occasions, the brightness of a nova may be sufficient to render it visible to the naked eye (e.g., Nova Cygni 1975, which attained to +2 magnitude, approximately the brightness of the North Star, Polaris).

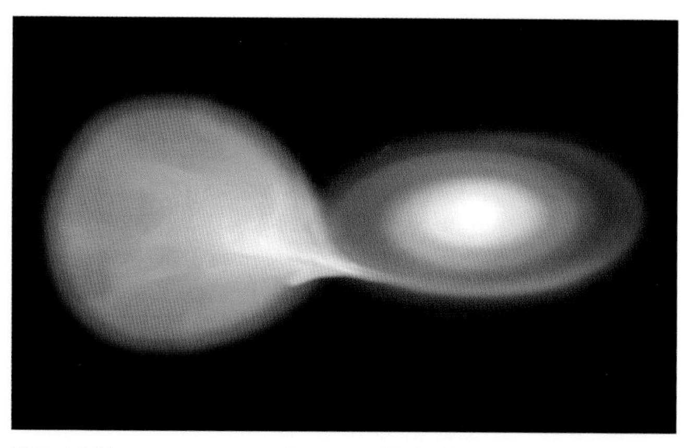

FIG. 4.2 How a nova occurs. Image credit: NASA/CXC/M. Weiss/Wikimedia Commons.

A supernova is an even more massive nuclear stellar explosion. There are two distinct categories: the first type (the binary-system supernova) is essentially a large and very bright nova that entails the destruction of the white dwarf, while the second type consists of a huge star that, having consumed all of its own nuclear fuel, implodes, causing a massive explosion of light and resulting in the star's disintegration. A supernova is a very rare phenomenon but, when one occurs, it is by far the brightest "star" in the sky. The light of a supernova may remain in the sky for months, indeed up to about two or three years.[58] A binary-system supernova may be up to 4,000,000,000 times brighter than the Sun (assuming a vantage point the same distance from each) during its peak.[59] Examples of binary-system supernovas include SN 1006 (in AD 1006), which became much brighter in the sky than Venus (reaching apparent magnitude -7.5) (and as large as half the Moon), and SN 1604 (Kepler's supernova in AD 1604), which appeared as bright as Jupiter at its peak (magnitude -2.5).[60] A huge-star supernova may be up to 1,000,000,000 times brighter than the Sun (assuming a vantage point the same distance from each).[61] An example of this kind of supernova is SN 1054 (the Crab Supernova) in AD 1054, the apparent brightness of which climaxed at about the maximum brightness of the International Space Station (magnitude -6).

Those who propose that the Star seen by the Magi was a nova or a supernova[62] suggest that such a phenomenon would have been a

[57] Moore, *Star of Bethlehem*, 73–74; see also Wikipedia, s.v. "Nova," http://en.wikipedia.org/wiki/Nova (last modified April 11, 2013).

[58] The AD 1006 supernova was visible at its brightest for 4 months in the spring before disappearing for a few months and then reappearing less bright in December for something like 2½ years. It was probably visible about 3 years after it was first observed (F. Richard Stephenson, "SN 1006: The Brightest Supernova," *Astrophysics and Geophysics* 51.5 [2010]: 27–32). According to Chinese records, the AD 1054 supernova lasted for 642 days. The AD 1572 supernova lasted well beyond 1 year, and the one in AD 1604 was tracked for a year by Kepler.

[59] Moore, *Star of Bethlehem*, 74.

[60] In astronomy a "magnitude" is a measurement of a star's apparent brightness. The scale is rather counterintuitive, since the brighter a star is, the smaller its magnitude value is (into minus numbers). A one unit increase is equivalent to a reduction of brightness by a factor of 2.51. A magnitude +1 star is 100 times brighter than a magnitude +6 star. A difference of 10 magnitudes would mean 10,000 times brighter; a difference of 15 magnitudes would mean 1,000,000 times brighter. The apparent magnitude of the Sun is -26.7, and that of the full Moon is -12.6; this means in effect that the Sun is 400,000 times brighter than the Moon.

[61] Ibid., 75.

[62] E. W. Maunder, "Star of the Magi," in *International Standard Bible Encyclopedia*, ed. James Orr, 5 vols. (Grand Rapids, MI: Eerdmans, 1939), 5:2848–2849; idem, *The Astronomy of the Bible: An Elementary Commentary on the Astronomical References*

celestial sign supremely worthy to mark a royal birth.[63]

There are, however, a few serious objections to the nova/supernova hypothesis.

First, there is no evidence that there was any nova or supernova in 6 BC or 5 BC. Therefore the theory is highly conjectural. Suitable candidates are lacking in the surviving astronomical records.[64] However, the fact that the records are not comprehensive means that this line of argumentation has limited weight. It is much more important that, as Guy Consolmagno emphasizes, supernovas leave remains in their wake—for example, the Crab Nebula is the remnants of a supernova in AD 1054. Unfortunately for the supernova hypothesis, no remnants of a supernova 2,000 years ago have ever been discovered.[65]

Some have claimed that the Chinese astronomical *hui-hsing*, recorded in the year 5 BC, was actually a nova.[66] However, most scholars concur that *hui-hsing* refers to a broom-star comet.[67] Normally the phrase refers to tailed comets. Moreover, the location in the sky of the *hui-hsing*, in Capricornus, suggests that it refers to a comet rather than a nova, for that constellation is not where one would normally expect to see a nova, since it is too far from the plane of the Milky Way galaxy, where most novas are found.[68] In addition, the *hui-hsing* of 5 BC did not heliacally rise. To argue that the *hui-hsing* is the Star, one is forced to claim that the Magi were confused—they believed that they were seeing a star's heliacal rising, in spite of the fact that the object was "far higher in the sky than could logically be expected for a first sighting of a star."[69]

Second, neither a nova nor a supernova moves within the framework of the fixed stars and constellations. Therefore neither phenomenon is capable of explaining the dramatic movement of the Star of Bethlehem, which shifted from the eastern morning sky to the southern evening sky within the space of a couple of months.

Third, an ordinary nova could not have done what Matthew states that the Star did—remain visible to the naked eye for over a year, go before the Magi to Bethlehem, or stand over a particular house in Bethlehem.

As for a supernova, while it could have remained visible for more than a year, its impressiveness would have steadily diminished over that time. Since we know that the Star's heliacal rising occurred the best part of a year after the first appearance, it would be difficult to understand why the Magi were so deeply impressed by the supernova at its rising. After all, not only would its brightness have been greatly reduced from the time of its first appearance, but also its placement in the heavens would, of course, have been exactly where it had been the whole time.

Fourth, ordinary novas were too common to secure the kind of attention and interpretation that the Star received.[70] Supernovas are more rare, at least in recent centuries (al-

of Holy Scripture (New York: Mitchell Kennerley, 1908), 393–400; R. S. McIvor, *Star of Bethlehem, Star of Messiah* (Toronto: Overland, 1998); A. J. Morehouse, "The Christmas Star as a Supernova in Aquila," *Journal of the Royal Astronomical Society of Canada* 72 (1978): 65–68; John Seymour and Michael W. Seymour, "The Historicity of the Gospels and Astronomical Events concerning the Birth of Christ," *Quarterly Journal of the Royal Astronomical Society* 19 (1978): 194–197.

[63] For example, James Mullaney, "The Star of Bethlehem," *Science Digest* 80 (December 1976): 61–65.

[64] Moore, *Star of Bethlehem*, 78.

[65] Guy Consolmagno, "Looking for the Star, or Coming to Adore?" *Thinking Faith*, http://www.thinkingfaith.org/articles/20101231_1.htm (last modified December 31, 2010).

[66] So, for example, Kidger, *Star of Bethlehem*, 234–246. We will take a closer look at the Chinese *hui-hsing* in chapter 6.

[67] See Donald K. Yeomans, *Comets: A Chronological History of Observation, Science, Myth, and Folklore* (New York: John Wiley, 1991), 361–362 and 367.

[68] In order to get around this problem, Kidger, *Star of Bethlehem*, 244–245, resorts to arguing that the Chinese data regarding the location of the *hui-hsing* is incorrect. He speculates that the nova actually appeared near the star Theta Aquilae, not far from Capricornus (246).

[69] So, for example, ibid., 281.

[70] Cf. Kenneth Boa and William Proctor, *The Return of the Star of Bethlehem: Comet, Stellar Explosion, or Signal from Above?* (New York: Doubleday, 1980), 81.

though note that there were two major ones in the eleventh century AD), but why would the Magi have interpreted one to signify the birth of the Jewish Messiah?

We are therefore very skeptical concerning the hypothesis that the Star was a nova or supernova.

METEORS HYPOTHESIS

Sir Patrick Moore was the most prominent champion of the hypothesis that the Star seen by the Magi in the east and then in Bethlehem was actually two separate shooting stars. Meteoroids are essentially debris, mostly from comets but also from some asteroids. The mostly pebble-size or smaller objects orbit the Sun until they are destroyed by it or crash into a planet. To become visible as shooting stars to human observers on Earth, the orbit of meteoroids must cross Earth's orbit when Earth is present. Such is the remarkable velocity of these bits of debris—up to 72 km per second!—that, upon striking Earth's atmosphere, they gradually disintegrate even as they excite the molecules of air, giving rise to glowing streaks of light.[71] Such streaks are classified as meteors or shooting stars. They may be short or long and vary greatly in brightness. Especially bright meteors are called fireballs. The very same bright meteor may be visible several hundred miles away.

While some meteors are associated with regular meteor showers, most are not. Those that are not are called sporadic meteors.

Moore believed that the Magi witnessed an especially spectacular meteor in their homeland.[72] On rare occasions meteors may be as bright as the Moon or even the Sun; Moore reckoned that one such especially bright meteor was involved in the Christmas story. This gloriously bright meteor, he imagined, rose in the east and traveled through the sky in a westward direction. Although the performance would have been rather fleeting, Moore suggested that the trail left by the fireball or bolide might have been visible for hours afterwards.[73]

Sir Patrick proposed that another gloriously bright meteor appeared, once again in the eastern sky and once again crossing the sky in a westward trajectory.[74] It is not altogether clear what was the purpose of this second fireball, whether to get the Magi to depart on their westward journey or to pinpoint the location of the child in Bethlehem at the end of their journey. Most likely, however, Moore was suggesting that the second meteor was doing the work of the Star when the Magi were in Bethlehem—unless Moore did mean this, his theory offers no explanation of the climactic appearance of the Star.[75]

Moore admitted that his hypothesis could not be proved, but, at the same time, insisted that it could not be disproved either and claimed that it fared better than other hypotheses.[76]

However, the hypothesis is easily refuted.

First, this view cannot explain how the Star could have first appeared one year or more before Herod issued the order to slaughter the infants of Bethlehem (Matt. 2:16; cf. v. 7b). If the first meteor occurred then, why did the Magi take so extraordinarily long to get to Judea?

Second, it is difficult to see how a fireball, regardless of how bright, could have prompted magi to think that an important birth had taken place, never mind that of the Messiah in Judea. Evidence that fireballs or bolides were ever interpreted by ancients as marking the birth of a king is notably lacking. As to the question of how the Magi

[71] Robert Lunsford, *Meteors and How to Observe Them* (New York: Springer, 2008), 2.
[72] Moore, *Star of Bethlehem*, 96.
[73] Ibid.
[74] Ibid.
[75] Cf. Kidger, *Star of Bethlehem*, 122.
[76] Moore, *Star of Bethlehem*, 99.

might have concluded that they should head to Judea, Moore suggested that the fireball pointed westward. But a westward-traveling fireball is hardly sufficient to identify Judea in particular as opposed to Syria, Greece, Italy, or Spain. Perhaps one could hypothesize that the fireball seemed to originate from a constellation that was suggestive of Judea or the Jews. However, Moore did not argue this, and ancients' ideas regarding the relationship between signs/constellations and particular geographical areas were not narrow enough to permit a firm identification with Judea.

Third, meteors are too fleeting to fulfill Matthew's description of the Star as having gone ahead of the Magi until it stood over the house where Jesus was. Moore conceded that no meteor could have done what the Star is said to have done in Matthew 2:9, but he tried to get around this problem by proposing that Matthew was simply exercising poetic license at this point in the narrative.[77] However, that Moore was forced to abandon a literal interpretation of Matthew's account, particularly at the point that the Gospel is giving its most detailed description of the Star's behavior, is a serious, indeed fatal, problem for his hypothesis.

Fourth, Matthew is quite explicit that it was the same "star" seen in the east and in Bethlehem. Moore's hypothesis, however, would mean that there were two meteors and hence two "stars." Certainly it defies belief that the Magi (and Matthew) would have regarded the Bethlehem meteor, coming a long time after the initial one, as identical to the meteor that they had previously seen zipping across the sky back in their homeland.[78]

We therefore conclude that the two meteors hypothesis is irreconcilable with the description of the Star in the Gospel of Matthew.

AN ORDINARY STAR: ALPHA AQUARII

In recent decades a few scholars, in particular Richard Coates and David Seargent, have postulated that the Star might have been an ordinary fixed star, in particular the star Alpha Aquarii.[79] This is a very ordinary star—it has a magnitude of +2.95,[80] which is not particularly bright. However, according to this hypothesis, what the star lacked in brightness it made up for in astrological significance. Fundamental to this hypothesis is the fact that Alpha Aquarii was known in Arabian tradition as "the lucky star of the king/kingdom,"[81] although it is conceded that it was not perceived to represent one particular royal figure.[82]

In order to explain why one particular annual morning rising prompted the Magi to embark on a journey westward to Judea, proponents of this view propose that some astronomical phenomenon in the relevant year must have seemed to invest the star with great significance. Seargent suggests that there may have been an alignment between a planet and this star.[83] Alternatively, he mentions the possibility that an important conjunction or a massing of planets at an astrologically significant moment might have engendered Alpha Aquarii's heliacal rising with special meaning that year.[84]

Seargent speculates that, in the wake of the triple conjunction of Jupiter and Saturn in Pisces of 7 BC, the rising of Alpha Aquarii in February of 6 BC would have been regarded as having special importance.[85] He claims that Pisces was the zodiacal sign of Israel and the

[77] Ibid., 98–99.
[78] Cf. Kidger, *Star of Bethlehem*, 122.
[79] Richard Coates, "A Linguist's Angle on the Star of Bethlehem," *Astronomy and Geophysics* 49.5 (October 2008): 27–32; and David Seargent, *Weird Astronomy* (New York: Springer, 2011), 276–281.
[80] Seargent, *Weird Astronomy*, 277.
[81] Ibid., 276–277.
[82] Ibid., 279.
[83] Ibid.
[84] Ibid., 278.
[85] Ibid., 280.

Messiah and that the "constellation" was associated with "change and new beginnings."[86]

According to this theory, during the Magi's short journey from Jerusalem to Bethlehem, sometime between July and September, this star would have moved in front of them until it crossed the meridian, its highest point ("culmination"), in the south. When an astronomical object culminates, it seems to pause for a while before descending in altitude to the west. Seargent deduces that this pause coincided with the arrival of the Magi in Bethlehem and seemed to pinpoint the precise location where Jesus and his mother were.[87]

However, this hypothesis is uncompelling.

First, we have no evidence to suggest that the name assigned to Alpha Aquarii by the Arabs was the same one attributed to the star by astrologers at the time of the Magi or that this star was of notable astrological significance in the first century BC.

Second, an overview of the history of the star in 7–5 BC shows that there was no conjunction involving the star, and no planetary massing in its vicinity. As for the idea that a Jupiter-Saturn conjunction in Pisces endowed the heliacal rising of Alpha Aquarii with special significance, one must ask why. The logic of such an association is obscure, to say the least.

Third, there is no foundation for the claim that Pisces was widely regarded as the zodiacal sign of Judea and its eschatological King, the Messiah, at the turn of the ages.[88]

Fourth, we lack a single scrap of evidence that the star Alpha Aquarii had any messianic significance or that its rising would have communicated that a divine figure had been born. There is no way that the ordinary rising of this unimpressive star could have prompted the Magi to embark on a long journey to Judea in search of the Messiah.

Fifth, this theory requires that the Magi took an inordinate amount of time to journey from their eastern homeland to Judea. At the same time, this hypothesis cannot explain why Herod asked when the Star had first appeared or why the Magi told him that it had first appeared between one and two years beforehand.

Sixth, an ordinary star cannot plausibly be regarded as having stood over one particular house, pinpointing it.

We can therefore safely eliminate this explanation of the Star of Bethlehem.

7–5 BC COMBINATION HYPOTHESIS

Mark Kidger has argued that the Star of Bethlehem was the combination of the triple conjunction of Jupiter and Saturn in 7 BC, the planetary massing of Mars, Jupiter, and Saturn within Pisces in 6 BC (a massing is a grouping of objects in the same celestial location), a pairing of the Moon and Jupiter in Pisces on February 20 in 5 BC, and a nova in the spring of 5 BC.[89] The nova was the catalyst for the Magi's journey to Judea, but, as an event that takes place once every 25 years or so, only had the impact on the Magi that it did because the celestial phenomena of 7 BC and 6 BC had primed them to expect a definitive final sign of the Messiah's birth.[90] Indeed Kidger believes that, in the wake of the nova in 5 BC, the Magi came to regard the triple conjunction of 7 BC as coinciding with the birth of the Messiah.[91]

As ingenious as Kidger's combination hypothesis is, it too fails to convince. Kidger,

[86] Ibid., 279.
[87] Ibid.
[88] The medieval Jewish scholars Bar Hiyya and Ibn Ezra, who maintained that Jupiter-Saturn conjunctions played an important determining role in Israel's history, did not believe that Pisces was Israel's sign, but rather Aquarius, one of the houses of Saturn. Abarbanel (in his *Wells of Salvation*) did invent some feeble connections between Pisces and Israel (see Roy A. Rosenberg, "The Star of the Messiah: Reconsidered," *Biblica* 53 [1972]: 106–107; and Hughes, *Star of Bethlehem Mystery*, 212), but he did not regard it as Israel's sign.
[89] Kidger, *Star of Bethlehem*, 198–275.
[90] Ibid., 258–259, 264–265.
[91] Ibid., 216.

clearly aware of the inadequacies of the triple conjunction and nova theories, believes that together they become strong. Unfortunately, joining together flawed hypotheses does not necessarily create a strong hypothesis. A good number of the objections we raised against the triple conjunction and nova/supernova theories apply against Kidger's combined view—e.g., no one in the first millennium BC believed that Pisces was associated with Judea and the Jews;[92] a nova cannot "stand" over a house and is not sufficiently unusual or spectacular to have prompted magi to make a long journey; the 5 BC *hui-hsing* is almost certainly a tailed comet; and 7 BC is too early to be Jesus's birth year.

In addition, the combination hypothesis of Kidger is overly complicated and, more importantly, incompatible with Matthew's account. It is very unlikely that each of the astronomical events mentioned by Kidger would have been interpreted as having a single, unified message regarding the birth of the Messiah in Judea. Furthermore, Kidger's hypothesis holds that the Star of Bethlehem was a selection of unrelated phenomena spread over 2 years, with, for example, the astronomical entity marking Jesus's nativity being the triple conjunction and the celestial guide to Bethlehem being a nova. However, the text of Matthew makes it clear that the Star was one single entity that appeared, heliacally rose, and went ahead of the Magi, finally standing over the house where the messianic child was.

We conclude therefore that Kidger's combination hypothesis is not a plausible contender for the Star of Bethlehem.

JUPITER IN 3–2 BC

Over the last few decades a number of scholars, judging that Herod the Great actually died in 1 BC rather than 4 BC, have proposed that the mystery of the Star of Bethlehem may be explained with reference to celestial phenomena in the years 3 and 2 BC.

Ernest L. Martin, in his book *The Star of Bethlehem: The Star That Astonished the World*,[93] pointed out that there was a triple occultation (or obscuring) of Regulus, the brightest star in Leo, by Jupiter in 3–2 BC—on September 14, 3 BC; February 17, 2 BC; and May 8/9, 2 BC. Regulus, to the ancient Babylonians, was The King (LUGAL). According to Martin, it was also the Messiah's Star, and the constellation in which it was found, Leo, was associated with Judah.[94] No bright star is closer to the path of the Sun and Moon than Regulus, and so, naturally, it is occasionally occulted by planets such as Venus and Jupiter. However, Martin proposed that the triple occultation of Regulus in 3–2 BC was especially susceptible to the interpretation that a moment of royal significance was about to dawn.[95] He was particularly impressed by the first occultation, since it occurred shortly after a conjunction between Jupiter the King Planet and Venus the Mother Planet. In his opinion the oval movement of Jupiter against the backdrop of the fixed stars during the period of the triple occultation was nothing less than a "crowning" of Regulus.[96]

As Martin saw it, the celestial events relating to the King Planet, the Messiah's Star, and Judea's constellation alluded strongly to Biblical traditions concerning the Messiah.

In addition to the triple occultation, Martin highlighted that on June 17, 2 BC, Jupiter reunited with Venus in Leo, seeming to most naked-eye observers in Babylon to form a single star. Since this was visible over the western horizon, it was natural for

[92] Contra ibid., 254; also 257.
[93] Ernest L. Martin, *The Star of Bethlehem: The Star That Astonished the World*, 2nd ed. (Portland, OR: Associates for Scriptural Knowledge, 1996), available at http://www.askelm.com/star.
[94] http://www.askelm.com/star/star004.htm (accessed March 26, 2014).
[95] Ibid.
[96] Ibid.

the Magi to interpret it as pointing toward Judea.[97]

Then on August 27 of 2 BC, during a massing of Mercury, Venus, Mars, and Jupiter in Leo, Jupiter and Mars had a conjunction.[98] Martin speculated that this phenomenon may have seemed to the Magi to announce the commencement of wars on the earth and that this would have been understood by them in connection with Old Testament prophecies that the messianic era would be ushered in by a cataclysmic war.[99]

For Martin, the Star of Bethlehem was Jupiter.[100] He suggested that on December 25 of 2 BC the King Planet, as it began a retrograde phase, seemed to pause in the belly of Virgo, the celestial maiden.[101] This apparent cessation of movement, according to Martin, would have been perceived to be astrologically significant.[102] Moreover, on that date, when the Magi, now in Judea, were observing the sky in the run-up to sunrise, Jupiter was culminating almost 70 degrees over the horizon in the south. Martin maintained that it is to this that Matthew was referring when he reported that the Star went before the Magi and then stood over the place where the child was (Matt. 2:9).[103] According to Martin, Jesus was born on September 11, 3 BC, when the Sun "clothed" Virgo and the Moon was under her feet (Rev. 12:1–2).[104]

Frederick A. Larson, an attorney in the United States, has recently popularized a very similar theory in a DVD called *The Star of Bethlehem*.[105] The major difference is that what Martin regarded as the moment of the birth of Jesus, Larson views as the moment of the conception.

As enthralling as the adventures of Jupiter in 3–2 BC were, the hypothesis of Martin, and Larson, must be rejected for a number of reasons.

First, the whole hypothesis rests on a very dubious redating of Herod's death to 1 BC. We have already seen some of the shortcomings of this chronological revisionism. For one thing, coins minted in Jerusalem by Herod's sons make it clear that they began their reigns in 4 BC.

Second, Matthew 2:9b strongly suggests that the Star was observed to stand over a particular house, not the town as a whole, leading the Magi to the exact place where the child was. Advocates of the 3–2 BC hypothesis are unable to offer a plausible explanation of what Matthew describes.

The Star's "standing" could not refer to Jupiter becoming stationary relative to the fixed stars immediately before beginning retrograde motion, because that is not detectable by the human eye in the short space of a few hours. Moreover, if the planet's becoming stationary were in Matthew's mind, that would mean that the Star's going in advance of the Magi, leading them forward, must refer to the movement of the planet relative to the fixed stars in the hours prior to its becoming stationary. But Jupiter's movement relative to the fixed stars during the last hours before it paused would have been negligible and similarly impossible to perceive visually.

At the same time, it should be realized that the Magi would have known well in advance about Jupiter's movements and would hardly have elected to base their itinerary on this.[106]

Third, it is peculiar that Martin identifies Jupiter as the Star of the Messiah, but regards the celestial event marking his birth as having

[97] Ibid.
[98] Ibid.
[99] Ibid.
[100] Ibid.
[101] Ibid.
[102] Ibid.
[103] Ibid.
[104] http://www.askelm.com/star/star006.htm (accessed March 26, 2014).
[105] Frederick A. Larson, *The Star of Bethlehem*, DVD, directed by Stephen Vidano (Santa Monica, CA: Mpower Pictures, 2006).
[106] Notably, the Magi did not set off back to their homeland as soon as Jupiter started its retrograde motion!

nothing to do with Jupiter. The celestial event marking Jesus's birth was, Martin claims, simply the Sun clothing Virgo while the Moon was under her feet (three days before the first of the three Jupiter-Regulus occultations).[107] The Gospel account, however, strongly connects the Star itself with the birth of the Messiah.[108]

Fourth, this theory is unable to offer a plausible chronology of events. According to Martin, the Magi saw the sign marking the Messiah's birth on September 11, 3 BC, and yet delayed departing for Judea more than 9 months until, after all three Jupiter-Regulus occultations, the King Planet had a conjunction with Venus on June 17 of 2 BC; or 11½ months, until the massing of planets on August 27, 2 BC! And then it took them 4 or 6½ months to travel there!

Fifth and most important, this hypothesis's interpretation of Jupiter's movements in 3–2 BC is not consistent with ancient astrological principles. For example, no magus would have regarded Leo as the constellation or sign of Judea, nor Regulus as the Messiah's star. Moreover, there is no basis for the claim that the movement of Jupiter relative to Regulus would have been paradigmed as "crowning."[109]

In spite of its recent increased popularity, therefore, the 3–2 BC hypothesis can be safely discarded.

SUPERNATURAL PHENOMENON

Many have assumed that the Star was miraculous in nature (at least as far as Matthew was concerned)[110] or merely a group visionary experience of the Magi.[111] Since the Star was, according to these hypotheses, a one-off phenomenon, there is no way that astronomical investigation in the twenty-first century could expect to make any progress in the quest for the Star.

The miraculous and visionary views are attractive since they can, of course, explain much of the data in Matthew's account, because their central claims, namely that there was a miraculous Star in 6–5 BC and that the Star was a subjective entity, are by their very nature impossible to falsify.[112] At the same time, they are positions of last resort[113]—that

[107] http://www.askelm.com/star/star008.htm (accessed March 26, 2014). Frederick Larson, in his *Star of Bethlehem* DVD, diverges from Martin at this point, suggesting that the Sun clothing Virgo and the Moon under her feet on September 11 of 3 BC simply marked Jesus's conception, with a very close conjunction of Jupiter and Venus on June 17 of 2 BC occurring at the point of Jesus's birth. If conception occurred on September 11, 3 BC, and the period between conception and birth was an average 266 days, birth would be expected on June 4, 2 BC. Larson suggests that the Magi started planning to travel to Judea in September of 3 BC, but actually departed only when they witnessed the conjunction of June 17, 2 BC (Frederick A. Larson, "Westward Leading," http://www.bethlehemstar.net/starry-dance/westward-leading [accessed March 26, 2014]). Larson subsequently states that, on December 25, 2 BC, Jupiter appeared to come to a halt relative to the fixed stars, because it was beginning retrograde motion. The Magi, making their way from Jerusalem to Bethlehem at this time, perceived that it was standing over the town where David had been born (Frederick A. Larson, "To Stop a Star," http://www.bethlehemstar.net/starry-dance/to-stop-a-star [accessed March 26, 2014]). That, of course, would mean that the Magi's trip lasted an absurdly long time—over 6 months! Moreover, quite why the conjunction of Jupiter and Venus in 2 BC, as stunning a sight as it undoubtedly was, would have been interpreted by Magi as a signal that the Messiah had been born, is unclear. Certainly there is no evidence that any ancient would have interpreted the conjunction to have this significance. Larson suggests that the Magi would have been surprised by the sight of Jupiter and Venus in conjunction. However, by the first-century BC astronomers were able to calculate the movements of the planets long in advance. As we have seen, the Babylonian astronomers even produced almanacs, containing their predictions for a given upcoming year. Therefore the Magi would not have been taken aback by the conjunction. As regards Jupiter's retrograde motion, that too would have been known about well before it happened. Further, a look at the Starry Night® Pro software that Larson uses reveals that the date when Jupiter "paused" was actually December 27/28, not, as Larson claims, December 25. In truth, the Magi would probably have been aware that Jupiter was still moving within the fixed stars and constellations on December 25.

[108] It is also worth pointing out that the third occultation occurred below the visible horizon.

[109] And, even if ancients could have regarded a planet's slow apparent loop motion in the sky as an act of "crowning," Jupiter was hardly crowning Regulus. Regulus was, after all, nowhere near the middle of the loop of the planet's celestial motion.

[110] So, for example, David L. Turner, *Matthew*, Baker Exegetical Commentary on the New Testament (Grand Rapids, MI: Baker Academic, 2008), 80, 86; Craig Blomberg, *Matthew*, New American Commentary (Nashville: Broadman, 1992), 65; cf. Hare, *Matthew*, 14; Luz, *Matthew 1–7*, 135.

[111] E.g., Ken Collins, "The Star of Bethlehem," http://www.kencollins.com/explanations/why-01.htm (last modified March 23, 2013); and F. Richard Stephenson, as cited by Fiona Veitch Smith, "Did the Star of Bethlehem Really Exist?" http://www.veitchsmith.com/2009/12/10/did-the-star-of-Bethlehem-really-exist/ (posted December 10, 2009).

[112] Cf. Donald A. Carson, "Matthew," in *Expositor's Bible Commentary*, rev. ed., ed. Tremper Longman III and David E. Garland, vol. 9 (Grand Rapids, MI: Zondervan, 2010), 111.

[113] Cf. Molnar, *Star of Bethlehem*, 16.

is, they are adopted only because the description of the Star in Matthew 2:9 is deemed to go beyond what could realistically be expected of normal astronomical phenomena.[114] In this book we shall demonstrate that there is no need to resort to the miraculous or visionary views to explain the behavior of the Star. When the relevant Biblical texts are correctly interpreted, everything that is said about the Star is seen to cohere with established astronomical facts.

There are a few key problems with the miraculous view.

First, the astronomical language—specifically "star" and "rising"—used in Matthew 2 is misleading if an astronomical phenomenon is not in view. It should be noted that it is not just Matthew (v. 9) who refers to a "star" that "rose," but the Eastern astronomers also (v. 2).

Second, the miraculous view leaves key questions unanswered: How did the Magi come to perceive that someone had been born, that he was King of the Jews, and that he was worthy of worship? The fact that the Magi traveled from their homeland to Judea in search of the Messiah suggests that they deduced from what the Star did prior to their departure that the Messiah prophesied in the Hebrew Scriptures had been born. However, the miraculous view cannot explain this without introducing elements that are absent from the narrative, such as that the Magi received other supernatural revelation too.

One version of the supernatural hypothesis is that "the guiding star was a guiding angel."[115] According to this view, the apocryphal *Gospel of the Infancy* (in particular, chapter 7) was correct in its interpretation of Matthew's Star when it said, "In the same hour there appeared to them an angel in the form of that star which had before guided them on their journey; and they went away, following the guidance of its light, until they arrived in their own country."[116]

However, Matthew, even in the nativity narrative, is not shy about referring explicitly to angels when they are involved in events. For example, Matthew 1:20 reports that "an angel of the Lord appeared to [Joseph] in a dream" to command him to marry Mary (cf. v. 24). Further, 2:13 states that "an angel of the Lord appeared to Joseph in a dream" to warn him to flee out of Herod's clutches to Egypt, and verses 19–20 record that "an angel of the Lord appeared in a dream to Joseph in Egypt" to call him to return to the land of Israel, because "those who sought the child's life are dead." In light of these explicit references to angels, it is implausible that Matthew, in 2:1–12, would have consistently spoken of the guidance of an angel as that of a star.

The proposal that the Star was observable only by the Magi and was visionary in nature is likewise flawed and makes for a very implausible reading of the text. This position makes much of the shock of Herod and the people of Jerusalem when they heard what the Magi were saying, interpreting it as evidence that the Judeans had not themselves seen the Star.

A strong case can be mounted against the visionary view.

First, Matthew at no point suggests that the Star was a subjective phenomenon. Elsewhere when a visionary experience is in view, there are clear indications that this is the case. Note that Matthew makes explicit in verses 12, 13, and 19–20 when subjective revelatory phenomena played a part in the narrative (also 1:20). However, in 2:1–18 both the Magi and

[114] For example, Kenneth D. Boa, "The Star of Bethlehem" (ThM thesis, Dallas Theological Seminary, 1972), 76.

[115] Dale C. Allison, "What Was the Star That Guided the Magi?," *Bible Review* 9.6 (1993): 24, and reprinted in *The First Christmas: The Story of Jesus' Birth in History and Tradition*, 25–31; cf. Michael J. Wilkins, *Matthew*, Zondervan Illustrated Bible Backgrounds Commentary (Grand Rapids, MI: Zondervan, 2002), 16; Craig Evans, *Matthew*, New Cambridge Bible Commentary (Cambridge: Cambridge University Press, 2012), 52–53.

[116] Translation from M. B. Riddle, "Arabic Gospel of the Infancy of the Saviour," in *The Ante-Nicene Fathers*, ed. Alexander Roberts and James Donaldson, 10 vols. (Grand Rapids, MI: Eerdmans, 1979), 8:406.

the narrator speak of the Star as an objective phenomenon, rising (v. 2 [the Magi's words] and v. 9 [the narrator's words]), going before them (v. 9 [the narrator's words]), and standing over the house (v. 9 [the narrator's words]). Indeed the Star is described in terms as objective as the Magi's journey to Judea and their entry into the house where Jesus and Mary were.

Second, the reaction of Herod and the people of Jerusalem to the report of the Magi does not make a lot of sense if the phenomenon was merely subjective—why would the Jewish people have been troubled by a subjective report from pagan astrologers from a foreign land, probably Babylon? And why would Herod have been convinced that the Magi were correct in claiming that the Messiah had been born? Besides, the text does not require that Herod and the people of Jerusalem were unaware of the Star prior to the Magi's arrival, but only that they did not appreciate its messianic significance.

We maintain, then, that the miraculous and visionary hypotheses are, by their very nature, positions of last resort that leave key questions unanswered. Moreover, they cannot explain satisfactorily all the data in the text. Matthew 2:1–12 strongly pushes the reader to accept that the Star was a literal and objective astronomical phenomenon that coincided with the birth of Jesus.

MYTHICAL (OR MIDRASHIC) HYPOTHESIS

Many scholars, including some astronomers and many liberal (as well as a number of conservative) theologians, do not accept Matthew's account of the Star as literally true; they do not believe that the birth of Jesus was really attended by an astronomical marvel.[117]

Cullen states that Matthew 2:1–12's narrative concerning the Magi and the Star can with confidence be categorized as mythical in nature.[118] This is a historical and literary judgment regarding Matthew's Gospel.

Some New Testament scholars are sympathetic to this view.

For example, Raymond E. Brown rejected the historical veracity of the account on the ground that it was marked by "intrinsic unlikelihoods. A star that rose in the East, appeared over Jerusalem, turned south to Bethlehem, and then came to rest over a house would have constituted a celestial phenomenon unparalleled in astronomical history; yet it received no notice in the records of the times."[119]

In addition, Ulrich Luz has claimed that the Magi story as a whole is historically improbable.[120] The narrative, he maintains, "does not conform to the laws of historical probability. The desperate questions of the interpreters demonstrate this: Why did Herod not at least send a spy along with the Magi? How could the whole population of Jerusalem, the scribes, and the unpopular King Herod be perplexed by the coming of the Messiah? The star also is not described realistically, i.e., as astronomically plausible."[121]

It is not uncommon for Matthew's infancy

[117] So, for example, Christopher Cullen, "Can We Find the Star of Bethlehem in Far Eastern Records?," *Quarterly Journal of the Royal Astronomical Society* 20 (1979): 153–159; Géza Vermes, *The Nativity: History and Legend* (London: Penguin, 2006), 22; and E. P. Sanders, *The Historical Figure of Jesus* (London: Penguin, 1993), 85.

[118] Cullen, "Can We Find the Star?," 158.

[119] Brown, *Birth of the Messiah*, 188. Cf. Marcus J. Borg and John Dominic Crossan, *The First Christmas: What the Gospels Really Teach about Jesus's Birth* (New York: HarperCollins, 2007), 182 (who reject as implausible that the Star led the Magi westward, then southward, and finally pinpointed the exact place of Jesus's birth); Steve Moyise, *Was the Birth of Jesus according to Scripture?* (Eugene, OR: Wipf & Stock, 2013), 51–54, 90 (who objects to the idea that the "sat-nav" Star could have guided the Magi to Jerusalem, then south to Bethlehem, and then "hover[ed] over" and pointed out a particular house). Rudolf Schnackenburg, *The Gospel of Matthew* (Grand Rapids, MI: Eerdmans, 2002), 22 (cf. 20), rejected the historicity of the Star because "stars do not move from north to southwest (the direction from Jerusalem to Bethlehem) and cannot indicate a precisely delimited location."

[120] Luz, *Matthew 1–7*, 132.

[121] Ibid. Cf. Kim Paffenroth, "The Star of Bethlehem Casts Light on Its Modern Interpreters," *Quarterly Journal of the Royal Astronomical Society* 34 (1993): 455–457.

narrative to be labeled "midrash" or "creative historiography."[122]

Obviously, if the cynicism of such theologians is justified, then the quest to identify the historical celestial event witnessed by the Magi is doomed to miserable failure.

However, the skepticism is unwarranted. We have already seen that the Gospels are ancient Greco-Roman biographies that seek to tell the story of Jesus in a historically faithful way, based on sources deemed reliable, and that the story of the Magi was regarded by Matthew as true and has strong historical credentials.[123] But here it is important to say two things in response to the claim that the account of the Magi and the Star in Matthew 2:1–18 is mythical.

First, the claim rests on astronomical ignorance. As this study will demonstrate, every element in Matthew's description of the behavior of the Star is consistent with established astronomical facts. We must appreciate that it was normal in antiquity to describe celestial phenomena from the vantage point of an Earth-bound observer. Even today, when an astronomer speaks of a Leonid meteor shower, he does so because the meteors seem to the human observer to radiate from the constellation Leo. Likewise, in the ancient world, astronomical sights could be described observationally. Therefore when we read of the Star moving in advance of the Magi to the place where Jesus was and indeed standing over it, this language almost certainly reflects the observational perspective of the Magi.

At the same time, while I will maintain that the astronomical phenomenon known as the Star of Bethlehem can be identified, were I unable to do so, this would certainly not justify cynicism regarding Matthew's account. The Star might conceivably have been a rare astronomical event with which modern astronomers are as yet unfamiliar.[124]

Second, as for the cynics' negative assessment of the non-astronomical aspects of Matthew's story, it must be said that these scholars display an astonishing lack of empathy and imagination.

Concerning the question of why Herod did not elect to send a spy or armed escort, the most natural explanation is the one suggested by the narrative: he saw no need to do so. Moreover, if the Magi had discovered that they were being watched, Herod's whole objective would have been jeopardized.[125] The king of Judea would have been certain that his simple plot was working—he knew that he had successfully hoodwinked the naive Magi into thinking that he too wanted to pay homage to the newborn Messiah. He had every reason to think that the Magi would return to him to let him know where precisely the infant was, if indeed they were able to discover him at all. Matthew tells us that the Magi would indeed have reported back to him, had they not been warned in a dream not to do so. In addition, Herod had a Plan B in the unlikely case that his targeted assassination of the messianic child failed: the massacre of Bethlehem's infants. In view of the fact that neither Plan A nor Plan B worked, Herod might well have wished that he had sent a

[122] Michael D. Goulder, *Midrash and Lection in Matthew* (London: SPCK, 1974), 3–46; Bernard P. Robinson, "Matthew's Nativity Stories: Historical and Theological Questions for Today's Readers," in *New Perspectives on the Nativity*, 113–115. Moyise, *Was the Birth of Jesus according to Scripture?*, 52, claims that Matthew was simply "offering a fulfillment of Scripture in narrative form." Although he acknowledges that a Scriptural background does not rule out the historicity of the events of the Nativity (ibid., 53), he nevertheless insists that "if their historicity is doubted on other grounds, then it offers additional support." Employing a hermeneutic of suspicion, he assumes that correspondences and parallels between Gospel narratives and ancient prophecies are generally best explained with reference to invention rather than historical fulfillment.

[123] Charles L. Quarles, *Midrash Criticism: Introduction and Appraisal* (Lanham, MD: University Press of America, 1998), 54, highlights that creative historiography was condemned by the early Christians, for example, in 1 Tim. 4:6–7; Titus 1:14; and 2 Pet. 1:16.

[124] As David H. Kelley and Eugene F. Milone, *Exploring Ancient Skies*, 2nd ed. (New York: Springer, 2011), 486, point out, the Star might have been a phenomenon that occurred once in human history 2,000 years ago, which, because it has not recurred, is beyond scientific analysis or identification.

[125] Cf. R. T. France, *The Gospel of Matthew*, New International Commentary on the New Testament (Grand Rapids, MI: Eerdmans, 2007), 74.

spy. But hindsight is 20/20, both for Herod and for modern critics.

With respect to the question of why Jerusalem and its religious establishment would have been rattled by the news that the Messiah had been born, the historical context readily explains this. Judea was ruled by Herod the Great, whom the people of Jerusalem knew to be capable not only of brutal atrocities against his enemies, but even of executing members of his own family. They understood that Herod would feel very threatened by the report that the Messiah had been born, since it would seem to imply that his own dynasty would soon be terminated. Therefore those in Jerusalem would have been certain that the king of Judea would take action against the messianic baby. Moreover, the Jerusalemites would have believed that, when the Messiah was full-grown, he would challenge the dynasty of Herod and the Romans. Such was the power of messianic expectation that the Jerusalemites might well have feared that a new revolutionary movement might be stirred up by the birth of this baby. At any rate, the Messiah's birth would have seemed to signal impending civil war and even all-out war with the Romans, with the city of Jerusalem being the main focus of the fighting. Therefore, although superficially surprising, the response of the people of Jerusalem to the news that the Messiah had been born makes excellent sense when the historical context is factored in.

Finally, the absence of any ancient astronomical record of the Bethlehem Star would constitute a problem only if we had reason to believe that the surviving astronomical records are comprehensive. Unfortunately, they are patchy at best.[126]

We maintain, therefore, that there is nothing historically implausible about Matthew's account of the visit of the Magi.[127]

Matthew portrays the Star as historical and expects his readers to treat it as such. He regards the Star, together with the journey of the Magi, as a major proof that Jesus was indeed the Messiah.

We have already highlighted a number of factors that favor the Star's historicity. On a general level, genre considerations, the importance of historical accuracy to the early Christians, the stability of the transmission of the Jesus tradition in the early church, Matthew's conservative handling of his sources, and the consistency of Matthew's portrayal of Herod with that of Josephus speak in favor of the account's historical reliability. With respect to the nativity narrative, the corroboration of many elements of the account by Luke, the lack of any plausible alternative explanation of the origin of the Magi story, and the positive presentation of the Magi argue strongly for historical reliability. So also does the fact that the account of the Magi and the Star has many traits that render it the sort of event that tends to be remembered accurately by eyewitnesses.

Ignatius, who ministered in the first century AD and knew the apostles and other prominent early Christians,[128] makes specific

[126] On the early Chinese, Korean, and Japanese records, see, for example, Thomas John York, "The Reliability of Early East Asian Astronomical Records" (PhD thesis, Durham University, 2003), available online at http://etheses.dur.ac.uk/3080/.

[127] It has been suggested that Matthew invented the story in the wake of the visit of a great entourage of the kings of Armenia and three other kingdoms (called "magi" in Pliny the Elder, *Natural History* 30.1.16–17) in AD 66 to Rome to honor Nero, in response to some astronomical sign (see Cassius Dio 63:1–7; Suetonius, *Nero* 13) (e.g., Francis Wright Beare, *The Gospel according to Matthew: A Commentary* [Oxford: Blackwell, 1981], 74–75; cf. Schnackenburg, *Gospel of Matthew*, 20). However, there is no evidence that Matthew was anything like the inventor of material that this hypothesis supposes. In truth, the historical incident in Nero's day actually confirms the plausibility of Matthew's account. As France (*Gospel of Matthew*, 64) put it, "it demonstrates that high-ranking eastern magi were willing and able to travel west for diplomatic reasons" (cf. Donald A. Hagner, *Matthew*, 2 vols., Word Biblical Commentary [Dallas: Word, 1993–1995], 1:25–26; see also the important essay on the historical nature of what Matthew writes concerning the Magi by Yamauchi, "Episode," 18–23).

[128] Ivor Bulmer-Thomas, "The Star of Bethlehem—A New Explanation—Stationary Point of a Planet," *Quarterly Journal of the Royal Astronomical Society* 33 (1992): 365–366, cavalierly dismisses Ignatius's evidence on the ground that someone writing more than 100 years later cannot be regarded as a dependable witness. However, (1) Ignatius was a contemporary of Matthew, and both seem to have been closely associated with Syrian Antioch; (2) it is widely believed that, in *To the Ephesians* 19:2–3, Ignatius was citing from a first-century hymn concerning the Star.

mention of the Bethlehem Star in chapter 19 of his letter *To the Ephesians*:

> A star shone in heaven [with a brightness] beyond all the stars; its light was indescribable, and its newness provoked astonishment. And all the other stars, together with the Sun and the Moon, formed a chorus to the star, yet its light far exceeded them all. And there was perplexity regarding from where this new entity came, so unlike anything else [in the heavens] was it.[129]

This is evidently a tradition independent of Matthew[130] and therefore supports the historicity of the Star.

Cullen seeks to undermine the historicity of the Star, asking how the Magi managed to arrive in Bethlehem at the correct, precise moment. He demands to know whether his fellow astronomers are claiming that this was merely a coincidence or something else, such as a miracle or correct astrology. Supposing that these explanations are self-evidently foolish, he pronounces that Matthew's account is historically implausible.[131] However, Cullen tips his hand at this point—he is rejecting the historicity of Matthew's account because he does not wish to accept the implication of it. Obviously, historical investigation should not operate in this way—one should begin by making historical judgments based on the evidence and then, and only then, ask the questions of why and how.

When one is on a quest for the historical Star of Bethlehem, one is duty-bound to adopt a sympathetic approach to Matthew 1:18–2:23, the only major narrative concerning the Star and the Magi that we have from the first century AD. Indeed the more a hypothesis is forced to diverge from a natural reading of Matthew's narrative, the less likely it proves itself to be. If Matthew is unreliable, then any hope of identifying the Star is dashed. But if he is reliable, then there is purpose to the quest, and the solution to the mystery will be able to explain naturally every aspect of Matthew's account.

CONCLUSION

In this chapter we have considered most of the major hypotheses concerning the Star seen by the Magi in the east and then again in Bethlehem. They have each been found wanting. We must turn next to consider the comet hypothesis.

[129] My translation.
[130] See William R. Schoedel, "Ignatius and the Reception of the Gospel of Matthew in Antioch," in *Social History of the Matthean Community: Cross-Disciplinary Approaches*, ed. David L. Balch (Minneapolis: Fortress, 1991), 155–156.
[131] Cullen, "Can We Find the Star?," 158.

5

"What Sudden Radiance from Afar?"

Introducing Comets

Having considered the other major proposals regarding the identity of the Star of Bethlehem, we must now turn to the last and most plausible of the main views, the comet hypothesis.

Unfortunately, much of the academic discussion concerning the comet hypothesis reflects a failure to come to terms with the history of comets and their interpretation, and with cometary astronomy generally. Therefore in this chapter we must pause briefly to become better acquainted with the phenomenon of comets. Armed with this knowledge, we shall in the next chapter be in the position to evaluate the two cometary hypotheses currently on the table, namely that the Bethlehem Star was Halley's Comet in 12 BC or that it was the 5 BC *hui-hsing* comet recorded by the Chinese. Also in that chapter we shall develop our own case that the Star was a great comet.

First, then, we need to familiarize ourselves with comets. Along with a meteor storm, a total solar eclipse, and the aurora borealis (Northern Lights), the appearance of a great comet is one of the chief celestial marvels that a human being can witness in his or her lifetime. A comet may be the largest,[1] longest, and brightest[2] object, not to mention the most unusual[3] one, in the sky. A great comet commands the attention of all humans.

THE CELESTIAL JOKERS

Comets are the celestial jokers,[4] mavericks that simply do not follow the normal rules of the heavens. They appear in the sky out of nowhere. They may venture into celestial territory where no planet can ever go, even into the polar regions. Whereas the stars and planets followed set schedules that the ancients were able to identify and on that basis determine seasons and make predictions, comets appeared to be completely random in

[1] Carl Sagan and Ann Druyan, *Comet* (New York: Pocket Books, 1986), 128.
[2] Only the Sun or a superbolide may be brighter.
[3] Apart from perhaps a great meteor or superbolide.
[4] Victor Clube and Bill Napier, *The Cosmic Serpent: A Catastrophist View of Earth History* (New York: Universe, 1982), 158.

their movements.[5] Sometimes a comet remains in one spot or region of the sky for weeks or indeed for its whole apparition;[6] at other times a comet races very quickly across large sections of the heavens.[7] Comets may suddenly change their apparent direction and behavior. Moreover, they frequently change their appearance. They can go from being barely visible to being brighter than the full Moon. These strangers may also grow in size to be much larger than the Sun and Moon, and may increase in length to the extent that they stretch across the whole sky.[8] Not only is each comet completely individual, but each apparition of a given comet is unique.

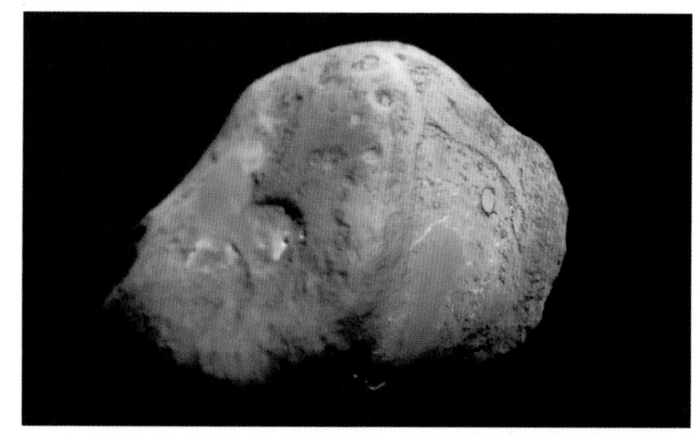

FIG. 5.1 The nucleus of Comet Tempel 1. It measures 7.6 x 4.9 km. Image credit: NASA/JPL/University of Maryland/Wikimedia Commons.

FIG. 5.2 The nucleus of Comet Hartley 2 as it degasses. The hyperactive comet is 2.5 km in diameter. This image was taken by NASA's EPOXI mission spacecraft during a fly-by on November 4, 2010. Image credit: NASA/JPL-Caltech/UMD/Wikimedia Commons.

WHAT IS A COMET?

A comet near the Sun consists of a coma, that is, a dense cloud of dust around a nucleus, and a tail. When far from the Sun, it is nothing more than a bare nucleus undetectable by the naked eye.

NUCLEUS

Just as a plum's flesh surrounds a pit, so a comet coma envelops an inner nucleus. This nucleus is essentially an icy[9] ball of dirt, dust,[10] and stones. Up close, a nucleus looks like an extraordinarily black and barren giant rock (fig. 5.1).[11] Nuclei may have distinctive shapes—for example, those of Comets 1P/Halley, 9P/Tempel 1 (fig 5.1), and 19P/Borrelly look like potatoes; those of Comets 8P/Tuttle and 103P/Hartley 2 (fig. 5.2) have been compared to pea-

[5] Cf. David Seargent, *The Greatest Comets in History: Broom Stars and Celestial Scimitars* (Berlin: Springer, 2009), 22–23.
[6] "Apparition" in astronomy refers to the time during which a comet is visible.
[7] Cf. ibid., 23.
[8] Cf. Jacques Crovisier and Thérèse Encrenaz, *Comet Science: The Study of Remnants from the Birth of the Solar System*, trans. Stephen Lyle (Cambridge: Cambridge University Press, 2000), 1.
[9] Decades ago it was believed that comets were dirty snowballs (so Fred L. Whipple, *The Mystery of Comets* [Washington, DC: Smithsonian Institution Press, 1985], 147–148), but now they are thought to be "icy dirtballs."
[10] As Nick James and Gerald North, *Observing Comets* (London: Springer, 2003), 24–25, point out, by "dust" astronomers do not mean domestic dust, but rather a variety of, among other things, magnesium-rich silicates, sulfides, and carbon.
[11] Crovisier and Encrenaz, *Comet Science*, 66. The nucleus of Halley's Comet has been described as "a pitch-black rock covered with mountains and valleys, . . . measured at 15.3 by 7.2 by 7.2 km" (Peter Jenniskens, *Meteor Showers and Their Parent Comets* [Cambridge: Cambridge University Press, 2006], 16).

nuts; and, from some angles, the nucleus of Comet 67P/Churyumov-Gerasimenko has the appearance of a rubber duck.[12]

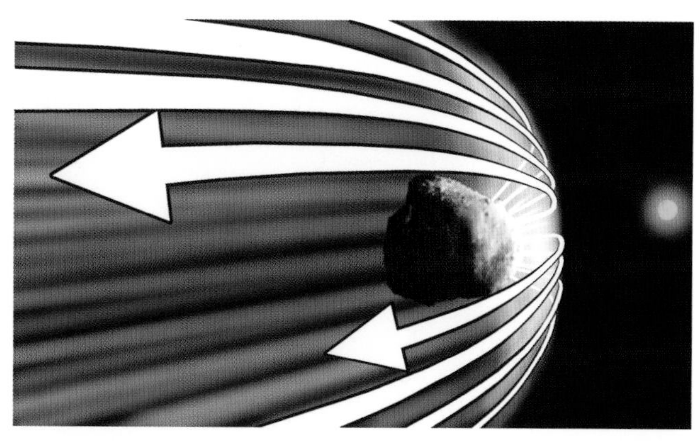

FIG. 5.3 How a comet develops a coma and tail. Image credit: Sirscha Nicholl.

Put simply, when a comet nucleus comes to within a certain distance from the Sun, it begins to react, its "ices" being converted directly from solids to gases. These gases rocket off the nucleus in jets toward the Sun and take with them dust and rock particles (fig. 5.2). The Sun's radiation pressure then pushes the debris back to form a coma and a dust tail (fig. 5.3).

How close a freezing cold comet nucleus must come to the Sun before it begins to "degas" varies greatly, based largely on the concoction of chemicals that make up the comet's "ices." If we use the distance from Earth to the Sun as a measure (= 1 "Astronomical Unit" or 1 AU), most comet nuclei react to the Sun's heat when they are within 3 AU (roughly 450 million km) of the Sun. However, some comets, like Hale-Bopp, which have a high proportion of exotic "ices" like carbon monoxide, begin producing fountains of gases some 20 AU from the Sun! When active comets are the same distance from the Sun as Earth, their nuclei have jets and are surrounded by a coma consisting of gas, dust, and rocks and they may have a gas tail and a dust tail. Comets are most productive when closest to the Sun, a point called their "perihelion" (from the Greek *peri*, meaning "near," and *helios*, meaning "Sun").

As a comet moves away from the Sun, it will react less and less to it and eventually will cease reacting at all.

Most comet nuclei are between 1 and 10 km in diameter,[13] although some are considerably larger than this. Hale-Bopp is reckoned to be about 40–70 km in diameter.[14] The Great September Comet of 1882 is believed to have had a nucleus of 50 km in diameter.[15] However, there are nuclei that are "giant" size. The asteroidal comet and centaur[16] 95P/Chiron has a nucleus 166–233 km in diameter.[17] Some long-period comets are extraordinarily large, 100 km in diameter or greater.[18] The sire of the sungrazer comet family is thought

[12] In 2014, Comet 67P was the subject of intensive study by the European Space Agency's Rosetta spacecraft. Then on November 12, 2014, the comet's nucleus became host to Rosetta's robotic lander called Philae. This little probe proceeded to take panoramic images of the nucleus surface, to examine the surface's electrical and mechanical properties, to take gas measurements, and to investigate the internal structure by means of low-frequency radio waves and laboratory testing of drilled sub-surface samples.

[13] Seargent, *Greatest Comets*, 5.

[14] See Zdenek Sekanina, "Statistical Investigation and Modeling of Sungrazing Comets Discovered with the Solar and Heliospheric Observatory," *Astrophysical Journal* 566.1 (2002): 582; Yanga R. Fernández, "The Nucleus of Comet Hale-Bopp (C/1995 O1): Size and Activity," *Earth, Moon, and Planets* 89 (2002): 3–25.

[15] Sekanina, "Statistical Investigation," 582.

[16] A centaur is a comet- or asteroid-like "minor planet" that orbits between Jupiter and Neptune.

[17] Consider also the size of other centaurs such as 5145 Pholus (about 200 km in diameter), 1995 SN55 (300 km), and 10199 Chariklo (260 km); and the size of the trans-Neptunian objects 1992 QB1 (160 km) and 1993 FW (175 km). It is believed that some centaurs may evolve into short-period comets.

[18] On giant comets, see Mark E. Bailey, S. V. M. Clube, G. Hahn, W. M. Napier, and G. B. Valsecchi, "Hazards Due to Giant Comets: Climate and Short-Term Catastrophism," in *Hazards Due to Comets and Asteroids*, ed. T. Gehrels (Tucson: University of Arizona Press, 1995), 479–533, esp. pp. 482–486; Mark E. Bailey, V. V. Emel'yanenko, G. Hahn, N. W. Harris, K. A. Hughes, K. Muininen, and J. V. Scotti, "Orbital Evolution of Comet 1995 O1 Hale-Bopp," *Monthly Notices of the Royal Astronomical Society* 281 (1996): 916–924; and S. V. M. Clube, F. Hoyle, W. M. Napier, and N. C. Wickramasinghe, "Giant Comets, Evolution, and Civilization," *Astrophysics and Space Science* 245 (1996): 43–60. By "giant" these authors mean a diameter of 100 km or larger.

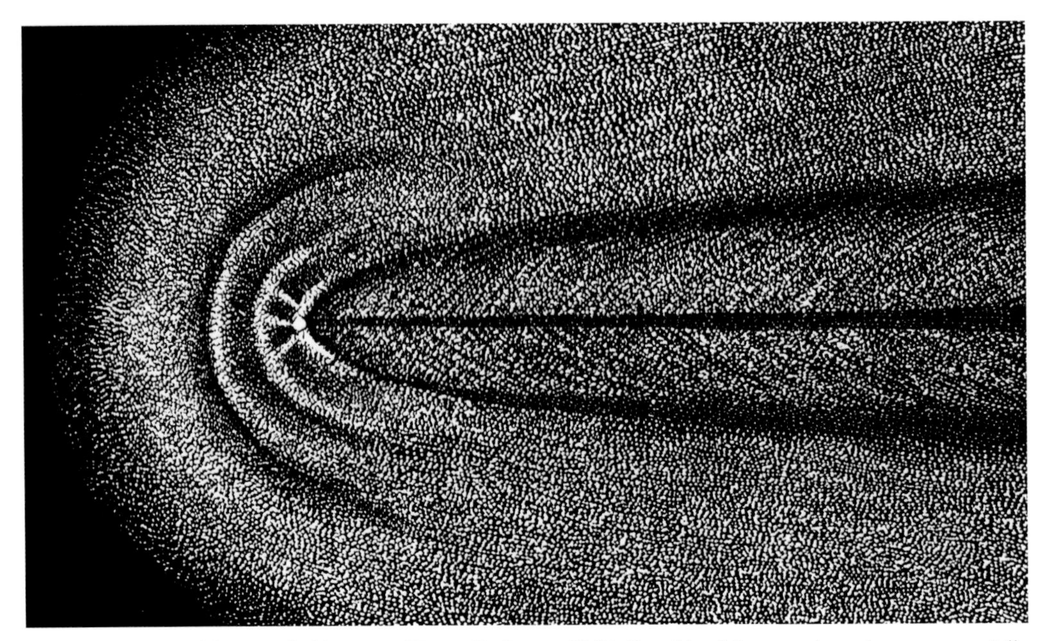

FIG. 5.4 A sketch of the remarkable coma of Donati's Comet, 1858. One side of the comet's nucleus was especially active whenever the nucleus's rotation brought it back into sunlight. It is this that gave rise to the strange haloes, or envelopes, on the sunward side of the coma. Image from George F. Chambers, *The Story of the Comets* (Oxford: Clarendon, 1909), plate 17 fig. 53 (opposite page 137).

to have been 100–120 km in diameter.[19] Sarabat's Comet of 1729 is believed to have been at least 100 km (some say 300 km).[20] Bailey et al. estimate that comets larger than 100 km cross Earth's orbit approximately once every 400 years and come within Jupiter's orbit once every 70 years or so.[21]

COMA

A comet's coma, or head, is the distinctive blotch of bright light on the sunward side of the comet. Usually it is the brightest part of the comet.

A coma is somewhat like an atmosphere around the nucleus, the innermost layer being denser and brighter and the outermost layer less dense and bright.[22] Immediately around the nucleus is the brightest region of the coma, known as the pseudonucleus.[23]

Comas may be anything from very small to mind-bogglingly massive. In space, the diameter of Hale-Bopp's coma is reckoned to have been twice the diameter of the Sun in 1996,[24] and Comet Holmes grew to be more than 5 times the diameter of the Sun in 2007–2008.[25]

[19] So, among many others, Brian G. Marsden, "The Sungrazing Comet Group," *Astronomical Journal* 72 (1967): 1170–1183; Ernst Julius Öpik, "Sun-Grazing Comets and Tidal Disruption," *Irish Astronomical Journal* 7 (March 1966): 141. A lot of work on the history of the sungrazers has been done in recent years by Zdenek Sekanina and P. W. Chodas—see, for example, "Fragmentation Hierarchy of Bright Sungrazing Comets and the Birth and Orbital Evolution of the Kreutz System. I. Two-Superfragment Model," *Astrophysical Journal* 607 (2004): 620–639; and "Fragmentation Hierarchy of Bright Sungrazing Comets and the Birth and Orbital Evolution of the Kreutz System. II. The Case for Cascading Fragmentation," *Astrophysical Journal* 663 (2007): 657–676. They suggest that the Kreutz progenitor's "maximum dimension must have been close to 100 km" ("Two-Superfragment Model," 635).
[20] Duncan Steel, *Rogue Asteroids and Doomsday Comets* (London: John Wiley, 1997), 126.
[21] Bailey et al., "Hazards Due to Giant Comets," 484–485.
[22] So Martin Mobberley, *Hunting and Imaging Comets* (Berlin: Springer, 2011), 22.
[23] For fascinating images of cometary structures in the vicinity of the nucleus, see Jürgen Rahe, Bertram Donn, and Karl Wurm, *Atlas of Cometary Forms: Structures Near the Nucleus* (Washington, DC: NASA, 1969).
[24] "Comet Hale-Bopp—Still Enormous!," http://www.eso.org/public/news/eso9933 (last modified June 29, 1999); Robert Burnham, *Great Comets* (Cambridge: Cambridge University Press, 2000), 101, 103; Zdenek Sekanina, "Activity of Comet Hale-Bopp (1995 O1) beyond 6 AU from the Sun," *Astronomy and Astrophysics* 314 (1996): 964.
[25] John F. Pane, "Comet 17P/Holmes," http://www.cs.cmu.edu/~pane/holmes (accessed July 1, 2014).

From the perspective of Earth, comas can look extremely large, because they really are huge in space and/or because they are so close to us.[26] Due to the fact that they made especially close approaches to Earth, the relatively small comets Hyakutake (with a nucleus about 4 km in diameter) in 1996 and Lexell in 1770 appeared respectively to be 5[27] and 6–8[28] times the diameter of the Sun. If the large comet Hale-Bopp had come closer to the Sun and Earth, it would have appeared to be something like 4 or 5 degrees in diameter when 0.1 AU away from Earth[29] and considerably larger if nearer still. Astronomers believe that Earth's skies hosted countless comets with huge, bright comas in the past.[30]

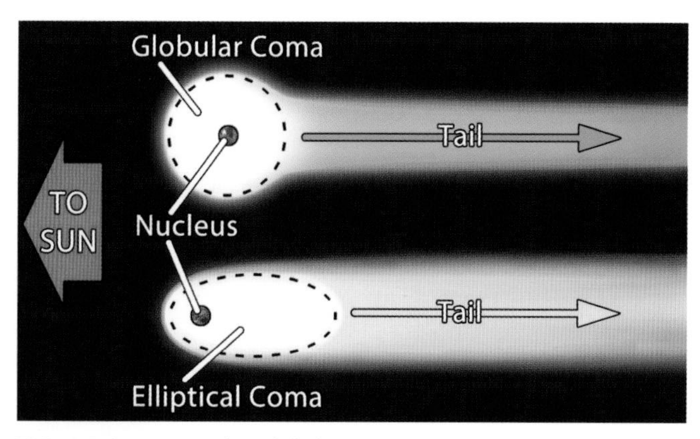

FIG. 5.5 A comet with a globular coma compared to a comet with an elliptical coma. Image credit: Sirscha Nicholl.

Generally speaking, with respect to shape, comet comas fall into two major categories. Some are more circular ("globular"), and others are oval (elliptical) (fig. 5.5), such as the comets Hale-Bopp and Ikeya-Zhang, or fan-shaped (parabolic), like Tebbutt's Comet of 1861.

Globular comas often have a blue or green hue, because they are more gassy than dusty.[31] Elliptical and parabolic comas often appear more yellowish, because they are dustier. The most spectacular comets in history, heavy dust-producing comets that have made particularly close passes by the Sun (coming within the orbit of Mercury), tend to have elliptical or parabolic comas.[32] Because of the Sun's intense radiation pressure, the dust expelled sunward by the nucleus is immediately forced back behind it, with the result that the nucleus is very close to the sunward side of the coma. These types of comas typically merge into their tails (see fig. 5.6).

TAILS

Basically there are two types of tails that comets may (or may not) have: (1) gas (sometimes called "plasma") tails, which are straight, bluish, structurally fine, and point almost exactly away from the Sun; and (2) dust tails, which are famously yellowish white and thick and may be curved.[33]

One might imagine that comet tails are like long loose hair flowing out from a motorcyclist's head as she accelerates along at breakneck speed.[34] However, comet tails always point in the direction opposite the Sun,

[26] At the coma's circumference the brightness decreases to the point that it is indistinguishable from the tail or the sky.
[27] N. D. James, "Comet C/1996 B2 (Hyakutake): The Great Comet of 1996," *Journal of the British Astronomical Association* 108 (1998): 161–162. Also Andreas Kammerer, personal email message to the author, October 30, 2012.
[28] Gary W. Kronk, *Cometography: A Catalog of Comets*, 6 vols. (Cambridge: Cambridge University Press, 1999–), 1:449.
[29] Andreas Kammerer, personal email message to the author, October 30, 2012. Cf. Mobberley, *Hunting and Imaging Comets*, 75: "maybe 5° across."
[30] See, for example, W. M. Napier, "Evidence for Cometary Bombardment Episodes," *Monthly Notices of the Royal Astronomical Society* 366 (2006): 977–982; Fred Schaaf, *Wonders of the Sky: Observing Rainbows, Comets, Eclipses, the Stars, and Other Phenomena* (Mineola, NY: Dover, 1983), 134–135.
[31] David Seargent, *Comets: Vagabonds of Space* (Garden City, NY: Doubleday, 1982), 26, 28.
[32] Ibid., 30.
[33] See John C. Brandt and Robert D. Chapman, *Introduction to Comets*, 2nd ed. (Cambridge: Cambridge University Press, 2004), 129, 148. A comet may also have a sodium, or neutral, tail, which fluoresces.
[34] Cf. Mobberley, *Hunting and Imaging Comets*, 7.

regardless of whether they are moving toward it or away from it (see figs. 5.7–8).

Gas tails are formed when the electrically charged gas particles that exploded toward the Sun from the nucleus are pushed by the solar wind directly back behind the nucleus. Because of this, gas tails are generally straight, narrow, and point directly away from the Sun. The sunlight causes them to fluoresce.

Dust tails are formed when small dust particles expelled from the comet's nucleus toward the Sun are pushed behind the coma by solar radiation pressure. Because dust particles travel more slowly than gas particles, they lag behind the comet, with the result that dust tails tend to be more curved and broad than gas tails and not directly anti-solar, but rather angled back toward the direction from which the comet has just come (fig. 5.7). The section of the dust tail farthest from the coma is less dense, less bright, and more curved. A tail's curvature and length are typically greatest just after the comet's closest encounter with the Sun, when it is traveling fastest and is at its most productive. In the case of a comet steeply inclined to the plane of Earth's orbit around the Sun (the "ecliptic"[35]; see fig. 5.24 and fig. 7.12), this curvature may be very apparent to Earth-dwellers, and the dust tail may seem wider and be easily distinguishable from the gas tail. However, where a comet's orbit is angled narrowly to the ecliptic plane, the dust tail will generally appear to Earth-dwellers to be narrower, straighter, brighter, and longer, and will combine with the gas tail to form a single tail.[36]

FIG. 5.6 A false-color image of Comet Ikeya-Zhang on March 18, 2002. Image credit: Bojan Dintinjana and Herman Mikuz, Črni Vrh Observatory, Slovenia.

Some comet tails are mindbogglingly large. Especially long and wide tails are associated with very productive comets that begin reacting to the Sun when far from it. Some hyperactive small nuclei (like that of Hyakutake) produce extraordinarily long tails. But the dream scenario for a long comet tail is a large, very volatile nucleus.

The longest known comet tails in space are Hyakutake's at 3.8 AU (570 million km),[37] that of Messier's Comet of 1769 at 3.5 AU (520 million km),[38] and that of the Great March Comet of 1843 at 2.15 AU (330 million km).[39] See fig. 5.9.

[35] The "ecliptic" is the apparent path of the Sun through the sky. Imagine that you are riding a horse on a carousel that moves in an anticlockwise direction—the horse is Earth and the central column is the Sun. The level you are on is "the ecliptic." Most of the planets go around the Sun at a similar angle to that of Earth. Now imagine that the carousel can be tilted up at different angles. Tilted up at 1 to 30 degrees, you are like a narrowly inclined comet going around the Sun. Tilted up (with seatbelts on!) at 45 degrees and 90 degrees, you are like a steeply inclined comet orbiting the Sun. The point is that "narrowly inclined" and "steeply inclined" reflect the perspective of someone used to going around the carousel at ground level (representing the ecliptic). By the way, a tilt of 0–90 degrees makes the orbit prograde, while a tilt of 90–180 degrees makes it retrograde. A tilt of 150–180 degrees is the retrograde equivalent of a prograde tilt of 0–30 degrees.

[36] Richard Schmude, *Comets and How to Observe Them* (New York: Springer, 2010), 144; Fred Schaaf and Guy Ottewell, *Mankind's Comet: Halley's Comet in the Past, the Future, and Especially the Present* (Greenville, SC: Furman University, 1985), 56; Seargent, *Greatest Comets*, 98.

[37] The Ulysses spacecraft unexpectedly came across the tail of Hyakutake on May 1, 1996 (see "Ulysses's Surprise Trip through Comet's Tail Puts Hyakutake in Record Books," a news feature on the Royal Astronomical Society website from the year 2000 [http://www.ras.org.uk/news-and-press/70-news2000/377-pn00-07, last modified May 8, 2012]).

[38] Seargent, *Comets: Vagabonds of Space*, 116; S. K. Vsekhsvyatskii, *Physical Characteristics of Comets* (Jerusalem: Israel Program for Scientific Translations, 1964), 130–131.

[39] Although we cannot be sure how long Comet McNaught (C/2006 P1) was (its tail was at least 1½ AU long), we know that the region of space affected by it was much greater than that affected by Hyakutake. Whereas it took 2½ days for the Ulysses spacecraft

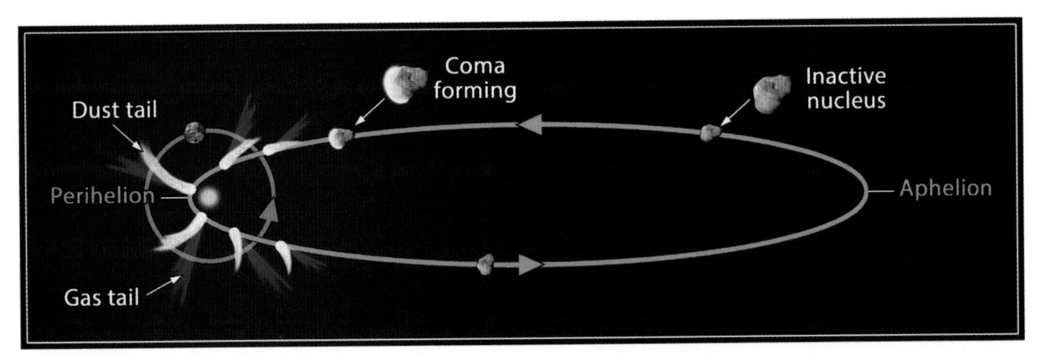

FIG. 5.7 Tail orientation at different stages of a comet's orbit. Image credit: Sirscha Nicholl.

As regards the apparent lengths of comets in history, the longest is a 300-degree one recorded in the year AD 905 by the Chinese (C/905 K1).[40] 200-degree comets were seen in AD 893[41] and 1618.[42] Tebbutt's Comet of 1861 was 120 degrees long, sufficient to extend two-thirds of the way across the dome of the sky.[43] Great comets in 1618 and 1769, as well as Halley's Comet in AD 837, were longer than 90 degrees, and many other comets in history peaked at 90 degrees (e.g., 1106, 1680, 1843, 1910 [Halley], 1996).[44] The apparent length of a comet is determined not just by the actual tail length but also by the comet's brightness, its distance from Earth, and the angle of the comet relative to the Sun and Earth.

FIG. 5.8 The orientation of Halley's Comet relative to the Sun as the Sun makes its way through the constellation of Leo the Lion in 1531. From Petrus Apianus, *Practica auff das 1532. Jahr.* Image credit: Crawford Library of the Royal Observatory, Edinburgh, Scotland.

Comet dust tails may also be very wide. The tail of Donati's Comet (1858; fig. 5.12) was as wide as the Big Dipper's handle (16 degrees),[45] while that of Tebbutt's Comet (1861; see fig. 5.21) was as broad as the distance between the top two stars in the Big Dipper's bowl (10 degrees).[46] Halley's Comet in AD 837 was as wide as four fingers at arm's length (7–8 degrees), and in 1066 it was as wide as the Big Dipper's bowl is high (5 degrees), as

to get through the shocked solar wind around Hyakutake, it took 18 days for it to get through the shocked solar wind around McNaught ("The Shocking Size of Comet McNaught," http://www.ras.org.uk/news-and-press/157-news2010/1782-the-shocking-size-of-comet-mcnaught; last modified April 13, 2010).

[40] David W. Pankenier, Zhentao Xu, and Yaotiao Jiang, *Archaeoastronomy in East Asia* (Amherst, NY: Cambria, 2008), 104–105.

[41] John Williams, *Observations of Comets, from B.C. 611 to A.D. 1640* (London: Strangeways & Walden, 1871), 52; Pankenier et al., *Archaeoastronomy in East Asia*, 103 (over 37 days it grew from 100 degrees to 200 degrees).

[42] Pankenier et al., *Archaeoastronomy in East Asia*, 247; Kronk, *Cometography*, 1:333–334.

[43] Among those who reported such extraordinary lengths for Comet Tebbutt at the end of June and start of July of 1861 was Johann Friedrich Julius Schmidt of Athens; see Gary Kronk, "C/1861 J1 (Great Comet of 1861)," http://cometography.com/lcomets/1861j1.html (last modified September 30, 2006).

[44] See Seargent, *Greatest Comets*, 45, 92, 111, 114, 209; Fred Schaaf, *Comet of the Century* (New York: Springer, 1997), 237, 246; Kronk, *Cometography*, 1:445; Gary Kronk, "C/1996 B2 (Hyakutake)," http://cometography.com/lcomets/1996b2.html (last modified September 30, 2006). Schaaf, *Comet of the Century*, 204–228, notes that there was a 90+-degree comet in 147 BC and that there were 100+-degree comets in AD 191, 287, 390, 418, and 891.

[45] Seargent, *Greatest Comets*, 133.

[46] Ibid., 140. Note too Seargent's comments regarding the width of Mithridates's Comet of 135 BC (p. 70).

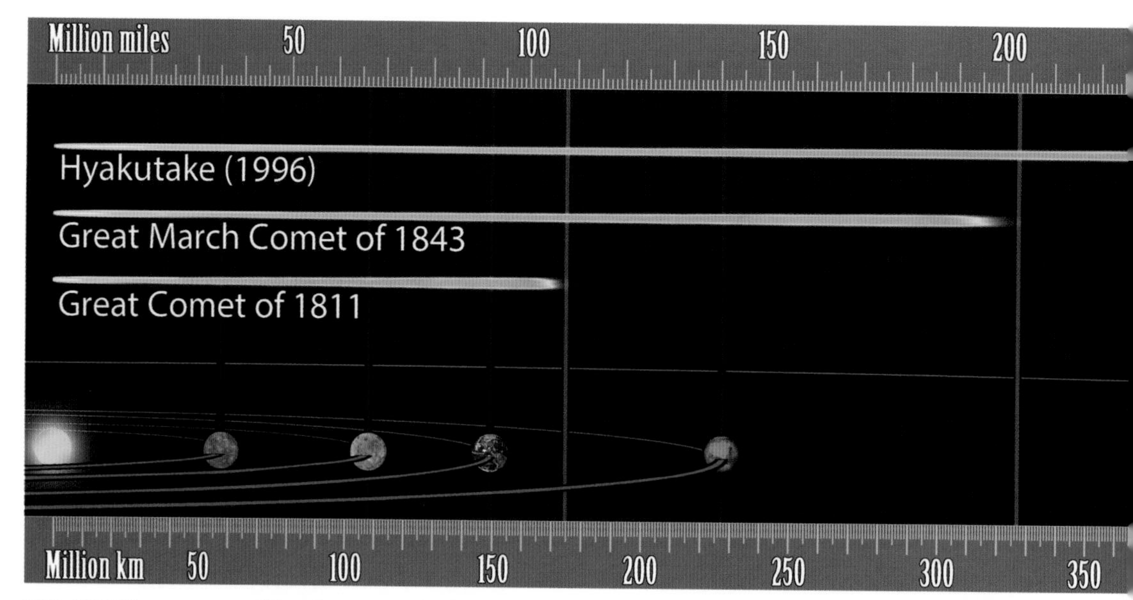

FIG. 5.9 Different comet tail lengths compared. The outermost planet is Jupiter. Image credit: Sirscha Nicholl. Here, as elsewhere, images of the planets are courtesy of NASA.

FIG. 5.10 The Great Comet of 1577 as seen from Prague. A broadside, "Von einem Schrecklichen und Wunderbarlichen Cometen" (Prague: Peter Codicillus, 1577). Image credit: Wickiana Collection, Graphische Sammlung und Fotoarchiv (Department of Prints and Drawings/Photo Archive), Zentralbibliothek Zürich (Zürich Central Library).

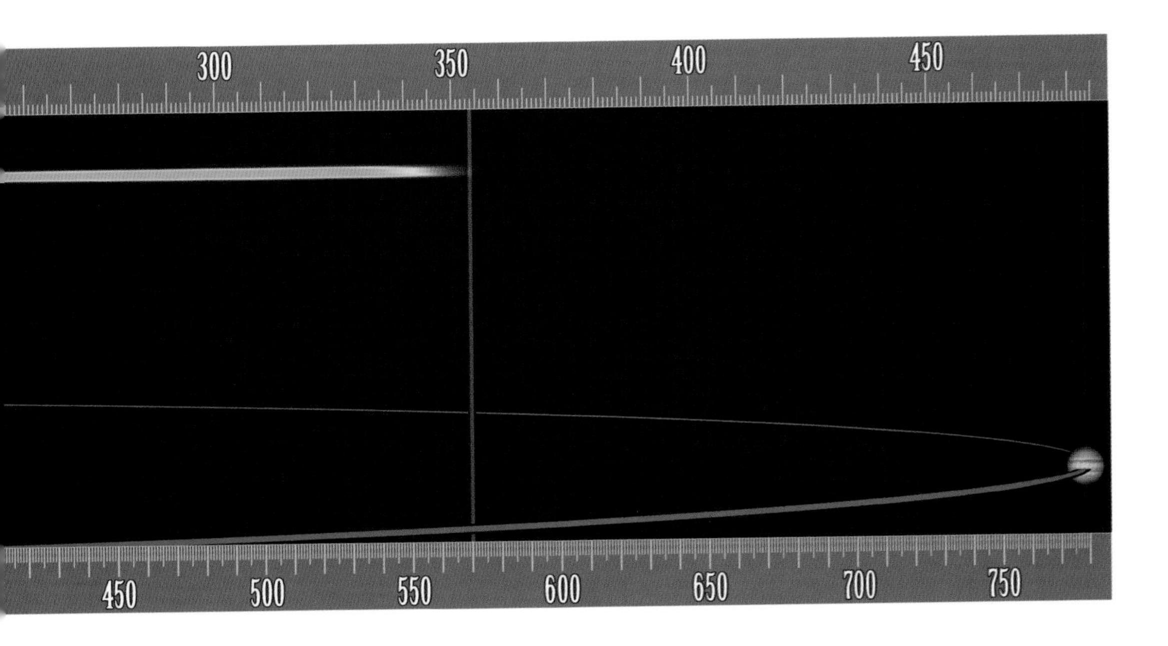

were comets in 1106 and 1471.[47] In 1965 the end of Comet Ikeya-Seki's tail was as wide as the distance between the 3 stars on Orion's belt (3 degrees).[48] In addition, the Great Comet of 1577 (fig. 5.10) was as wide as the Andromeda Galaxy (2.5 degrees).[49] The Great Comet of 1680 (figs. 5.15; 6.7; 10.19) was described as being "like a very wide belt" that stretched from one side of the horizon to the other with little difference in its breadth[50] (2 degrees).[51]

It is surely one of the most remarkable wonders of our solar system that relatively small bodies like comet nuclei may give rise to tails so astonishingly long and wide.

ANTITAIL

Occasionally, comets may have a thin "minitail" that points toward the Sun—the so-called antitail.

FIG. 5.11 An engraving of Coggia's Comet (C/1874 H1) as seen from the Pont Neuf by Charles La Plante, 1874. Image credit: Patrick Moore Collection, www.patrickmoorecollection.com.

[47] Ibid., 45, 47, 92, 103.
[48] Ibid., 221.
[49] Ibid., 107.
[50] Casimiro Diaz in Manila (The Philippines) as cited by Kronk, *Cometography,* 1:372. What Diaz writes suggests that the 1680 comet was about 180 degrees in length! It is possible that Diaz, in the Philippines, saw the comet around the time of its peak (i.e., on December 11–16 or 19–23), when Europeans were unable to see it.
[51] Richard Hooke as cited by Kronk, *Cometography,* 1:370; and Seargent, *Greatest Comets,* 114. Comet Skjellerup-Maristany in 1927 was similarly broad (ibid., 151).

FIG. 5.12 Comet Donati on October 5, 1858. From E. Weiß, *Bilderatlas der Sternenwelt* (Esslingen: J. F. Schreiber, 1888). Image credit: Wikimedia Commons.

This happens, from the Earth-dweller's vantage point, when a comet's orbit takes it close to the plane of Earth's orbit. The antitail consists of larger particles of dust recently expelled from the comet that are not so easily pushed back behind the comet and remain in the same orbital plane as the coma. When Earth moves near to the plane of the comet's orbit, observers can see not just the main part of the tail on the side of the comet away from the Sun, but also some of this larger material still remaining close to its orbital plane, which appears to poke out in front of the coma on the sunward side.[52] An antitail may also be seen when Earth's position relative to a productive comet is such that the rear part of the comet's tail is seen, in projection, on the sunward side of the coma.[53] Antitails may take the form of a fan or a spike, depending on the position of the comet relative to the ecliptic plane.[54] Comet Arend-Roland on April 22, 1957, had a fan antitail and on April 25 a spike (fig. 5.13).

The dream scenario for antitail development is a narrowly inclined and very productive comet making one of its first perihelion visits.[55] For example, Comet Lulin in 2009 had an inclination of 178.4 degrees, that is, less than 2 degrees from the ecliptic plane, and sported an antitail spike for two whole months,[56] although most of the time this was not visible to the naked eye (fig. 5.14).

VISIBILITY

As comets approach the Sun, they produce more fluorescent gas and sunlight-reflecting

[52] For more, see Brandt and Chapman, *Introduction to Comets*, 136–141.
[53] James and North, *Observing Comets*, 28, 30 fig. 2.6.
[54] Brandt and Chapman, *Introduction to Comets*, 137.
[55] Mobberley, *Hunting and Imaging Comets*, 9.
[56] Ibid., 168.

dust. The closer a comet gets to the Sun, the larger and brighter the comet becomes in space (though not necessarily to those observing from Earth).[57] If the brightness increases sufficiently, a comet may become visible to the naked eye.

Active comets can be seen with the naked eye in a clear, dark sky when their astronomical magnitude attains to approximately +3.4 to +4.[58]

Comet brightness is dependent on many factors, such as the comet's size, its productivity, and its proximity to Earth.[59]

It is very important to differentiate between apparent magnitude, the brightness as it appears to a human observer on Earth at a given point in time, and absolute magnitude, which is a rough measure of the intrinsic brightness of a celestial body. In order to calculate a comet's intrinsic brightness, astronomers work out how bright it would be if it were exactly 1 AU from both Earth and the Sun.[60] A comet may be intrinsically bright, but, owing to its distance from Earth and the Sun, never display its full glory to human observers. Hale-Bopp's maximum apparent magnitude was -1, but, since its absolute magnitude was around -1, had it arrived at perihelion four months earlier, when it would have been much closer to Earth, its apparent magnitude would have been about -6 to -10, sufficient to make it visible in the daytime.[61] Tycho's comet of 1577 seems to have had an absolute magnitude of -1.8,[62] and may have had an apparent brightness greater than magnitude -8.[63] Sarabat's Comet of 1729 had a rather unimpressive maximum apparent magnitude (certainly no greater than +2.6), since it never came closer to the Sun than 4 AU. However, its absolute magnitude was between -3 and -6.[64] Had it made a close pass by the Sun and/or Earth, it would

FIG. 5.13 Comet Arend-Roland with its long antitail. On April 25, 1957, the antitail was 12 degrees long. Image credit: Armagh Observatory.

[57] However, as the comet gets closer to the Sun, the Sun actually constricts the coma, causing it to shrink.

[58] F. Richard Stephenson, Kevin Yau, and Hermann Hunger, "Records of Halley's Comet on Babylonian Tablets," *Nature* 314 (April 18, 1985): 587–592; Kevin Yau, Donald Yeomans, and Paul Weissman, "The Past and Future Motion of Comet P/Swift-Tuttle," *Monthly Notices of the Royal Astronomical Society* 266 (1994): 314. In a rural location with clear, dark skies, the faintest ordinary star visible to the naked eye is generally sixth magnitude (up to +6.5).

[59] Mark Littmann and Donald K. Yeomans, *Comet Halley: Once in a Lifetime* (Washington, DC: American Chemical Society, 1985), 111.

[60] It should be appreciated that the absolute magnitude of a comet is not constant, and may change during a single apparition, particularly around the time of perihelion.

[61] Gary W. Kronk, personal email message to the author, September 12, 2012 ("something like -8 to -10!"); Mobberley, *Hunting and Imaging Comets*, 74–75 (-5 to -6). Andreas Kammerer reckons that if Hale-Bopp's perihelion distance had been 0.1 AU and its perigee distance 0.1 AU, its apparent magnitude would have been -12 and -8 respectively (personal email message to the author, October 30, 2012).

[62] Mobberley, *Hunting and Imaging Comets*, 34–35; Vsekhsvyatskii, *Physical Characteristics*, 106; Kronk, *Cometography*, 1:320.

[63] Seargent, *Greatest Comets*, 107.

[64] -3 is the value of absolute magnitude favored by Vsekhsvyatskii, *Physical Characteristics*, 51, 124; Donald K. Yeomans, *Comets: A Chronological History of Observation, Science, Myth, and Folklore* (New York: John Wiley, 1991), 160–161; and Kronk, *Cometography*, 1:396. F. G. Watson, *Between the Planets*, rev. ed. (Cambridge, MA: Harvard University Press, 1956), 62, opts for something closer to -6. If we assume the orbit preferred by Kronk (*Cometography*, 1:396) and an (average) brightness slope (n) of 4, then an absolute magnitude of -6 would mean that the comet was first spotted at apparent magnitude +2.6 on August 1, 1729, and last observed at apparent magnitude +4.1 "in slight twilight" (Kronk, *Cometography*, 1:396) on January 21, 1730. (For more on "brightness slope," see chapter 9.) With an absolute magnitude of -3, the comet would have been spotted first at apparent magnitude +5.6 and last seen at +7.1. All in all, it seems most likely that Sarabat's Comet was first spotted at apparent magnitude +3 to +4 and had an absolute magnitude of -4.6 to -5.6 (if first seen at apparent magnitude +3.4, its absolute magnitude would have been -5.2).

FIG. 5.14 Comet Lulin: A photograph taken on January 31, 2009. You can see the sunward antitail to the left of the coma. Image credit: Joseph Brimacombe, Coral Towers Observatory, Cairns, Australia/Wikimedia Commons.

FIG. 5.15 The Great Comet of 1680 over Rotterdam as painted by Dutch artist Lieve Verschuier. Image credit: Collection Museum Rotterdam, inv. no. 11028, http://collectie.museumrotterdam.nl/objecten/11028-A.

almost certainly have been as bright as the full Moon, probably hundreds of times brighter, sufficient to enable Earth-dwellers to read a newspaper at midnight.[65] The giant parent of the Kreutz sungrazing family of comets, responsible for many of the greatest historical comets (e.g., 1843, 1882, and 1965), might well have been -5 in absolute magnitude.[66]

The greatest apparent magnitude values among comets belong to the Great Comet of 1680 (see fig. 5.15), which at -18 was 100 times brighter than the full Moon (-12.6), the Great September Comet of 1882, which was -17[67] or -15 to -20,[68] and the twentieth century's brightest comet, Ikeya-Seki of 1965, which was -15.[69] Other noteworthy apparent magnitudes include the Great Comet of 1577 and the Great Southern Comet of 1865, both of which reached -8.[70] The Great Comet of 1744 peaked at magnitude -7, the Great March Comet of 1843 at between -7 and -10, and Comet Skjellerup-Maristany of 1927 at

[65] Sagan and Druyan, *Comet*, 131–132.
[66] Zdenek Sekanina, as cited by Schaaf, *Comet of the Century*, 73.
[67] So Wikipedia, s.v. "Great Comet of 1882," http://en.wikipedia.org/wiki/Great_Comet_of_1882 (last modified February 26, 2013).
[68] Gary W. Kronk, *Comets: A Descriptive Catalog* (Hillside, NJ: Enslow, 1984), 69.
[69] John E. Bortle, "The Bright-Comet Chronicles," *International Comet Quarterly* (1998), http://www.icq.eps.harvard.edu/bortle.html (accessed March 26, 2014).
[70] Andreas Kammerer, "Analysis of Past Comet Apparitions: C/1995 O1 (Hale-Bopp)," http://kometen.fg-vds.de/koj_1997/c1995o1/95o1eaus.htm (last modified June 26, 2007).

-6 to -9.[71] The stunning Comet McNaught in 2007 climaxed at -5.5.[72] These are all part of an elite group of daytime comets. Few comets attain to a magnitude more impressive than Venus's, and most of those that do are too close to the Sun to be easily visible.[73] At perihelion Comet Lovejoy, the Great Christmas Comet of 2011 (fig. 5.16), attained to an apparent magnitude akin to that of Venus (-4), but, because of how near it was to the Sun at the time, it was not visible to the naked eye.[74]

Even bright comets usually have to get at least 8–10 degrees from the Sun before their comas are clearly observable (e.g., Kirch's Comet of 1680). On rare occasions, however, especially bright comets may be detected in broad daylight, even when they are very close to the solar disk, by those who are enjoying clear skies and who block out the Sun, using a wall or their hand. For example, the Great Comet of 1843 and Ikeya-Seki of 1965 could both be seen within a few degrees of the Sun.[75] Comet Skjellerup-Maristany was detectable by the naked eye when only 5 degrees from the Sun in 1927, Comet McNaught in 2007 when it was just 5½ degrees away (fig. 5.19),[76] and Comet West in 1975 when it was 6½ degrees from the Sun.[77]

FIG. 5.16 Comet Lovejoy as seen from the International Space Station on December 22, 2011. Image credit: NASA/Dan Burbank/Wikimedia Commons.

It is from comets that are visiting the outer and inner solar system (i.e., the entire region from Neptune to the Sun) for the first time or, at any rate, one of the first times, and are loaded with volatile chemicals, that the brightest comets tend to be drawn. Comet Hale-Bopp was so rich in volatiles that it became visible to the naked eye (on May 20, 1996, at a magnitude of +6.7) some 10½ months prior to perihelion and remained visible for a total of at least 18 months. Even though its peak magnitude was -1, it remained brighter than magnitude 0 for some 8 weeks, the longest of any comet on record.[78]

Generally, you can calculate a comet's apparent magnitude if you know its absolute magnitude (intrinsic brightness) and its location with respect to the Sun and Earth. However, comet brightness is not always predictable.[79]

[71] See "Brightest Comets Seen since 1935," *International Comet Quarterly*, http://www.icq.eps.harvard.edu/brightest.html (accessed March 26, 2014); Joe Rao, "The Greatest Comets of All Time," SPACE.com (January 19, 2007), http://www.space.com/3366-greatest-comets-time.html (accessed March 26, 2014); and Seargent, *Greatest Comets*, 150, 208–224.
[72] "Brightest Comets Seen since 1935."
[73] Seargent, *Comets: Vagabonds of Space*, 34.
[74] Karl Battams, "The Great Birthday Comet of 2011," http://sungrazer.nrl.navy.mil/index.php?p=news/birthday_comet (last modified December 28, 2011).
[75] Kronk, *Cometography*, 2:130–133. Other instances of especially bright comets being spotted relatively close to the Sun can be found in Seargent, *Greatest Comets*, 113, 144, and 212; and Kronk, *Cometography*, 2:506–507.
[76] Joseph N. Marcus, "Forward-Scattering Enhancement of Comet Brightness. II. The Light Curve of C/2006 P1 (McNaught)," *International Comet Quarterly* 29 (2007): 119, 128.
[77] Gary W. Kronk, "C/1975 V1 (West)," http://cometography.com/lcomets/1975v1.html (last modified October 3, 2006).
[78] "Comet Hale-Bopp: The Great Comet of 1997," an article on NASA's Stardust website, http://stardust.jpl.nasa.gov/science/hb.html (last modified November 26, 2003).
[79] Crovisier and Encrenaz, *Comet Science*, 31.

FIG. 5.17 Comet Holmes during its outburst in 2007. Image credit: John Buonomo/Wikimedia Commons.

Comets may underperform. Some new comets, like Comet Kohoutek of 1973 and Comet ISON of 2013, promise to be magnificent when first observed but end up failing to live up to expectations. When first discovered, many astronomers predicted that at perihelion ISON would be one of the most glorious comets in human history—not only would it have an apparent magnitude of -17, but it would also put on a spectacular, once-in-a-millennium show for Earth's inhabitants. However, in light of the comet's failure to brighten as expected over the following months, peak magnitude forecasts were downgraded to -6. Tragically, ISON did not attain even to that brightness level or put on any kind of notable display; in fact, half an hour before perihelion, exposed to the full force of the Sun's powerful gravity and heat, the comet completely disintegrated.

However, sometimes comets end up brighter than simple predictions based on their intrinsic brightness might suggest. Comets, particularly but not only those new to our part of the solar system, may on occasion undergo sudden bursts of brightness that increase their brightness by between 6 and 100 times (= 2–5 magnitudes) or even 1,000 times (= 7.5 magnitudes) for 3–4 weeks or more.[80]

The comet most famous for its outbursts is Comet Holmes (fig. 5.17). Within 42 hours in October 2007 it grew incredibly and went from being magnitude +17, "a thousand times too faint even to be seen with binoculars," to magnitude +2.7, which was almost enough to make it one of the top 100 brightest "stars" in the sky.[81] That is a half-a-million-fold intensification of brightness.

Another comet that had an outburst was Halley's Comet, on December 12, 1991, when it was 14.3 AU from the Sun—it became 300 times brighter. In the year 2000, comets C/1999 S4 (LINEAR) and 73P/Schwassmann 3 also underwent major outbursts.

These outbursts occurred because fresh ice and dust were suddenly being released from the comet nucleus. In the case of the outbursts in the year 2000, what happened marked a key stage in the process of the comets' splitting into pieces.[82] As regards the case of comets Holmes and Halley, the nuclei may have been hit by asteroids, and/or gas pockets within the nuclei may have exploded.[83]

Curiously, in the months after it was first observed, "rather than flaring in brightness from a single [outburst] or infrequent outbursts, Hale-Bopp seemed to be puffing them

[80] Cf. Brandt and Chapman, *Introduction to Comets*, 258; Littmann and Yeomans, *Comet Halley*, 74.

[81] "Comet in Major Outburst Now Visible in Evening Sky," an article on the Armagh Observatory's website in 2010, http://star.arm.ac.uk/press/2007/cometholmes (last modified October 10, 2012).

[82] We shall consider comet fragmentation events later in this chapter.

[83] On these and other possibilities, see Sergei I. Ipatov, "Cavities as a Source of Outbursts from Comets," in *Comets: Characteristics, Composition, and Orbits*, ed. Peter G. Melark (Hauppauge, NY: Nova Science, 2011), 101–112.

out one after another like a locomotive."[84] In particular, it had a series of outbursts separated by about 2 to 3 weeks.[85]

Moreover, comets, particularly medium or large dusty ones,[86] are subject to brightness boosts when they come close to the imaginary line that cuts through Earth and the Sun, due to the forward-scattering or backscattering of the Sun's light. To put it simply, as a comet moves closer to the zone between the Sun and Earth, the Sun's light hits the small dust particles of the coma and tail and is scattered forward. The result is that the coma and the dust tail are subject to an increasingly large spike in brightness. It is just like when a spider's web is between you and the Sun—the web suddenly becomes stunningly visible as the sunlight strikes it and is scattered forward in your direction. See fig. 5.18.

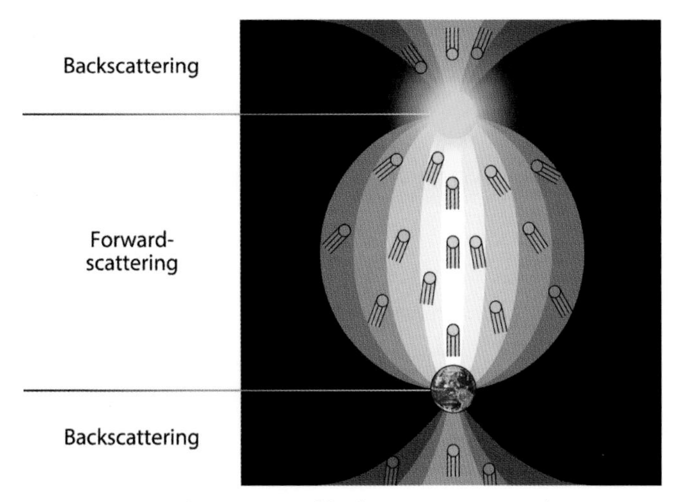

FIG. 5.18 Forward-scattering and backscattering. Comets that move into the zone between Earth and the Sun or move "behind" Earth or the Sun are boosted in their brightness. This is because the sunlight is scattered forward and backward (toward Earth) by the comet's dust. The closer the comet gets to the Earth-Sun axis, the greater the brightness boost is. Image credit: Sirscha Nicholl.

Where lines extending to the comet from the Sun and from Earth converge at an angle (= "phase angle") of 90 degrees or more, this effect is notable. At 150 degrees the comet is 2.5 magnitudes brighter than at 90 degrees.[87] At 166.5 degrees, the brightness boost is 5 magnitudes (a hundredfold boost) greater than at 90 degrees.[88] In extreme cases the effect can result in a boost of more than 7 magnitudes, as happened when Comet Skjellerup-Maristany had a phase angle of 173.5 degrees on December 15, 1927.[89] Due to forward-scattering, Comet McNaught in January 2007 was boosted by 2–3 magnitudes and so became visible during the daytime[90] (fig. 5.19) and Comet Tebbutt in 1861 became so bright that it cast shadows on the walls of Athens Observatory at night.[91]

Where a comet is on the other side of Earth from the perspective of the Sun or the other side of the Sun from the perspective of Earth, the light that it gives off is backscattered, that is, reflected back off the larger dust particles. When such a comet is positioned at an angle close to the Sun-Earth axis (a phase angle of between 30 degrees and 0), it will experience a brightness boost of up to 1 magnitude.[92]

Another factor affecting a comet's apparent brightness is that the more extended the coma is, the larger the "surface area" is over which its brightness is distributed. All other

[84] Schaaf, *Comet of the Century*, 284.
[85] Ibid.
[86] Joseph N. Marcus, "Forward-Scattering Enhancement of Comet Brightness. I. Background and Model," *International Comet Quarterly* 29 (2007): 61.
[87] Ibid., 58.
[88] Ibid.
[89] Ibid., 56.
[90] Marcus, "C/2006 P1 (McNaught)," 119.
[91] Marcus, "Background and Model," 40, 62.
[92] Ibid., 59.

things being equal, this means that the larger a coma is, the duller it will appear. For any object to be visible in the night sky or the day sky, it must be brighter than the sky.[93]

Finally, the terrestrial circumstances of the observer, particularly local atmospheric conditions and the lay of the land, are a key factor determining which parts of a cometary apparition may be visible to observers. For example, most Europeans missed daytime sightings of the Great Comet of 1843 because of widespread cloudy conditions.[94]

FIG. 5.19 Comet McNaught (C/2006 P1) seen from Lawlers Gold Mine in Western Australia on January 20, 2007. Image credit: Sjbmgrtl/Wikimedia Commons.

VARIABILITY

Comets are remarkably varied not only in celestial route, coma and tail size, and brightness, but also in shape and color, and in the duration of their apparitions.

With respect to shape, Pliny the Elder listed ten different types of comets, ranging from the bearded to the sword and from the horn to the burning torch (compare fig. 5.20).[95] In their book *Comet*, Carl Sagan and Ann Druyan offer what they call a "bestiary" of comets, with images of comets that look like a fan or horse's mane, a fountain, a tall glass, a syringe, an angel, a fetus or rabbit, a lighthouse or ball-point pen, an arrow, and a human.[96] A fan-shaped (parabolic) coma, like Tebbutt's Comet of 1861, may look like an angel, and an oval (elliptical) coma, like that of Hale-Bopp, Hyakutake, or Ikeya-Zhang, may look like a fetus in the fetal position or a baby in swaddling clothes.

Regarding color, comets are also diverse—they are often silvery grey, but dustier comets tend to have a yellowish hue (e.g., Comet

[93] For this reason, astronomers refer to "surface brightness," the magnitude of 1 square arcsecond (1/3600 degree) of any given astronomical object. The surface brightness of a comet that is 1 square arcminute (an arcminute is 60 arcseconds or 1/60 degree) will be 10,000 times, and hence 10 magnitudes, brighter than if it were 100 square arcminutes. In ideal observing conditions, such as would generally have prevailed in ancient Babylon, the surface brightness of the night sky is +22 magnitudes per square arcsecond and that of the clear daytime sky approximately +4 (an overcast daytime sky's surface brightness is about +6). To be seen, an object needs to have a surface brightness greater than the background sky. The Milky Way Galaxy, with its surface brightness of +21, can be seen only in very clear night skies. The full Moon has a surface brightness of approximately +3.6 and so is brighter than a clear daytime sky and hence may be clearly visible even when the Sun is present. Venus, with its surface brightness of +1.9, is also detectable during the day in clear skies to those with excellent eyesight who know exactly where to focus their vision. However, due to its small size, it is more difficult to spot than the Moon. The Sun's surface brightness is -10.7, Jupiter's +5.7, Saturn's +5.9, and Mars's +3.9 (see Roger Nelson Clark, *Visual Astronomy of the Deep Sky* [Cambridge: Cambridge University Press, 1990], 11 table 2.3; Mike Luciuk, "Astronomical Magnitudes: Why Can We See the Moon and Planets in Daylight?," 7 [http://www.asterism.org/tutorials/tut35%20Magnitudes.pdf (last modified April 25, 2013)]; Paul Schlyter, "Radio and Photometry in Astronomy," http://stjarnhimlen.se/comp/radfaq.html [last modified April 13, 2010]; also "Planetary Photo Techniques," a webpage of Galactic Photography, http://www.galacticphotography.com/astro_Planetary_technique_3.html [last modified September 9, 2012]). Venus, Jupiter, Saturn, and Mars become clearly visible around sunset, when the sky's surface brightness is dimming. When the Sun is just 5 degrees above the horizon, a clear sky's brightness at its zenith is +6.5; 15 minutes after sunset it is +13—at this point one can see stars that have an apparent stellar magnitude (note: not "surface brightness") of +3.5 (Clark, *Visual Astronomy of the Deep Sky*, 16). In ideal dark conditions at night, when the sky's surface brightness is +22, naked-eye observers can generally see stars up to +6.5 apparent stellar magnitude (note: not "surface brightness"). The surface brightness of the middle "star" of Orion's sword, the Orion Nebula, is +17 magnitudes per square arcsecond, and it is visible to the naked eye at night, even in light-polluted skies.

[94] Kronk, *Cometography*, 2:130.

[95] See Yeomans, *Comets*, 11–14.

[96] Sagan and Druyan, *Comet*, 173–187.

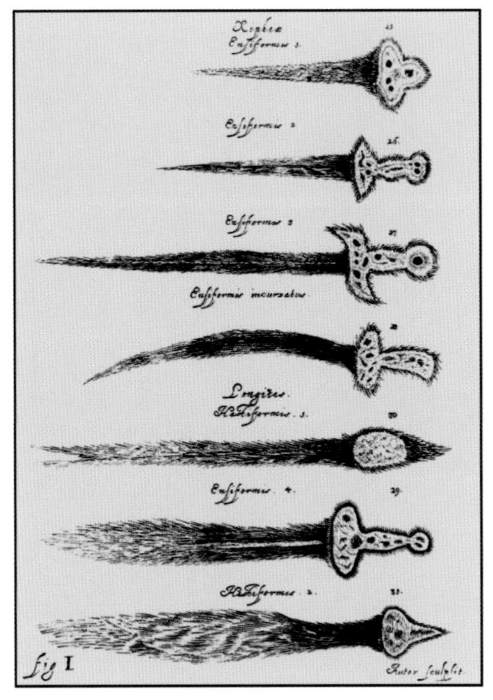
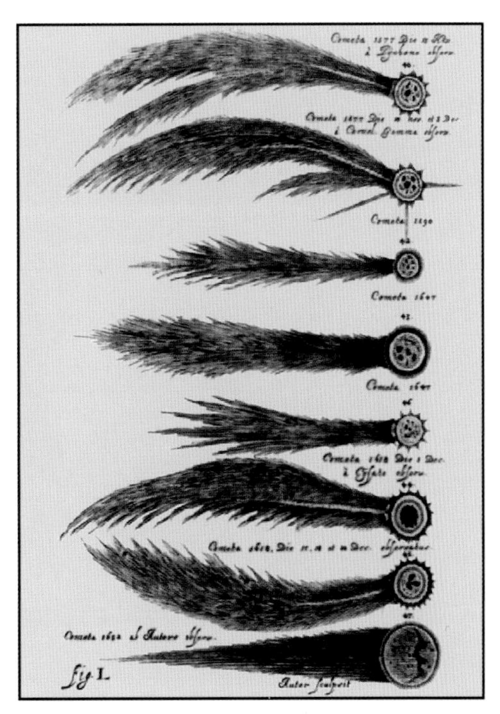

FIG. 5.20 Images of comets from *Cometographia*, by Johannes Hevelius of Danzig, 1668. Image credits: NASA/JPL (left); Library of Congress (right). Image enhancement: Sirscha Nicholl.

West in 1975–1976)[97] and gassier ones a bluish-green hue[98] (e.g., Comet Hyakutake). Single comets may change their color during the course of their apparition: for example, Comet Tebbutt was variously portrayed as white, golden, silver, "bluish-green," "greenish-blue," "greenish-yellow."[99] See. fig. 5.21.

Concerning the duration of cometary apparitions, a glance at the historical records demonstrates that they may be extremely brief or extraordinarily long (lasting well in excess of a year) or anything in between.[100]

COMET ORBITS

Apollonius of Myndus, who studied with the Babylonians,[101] the forefathers of modern comet astronomy, claims that they believed that comets were reckoned to be in the same general category as the planets, albeit with eccentric orbits (which, nevertheless, could be calculated).[102] Seneca, too, realized that comets were objects orbiting in the highest heavens that were not confined to the zodiac, in contrast to the planets.[103]

Whereas the orbits of the planets are narrowly inclined to the ecliptic plane and are all prograde (i.e., counterclockwise from the vantage point of Earth's north pole), cometary orbital planes can be inclined at any angle to the ecliptic and hence be either prograde or retrograde (clockwise).[104] In addition, comet orbital planes may be oriented at any angle,

[97] See Seargent, *Greatest Comets*, 154.
[98] George F. Chambers, *The Story of the Comets* (London: Clarendon, 1909), 8–9.
[99] See Seargent, *Greatest Comets*, 141; Kronk, *Cometography*, 2:294, 295–296, 298, 299.
[100] Josephus, *J.W.* 6.5.3 (§289).
[101] Seneca, *Natural Questions* 7.3.2–3.
[102] Ibid., 7.17–18. See Mark E. Bailey, Victor M. Clube, and William M. Napier, *The Origin of Comets* (Oxford: Pergamon, 1990), 10–11; and Yeomans, *Comets*, 8.
[103] Seneca, *Natural Questions* 7.
[104] Brandt and Chapman, *Introduction to Comets*, 17. If the inclination is less than 90° to the ecliptic, the comet is prograde; if the inclination is more than 90° the comet is retrograde.

FIG. 5.21 The Great Comet of 1861 (C/1861 [Tebbutt]). From E. Weiß, *Bilderatlas der Sternenwelt* (Esslingen: J. F. Schreiber, 1888). Image credit: Wikimedia Commons.

and the orbit itself may be positioned in any direction within the plane.

How long a comet takes to complete one revolution and how far away from the Sun it goes depend on the comet's eccentricity[105] (the shape of its orbit) and how close it gets to the Sun at perihelion.

As regards eccentricity, while some comets, like the major planets, have almost circular orbits, most comets have orbits that are elongated ovals (ellipses), and some have evolved into orbits that are hyperbolic, meaning that they may never return to the inner solar system (fig. 5.22). Short-period orbits are less elongated than long-period orbits.

With respect to how near comets come to the Sun, at one extreme we have the sungrazers and at the other extreme we have comets 167P/CINEOS and C/2003 A2 (Gleason), which do not come closer than 11.8 AU and 11.4 AU respectively to the Sun.

As regards the orbital periods of comets, they may be as short as Encke's 3.3 years or in the hundreds of millions of years.[106]

As far as Earth-dwellers are concerned, the impressiveness of a cometary apparition is heavily dependent on the time at which the comet arrives at perihelion, since that determines where Earth is on its orbit and hence the perspective humans will enjoy. On the one hand, if the comet's orbit is synchronized ideally with Earth's, it is possible for Earth-bound observers to have front-row seats for the cometary spectacle both before and after perihelion. However, if a comet is out of sync with Earth, as was Hale-Bopp, humans can only watch the action from their seats near the back row.

[105] "Eccentricity" is the degree to which a celestial body's orbit deviates from perfect circularity—the eccentricity of a circle is 0; the more stretched the oval is, the higher the eccentricity is, up to 1 (elliptical); an eccentricity of 1 (parabolic) or above (hyperbolic) means that the object is incapable of completing an orbital revolution.

[106] The eccentricity and perihelion distance also determine the range of a comet's velocity.

LONG-PERIOD AND SHORT-PERIOD COMETS

Astronomers like to divide comets into two major groups: long-period and short-period ones. Long-period comets take more than 200 years to complete a single orbit, while short-period comets have an orbital period of 200 years or less. In terms of performance and productivity, long-period comets are wild and powerful in contrast to short-period comets, which are tame and weak.[107]

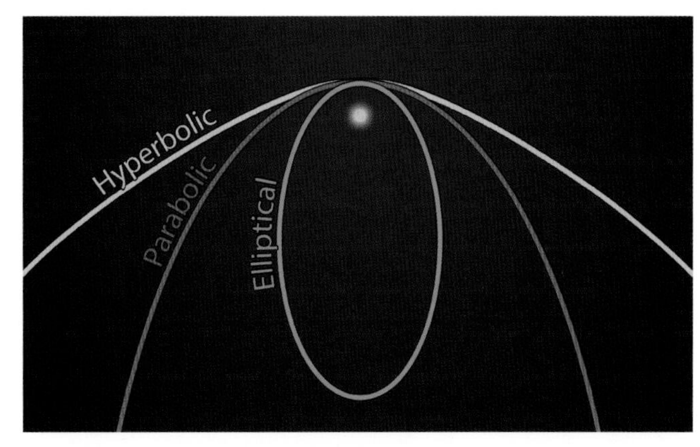

FIG. 5.22 Cometary orbits: elliptical, parabolic, and hyperbolic. Image credit: Sirscha Nicholl.

LONG-PERIOD COMETS

Many of the individual comets observed over the last few hundred years have been in extraordinarily long orbits with periods in the hundreds, thousands, and even millions of years. Indeed almost all the great comets in history are long-period comets.

Long-period comets have no particular preference for the plane of Earth's orbit, with many cutting across it at very sharp angles. They may have any inclination from 0° to 180°. Accordingly, they may journey around the Sun in a prograde (counterclockwise) or retrograde (clockwise) fashion. They are widely considered to originate in a nearly spherical region extending halfway to the nearest star, known as the Oort Cloud.

These long-period comets, hailing from the deep and dark reaches of the solar system, far from the Sun's light and warmth, have the potential to become notably bright as they approach the Sun, because they contain a high concentration of volatiles.[108] They are visible for longer and at greater distances than short-period comets. Some are large, and a small percentage are giant size.

With respect to their constitution, long-period comets also tend to be more fragile and susceptible to fragmentation.[109]

The brightest long-period comets in history tend to be ones that have made close approaches to the Sun.[110] Some, known as sungrazers, come perilously close to the Sun at perihelion. From their population come many of the brightest and most spectacular comets in history.[111] One example of a sungrazer is Comet Ikeya-Seki, which has a period of 877 years and in October 1965 came as close as 0.008 AU to the Sun (by comparison, the Moon is 0.00257 AU from Earth). Some sungrazers clearly belong to the same family. For example, the Kreutz Sungrazing Group consists of comets with high-inclination and 600- to 1,100-year orbits.[112] One member of this group, the Great September Comet of 1882, became a stunning daytime comet around perihelion time, partly because it fragmented when exposed to the raw pressure of the Sun's gravitational pull and its

[107] Cf. Guy Ottewell as cited by Schaaf, *Comet of the Century*, 197.
[108] Steel, *Rogue Asteroids and Doomsday Comets*, 36.
[109] H. F. Levison, A. Morbidelli, L. Domes, R. Jedicke, P. A. Wiegert, and W. F. Bottke, Jr., "The Mass Disruption of Oort Cloud Comets," *Science* 296 (2002): 2212–2215.
[110] Jenniskens, *Meteor Showers*, 72.
[111] For an overview of the greatest sungrazers, see Seargent's *Greatest Comets*, 191–224, and his *Sungrazing Comets: Snowballs in the Furnace* (Kindle Digital book, Amazon Media, 2012).
[112] Jenniskens, *Meteor Showers*, 423–427.

FIG. 5.23 A sketch of the Great September Comet of 1882, as seen from Cairo, Egypt. It peaked at apparent magnitude -17 or -15 to -20, and was seen in broad daylight. Image credit: *The Graphic* (November 4, 1882): 477.

fierce heat (see fig. 5.23). The Kracht Group consists of comets that come as close as 0.047 AU from the Sun and have a relatively low inclination (roughly 13.4 degrees). The Great Comet of 1680 was not related to either of these groups of comets, but nevertheless came to within 0.0062 AU of the Sun.

Passing this close to the Sun may prove catastrophic for a comet. Comet ISON (which had a perihelion distance of 0.0124 AU), for example, did not survive the Sun's scalding and gravitational pull as it made its way around the solar sphere on Thanksgiving Day 2013, but simply disintegrated. Many sungrazers share this fate.[113] Those that survive the close encounter with the Sun do so because they are relatively large (over 2 km in diameter) and structurally sound and because they are hurtling so fast at that point in their orbit—the effect is similar to the rapid movement of a finger through the flame of a candle.[114] As Fred Schaaf comments, "If the Moon orbited Earth at such a speed, we would see it complete its orbit and go through its entire set of phases in less than an hour. If Earth traveled around the Sun at this velocity, each season would last about 3 days, and the year would complete in less than 2 weeks. . . . No other enduring, discrete, macroscopic object in our solar system travels anywhere near so fast."[115]

SHORT-PERIOD COMETS

Short-period comets spend much or all of their time in the inner solar system. They include a wide range of comets and may be subdivided into two main categories, Halley-type (more than 20-year periods but less than or equal to 200-year periods) and Jupiter-type (20-year periods or less).

[113] Schmude, *Comets and How to Observe Them*, 17.
[114] Ibid., 17–18.
[115] Schaaf, *Comet of the Century*, 71.

Halley-type comets are, of course, named after Halley's Comet. This comet has been observed since 239 BC and perhaps even earlier. Its orbital period has consistently been 75–80 years. Like long-period comets, Halley-type comets may travel around the Sun in either a prograde or a retrograde revolution. However, in general they are less steeply inclined than long-period comets.

Most Jupiter-family comets are very narrowly inclined to the ecliptic and are prograde,[116] and have orbits that are greatly indebted to Jupiter's gravitational influence. In short, Jupiter, acting like a pinball flipper,[117] is able to fling narrowly inclined, prograde, long-period comets that make close passes by it into short orbits. Thereafter the orbits of these "captured" comets are perturbed by frequent close encounters with Jupiter.

Within the Jupiter family is the group of Encke-type comets. The shape and orientation of these comets' orbits around the Sun are modified by long-range interactions with Jupiter, but their orbits no longer bring them into close approaches to Jupiter.[118] Indeed their entire orbits are entirely within Jupiter's orbit.[119]

Over time, it seems that a comet's nucleus may form a crust that seals its remaining volatiles inside.[120] As a consequence, it may cease to react to the Sun's heat and therefore no longer develop a coma and tails. Since the nucleus itself is darker than freshly laid asphalt, it ceases to be visible to naked-eye observers on Earth and is liable to be mistaken for an asteroid.[121] Astronomers had classified 4015 Wilson-Harrington an asteroid but then discovered some old images from decades beforehand that revealed that it had formerly sported a gas tail.[122] Comets that cease reacting to the Sun may be either dormant (that is, generally inactive, but occasionally flaring into life for a limited period when some of their volatiles are freshly exposed) or extinct (that is, devoid of volatiles and therefore never reacting to the Sun). Comet Encke, with its orbital period of 3.3–3.5 years, burst to life in 1786, but scholars have been unable to find a single reference to it in the historical records stretching back over two millennia,[123] most likely because it was dormant most, if not all, of that time.[124] The comet is now, it would seem, in the last decades of its current phase of activity.

It should also be noted that, due to fragmentation, each Jupiter-family comet is probably little more than a small kernel of a larger original progenitor comet.[125]

ORBITAL ELEMENTS

The orbit of a comet at any particular point in time, and therefore its place within the dome of the sky, may be fully known if we have six pieces of technical information, known as the orbital elements. The six elements are the closest distance that the object comes to the Sun (perihelion distance) in AU (q) and the time when this occurs (T), the eccentricity

[116] See Sagan and Druyan, *Comet*, 96.
[117] Carolyn Sumners and Carlton Allen, *Cosmic Pinball: The Science of Comets, Meteors, and Asteroids* (New York: McGraw-Hill, 2000), 3.
[118] So Jenniskens, *Meteor Showers*, 130–132; Victor Clube and Bill Napier, *Cosmic Winter* (Oxford: Blackwell, 1990), 148.
[119] See Jenniskens, *Meteor Showers*, 133, on 2P/Encke.
[120] David Levy, *Comets: Creators and Destroyers* (New York: Simon & Schuster, 1998), 29; Yeomans, *Comets*, 353; Crovisier and Encrenaz, *Comet Science*, 67.
[121] Yeomans defines an asteroid (minor planet) as "an interplanetary body that formed without appreciable ice content and thus never had, or can have, cometary activity" (Yeomans, *Comets*, 352).
[122] Stephen J. Edberg and David H. Levy, *Observing Comets, Asteroids, Meteors, and the Zodiacal Light* (Cambridge: Cambridge University Press, 1994), 29.
[123] Brian G. Marsden and Zdenek Sekanina, "Comets and Nongravitational Forces. VI. Periodic Comet Encke 1786–1971," *Astronomical Journal* 79 (1974): 418; Fred L. Whipple and S. E. Hamid, "A Search for Encke's Comet in Ancient Chinese Records: A Progress Report," in *The Motion, Evolution of Orbits, and Origin of Comets*, ed. Gleb Aleksandrovich Chebotarev, E. I. Kazimirchak-Polonskaia, and B. G. Marsden (Dordrecht, Netherlands: Reidel, 1972), 152–154.
[124] Steel, *Rogue Asteroids and Doomsday Comets*, 27–28; D. J. Asher and S. V. M. Clube, "An Extraterrestrial Influence during the Current Glacial-Interglacial," *Quarterly Journal of the Royal Astronomical Society* 34 (1993): 489.
[125] Jenniskens, *Meteor Showers*, 126.

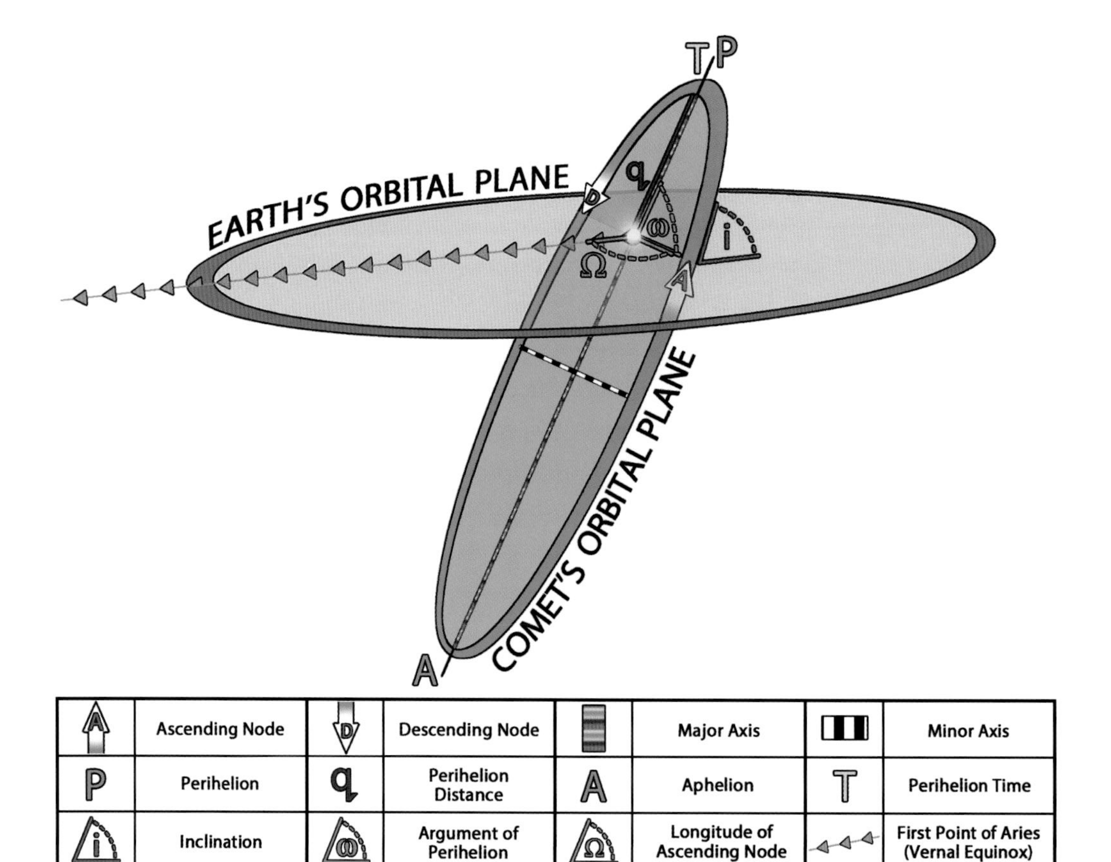

FIG. 5.24 The orbital elements of a comet. Earth's orbital plane is known as the ecliptic. Image credit: Sirscha Nicholl.

(e) of the cometary orbit, the inclination of the plane of that orbit relative to the plane of the ecliptic (i), the point where the cometary orbit crosses the plane of the ecliptic as the comet moves from the south to the north (the longitude of the ascending node) (Ω), and the angular distance from there to the perihelion point (the argument of perihelion) (ω).[126]

You can insert these pieces of information with respect to any comet into planetarium software such as Starry Night® Pro,[127] Redshift,[128] or Project Pluto's Guide[129] and follow the orbital course of the comet.

In order for the six orbital elements to be calculated approximately, at least three good-quality observations of a cometary apparition must be made. The more observations on which a cometary orbit is based and the

[126] See Brandt and Chapman, *Introduction to Comets*, 67; and Crovisier and Encrenaz, *Comet Science*, 17. Imagine a tablet computer laptop with a swivel screen and an oval piece of paper placed over the screen, affixed to it at one point in the center of the bottom of the screen. The keyboard is resting on the horizontal level, which represents the ecliptic plane on which Earth orbits the Sun. When you open the laptop and lift the screen to a certain angle, whether 45 degrees, 90 degrees, or 135 degrees, you are changing the screen's *inclination*. When you then turn the swivel screen around, it represents the changing of the *longitude of the ascending node*. Now twist the picture affixed to the screen around on its pivot—this represents changing the *argument of perihelion*. Now, suspending reality for a moment, imagine that the screen is monstrous, extending long in every direction, and that the similarly massive picture affixed to the screen is a giant oval. The longer the oval, the higher the *eccentricity* is; the more circular it is, the lower the eccentricity. The pivot point stands for the Sun. The point of the oval that is closest to the pivot represents perihelion, and the distance between them corresponds to the *perihelion distance*. Finally, if you run your finger along the edge of the oval, the moment when your finger (symbolizing the comet) is nearest the pivot (signifying the Sun) denotes the *time of perihelion*.

[127] Starry Night® Pro 6.4.3.

[128] Redshift 7, United Soft Media Verlag GmbH, Thomas-Wimmer-Ring 11, D-80539 Munich, Germany, http://www.redshift-live.com.

[129] Project Pluto, Guide 9.0, 168 Ridge Road, Bowdoinham, ME 04008, http://www.projectpluto.com.

longer the length of time they span, the more accurate the orbital elements will be.[130]

Whenever a set of orbital elements is determined, they may remain valid for only a relatively brief window of time, becoming increasingly unreliable as one moves forward or backwards in time due to gravitational and, to a lesser extent, nongravitational effects. We shall now consider these two factors briefly in turn.

GRAVITATIONAL EFFECTS

When a comet comes close to a planet either on the way toward or away from the inner solar system, it is gravitationally perturbed. This can have a significant effect on its orbit. Jupiter has the greatest gravitational pull in the solar system and so is the chief "bully" of the comet population.[131] For example, it was an encounter with Jupiter in April 1996 that changed the period of Hale-Bopp from 4,269 years to 2,534 years.[132] Saturn is likewise capable of seriously perturbing comets that venture too close. Other planets such as Uranus and Neptune may also act to change the speed and trajectory of a comet.

Jupiter may also throw comets out of the solar system altogether, or cause comets to split or disintegrate. In one particularly famous case, that of D/1993 F2 (Shoemaker-Levy 9), the comet, after getting trapped in an ever-decreasing short-period orbit around Jupiter, was split into pieces under tidal forces when it made a close approach to the gas giant in 1993. Then the resultant objects spread out around the orbit and collided with Jupiter the next time they passed it, in July 1994. Working out the effect of planetary perturbations on a given comet's orbit is a complex business, undertaken only by those who specialize in the field of solar system dynamics.

NONGRAVITATIONAL EFFECTS

The most important nongravitational effect on a comet is outgassing. Comet Encke's orbital period shortened from 3.5 years at the end of the eighteenth century to 3.3 years in the 1970s. From that point it stabilized. The reason for 2P/Encke's acceleration was that its spin axis was tilted in such a way that the nucleus rotated in a direction opposite to its orbital motion.[133] As a result, the rocket effect of its outgassing sped the comet up.[134]

FRAGMENTATION AND DESTRUCTION OF COMETS

Comets are relatively fragile objects and sometimes break up, whether due to the explosive release of internal pressure, collisions with small solar system bodies, and/or the gravitational pull of Jupiter or the Sun. This can result in boulders (e.g., Hyakutake) or significant fragments (e.g., 73P/Schwassmann–Wachmann) being thrust away from the nucleus or the wholesale disintegration of the nucleus (e.g., C/1999 S4).[135]

As we have already seen, some comets climax their career by being swallowed up by the Sun or by colliding with a planet.

COMETS AND METEOROID STREAMS

Comets are responsible for most meteor showers and meteor storms. Due to outgassing and/or fragmentation, comets deposit along their orbital course a stream of dust particles, stones, and some boulders.

As soon as each dust particle has been ejected from the nucleus, it orbits the Sun in its

[130] Schmude, *Comets and How to Observe Them*, 8.
[131] Cf. Seargent, *Comets: Vagabonds of Space*, 85.
[132] Data from NASA's JPL Small-Body Database Browser, http://ssd.jpl.nasa.gov/sbdb.cgi?sstr=Hale-Bopp (accessed May 3, 2014). Hale-Bopp passed within 0.77 AU of Jupiter in April 1996, and this altered its path.
[133] Whipple, *Mystery of Comets*, 149.
[134] Schmude, *Comets and How to Observe Them*, 25–26. On other nongravitational factors, see Donald K. Yeomans and Paul W. Chodas, "Predicting Close Approaches of Asteroids and Comets to Earth," in *Hazards Due to Comets and Asteroids*, ed. T. Gehrels (Tucson: University of Arizona Press, 1995), 241.
[135] Jenniskens, *Meteor Showers*, 378; cf. Clube and Napier, *Cosmic Winter*, 140.

FIG. 5.25 How meteoroid streams may give rise to meteor showers and storms. The meteoroids in a meteoroid stream are crossing Earth's orbit when Earth is present. This image is not to scale. Image credit: Sirscha Nicholl, using an image of Earth at night from NASA Earth Observatory.

own path, which is, naturally, almost identical to that of the parent comet. As a result, there are ribbons of particles and little stones journeying around the Sun on similar orbits, subject to the effects of the planets' gravitational pull. Over time the orbits of these particles, or meteoroids, evolve and the meteoroids spread out, so that the ribbons become convoluted and contorted.[136] These ribbons (or groups of ribbons) of meteoroids on evolving orbits are called meteoroid streams. The primary way we come to discover their existence is when they cross the plane on which Earth orbits, about 1 AU away from the Sun, and Earth passes through them (fig. 5.25).

Essentially, meteors are bright streaks in the night sky that occur when meteoroids crash into Earth's upper atmosphere. The meteoroids begin to heat up when encountering resistance at 130–150 km altitude. By the time they have reached 100 km they are bright enough to be visible on Earth as "shooting stars." By 75 km, most have disappeared. Larger meteoroids, however, are able to penetrate further through the atmosphere and tend to produce brighter, bigger meteors. Especially bright meteors are called fireballs; their brightness will be equal to or greater than that of the planets Jupiter or Venus.[137]

When the trails of a number of meteors seem to point back to one particular point in the sky, called a radiant, they are regarded as constituting a "meteor shower." Whenever astronomers judge that more than 1,000 meteors would have been observable per hour in a dark sky had the radiant been at the zenith (the imaginary point immediately above an observer's head), they classify the phenomenon as a "meteor storm."

During the great meteor storm of 1833, hundreds of thousands of meteors seemed to radiate from the head of Leo the lion, causing many observers in North America to imagine that the stars were falling from the sky, that the heavens were on fire, and that the world was coming to an end. This and other Leonid meteor storms before and since then were due to a dense section of meteoroids along the meteoroid stream parented by the Halley-type comet Tempel-Tuttle. Meteor storms radiating from the constellation Andromeda in 1872 and 1885 were obviously linked to the splitting and disintegration of the Jupiter-family Comet Biela over the previous few decades. Meteor showers and storms are usually related to Jupiter-family or Halley-type comets. However, some meteor outbursts have been identified with long-period comets with orbital periods of less than 10,000 years that come within 0.1 AU of Earth.[138]

[136] Jenniskens, *Meteor Showers*, 29.
[137] See Brandt and Chapman, *Introduction to Comets*, 331.
[138] See Jenniskens, *Meteor Showers*, 81–86, 172–200; Esko Lyytinen and Peter Jenniskens, "Meteor Outbursts from Long-Period Comet Dust Trails," *Icarus* 162 (2003): 443–452.

Most meteors, however, are "sporadic," that is, they are not associated with any particular shower. Many undoubtedly did in the distant past belong to clearly defined meteor showers, but it has now been so long since they were last replenished with fresh debris that the streams are too diffuse to be identified as showers.

FIG. 5.26 The radiant of the Andromedid Meteor Storm on November 27, 1872. Chromolith by F. Méneux. From Amédée Guillemin, *Le Ciel, notion élémentaire d'Astronomie physique* (Paris: Librairie Hachette et Cie, 1877), 623.

GREAT COMETS

Some comets set themselves apart from the majority by virtue of their sheer magnificence—their brightness and largeness and/or length seen against the backdrop of a dark sky. Such comets are classified as "great comets."

There is widespread agreement regarding the attributes that render a comet "great."[139]

FIG. 5.27 A painting of the magnificent Great Comet of 1843. From Amedée Guillemin, *The World of Comets*, ed. and trans. James Glaisher (London: Sampson Low, Marston, Searle, & Rivington, 1877), opposite page 152. Image credit: Patrick Moore Collection, www.patrickmoore collection.com.

First, a comet may make a close pass by the Sun. Chief among those attaining to greatness primarily because they came close to the Sun are the Kreutz Sungrazers. For example, the sungrazing Great March Comet of 1843 (see figs. 5.27–28) came to within 0.005 AU of the Sun (about 130,000 km from the Sun's surface),[140] traveling so fast (560 km a second) that it made its way three-fourths of the way around the Sun in less than 12 hours,[141] and attaining to -7 to -10 magnitude.

Second, a comet may make a close approach to Earth. The closer a comet comes to our planet, the brighter it seems to human observers—even normally dim objects can become remarkably bright when they make close passes by Earth. Among the great comets notable for coming near to Earth is the

[139] David Hughes as cited by Burnham, *Great Comets*, 51 (see also Burnham's comments on p. 70); John. E. Bortle, "Great Comets in History," *Sky and Telescope* 93.1 (1997): 44; Seargent, *Greatest Comets*, vii, 78; Donald K. Yeomans, "Cometary Astronomy," in *History of Astronomy: An Encyclopedia*, ed. John Lankford (New York: Routledge, 1996), 159 (and "Great Comets in History," http://ssd.jpl.nasa.gov/?great_comets [posted April 2007]).

[140] Mobberley, *Hunting and Imaging Comets*, 46.

[141] Isaac Asimov, *Asimov's Guide to Halley's Comet: The Awesome Story of Comets* (New York: Walker, 1985), 56.

Great Comet of 1861, which came as close as 0.13 AU to Earth, and Comet Hyakutake, which came to within 0.1 AU of Earth on March 25, 1996.

Third, a comet may develop a large, bright coma. Since large nuclei tend to have more volatiles to react to the Sun and hence become more active, they usually form bigger comas. Among those considered great primarily because of the large size of their comas are the Great Comet of 1811 (see figs. 5.29–31) and Hale-Bopp in 1996–1997, neither of which came close to the Sun or Earth.

Fourth, a comet may sport an eye-catchingly long tail. A great comet will generally have a tail that is 10 degrees or more in length. The Great Comet of 1618 (C/1618 W1) and Tebbutt's Comet of 1861 are two striking examples of comets rendered great because of their impressively long tails, 104 and 120 degrees respectively.

Fifth, a comet should offer the general public in the more populated northern hemisphere good viewing opportunities, preferably in a dark sky (at least 16–20 degrees from the Sun) in the hours after sunset, and should capture the public's attention. In addition, the comet should be clearly visible to the naked eye for a significant period of time.

Almost all great cometary apparitions are associated with long-period comets. The only short-period comet whose apparitions may be considered "great" is Halley's Comet.

FIG. 5.28 An impression of the 1843 Comet by D. A. Hardy. Image credit: The Patrick Moore Collection, www.patrickmoorecollection.com.

ANCIENT COMETARY RECORDS

While the places of the planets and Sun and Moon could be established by the ancients by means of calculation, the unpredictability of comets meant that they had to be observed and careful records of each stage of their apparition kept. We are fortunate that the Babylonians and Far Easterners made astronomical observations on a regular basis and over a long period. Their reason for doing so was largely their belief in astrology.[142]

BABYLONIAN RECORDS

The Babylonians kept astronomical diaries, namely detailed records of their celestial observations, for many centuries, from the middle of the eighth century BC right up to the first century AD. These records refer to a comet as a *sallammu* or *sallummu*.[143] Unfortunately, no Babylonian records of comets survive from the fifty years either side of the birth of Jesus.[144] Only nine Babylonian comet records have survived, and they only in fragmentary form, mentioning comets in 234, 210, 164, 163, 157, 138, 120, 110,

[142] Hermann Hunger, F. Richard Stephenson, C. B. F. Walker, and K. K. C. Yau, *Halley's Comet in History* (London: British Museum, 1985), 10.
[143] F. Richard Stephenson, "The Ancient History of Halley's Comet," in *Standing on the Shoulders of Giants*, ed. Norman Thrower (Berkeley: University of California Press, 1990), 243, who mentions private correspondence with Hermann Hunger on this issue.
[144] Ibid., 245–248; Hunger et al., *Halley's Comet in History*, 18–40.

and 87 BC.[145] The 164 BC and 87 BC comets are believed to be apparitions of Halley's Comet.[146]

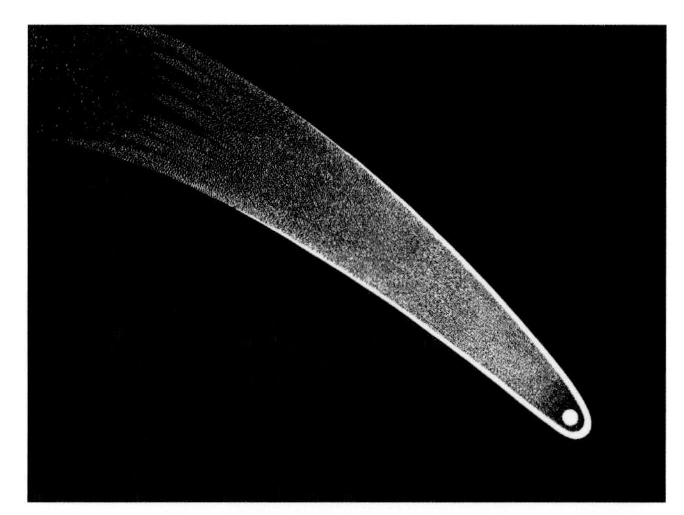

FIG. 5.29 The Great Comet of 1811, with its prominent pseudonucleus, as drawn by William Henry Smyth. From George F. Chambers, *The Story of the Comets* (Oxford: Clarendon, 1909), 130 fig. 47.

It is worth briefly overviewing some of these Babylonian cometary records, preserved on surviving fragments of cuneiform texts, to get a sense of what the Babylonian astronomers tended to take note of regarding comets. Concerning the 234 BC comet, they noted that it was first observed in the east in the last watch of the night sometime in January/February. With respect to the 210 BC comet, they recorded that it first appeared along the ecliptic in Scorpius, with its tail pointing eastward, in June/July. Regarding the 164 BC comet (Halley's), the Babylonians took note of the fact that the comet had appeared in the area of the Pleiades and Taurus and then, with its tail 7½ degrees long and oriented northward, moved to about 2½ degrees from Jupiter within Sagittarius. One cuneiform fragment noted that the 163 BC comet had a southward-oriented tail and was located 1½ degrees above the star α (Alpha) Coronae Borealis on September 5; another, more damaged fragment compares the comet's location to that of the same star and identifies the time as the first watch of the night. Only the date of the 157 BC comet, in October/November, has survived. With reference to the 138 BC comet, the Babylonians noted where and when the comet became stationary, and the zodiacal location (Libra) and date (May 28) of the comet's heliacal setting (its final visible setting in the evening in the run-up to a period when it is invisible due to its closeness to the Sun) in the west. With respect to the 120 BC comet, fragments of several records have survived, detailing:

FIG. 5.30 The Great Comet of 1811. A painting by C. H. R. Schreiber, "Komet von 1811 der Burg Katz," in *Bibliothek des allgemeinen und praktischen Wissens*, vol. 4, ed. Emanuel Müller-Baden (Berlin: Deutsches Verlagshaus Bong, 1912). This was one of a number of comets that developed a coma that appeared larger than the Sun or Moon. Source: Wikimedia Commons.

[145] Stephenson, "Ancient History of Halley's Comet," 244.
[146] Hunger et al., *Halley's Comet in History*, 10.

FIG. 5.31 The Great Comet of 1811 as seen on October 15, 1811, from near Winchester, England. An engraving by H. R. Cook based on a drawing by Abraham Pether. Image credit: The Patrick Moore Collection, www.patrickmoorecollection.com.

(1) the celestial location of the comet on May 18; (2) where and when it became stationary (May 20); (3) that its tail pointed southward on June 16; and (4) that the comet was seen on July 13 at the beginning of the night and that it had previously appeared in Aries in the east on day 29 of an earlier Babylonian month (probably equivalent to May 18). Concerning the comet of 110 BC, the Babylonians noted, on November 23, the part of the sky—the east—where the comet was and indeed its precise celestial location and the orientation of the tail—pointing westward. A later record recalled that initial observation and mentioned that the comet had subsequently migrated into the northern sky (or perhaps simply north of some celestial entity). Finally, with regard to the 87 BC (Halley's) comet, the Babylonians recorded that it was seen in the first part of some night in July/August and commented on the rate of its motion through the sky ("day beyond day one cubit") and the northwestern direction of the 10-degree tail.[147]

From these surviving Babylonian records, we can get a good idea of what typical Babylonian cometary records looked like. Stephenson deduces from the Babylonian evidence that the Babylonian astronomers included entries concerning comets at particular stages of a comet's apparition—at the first observation of the comet, at its heliacal setting, at its heliacal rising, whenever the comet (i.e., coma) became stationary relative to the fixed stars and constellations, and when the comet last appeared.[148] Each entry after the initial one made a summary reference back to a previous entry concerning the comet.[149] The Babylonians commented on the comet's locations within the sky generally (east, west, north, equatorial zone, south) and specifically within the constellations, and

[147] Information concerning each of these cometary observations from Babylon can be found in Kronk, *Cometography*, 1:7–18. On the 164 BC and 87 BC comets (Halley's Comet), see Hunger et al., *Halley's Comet in History*, 18–40.
[148] Stephenson, "Ancient History of Halley's Comet," 244.
[149] Ibid.

the direction and length of the tail.[150] Moreover, Stephenson points out that they obviously knew that the very same comet could be observed both prior to conjunction with the Sun and after it.[151] Interestingly, the surviving Babylonian cometary records seem to reflect a bias toward comets that are located within the zodiac (the band of sky through which the Sun, Moon, and planets seem to traverse).

Strangely, Diodorus Siculus, *Bibliotheca* 15.50.3, reports that the Babylonians were believed to be able to predict cometary apparitions accurately on the basis that comets complete cycles through age-long movements in appointed courses. Hermann Hunger et al. highlight that the logical deduction of what Diodorus Siculus claims is that the Babylonians compiled extensive tables of data regarding comets, as they did for eclipses.[152] It is not unlikely that there were, included in these records, cometary observations dating back as far as the eighth century BC.

As we have already seen, Babylonian records of Halley's Comet in 164 BC and 87 BC have survived, the latter only in fragmentary form. Concerning the Halley's apparition of 12 BC, Hunger et al. comment that, although the Babylonian astronomers no doubt analyzed the comet very closely, no trace of their records has survived.[153]

CHINESE RECORDS

A major boon for students of the history of astronomy is the surviving records of celestial phenomena kept in the Far East, particularly in China and Korea. They provide an invaluable collection of cometary observations all the way back to the first millennium BC.

For our purposes, however, the records we have from Korea are few and far between and of questionable reliability, and those from ancient China, largely preserved in a multivolume historical work known as the *Han shu*,[154] while more numerous and reliable, are far from complete.

There can be little doubt that the Chinese astronomers at the end of the first century BC observed many, if not most, naked-eye comets. However, a large number of these reports were not available to the writers of the *Han shu*, either because they were not made the subjects of memorials to the emperor and therefore were not formally included in the court register, or because they had been lost in the subsequent period. Of the records of cometary apparitions that *were* available to the historians, they elected to use only certain ones that were in accord with their ideological reading of the larger narrative of the Former Han dynasty. The criteria for inclusion rather than exclusion were not the brightness or coma/tail size, but rather the perceived astrological significance of the comet and especially the usefulness of the apparition in advancing the historian's ideological agenda.[155]

As John T. Ramsey and A. Lewis Licht put it, "Our extant sources clearly reflect but a small fraction of the records that were once kept by the imperial Astronomical Observatory, and the records that do survive are far less detailed and complete than the accounts from which they were drawn. We should not, therefore, conclude from the absence of a report in the Chinese sources available to us today that the sighting of a given comet

[150] Ibid.
[151] Ibid.
[152] Hunger et al., *Halley's Comet in History*, 18.
[153] Ibid., 53.
[154] The *Han shu*, or *The History of the Former Han Dynasty*, was composed in the late first and early second century AD and completed in AD 111.
[155] See David W. Hughes, "The Magnitude Distribution, Perihelion Distribution, and Flux of Long-Period Comets," *Monthly Notices of the Royal Astronomical Society* 326 (2001): 515–516. Also Thomas John York, "The Reliability of Early East Asian Astronomical Records" (PhD thesis, Durham University, 2003), 64, 67–68, 78, 121; available online at http://etheses.dur.ac.uk/3080/. See my appendix 1 on the Chinese records.

was not necessarily made from China."[156]

HOW COMPREHENSIVE ARE THE SURVIVING ANCIENT RECORDS?

Unfortunately, even when we combine all the records for any given ancient period, they fall far short of a comprehensive list of visible comets.

In his study of how many naked-eye cometary apparitions there would ordinarily be each century, Licht calculated that there would be an average of 87.[157] Ramsey and Licht wrote about the situation in the first century BC:

FIG. 5.32 Comet Ikeya-Zhang (on March 20, 2002), which now has an orbital period of 367 years, visited the inner solar system in 1661 and 2002. Image credit: The High School Astronomy Class at Alssundgymnasiet Sønderborg, Denmark, www.astronomy-ags.dk.

> To take the first century B.C. as an example, Hasegawa lists 34 naked-eye comets seen during those hundred years, out of which 16 were seen by only the Romans and/or Greeks, 15 by only the Chinese and/or Koreans, and 3 by observers both in the Mediterranean and in the Far East. However, based upon the statistics in Hasegawa's catalogue for recent centuries, for which our records are fullest, there should have been as many as 87 naked-eye comets visible during the first century B.C.[158]

According to the analysis of Hughes, the rate of naked-eye-observed long-period comets has been rather consistent over the past 2,000 years at approximately 81–99 per century.[159] For the period 50 BC to AD 50 we have surviving records of at most one-third of all visible comets.

Consequently, any investigation of ancient comets must bear in mind that relatively few records have been preserved. As Ramsey and Licht highlight, "The vast majority of comets have come and gone without leaving a trace."[160]

ANCIENT VIEWS OF COMETS

The influential Greek philosopher Aristotle believed that comets were merely meteorological phenomena, the product of the earth's warm exhalations rising to the highest part of the sphere of air and indeed to the border of the sphere of fire, where they ignited. Thereafter they were transported around the region of the upper air.[161]

Seneca, following the Babylonians, had

[156] John T. Ramsey and A. Lewis Licht, *The Comet of 44 B.C. and Caesar's Funeral Games* (Oxford: Oxford University Press, 1997), 109.

[157] Ibid., 116, also 73, 109; A. Lewis Licht, "The Rate of Naked-Eye Comets from 101 BC to 1970 AD," *Icarus* 137 (1999): 355–356.

[158] Ramsey and Licht, *Comet of 44 B.C.*, 116.

[159] David W. Hughes, "Early Long-Period Comets: Their Discovery and Flux," *Monthly Notices of the Royal Astronomical Society* 339 (2003): 1103–1110. He notes that while no single nation's astronomers ever actually recorded quite this many, the combined records of astronomers in the Far East, the Middle East, North Africa, and Europe between AD 1335 and 1600 totaled about 75 per century (pp. 1103, 1109). Between AD 1 and 1600 the total number of comet records per century increased from 37 (AD 100–810) to 53 (AD 810–1355) to 75 (AD 1355–1600) (pp. 1106–1107). Hughes concludes that the relatively low number of cometary records per century before AD 500 is evidence that the data that survives from that early period is incomplete (1108).

[160] Ramsey and Licht, *Comet of 44 B.C.*, 111–112.

[161] *Meteorologica*, esp. parts 4–7.

more sensible ideas concerning comets. He accurately predicted that "one day a human will be able to show the regions in which comets have their orbits, why their course is so far from the other stars, what size they are, and of what they are constituted" (Seneca, *Natural Questions* 7.26.1).[162] Unfortunately, it was Aristotle's atmospheric view of comets that dominated Western thinking about comets until the modern era.

These "hairy stars," as the Egyptians and Greeks thought of comets, were widely regarded by the ancients as heralds of important events on Earth.

The Romans, particularly the ruling elite, could assign great significance to the appearance of comets. For example, Suetonius, *Nero* 36, relates how Nero, in the middle of the first century AD, responded to a comet during his latter years: "It chanced that a comet had begun to appear on several successive nights, a thing which is commonly believed to portend the death of great rulers. Worried by this, and learning from the astrologer Balbillus that kings usually averted such omens by the death of some distinguished man, thus turning [the omens] from themselves upon the heads of the nobles, he resolved on the death of all the eminent men of the State."[163]

Rulers from across the ancient world, from Europe to the Near East to the Far East, were acutely aware of the need to be kept abreast of any strange astronomical phenomena, in particular eclipses and comets, that might augur ill for them, their dynasty, or their kingdom. They also knew that celestial signs, and any unfavorable interpretations of them by court astrologers that might leak out, risked empowering enemies inside and/or outside their realm, because to those eager for change, comets could be important portents of hope.

Even Aristotle and Seneca believed that cometary apparitions were harbingers of disaster.

Because bright comets have orbits that may cut across the ecliptic at any angle and hence are frequently located well away from the zodiac, they often did not fit neatly into the standard Babylonian and Greco-Roman system of astrological interpretation focused on the zodiac. A different interpretive approach was therefore called for in the case of comets.

Pliny the Elder (*Natural History* 2.22–23) set out principles by which cometary apparitions could be interpreted; by taking note of the comet's appearance, its placement within the sky, and the way in which its tail pointed, one could figure out the nature and geographical location of the doom of which the comet was warning:

> It is thought important to take note of the direction in which [the comet] shoots, the star from which it receives its influence, what it looks like, and in what places [in the sky] it shines. If it looks like a flute, it is an omen regarding the art of music; if it appears in the private parts of constellations, it is an omen for immoral behavior; it portends genius and erudition if it forms an equilateral triangle or a rectangular quadrilateral in relation to some of the fixed stars; and it portends poisonings if it appears in the head of either the northern or the southern Serpent.[164]

Ptolemy held a view similar to Pliny, stating that the shape, the zodiacal constellation in which the coma appeared, the direction of the tails, the timing and duration of the apparition, and the position of the comet

[162] My translation.
[163] Translation from Suetonius, *The Lives of the Twelve Caesars*, vol. 2, ed. and trans. J. C. Rolfe, rev. ed., Loeb Classical Library (Cambridge, MA: Harvard University Press, 1914), 151.
[164] My translation.

relative to the Sun were important clues for determining the meaning and target audience of a comet.[165]

From Pliny and Ptolemy we get a good idea how the astrological system of interpretation with regard to comets worked. When one factors in that comets often move around through different parts of the sky, one begins to appreciate the number and complexity of messages that astrologers might divine from them.

Based on what Pliny and Ptolemy wrote, we can set forth a series of questions that astrologers in the Greco-Roman and probably Babylonian environment around the turn of the ages would have asked in order to determine a comet's meaning and significance:[166]

1. Where in the sky did the comet first appear?
2. What did the comet as a whole look like? For example, did it seem similar to a beam of wood, a sword, a javelin, a flute, a trumpet, a horn, a torch, a beard, a mane, a goat, a discus, a jar, or a cask? What was its color? Shape? Size? Movement? Brightness?
3. What did the comet's coma (or head) look like?
4. In what zodiacal constellations was the comet seen?
5. Where within constellations did it appear? The private parts of a human constellation figure such as Hercules, Orion, or Andromeda? The head of one of the celestial serpents? Etc.
6. In what direction did the comet seem to point?
7. What events in the heavens or on the earth coincided with the comet's appearance? For instance, a comet that occurred at the time of an eclipse or the commencement of a new ruler's reign was liable to be interpreted with reference to that event.
8. Where in the sky did the comet seem to pause?
9. How long did the cometary apparition last?
10. With which stars was the comet in conjunction? How did it relate to established celestial entities like the stars, the planets, the Moon, and the Sun?
11. What was the position of the comet relative to the Sun? This revealed the beginning of the augured events: if it was to the west, it meant the onset of the prophesied woes was delayed; to the east, it meant the onset was imminent (Ptolemy).

Of course, many of these principles for the interpretation of comets would have been adopted well beyond the circle of astrologers. One did not need to buy into an astrological system of interpretation to conclude that a serpent-shaped comet in a serpentine constellation might be bad news for

[165] Ptolemy, *Tetrabiblos: Or Quadripartite*, trans. Frank Egleston Robbins, Loeb Classical Library (Cambridge, MA: Harvard University Press, 1940), 193, 195:

> We must observe . . . for the prediction of general conditions, the comets which appear either at the time of the eclipse or at any time whatever; for instance, the so-called "beams," "trumpets," "jars," and the like, for these naturally produce the effects peculiar to Mars and to Mercury—wars, hot weather, disturbed conditions, and the accompaniments of these; and they show, through the parts of the zodiac in which their heads appear and through the directions in which the shapes of their tails point, the regions upon which the misfortunes impend. Through the formations, as it were, of their heads, they indicate the kind of the event and the class upon which the misfortune will take effect; through the time which they last, the duration of the events; and through their position relative to the sun likewise their beginning; for in general their appearance in the orient [the east] betokens rapidly approaching events and in the occident [west] those that approach more slowly.

See Yeomans, *Comets*, 14–16, on Ptolemy's view of comets.

[166] Sara Schechner Genuth, *Comets, Popular Culture, and the Birth of Modern Cosmology* (Princeton, NJ: Princeton University Press, 1997), 55, speaking of seventeenth-century Western astrologers, comments that "There was a lot of latitude in the iconographic technique, and the astrologer could put a positive spin on his predictions if he was so inclined." Her statement applies equally to ancient astrologers.

a ruler or dynasty, or to believe that a large and bright sword-like comet hanging over a city was an omen of judgment against that city. Josephus, the Jewish historian, made it clear that he believed that a cometary sword standing over Jerusalem in AD 65–66 was a powerful omen auguring the destruction of the city in AD 70.

Although comets were often perceived to be negative omens in the ancient world, on many occasions they were interpreted positively, as we shall see in the following chapter.

COMETS IN THE BIBLE?

In the Biblical tradition, comets seem to be regarded as capable of functioning as messengers from God. Genesis 1:14–15 appears to have comets chiefly in mind when it refers to "lights" that are "for signs":

> God said, "Let there be lights in the expanse of the heavens to separate the day from the night. And let them be for signs and for seasons, and for days and years, and let them be lights in the expanse of the heavens to give light upon the earth." And it was so.

No member of the starry host is more equipped to communicate meaning ("for signs") than the comet. As comets move across the sky through the constellations, they are susceptible to being interpreted as conveying all kinds of messages, even complex ones, to human observers. While the Hebrew Bible has a very negative view of astrology, it does hold out the possibility that God could at certain times communicate messages through comets.

An important example is found in Numbers 24:17, a text we shall explore in detail in chapter 8. Here we find a striking prophecy from the Mesopotamian seer Balaam: "I see him, but not now; I behold him, but not near: a star shall come out of Jacob, and a scepter shall rise out of Israel." Balaam's mysterious reference to a celestial body identified as "a star" and "a scepter" (note the synonymous parallelism) is most naturally interpreted as speaking of a long-tailed comet. He is envisioning the Messiah as a cometary scepter, strongly intimating that his coming will be attended by a great comet. Balaam seems to imply that this comet would do something extraordinary in connection with its rising to announce the Messiah's coming. In particular, the oracle appears to prophesy that the comet would look like a scepter at a key stage of its apparition.

A second prophetic oracle that declares that a cometary apparition would convey a positive message from God is found in Isaiah 9:2, another verse that we shall examine closely in chapter 8: "The people who walked in darkness have seen a great light; those who dwelt in a land of deep darkness, on them has light shone." Isaiah's description is strongly suggestive of a great comet. "Deep darkness" most naturally refers to the time when the Sun and the Moon are absent from the sky. "Great light" implies that the celestial body was, like the two great lights of Genesis 1 (the Sun and the Moon), large and bright. The only large and bright object with a steady beam that can light up the night in the absence of the Moon is a great comet. Isaiah is therefore, it would seem, prophesying that an extraordinary comet that shone during the deep darkness of night would be the heavenly signal of the coming to fulfillment of God's plan of salvation through the Messiah. Like Balaam, Isaiah portrays the Messiah himself in terms of the comet that coincided with his birth.

We can see, therefore, that the Hebrew Bible assigns comets a positive, hopeful role in salvation-history.

CONCLUSION

Comets have fascinated and mystified humans since the dawn of civilization. They are remarkably diverse icy balls of dirt and dust. To ancient sky observers, they turned the heavens into something akin to a celestial movie screen conveying divine messages. In ancient thought, they were often understood to be negative portents, but could also be interpreted positively. The Hebrew Bible reflects a positive view of comets, even prophesying that one will appear to signal the Messiah's coming, and portraying him in terms of it. The question to which we must now turn is whether a comet might have performed the role of the Star in the story of the Nativity.

6

"A Stranger midst the Orbs of Light"

The Star as a Comet

Having evaluated the other major hypotheses put forward to explain the Bethlehem Star in chapter 4, and having then introduced comets in chapter 5, we must now turn to the key question: Could the Star of Bethlehem have been a comet?[1]

In this chapter we will consider the two comet proposals currently on the table—the Halley's Comet and the 5 BC Comet theories—before making a more robust case for the view that the Star is to be identified as one of the icy balls of dirt and dust zipping around the solar system in eccentric orbits.

A BRIEF HISTORY OF THE COMET HYPOTHESIS

The comet hypothesis has a long history, stretching back into the first three centuries of the Christian era. At the start of the second century, Ignatius, probably drawing on an established and authoritative hymn from the first century, referred to the Star in terms that were strongly suggestive of a comet. According to this tradition, the Star was new, "brighter than all the stars," and provoked astonishment because it was so unlike anything else in the heavens (*To the Ephesians* 19:2).

The Protevangelium of James (*Gospel of James*) dates to around AD 150. According to it, in response to Herod's question regarding what celestial sign the Magi saw that related to the newborn King, the Magi answered, "We saw an immense star [*astera pammegethē*] shining among these stars and causing them to become dim, so that they no longer shone; and we knew that a king had been born in Israel" (21:2–3).[2] As Olson and Pasachoff point out, here, "the Protevangelium of James describes the strange new star in language only befitting a comet."[3]

Then, in the first half of the third century,

[1] One prominent recent advocate of the comet hypothesis is the host of the BBC series *Wonders of the Solar System* and *Wonders of the Universe* and author of the accompanying books, Brian Cox. His opinion on the matter was given in the documentary "Star of Bethlehem: Behind the Myth," produced by Atlantic Productions in London and shown on the BBC in the UK in 2008 and on ABC in Australia in 2009. I am grateful to Atlantic for generously sending me a complimentary DVD copy of the production.

[2] My translation of the Greek text in Emile de Strycker, *La forme la plus ancienne du Protevangile de Jacques* (Brussels: Société des Bollandistes, 1961), 168–170.

[3] Roberta J. M. Olson and Jay M. Pasachoff, "New Information on Comet Halley as Depicted by Giotto Di Bondone and Other Western Artists," in *20th ESLAB Symposium on the Exploration of Halley's Comet: Proceedings of the International Symposium, Heidelberg, Germany, 27–31 October*, vol. 3 (Noordwijk, Netherlands: European Space Agency, 1986), C207.

Origen made explicit his conviction that the Star was a comet, in *Contra Celsum* 1.58–59:

> We consider that the star that was seen "at its rising" was a new star, and not like any of the normal celestial bodies, either those in the fixed sphere [above] or those in the lower spheres. Rather, it should be reckoned with the celestial bodies which occur from time to time, known as "hairy stars," "beams,"[4] "beards," "wine-jars," or any other such name by which the Greeks like to describe their various forms. We establish this point in the following way:
>
> It has been noticed that at the occurrence of momentous events and at the most profound transitions on earth, stars of this kind appear, announcing changes in dynasties or the breaking out of wars, or the occurrence of some phenomenon in the human realm that shakes affairs on earth. We have read in the book called *Concerning Comets* by Chaeremon the Stoic that at times comets have appeared when good events were about to occur. He has given an account of such occurrences. If, then, at the commencement of new dynasties or on the occasion of other momentous events on earth, the so-called "hairy star" or some similar body [namely, the "beam," "beard," "wine-jar," or some other type of comet] appears, why would it be a great surprise that a star should have appeared at the birth of the one who was going to introduce new ideas to the human race and to reveal his teaching not only to Jews, but also to Greeks, and to many barbarian nations in addition? Now I would point out with respect to comets that there is no prophecy about comets in circulation stating that such and such a comet would appear at the rise of a particular kingdom or at a particular time. However, the star which appeared at Jesus' birth had been prophesied by Balaam, recorded by Moses, when he said: "A star shall appear out of Jacob, and a man shall rise up out of Israel."[5]

Probably no one has done more to promote the comet hypothesis than Giotto di Bondone (1266/1267–1337), the medieval Italian artist. At the turn of the fourteenth century he painted a fresco entitled "The Adoration of the Magi" in Padua's Arena Chapel, in which he portrayed the Star of Bethlehem as a comet (fig. 6.1). Many have thought that he was inspired by the apparition of Halley's Comet in 1301, although that is questionable.[6] In memory of this depiction of the Magi's Star, the European

[4]The standard rendering of English translations (e.g., Alexander Roberts, James Donaldson, and A. Cleveland Coxe, *The Ante-Nicene Fathers Vol. IV: Translations of the Writings of the Fathers Down to A.D. 325* [New York: Scribner, 1926], 422; Henry Chadwick, *Origen: Contra Celsum* [Cambridge: Cambridge University Press, 1965], 53–54) at this point, "meteors," is clearly inappropriate. After all, Origen makes it clear in the subsequent context that he is speaking only of comets, and it is well known and established that cometary apparitions may take a multitude of forms—see Carl Sagan and Ann Druyan, *Comet* (New York: Pocket Books, 1986), 157–187. "Beam [of wood]" here (the most common meaning of the Greek term and of its Latin rendering) is evidently a type of comet—one with a long, straight tail. Meteors do not have anything like the same multiplicity of forms.

[5]My translation of the Greek text. The fourth-century theologian Ephrem the Syrian wrote concerning the Star of Bethlehem (Ephraem Syrus, *Opera Syriaca* [Rome: Vatican, 1740], 4), "A star shone forth suddenly with preternatural light, less than the sun and greater than the sun. It was less than the sun in manifest light; it was greater than it in secret strength by reason of its mystery. A star in the east darted its rays into the house of darkness" (as cited by J. B. Lightfoot, *Apostolic Fathers, Pt. II. S. Ignatius. S. Polycarp. Revised texts, with Introductions, Notes, Dissertations, and Translations* [London: Macmillan, 1885], 82). In addition, the Byzantine scholar John of Damascus (*Exposition of the Orthodox Faith*, book 2, chapter 7) wrote, "It often happens . . . that comets arise. . . . They are not of the stars that were made in the beginning, but are formed at the same time [as they arise] by divine command and again dissolved. And so not even the star which the Magi saw at the birth of the Friend and Saviour of Man, our Lord, who became flesh for our sake, is of the number of those that were made in the beginning" (John of Damascus, *Exposition of the Orthodox Faith*, trans. S. D. F. Salmond, A Select Library of Nicene and Post-Nicene Fathers of the Christian Church, Second Series, Volume 9 [Oxford: J. Parker, 1899], 24).

[6]David W. Hughes, Kevin K. C. Yau, and F. Richard Stephenson, "Giotto's Comet—Was It the Comet of 1304 and Not Comet Halley?," *Quarterly Journal of the Royal Astronomical Society* 34 (1993): 21–32, argue that the inspiration was closer to the time that Giotto painted the scene—1304. They suggest that it was the naked-eye comet C/1304 C1, which appeared for 74 days, from February 3 to April 18, and had a shorter tail than Halley's Comet in 1301. Of course, it is also possible that no particular comet was in Giotto's mind.

Space Agency named the robotic spacecraft sent to explore Halley's Comet, as well as the whole mission, *Giotto*.[7]

In recent decades, advocates of the comet hypothesis have fallen into two groups: those who identify it as Halley's Comet,[8] and those who maintain that it is the 70+-day "broom star comet" of spring 5 BC mentioned in Chinese astronomical records.[9]

FIG. 6.1 Giotto's "Adoration of the Magi" fresco in Padua—notable for its picture of the comet, often dubiously identified with Halley's 1301 apparition. Image credit: The Athenaeum.

THE HALLEY'S COMET THEORY EVALUATED

The view that the Star was Halley's Comet is often the only version of the comet hypothesis that is evaluated by those criticizing the position.[10]

Halley's Comet (officially designated 1P/Halley) has historically been one of the most consistently impressive comets. For more than 2,000 years it has faithfully returned every 75–80 years, gracing our skies as recently as 1985–1986. As mentioned in the last chapter, the Chinese *Han shu* preserved the record of a comet that was regarded as of great astrological significance: in August of the year 12 BC a "bushy star comet" was observed near Canis Minor. It remained in the skies for some 56 days before finally disappearing, last being seen between Ophiuchus and Scorpius.[11] It is generally accepted that this was an apparition of 1P/Halley.

[7] David Ritchie, *Comets: Swords of Heaven* (New York: New American Library, 1985), 11, notes that many philosophers in the time of Giotto and earlier regarded the Star as a comet (including the Genoese historian and theologian Jacobus de Veragine, author of *The Golden Legend*).

[8] J. Edgar Bruns, "The Magi Episode in Matthew 2," *Catholic Biblical Quarterly* 23 (1961): 54; Robert S. Richardson, "The Star of Bethlehem—Fact or Myth?," *The Griffith Observer* 22 (December 1958): 163–164; Arthur Stenzel, *Jesus Christus und sein Stern* (Hamburg: Verlag der Astronomischen Korrespondenz, 1913), 73; Jerry Vardaman, "Jesus' Life: A New Chronology," in *Chronos, Kairos, Christos*, ed. Jerry Vardaman and E. M. Yamauchi (Winona Lake, IN: Eisenbrauns, 1989), 66, 78 table 4; H. W. Montefiore, "Josephus and the New Testament," *Novum Testamentum* 4 (1960): 140–146; Nikos Kokkinos, "Crucifixion in A.D. 36," in Vardaman and Yamauchi, *Chronos, Kairos, Christos*, 158; A. I. Reznikov, "La comète de Halley: une démystification de la légende de Noël?," *Recherches d'astronomie historique* 18 (1986): 65–68; James Fleming in *The Advertiser* (December 21, 1985), as referenced by P. A. H. Seymour, *The Birth of Christ: Exploding the Myth* (London: Virgin, 1998), 102; William Phipps, "The Magi and Halley's Comet," *Theology Today* 43 (1986–1987): 88–92.

[9] Colin J. Humphreys, "The Star of Bethlehem—A Comet in 5 B.C.—And the Date of the Birth of Christ," *Quarterly Journal of the Royal Astronomical Society* 32 (1991): 389–407; idem, "The Star of Bethlehem, a Comet in 5 B.C., and the Date of the Christ's Birth," *Tyndale Bulletin* 43 (1992): 31–56; idem, "The Star of Bethlehem," *Science and Christian Belief* 5 (1995): 83–101; Duncan Steel, *Eclipse* (London: Headline, 1999), 20–21. The 5 BC comet hypothesis gets a mention by Gary W. Kronk in his monumental work, *Cometography: A Catalog of Comets*, 6 vols. (Cambridge: Cambridge University Press, 1999–), 1:26.

[10] E.g., Raymond Brown, *The Birth of the Messiah: A Commentary on the Infancy Narratives in the Gospels of Matthew and Luke*, 2nd ed. (New York: Doubleday, 1993), 171–172; Donald A. Carson, "Matthew," in *Expositor's Bible Commentary*, rev. ed., ed. Tremper Longman III and David E. Garland, vol. 9 (Grand Rapids, MI: Zondervan, 2010), 111; Simo Parpola, "The Magi and the Star: Babylonian Astronomy Dates Jesus' Birth," in *The First Christmas: The Story of Jesus' Birth in History and Tradition*, ed. Sara Murphy (Washington, DC: Biblical Archaeology Society, 2009), 15.

[11] Ho Peng-Yoke, "Ancient and Mediaeval Observations of Comets and Novae in Chinese Sources," *Vistas in Astronomy* 5 (1962): 127–225, catalog number 61; Donald K. Yeomans, *Comets: A Chronological History of Observation, Science, Myth, and Folk-*

The Chinese record of Halley's Comet in 12 BC was the most exhaustive comet report up to the sixth century. It reveals that it was seen from August 26, when it was in Gemini in the eastern morning sky, to October 20, when it was in Scorpius, low in the western sky. Significantly, the comet came to within 0.16 AU of Earth on September 9, permitting Earth-dwellers to see an approximately 25-degree tail around that time, although unfortunately this was a month before perihelion (October 10) and so the comet's brightness was only first magnitude. The comet, with its striking tail, was standing up near-vertically over the western horizon from September 11 to 15.

FIG. 6.2 Halley's Comet on June 6, 1910. A photograph taken by the Yerkes Observatory and published in an article in the New York Times on July 3, 1910. Image credit: Wikimedia Commons.

There are, then, a couple of interesting points of comparison between this apparition of Halley's Comet and the Star of Bethlehem. First, Mark Littmann and Don Yeomans point out that, just as the Star of Bethlehem was seen in the eastern sky and then led the Magi westward to Jerusalem, so also, viewed from the Middle East, Halley's Comet beamed brightly over the eastern horizon at the start of September and then over the western horizon later in the month.[12] Second, for a few days in mid-September, the comet, with its sizable tail, could readily have been described as standing over the western horizon.

The temptation to identify Halley's Comet in 12 BC with the Star of Bethlehem has proved too great for some. James Fleming tried to support this hypothesis by arguing that there was a census in 12 BC that would qualify as Quirinius's first census (Luke 2:2), which brought Joseph and Mary to Bethlehem for the birth.[13] According to Nikos Kokkinos, the Magi departed for Judea shortly after seeing the comet for the first time and arrived in Jerusalem "by mid-September 12 B.C."[14]

However, upon close examination this hypothesis quickly disintegrates.

First and foremost, a date of 12 BC is

lore (New York: John Wiley, 1991), 367. Some believe that this apparition of Halley's Comet is also mentioned by Cassius Dio as presaging the death of the Roman General Agrippa (54.29): "The star known as comet hung (*aiōrētheis*) for many days over (*huper*) the City [of Rome] and finally was broken up into torches" (my translation) (so, for example, Kronk, *Cometography*, 1:25). However, the comet described by Cassius Dio is unlikely to have been Halley's Comet. A number of comets have famously split and/or disintegrated: Aristotle's Comet of 373–372 BC (Ephorus as cited by Seneca, *Natural Questions* 7.16.2); Comet Biela in the mid-nineteenth century; the Great September Comet of 1882; and Comet West in 1976 (cf. James C. Watson, *A Popular Treatise on Comets* [Philadelphia: James Challen & Son, 1861], 81). The fact that Dio's comet was observed to split suggests that it was visible for a long time. Moreover, the extraordinary performance of the comet—hanging over Rome—is partly explained because fragmenting comets typically release extraordinary quantities of dust and therefore become brighter, larger, and longer. It is not uncommon for bright long-period comets to appear around the time of the return of Halley's Comet (e.g., 1910) (see F. Richard Stephenson, "The Ancient History of Halley's Comet," in *Standing on the Shoulders of Giants*, ed. Norman Thrower [Berkeley: University of California Press, 1990], 234; David Hughes, "Apian's Woodcut and Halley's Comet," *International Halleywatch Newsletter* 5 [1984]: 24–25).

[12] Mark Littmann and Donald K. Yeomans, *Comet Halley: Once in a Lifetime* (Washington, DC: American Chemical Society, 1985), 10–11.

[13] James Fleming, as cited by Seymour, *Birth of Christ: Exploding the Myth*, 102.

[14] Kokkinos, "Crucifixion in A.D. 36," 162.

impossibly early for the birth of Jesus. Those who argue for the 12 BC Halley's Comet apparition end up having to undertake a radical revision of the chronology of Jesus's life. Jerry Vardaman, for example, dates the commencement of Jesus's ministry to AD 15 and the crucifixion to AD 21.[15] This naturally raises a whole raft of serious problems, for example, with respect to Luke 3:1–3 (John the Baptist began his ministry in "the fifteenth year of the reign of Tiberius," who came to the throne in AD 14); John 2:20 ("it took/has taken[16] forty-six years to build this temple"); the established chronology of Pontius Pilate (AD 26–36); and the history of the early church. The evidence strongly favors the view that Jesus was born in 6 or 5 BC and then died in AD 30 or 33. The fact that Halley's Comet appeared in 12 BC means that it simply could not have been the Star.

FIG. 6.3 Halley's Comet in 1066 as portrayed on Scene 32 of the Bayeux Tapestry. Image credit: Wikimedia Commons.

Second, unlike the Star seen by the Magi, Halley's Comet, although it did appear in the east, did not at any stage rise heliacally (namely, after being invisible because of proximity to the Sun) in 12 BC.

Third, this hypothesis fails to offer a credible explanation for the behavior of the Magi. One must assume that the Magi and their predecessors had seen a number of comets, including bright ones like Halley's, come and go without undertaking major treks in search of a newborn King of the Jews. Why, then, would they have responded any differently in 12 BC? The Chinese were deeply impacted by this apparition of Halley's, but they had a completely different way of reading the heavens than the Babylonians and Greeks. What might have caused the Magi to think that a new king had been born and that he was Jewish and divine? An examination of Halley's behavior in 12 BC on astronomical software fails to shed any light on the issue.

Fourth, Halley's Comet did not remain visible for anywhere near as long as the Star of Bethlehem. Halley's 56 days do not compare well with the Star's minimum apparition of one year.

Fifth, the chronology of the Magi's journey set out by Kokkinos allows less than 3 weeks (from August 26, when the comet attained to naked-eye visibility, to mid-September) for the Magi to travel to Judea. Even if they had traveled an average distance of 20 (indeed 28!) miles per day, they would not have completed the journey from Babylon that quickly.

Sixth, at no stage did Halley's Comet

[15] Vardaman, "Jesus' Life," 78 table 4.
[16] The Greek text can be rendered either way. Those dating Jesus's birth to 12 BC cannot make sense of either translation.

during its 12 BC apparition ever appear in the south. It first appeared in the east before slowly moving to the north and then to the west, at which point it disappeared from the sky. This is in contrast to the Bethlehem Star, which was seen in the southern sky, when it led the Magi from Jerusalem to Bethlehem.

It is clear that the Halley's Comet hypothesis is fundamentally flawed and untenable. It is puzzling that, of the entire comet population, so many scholars discussing the Star of Bethlehem have zeroed in on it.[17] As impressively bright as it has been over recent millennia, Halley's is scarcely the only great comet in history. Indeed, long-period comets, like the Great Comets of 1680, 1843, 1880, 1882, 1910, 1996, and 1997, are often even more stunning than Halley's (particularly in its 12 BC apparition)—appearing brighter, sporting larger comas and/or greater tails, and remaining visible for longer.

THE CHINESE COMET OF 5 BC THEORY EVALUATED

In an excellent contribution to the Star of Bethlehem debate, Professor Sir Colin Humphreys has offered an alternative cometary theory.[18] He points out that, according to the Chinese astronomical records, there was a *hui-hsing* in the constellation Capricornus that remained visible for just over 70 days, beginning in March/April of the year 5 BC.[19] The record is as follows:

> Second year of the *Ch'ien-p'ing* reign period of Emperor Ai of the Han dynasty, second month [March 10–April 7, 5 BC], a *hui-hsing* appeared at *Ch'ien-niu* for over 70 days.[20]

On the understanding that the *hui-hsing* refers to "a broom star comet" (as noted in chapter 4), Humphreys suggests that this comet, which falls within the plausible time frame for the birth of Jesus and was present long enough to permit the Magi to travel to Judea, played the part of the Star of Bethlehem.[21] Sir Colin takes the view that what brought the Magi west was the combination of the *hui-hsing* in 5 BC, the planetary massing in 6 BC, and the triple conjunction of 7 BC.[22] He suggests that the triple conjunction and planetary massing convinced the Magi that the birth of the mighty messianic King of Israel would take place in the near future. "The scene was set: their expectations were aroused for a third sign which would indicate that the birth of the king was imminent."[23] The third sign was the comet in 5 BC, which told them that the King had now been born.[24] In Humphreys's view, it was this final celestial wonder, the comet, that was the Star seen twice by the Magi.[25]

[17] Some scholars, such as Brown, *Birth of the Messiah*, 172, believe that the 12 BC Halley's Comet apparition may have played a key role in the development of the Magi narrative: "It is possible that the appearance of Halley's comet in 12 B.C. and the coming of foreign ambassadors two years later to hail King Herod on the occasion of the completion of Caesarea Maritima have been combined in Matthew's story of the star and the magi from the East." That is a rather far-fetched and naive proposal. Inexplicably, Brown fails to devote any attention to the particulars of the apparition of Halley's Comet in 12 BC to discover the extent to which it was consistent with Matthew's striking portrayal of the Star.

[18] Humphreys, "Star of Bethlehem—A Comet in 5 B.C.," *Quarterly Journal of the Royal Astronomical Society*, 389–407; idem, "Star of Bethlehem, a Comet in 5 B.C.," *Tyndale Bulletin*, 31–56; idem, "Star of Bethlehem," *Science and Christian Belief*, 83–101. It should be appreciated that this position has a history that goes back centuries before Humphreys. One prominent cometary astronomer who has argued for this view is Duncan Steel in his *Marking Time: The Epic Quest to Invent the Perfect Calendar* (New York: John Wiley, 2000), 324–332.

[19] Humphreys, "Star of Bethlehem, a Comet in 5 B.C.," *Tyndale Bulletin*, 42–44; Ho, "Ancient and Mediaeval Observations," catalog number 63.

[20] Translation adapted from David H. Clark, John H. Parkinson, and F. Richard Stephenson, "An Astronomical Re-Appraisal of the Star of Bethlehem—A Nova in 5 BC," *Quarterly Journal of the Royal Astronomical Society* 18 (1977): 444; and David W. Pankenier, Zhentao Xu, and Yaotiao Jiang, *Archaeoastronomy in East Asia* (Amherst, NY: Cambria, 2008), 23–24.

[21] Humphreys, "Star of Bethlehem, a Comet in 5 B.C.," *Tyndale Bulletin*, 42.

[22] Ibid., 45–47; cf. Montefiore, "Josephus," 140–146; Jack Finegan, *Handbook of Biblical Chronology* (Peabody, MA: Hendrickson, 1998), 313–319 §§537–549; Steel, *Eclipse*, 20–21.

[23] Humphreys, "Star of Bethlehem, a Comet in 5 B.C.," *Tyndale Bulletin*, 46.

[24] Ibid., 47.

[25] Ibid., 45–47.

Unfortunately, there are a number of problems with this hypothesis.

First, while Matthew 2:2 reports that it was what the Star did in connection with its heliacal rising that prompted the Magi to travel to Judea, we have no reason to believe that the comet observed by the Chinese in Capricornus ever heliacally rose. Humphreys argues that what the 5 BC comet did to impress the Magi, it did at the beginning of its apparition, coinciding with the period shortly after perihelion, when the comet would have been at its most impressive.[26] However, although it is true that this comet was at this time in the eastern (technically, southeastern) sky in the hours before dawn, it was way too far from the Sun to have been regarded as heliacally rising. The comet became visible in Capricornus in March/April, but the stars of Capricornus had heliacally risen more than a month beforehand.

Second, the fact that the comet of 5 BC remained observable for just over 70 days constitutes a major problem, for Herod's order that one-year-old infants be slain was based on the fact that the Star had first appeared at least a year beforehand (vv. 7, 16). Of course, Humphreys, although he insists that the birth of Jesus coincided with the comet's appearance, attempts to get around this problem by postulating that, when the Magi told Herod that they had first seen the Star "two years ago," they were referring to the non-cometary phenomena.[27] However, this simply does not work. After all, Herod asks the Magi not when the first of a number of signs occurred, but "when the star had appeared" (v. 7). And we must assume that the Magi answered his question straightforwardly, speaking of the first appearance of the very same Star that they had reported seeing rise in verse 2. It should also be noted that, according to verse 2, what underlay the Magi's journey to Judea was one thing: their observation of a "star" that they interpreted to be the Jewish Messiah's. Had the apparition of the Star alone been inadequate to get them journeying westward, then we would have expected the Magi to speak of "signs" rather than the appearance of a "star."

Third, there was obviously something extraordinary about the Star of Bethlehem that prompted the Magi to conclude at the time that it was signaling the Messiah's birth, and to be so sure of this that they traveled 550 miles to Jerusalem in search of the newborn King. However, Humphreys's hypothesis is unable to offer any explanation of why the comet itself was regarded as remarkable or interpreted in messianic terms. One of the few things we know about the Chinese *hui-hsing* is that it appeared in Capricornus, an area of sky with no obvious connection to the Jewish people.

Fourth, there is no evidence that this comet appeared in the western or southern evening sky. Indeed there is reason to wonder if the comet ever left Capricornus. As David H. Clark, John H. Parkinson, and F. Richard Stephenson highlight, a *hui-hsing* in the Chinese records is usually a tailed comet and, when reference is made to one, there is generally some mention of the object's movement.[28] These scholars point out that the Chinese record of the 12 BC apparition of Halley's Comet was probably made by the same astronomers who made the 5 BC record.[29] While the Chinese detailed the 12 BC comet's movements over 150 degrees through more than 10 asterisms, they make no mention of any motion with respect to the 5 BC entity.[30]

[26] Ibid., 36.
[27] Ibid., 48, proposes that the Magi informed Herod of the triple conjunction in 7 BC, the planetary massing of 6 BC, and the comet "about one month previously."
[28] Clark, Parkinson, and Stephenson, "Astronomical Re-Appraisal," 444.
[29] Ibid.
[30] Ibid.

It is, however, possible that this Chinese record is incomplete and that the comet was only in Capricornus at the point when it was first discovered.[31] Certainly it is unfair to compare this comet record to that of Halley's Comet in 12 BC, which was perceived to be of extraordinary astrological significance and therefore was included in unusual detail in the *Han shu*.

Nevertheless, there are many examples from 110 BC to AD 100 that detail movement across the constellations.[32] And there are examples where the comet's location in the sky is omitted[33] or is put in more general terms.[34] Quite simply, it was not normal during this period for the Chinese to specify only the initial location of the comet.

It is therefore not at all certain that the *hui-hsing* in 5 BC moved beyond the constellation of Capricornus during the 70+ days of its apparition.[35] Relative celestial stability would be unusual for a broom star cometary apparition, but it is possible when a formerly dormant comet experiences a major outburst due to a fragmentation or splitting event a few weeks after perihelion, when a tail is capable of becoming visible from Earth and its movement within the starry sky might be constrained to one small region of the sky. Had such an outburst happened farther away, it probably would have been described as a *po*, the Chinese term for a tailless comet (a bushy star comet).[36] When Comet Holmes, located far from Earth and the Sun, had its magnificent 2007 outburst (fig. 5.17), it remained visible to the naked eye for more than 3 months.

Of course, if the 5 BC comet did not move from Capricornus and hence the morning sky, that would disqualify it from being the Bethlehem Star. After all, the Star of Bethlehem clearly migrated across a broad swath of the starry heavens from the eastern morning sky to the southern evening sky in the space of only a few months. A comet confined to Capricornus would not have been able to guide the Magi from Jerusalem to Bethlehem or to stand over a particular house in Bethlehem.

Even if we were to assume that the Chinese record is simply indicating that the comet began its apparition in Capricornus before moving on from there, we have no basis for thinking that it made its way to the southern evening sky, acted in such a way as to guide the Magi toward Bethlehem, or "stood over" one particular house. The Chinese record furnishes very little information about the comet in question. In light of the great ambiguity, we should be very cautious about jumping to the conclusion that this comet was the Bethlehem Star. The mere fact that we have a record of a comet in 6–5 BC does not constitute a firm foundation for identification.

This version of the comet hypothesis is regrettably unsatisfying, therefore. If the Star of Bethlehem was indeed a comet, we must look elsewhere.

A STRONGER CASE FOR THE STAR OF BETHLEHEM BEING A COMET

In spite of the weaknesses of the Halley's Comet and 5 BC Comet hypotheses, the evidence that the celestial phenomena that the Magi witnessed were caused by a comet is overwhelming. I shall now set out a new, stronger case for identifying the Star of Bethlehem as a great comet.

[31] Gary W. Kronk, in a personal email message to the author (October 23, 2011), wrote, "Although many Chinese records contain an incredible amount of detail as to a comet's motion, many can be found where only the location of the discovery is given, sometimes being a direction and sometimes a Chinese constellation. . . . The initial location was very important to these astrologers/astronomers."

[32] See, for example, Ho, "Ancient and Mediaeval Observations," catalog numbers 44, 49, 59, 61, 68, 73, 76, 81, 82, 83, 86.

[33] See, for example, ibid., number 65.

[34] See, for example, ibid., numbers 48, 55, 69.

[35] This has naturally caused some scholars to wonder if this really was a comet (see Clark, Parkinson, and Stephenson, "Astronomical Re-Appraisal," 443; Kokkinos, "Crucifixion in A.D. 36," 160n92).

[36] On the nomenclature, see Yeomans, *Comets*, 361.

FIG. 6.4 Comet Hale-Bopp as seen from Tierra del Sol in San Diego County in April, 1997. Image credit: Michael K. Fairbanks, DPM La Mesa, CA. Image source: Wikimedia Commons.

First, the simple fact that the Star was a bright light that suddenly appeared in the heavens can be explained only with reference to a meteor, nova, supernova, or comet.[37] When we also consider that the Star was visible for at least 1–2 years, the possible identifications are narrowed down to two—a supernova or a large, productive, long-period comet like Hale-Bopp. Hale-Bopp became visible to the naked eye 10½ months before perihelion and maintained its naked-eye visibility for a total of 18 months. If the Star was a very large comet as great as or even greater than Hale-Bopp (fig. 6.4), it would not be surprising if it was initially spotted many months before perihelion and remained visible for longer than one year. The comet hypothesis does not need to introduce other astronomical phenomena into the picture to explain Herod's decision to kill infants in their first and second years. It can simply accept Matthew's claim that the upper age threshold was set according to the time of the Star's first appearance that Herod had ascertained from the Magi (2:7, 16).

Second, no celestial entity at or around the time of its heliacal rising was more capable of impressing and surprising an ancient astronomer than that of a great comet at perihelion.[38] According to Matthew 2:2, the Magi asked the people of Jerusalem, "Where is he who has been born the king of the Jews? For *we saw his star at its rising* and have come to worship him."

The heliacal risings of stars and planets were certainly not particularly visually impressive, nor were they in any way unexpected. By the first century BC, Babylonian astronomers were quite proficient in calculating in advance when most celestial phenomena would occur, including the heliacal risings of stars and planets. This kind of information was included in their almanacs. Moreover, the Magi knew from experience what the heliacal rising of the different planets and stars looked like—the entity faintly appeared over the horizon at dawn, immediately before its light was extinguished by the rising Sun's light. Then, over the following days, it steadily appeared earlier and earlier and in a darker sky. It is very unlikely that the Magi would have been profoundly impacted by what they *saw* the Star do at the time of its heliacal rising if it was predictable or unexceptional.[39]

[37] Cf. Werner Keller, *The Bible as History*, rev. ed. (London: Hodder & Stoughton, 1980), 360.

[38] When the Babylonians and other ancients spoke of a comet's heliacal rising or setting, they had in mind first and foremost its coma or head.

[39] Even when a heliacal rising of a planet or star had astrological or societal importance, it was not the seeing of it that mattered, since such events were calculated in advance. The fact that the Magi explicitly state that they had traveled to Judea because they had *seen* what the Star had done at its heliacal rising suggests that it was observations rather than calculations that had had the

The Magi evidently observed the Star doing something extraordinary in connection with its heliacal rising many months after its first appearance. This makes excellent sense if the Star at its rising was a great comet that ventured very close to the Sun at perihelion. Such comets, depending on their orbit and the place of Earth on its orbit, may emerge from below the eastern (e.g., Halley's Comet in AD 66, and the Great Comets of 1147 and 1689) or western (e.g., the Great Comets of 1680, 1843, 1882, and 2007) horizon as they separate from the Sun in the aftermath of perihelion. When a comet does this, it is a heliacal rising.[40] Such risings may be very striking, because comets are at their most impressive around perihelion time—they are at their brightest, longest, largest, and fastest. Indeed no other entity's heliacal rising can compare in majesty to that of a great comet speeding away from the Sun immediately after its closest encounter with it. In the case of the Star, its early first appearance reveals that it was large and intrinsically very bright and hence had the capacity to put on an extraordinary display at its heliacal rising.

Moreover, the heliacal rising of a great comet would have been an unpredictable and surprising astronomical event. Ancient astronomers could not have confidently predicted what a comet would do in the heavens. Even aside from comets' maverick movements in the heavens, they were uniquely capable of springing surprises on human observers—for example, because of changes in the coma's size, brightness, or form, changes in the tail's length, shape, and orientation, and/or the sudden appearance of an antitail. No astronomical entity had a greater capacity to amaze ancient observers at its heliacal rising than a great comet. This renders a great comet by far the most natural celestial candidate for the role of the Star that rose in the east.

Third, the fact that the Star moved rapidly within the framework of the fixed stars and constellations is explicable only if it was a comet. Aside from meteors and the Moon, no celestial entity other than a comet in the inner solar system is capable of covering the celestial territory that the Star did over such a short time frame.[41]

The Magi saw the Star do something special in connection with its heliacal rising in the eastern sky. Some 30–40 days after this sign in the east was completed, they saw the Star appear in the southern sky, going before them in the direction of Bethlehem from Jerusalem. Comets, particularly those with a small perihelion distance, that are making their final approach to the Sun or receding from it may progress quickly through a significant portion of the sky. A comet with orbital elements within certain ranges could have heliacally risen in the eastern sky and, shortly thereafter, shifted to the western evening sky, and then migrated to the southern evening sky on schedule to usher the Magi to Bethlehem and point out the location of the house where baby Jesus was.

Prograde and retrograde comets with relatively narrow inclinations and with small perihelion distances may, around perihelion time, shift over a fairly short period of time from the western evening sky to the eastern morning sky and/or from the eastern morning sky to the western evening sky.

If the comet's perihelion is the other side of the Sun from Earth's perspective, a prograde comet with fairly low inclination would be on the east side of Earth (and hence in the eastern sky) shortly before perihelion and on the west side of Earth (and hence in the western sky)

decisive influence on them. Indeed it may well imply that they had been unable to determine reliably or precisely in advance of the Star's heliacal rising, based solely on mathematical calculations and/or past experience, what ended up happening. Accordingly, that the Magi saw the Star may reveal that they had been surprised by what it did at its rising or at least had been uncertain about what it would do.

[40] When a bright comet is positioned just above the Sun at dawn or sunset (usually with the tail upwards), it generally means that the comet is very near the Sun in outer space (Seargent, *Greatest Comets*, 23, 175).

[41] Cf. Steve Moyise, *Was the Birth of Jesus according to Scripture?* (Eugene, OR: Wipf & Stock, 2013), 51n6: "The swift movement across the sky of a comet comes closest to what Matthew describes."

after perihelion. If, however, that comet on its approach to the Sun passes through the Earth-Sun line, it would be on the west side of Earth before that point, then the east side, and finally, after perihelion, the west side again. If the comet cuts through the Earth-Sun line around perihelion time, the pattern would be west-east, and, if it does so after perihelion, then the pattern would be east-west-east.

On the other hand, if a retrograde comet narrowly inclined to the ecliptic passes perihelion on the other side of the Sun from Earth's perspective, it would appear in the west before perihelion and in the east after it. If such a retrograde comet cuts through the Earth-Sun line before perihelion, the pattern would be east-west-east. If it does so after perihelion, the pattern would be west-east-west.

The Star of Bethlehem's dramatic rapid movement within the framework of the fixed stars therefore strongly favors its identification as a comet.

Fourth, the Hebrew Bible (Old Testament) associated the Messiah's birth with a comet, most notably, as we have seen, in Numbers 24:17–19, where the seer Balaam prophetically declared, "I see him, but not now; I behold him, but not near: a star shall come out of Jacob, and a scepter [*shbt*] shall rise out of Israel; it shall crush the forehead of Moab and break down all the sons of Sheth. Edom shall be dispossessed; Seir also, his enemies, shall be dispossessed. Israel will do[42] valiantly. And one from Jacob shall exercise dominion and destroy the survivors of cities!" In light of the parallelism of "star" and "scepter," the identification of the astronomical entity heralding the birth of the Messiah as a comet is most natural. The Babylonian Talmud tractate *Berakhot* 58b strengthens this interpretation, for in it Rabbi Samuel refers to a comet (in Aramaic) as "a scepter [*shbyt*] star."[43] Origen correctly concluded that Balaam's oracle about the star in Numbers 24:17 was prophesying that a cometary apparition would mark the birth of the Messiah. That Balaam was speaking of a long-tailed comet scepter is accepted by a good number of modern scholars[44] and is represented in the *New English Bible* and the *Revised English Bible* ("a comet [will] arise from Israel" [NEB, REB]). Significantly, it is now widely accepted that when the Magi in Matthew 2:2 speak of having seen "his star at its rising," they are alluding to this ancient oracle of the Mesopotamian seer, implicitly claiming that what Balaam prophesied about the scepter-star has recently come to fulfillment.

Another important oracle that prophesied that the Messiah's nativity would be attended by a comet was Isaiah 9:2: "The people who walked in darkness have seen a great light; those who dwelt in a land of deep darkness, on them has light shone." That the Messiah's birth is in view is clear from verses 6–7. As we saw in the last chapter, the only plausible identification of the great celestial light that would shine in the deep darkness is that it is a great comet.

Fifth, if the star were a comet, this would shed light on the peculiar behavior of the

[42] I have rendered the Hebrew "will do" (rather than "is doing"), since the entire content of these verses is manifestly prophetic, speaking of the distant future.

[43] It is, however, interesting to observe that a number of scholars, including W. Staerk, *Die jüdische Gemeinde des Neuen Bundes in Damaskus* (Gütersloh: C. Bertelsmann, 1922), 28, 65; Berend Gemser, "Der Stern aus Jacob (Num. 24.17)," *Zeitschrift für die Alttestamentliche Wissenschaft* 43 (1925): 301–302; Jacob Milgrom, *Numbers*, The JPS Torah Commentary (New York: Jewish Publication Society, 1992), 207–208; and Thomas F. McDaniel, "Problems in the Balaam Tradition," http://tmcdaniel.palmerseminary.edu/Balaam.pdf (accessed August 1, 2014), maintain that *Berakhot* 58b's *shbyt* is related to the Akkadian cognate *šibṭu* and means "comet" or "meteor." If so, the use of the Hebrew word *shbt* in Num. 24:17 might be double entendre, with Balaam referring simultaneously to a scepter and a comet.

[44] For example, Staerk, *Die jüdische Gemeinde*, 28, 65; Gemser, "Der Stern aus Jacob," 301–302; I. Zolli, "Il significato di '*shēbheṭ*' nel Salmo CXXV," *Atti del XIX congress Internazionale degli Orientalisti* (Rome: G. Bardi, 1938), 459; Sigmund Mowinckel, *He That Cometh: The Messiah Concept in the Old Testament and Later Judaism* (New York: Abingdon, 1954), 12, 13, 68, 102; Hans-Jürgen Zobel, "šēḇeṭ," in *Theological Dictionary of the Old Testament*, vol. 14, ed. G. J. Botterweck, H. Ringgren, and H.-J. Fabry (Grand Rapids, MI: Eerdmans, 2004), 305; Joseph A. Fitzmyer, *The One Who Is to Come* (Grand Rapids: Eerdmans, 2007), 30n21; McDaniel, "Problems in the Balaam Tradition." Cf. Baruch A. Levine, *Numbers 21–36*, Anchor Yale Bible (New Haven, CT: Yale University Press, 2000), 190, 199–201 ("meteor").

Star at the climax of its apparition, coinciding with the final phase of the Magi's journey.

Notably, according to Matthew 2:9, the Star seemed to go ahead of the Magi, first to Bethlehem. Thereafter the Star went through a phase of "coming" (recall our discussion in chapter 3). Then, presumably within a half-dozen hours or so of their arrival in Bethlehem, the Star was stationed over one particular house, standing over it and thereby pinpointing it as the place where the messianic child and his mother were located. The Star's going on ahead of the Magi to Bethlehem, in the south, when it was at its culmination (highest point), would have made it seem that it was moving on a basically horizontal plane. The "coming" seems to indicate that the Star's movement thereafter was a downward one. The Star's standing over a particular house strongly implies that it was setting essentially upright behind the structure, from the Magi's perspective, looking like it was about to enter it.

A bright-tailed comet that appears at sunset reasonably high in the sky and far from the western horizon is better qualified to serve as the Magi's celestial guide from Jerusalem to the Messiah's house than any other astronomical entity. Some comets are so close to Earth that they can move at considerable speed through the sky over a matter of hours. But a comet does not need to be moving noticeably against the backdrop of the fixed stars and constellations to function as a guide. A comet's brightness and size make it stand out in the sky and draw attention to its movement through the dome of the sky as it follows its normal daily course through the heavens. A tailed comet moving toward the south or south-southwest, the same direction as the Magi were heading as they went from Jerusalem to Bethlehem, might readily have been perceived by them to be traveling in front of them. No star or planet moving in the same direction as a traveler can match the impression made by a bright long-tailed comet at a reasonable altitude.

Likewise, a long-tailed comet's descent as it moves from the meridian in the south to the western horizon is much more dramatic than that of any other celestial entity, because to observers the whole orientation of the comet is radically transformed as it descends.

Further, the setting of a tailed comet on the western horizon is uniquely qualified to be perceived to "stand over" a particular house.[45] For examples of comets "standing" over places, we need only turn to Roman historians Josephus and Cassius Dio. Using the same verb as Matthew for "stand," Josephus referred to "a star, resembling a sword, which stood over the city [of Jerusalem]."[46] In a

[45] Kokkinos, "Crucifixion in A.D. 36," 160; Humphreys, "Star of Bethlehem, a Comet in 5 B.C.," *Tyndale Bulletin*, 37–38.

[46] Josephus, *J.W.* 6.5.3 (§289). The sword-like comet has been identified as 1P/Halley in AD 66 by R. M. Jenkins, "The Star of Bethlehem and the Comet of AD 66," *Journal of the British Astronomical Association* 114 (2004): 336–343 (http://www.bristolastrosoc.org.uk/uploaded/BAAJournalJenkins.pdf). If Josephus is referring to two different comets, one like a sword "and" one that lasted for a year, it is possible that the sword-like comet is referring to Halley's Comet in AD 66. However, if the Flavian historian is referring to a single comet that at one stage of its year-long apparition looked like a sword ("a star . . . , even a comet . . ."), then he was certainly not thinking of Halley's Comet, since it was visible for less than 2½ months, from January 30 to April 11 (unless we charge Josephus with an unbecoming exaggeration regarding a detail the truth of which Vespasian and Titus and many readers would have known well). Unfortunately, there is some debate regarding the date(s) of Josephus's comet(s). Humphreys, "Star of Bethlehem, a Comet in 5 B.C." *Tyndale Bulletin*, 37–38, reckons that the sword-like comet Josephus mentions is the one mentioned in Tacitus, *Ann.* 15.47.1, as occurring in the year AD 64. However, it is probably best to see Josephus's comet(s) as occurring in AD 65–66, since he is writing about omens in the run-up to the outbreak of the Judean War in AD 66, and since the immediately succeeding context refers to an omen dated to the spring of AD 66.

According to the fourth-century AD Pseudo-Hegesippus (*On the Ruin of the City of Jerusalem*), in the run-up to the Judean War there was one comet that lasted a year and looked like a sword and indeed was so bright in early spring that it shone on the temple and altar for half an hour each night through Passover week. Josephus also mentions this strange light, but does not make explicit any link to the comet or give any indication that it occurred on successive nights. Pseudo-Hegesippus also claims that there was great division regarding how to interpret the comet—some regarding it as heralding freedom but others perceiving it to announce war (see Wade Blocker's translation, http://www.tertullian.org/fathers/hegesippus_05_book5.htm [last modified November 25, 2005]). Josephus mentions only his own interpretation—that it was an omen of judgment. Pseudo-Hegesippus states that the year-long cometary apparition occurred "before the people dissociated themselves from the Romans" (5.44). That would suggest that the comet occurred in AD 65–66, climaxing in the spring of AD 66, shortly after Halley's Comet. Whether Pseudo-Hegesippus is

FIG. 6.5 The Great Comet of 1881 (C/1881 K1 [Tebbutt]) as observed over the northern horizon on June 25/26, 1881: a chromolithograph in *The Trouvelot Astronomical Drawings* by Étienne Léopold Trouvelot, 1881. Image credit: University of Michigan Library/MLibrary Digital Collections. http://creativecommons.org/licenses/by/3.0/.

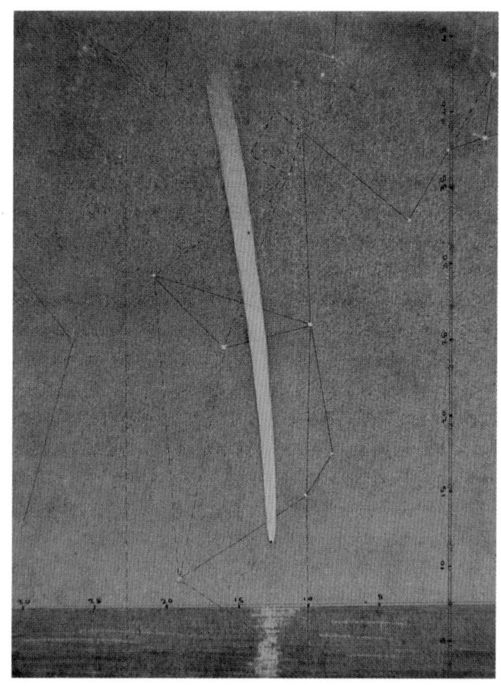

FIG. 6.6 The Great Comet of 1843, with its stunning tail, as it set over the western horizon. An observational drawing by William Clerihew, who viewed the comet from the southern waters of the Bay of Bengal. Image credit: Royal Astronomical Society Library, London (Ms. Add. 183/3, 186 x 135 mm).

similar vein, Cassius Dio wrote of a "comet" hanging (or raised up) over (*aiōrētheis huper*) the city of Rome for several days in 12 BC, which would seem to be essentially the same kind of phenomenon that Josephus was describing.[47]

As astronomer Chandra Wickramasinghe points out, only a tailed comet can appear to "stand over" a place, its long upward-streaking tail seeming to point downwards to one particular location on the earth below[48] (see figs. 6.5–7).

Similarly, New Testament scholar Craig Keener writes, "Without a tail extending as suggested in some modern artistic portrayals of the event, a celestial light could have

simply interpreting Josephus freely or is drawing on other sources we do not know. Regardless, it is possible that his understanding of the events is essentially correct.

Like Jenkins, Phipps, "Magi and Halley's Comet," 88, suggests that Matthew invented the Star of Bethlehem under the inspiration of the dramatic AD 66 apparition of Halley's Comet. However, it is surely methodologically sounder to consider whether a different comet, one at the time of Jesus's birth, inspired Matthew's account than to propose that it was one sometime around when Matthew wrote (before or after AD 70). Certainly, Halley's Comet in AD 66 did rise heliacally in the east, did make it to the southern sky by mid-March, and was able to set in the west from about March 24 through to April 11 (at which point even the trained Chinese astronomers could not see it any more). To that extent it bore some similarities to the Christ Comet as described by Matthew. However, by the time Halley's Comet eventually was able to set in the west in a dark sky at the close of the apparition, the comet was very faint (about fourth magnitude) and rapidly fading, and its tail was exceedingly short (1–2 degrees long, according to Project Pluto's Guide 9.0) and setting at too sharp an angle to be regarded as "standing over" a place. It is really difficult to believe that any non-astronomers like Matthew were still keeping track of the comet at this stage, when it was so faint. Further, the Star of Bethlehem probably went from the east to the south via the west, unlike Halley's Comet in AD 66. Therefore Matthew's description is much too different from the AD 66 Halley's apparition to have been inspired by it. Moreover, as we highlighted in chapters 2–3, Matthew was not an inventor of stories.

[47] Cassius Dio 54.29.8.

[48] As cited by Steffan Rhys, "Star of Bethlehem Comet Theory," http://www.walesonline.co.uk/news/wales-news/2008/12/22/star-of-bethlehem-comet-theory-91466-22528488 (accessed March 26, 2014). This counters the claim of Kenneth Boa and William

FIG. 6.7 A watercolor of the Great Comet of 1680 over Beverwijk on December 22, 1680, by Atlas van Stolk (Rotterdam). The tail is sticking upwards because the comet is immediately "above" the Sun, which has just disappeared below the horizon (the Sun is less than 10 degrees from the comet). Image credit: Atlas Van Stolk Collection, Rotterdam, www.collectie.atlasvanstolk.nl (inventory no. 11283).

FIG. 6.8 A wood engraving of the Great Southern Comet of 1880 "standing up" over the horizon, as viewed from Melbourne Observatory on February 16, 1880. From *The Illustrated Australian News*. The comet grew a tail as long as 75 degrees, although it was only about third magnitude at the time. Image credit: State Library of Victoria (accession no.: IAN16/02/80/17).

pointed them only in the most general way or by symbolic means."[49]

For a comet to develop a long tail in the run-up to perihelion and particularly in the period after it,[50] due to the increase of degassing, is, of course, normal.

In addition, it is common for comets to set coma-down and tail-up, because their tail always points away from the Sun. Depending on the location of the comet with respect to the Sun and Earth and on the season, it is perfectly possible for a comet's tail to stand up vertically or near-vertically.

A long-tailed comet might well appear to be pinpointing one particular structure, if that structure is located on the visible horizon and the observer is located on the directly opposite side of the structure to the comet (figs. 6.8–10).

In light of the Magi's prior history with the Star and the fact that it had seemed to have ushered them to Bethlehem, one can well understand why they would have interpreted the "standing" of this comet while they were in Bethlehem as revealing the precise location of the messianic baby.

Ironically, Boa claims that "A comet must be ruled out because it could not move on before the magi until it came and stood over where the Child was (Matt. 2:9)."[51] In fact, he could not be further from the truth, for the *only* celestial entity that could do this is a comet!

Proctor, *The Return of the Star of Bethlehem: Comet, Stellar Explosion, or Signal from Above?* (New York: Doubleday, 1980), 74: "The Gospel says that the Star stood over the place where the child was—and that would have been an extremely hard trick for any comet."

[49] Craig S. Keener, *The Gospel of Matthew: A Socio-Rhetorical Commentary* (Grand Rapids, MI: Eerdmans, 2009), 104.

[50] Because time is required for the dust particles to make their way to and along the tail (Littmann and Yeomans, *Comet Halley*, 58).

[51] Kenneth D. Boa, "The Star of Bethlehem" (ThM thesis, Dallas Theological Seminary, 1972), 66; cf. 35–36.

FIG. 6.9 A wood engraving by Martin Ebsworth of Comet Wells of 1882 (C/1882 F1 [Wells]), viewed from Australia on June 15, 1882. This image gives us a good idea how a setting comet can seem to highlight a specific point along the horizon. Image credit: State Library of Victoria, Australia, http://www.slv.vic.gov.au.

Sixth, comets by their movements within and among constellations are exceptionally well placed to convey significant and even complex messages to those equipped with the knowledge and paradigms to understand them. Simply put, these eccentric astronomical entities are capable of turning the fixed stars and constellations into a giant noticeboard. It is evident that what the Magi observed in the eastern sky back in their homeland was extraordinarily powerful and full of meaning. So hard-hitting was it that they embarked on a long journey, indeed a pilgrimage, to Judea. From the heavenly scene they witnessed, the Magi perceived an incredible amount of information: that someone had been born, that this person was a king, and that he was divine. Among astronomical phenomena, a cometary apparition is uniquely able to communicate such a complex set of ideas.

Seventh, the Star was clearly unique. The Magi evidently developed an extraordinary emotional bond with the Star, convinced that it was communicating especially to them, personally charging them to go to Judea and worship the newborn Messiah, and even guiding them there. They had obviously not seen the Star before, nor observed a celestial performance quite like it. The sheer uniqueness of the Star of Bethlehem's apparition is most naturally explained if it was a comet. Because comets are so diverse in size, intrinsic brightness, shape, color, chemical makeup, and behavior, and because they orbit the Sun at different speeds, angles, and distances from the solar disk and make their passes by the Sun at different times of the year, no two comets ever put on the same celestial show. Even individual periodic comets do not repeat the same display, since they are subject to gravitational and nongravitational forces that alter their orbit, and since their returns do not occur at the same precise time of the year as their previous visit. Therefore the unique nature of a cometary apparition fits well with the unique nature of the Star of Bethlehem.

Eighth, the fact that comets in the ancient world were often interpreted to augur regime change and regarded as threatening to the ruling establishment may help elucidate the negative responses of Herod the Great and the people of Jerusalem to the report of the Magi regarding the Star (Matt. 2:3).

Comets were often interpreted by the ruling elite as distinctly bad news, portending the death of a ruler.[52] The death of the Emperor Claudius was presaged by a comet.[53] As we have seen, when a comet appeared during Nero's reign, he was so fearful that it was an omen of his death that he consulted his astrologer Balbillus, who recommended that he execute prominent subjects to redirect the wrath of the gods.[54] Nero then embarked on a brutal massacre of the nobility that ironically ended up bringing about his own deposition.[55] Moreover, according to Cassius Dio,[56] the Emperor Vitellius's death was announced by a number of different celestial phenomena, including "a comet star." Cassius Dio also records that Vespasian's death in AD 79 was presaged by "the comet star which was seen for a considerable period" and which Vespasian tried to play down by insisting that the comet's long hair meant that it was an omen not for him, since he was bald, but rather for the long-haired Parthian king.[57] In the case of the Magi's Star, however, the threat to Herod was based not simply on the fact that a comet had appeared but on the fact that the comet clearly represented the Messiah and indicated that he had now been born.

FIG. 6.10 A lithograph by Charles Piazzi Smyth:* "The Great Comet of 1843 as seen at the Cape of Good Hope on March 4th in the evening (34 S Lat.)." Image credit: © The Royal Society (image number: RS.10050).

NOTE: *On Charles Piazzi Smyth's fascinating career, see Brian Warner, *Charles Piazzi Smyth, Astronomer-Artist: His Cape Years, 1835–1845* (Cape Town: A. A. Balkema, 1983); and Roberta J. M. Olson and Jay M. Pasachoff, *Fire in the Sky: Comets and Meteors, the Decisive Centuries, in British Art and Science* (Cambridge: Cambridge University Press, 1998), 201–209.

The negative response of the people of Jerusalem may possibly also have been partly due to a conviction that comets were often portents of disaster and partly due to their belief that this particular comet was announcing ultimate regime change—the end of Herod's dynasty and even the end of the Roman empire as the Messiah overthrew his enemies and established his kingdom on the earth.[58]

Ninth, we know that the Babylonians were very interested in comets. They made records of them in their astronomical diaries. In these records they were particularly concerned to note the time and details of a comet's first appearance and key moments of the apparition such as its heliacal setting and

[52] See especially the comment by Suetonius in *Nero* 36.1—a comet, or hairy star, is "a thing that is popularly perceived to portend the demise of great dignitaries." Cf. Tacitus, *Ann.* 14.22.

[53] Suetonius, *Claudius* 46.

[54] Suetonius, *Nero* 36.1.

[55] Nicholas Campion, *A History of Western Astrology: Volume 1, The Ancient World* (London and New York: Continuum, 2008), 239.

[56] Cassius Dio 65.8.

[57] Ibid., 66.17 (cf. Suetonius, *Vespasian* 23). Translation from *Dio's Rome, Volume 5*, trans. Herbert Baldwin Foster (New York: Pafraets, 1906), 141.

[58] The Hebrew and Aramaic Scriptures prophesied that the eschatological turn of the ages would see severe tribulation in the world and more particularly in Jerusalem (see, for instance, Psalms 2, 110; Daniel 2, 7, 11–12; Zechariah 14). The people of Jerusalem knew that neither Herod nor his sons, nor their Roman overlords, would quickly and quietly stand aside to let the Messiah take his place as King of kings.

rising. As Stephenson writes, "Even the existing descriptions [of comets by the Babylonians] are fragmentary, but they still enable us to infer the characteristic features of a typical Babylonian cometary account. Whether in the daily reports or the monthly summaries, records of comets were apparently entered only on the following occasions: (1) first sighting; (2) heliacal setting; (3) heliacal rising; (4) any stationary points; and (5) last visibility."[59] In light of this, it is noteworthy that in Matthew 2 the Magi are portrayed as having made a record of the first appearance of the Star and its heliacal rising.

Tenth, in the ancient Near East around the turn of the ages, a cometary apparition was sometimes interpreted as a sign of a great ruler's birth. Justinus (writing in the second century AD, but no doubt reflecting opinion from the second and first centuries BC) emphasized that the "future excellency" of Mithridates VI Eupator (134–63 BC), king of Pontus and Armenia Minor and formidable foe of Rome, was signaled in the heavens on the occasion of his birth in 135 BC by a brilliant, fiery, long-tailed comet that lasted for 70 successive days and at one stage took four hours to rise and set.[60] There is therefore precedent for the idea that a great comet could function as a celestial announcement of the birth of a great leader. Since the Star of Bethlehem was a heavenly sign marking the birth of the messianic King of the Jews, a comet is an ideal candidate for the role.

Eleventh, the proposal that the Star of Bethlehem was a comet has in its favor that it is the earliest explicit identification of the Star of which we know and is the only astronomical explanation of the Star that is suggested by Christian descriptions of the Star in the first three centuries AD. We have already considered evidence for the identification of the Star as a comet from Origen, and we have seen that Ignatius and the *Protevangelium of James* spoke of the Star in terms that seem to require that it was a comet.[61]

These arguments form a formidable case that the Star seen by the Magi in their eastern homeland and in Bethlehem in 6–5 BC was a great comet.

We can, however, go further than this.

A LONG-PERIOD, RETROGRADE COMET NARROWLY INCLINED TO THE ECLIPTIC

The most plausible candidate for the role of the Star of Bethlehem is assuredly a long-period comet. Moore observed that short-period comets can be safely discounted, because only rarely do they attain to naked-eye visibility, and even then they are faint and fuzzy.[62] His point is valid: some short-period comets, although dormant for long stretches of time, reactivate to such an extent that they cross the threshold of naked-eye visibility for brief seasons, sometimes just for single apparitions (e.g., Comet Helfenzrieder), but other times for up to a few hundred years, with the intensity of their brightness fading over this time (e.g., Comet Encke). But even these comets are generally difficult for the naked eye to make out.

[59] Stephenson, "Ancient History of Halley's Comet," 244.

[60] Justinus, *Epitome of the Philippic History of Pompeius Trogus* 37.2.1–3. See Justinus, *Epitome of the Philippic History of Pompeius Trogus*, trans. J. C. Yardley (New York: Oxford University Press, 1994), 234. In addition, R. A. Hazzard ("Theos Epiphanes: Crisis and Response," *Harvard Theological Review* 88 [1995]: 426–427; and *Imagination of a Monarchy: Studies in Ptolemaic Propaganda* [Toronto: University of Toronto Press, 2000], 185) has made the case that, when Ptolemy V was proclaimed Theos Epiphanes in 199/198 BC, it was claimed by the court that Zeus had announced the Ptolemaic ruler's greatness in advance by sending two comets, one at the time of his birth (210 BC) and one at the time of the boy-king's accession (204 BC). Hazzard maintains that images of stars/comets on coinage from that period reflect this.

[61] *Sib. Or.* 1.323–324 speaks of the Star as coming from the east, brightly shining even in the middle of the day. Similarly, *Sib. Or.* 12:30–33 describes it as a celestial body extremely like the Sun, that shines so brightly that it can be clearly seen at noon. These passages, from the second or third century AD, could only be referring to a large daytime comet in the same league of brightness as Ikeya-Seki of 1965 and the Great September Comet of 1882. The Gnostic writer Theodotus (second century AD) speaks of "a strange and new star" (*Excerpta ex Theodoto* 74).

[62] Patrick Moore, *The Star of Bethlehem* (Bath, England: Canopus, 2001), 67.

At the same time, we should bear in mind that Jupiter-family comets are capable of remarkable outbursts of activity that increase their level of brightness dramatically for a brief time, from a day up to a few months (e.g., Comet Holmes). Furthermore, a short-period comet is capable of becoming a striking object in the night sky if it makes a very close pass by Earth[63] (e.g., Comet Lexell in June/July of 1770). In addition, a number of astronomers believe that many short-period comets which in the modern era have a weak magnitude were probably brighter a few millennia ago.[64] It is sometimes suggested that a given comet may have been brighter in the past by 1 magnitude per millennium,[65] although this is not a reliable measure.

However, we do well to remember that virtually all the great historical comets, with the notable exception of Halley's Comet, have been long-period comets that had made relatively few passes into the inner solar system and hence were still loaded with volatiles.

It is long-period comets whose closest approaches to the Sun are within the orbit of Mercury (at its farthest, just under half the Earth-Sun distance) that are most likely to become bright, large, and long, and hence may put on a compelling celestial display.[66] The fact that the Star of Bethlehem was visible for so long reveals that it was intrinsically extremely bright and indeed very large (like Hale-Bopp in 1996–1997, and Sarabat in 1729–1730), meaning that it was a long-period comet. That the tail was very long is suggested not just by the implicit claim that Numbers 24:17's oracle of the cometary scepter was fulfilled (Matt. 2:2; Rev. 12:5; see also chapter 8 below), but also by the Star's going ahead of the Magi to Bethlehem and then standing over the house where Jesus was. Long tails are characteristic of long-period comets. Moreover, the awesome nature of the heliacal rising of the Star favors a productive comet with a close perihelion distance, which likewise strongly favors a long-period comet.

Not only was the Star evidently a large, intrinsically bright, long-period comet with a perihelion distance within Mercury's orbit, but it was also almost certainly narrowly inclined to the ecliptic plane on which Earth orbits the Sun. The fact that the comet, obviously long, stood vertically or near-vertically over the western horizon at the conclusion of the Magi's journey strongly favors this.[67] We are privileged to have a number of glorious images of historical comets standing up over the western horizon. In each case the comet was located in a zodiacal constellation and was setting in approximately the same location as the Sun—the tail pointing upwards because tails are always oriented away from the Sun. The fact that the Star at its climactic appearance was probably first seen in the south-southeast and then crossed the meridian (in the south) and finally set upright (Matt. 2:9, 11) seems to indicate that the scene occurred between October and December. At that time the angle of the ecliptic midway through the night, from the perspective of someone in the ancient Near East, was nearly vertical. Further, a comet that "stands" is most likely a straight-tailed comet. This, coupled with the fact that the Star was regarded as having fulfilled Numbers 24:17, which prophesied that a straight-tailed cometary scepter would signal the Messiah's coming, favors a narrowly inclined comet. When

[63] Fred Schaaf, *Comet of the Century* (New York: Springer, 1997), 39.

[64] F. L. Whipple and S. E. Hamid, "A Search for Encke's Comet in Ancient Chinese Records: A Progress Report," in *The Motion, Evolution of Orbits, and Origin of Comets*, ed. Gleb Aleksandrovich Chebotarev, E. I. Kazimirchak-Polonskaia, and B. G. Marsden (Dordrecht, Netherlands: Reidel, 1972), 152, claimed that the hypothesis that the brightness of a comet steadily decreases over time was supported by the history of observations, basic logic, and the icy nucleus model.

[65] Yeomans, *Comets*, 344–345; one exception is 109P/Swift-Tuttle, which seems to have maintained the same absolute magnitude for at least the last two millennia (Yau et al., "Past and Future Motion," 314).

[66] Schaaf, *Comet of the Century*, 40.

[67] That the Star's standing as it set is suggestive of a comet located on the ecliptic plane was first pointed out to me by Prof. Mark Bailey, Director of the Armagh Observatory, in April of 2011.

a comet orbits on or near the plane of Earth's orbit, the sharp curvature of the tail in outer space is not apparent to Earth-dwellers—the tail appears to be straight.

Moreover, as we saw above, the fact that what most deeply impressed the Magi was what the Star did around the time of its heliacal rising is most naturally explained if it was a great comet that ventured very close to the Sun. Comets making close passes by the Sun are generally located within the zodiacal constellations at the time. So it would seem that the comet was in the zodiacal band of sky not only when it seemed to stand over the house where Jesus was in Bethlehem, but also some weeks earlier, when it was at perihelion. This would seem to confirm that the comet was indeed narrowly inclined to the ecliptic.

At the same time, a comet in the zodiacal band would have been perceived to have greater astrological significance than one outside it. The primary task of astrologers at the time of the birth of Jesus was formulating horoscopes based upon a person's zodiacal birth sign, that is, the zodiacal sign that was heliacally rising at the point of birth. Considering that the Magi interpreted what the Star did in the eastern sky as disclosing the birth of a great leader, it would obviously make most sense if the Star was within a zodiacal constellation/sign.

Along these same lines, the discovery of the comet by the Magi the best part of a year, or more, before perihelion, when the comet would have been very dim, is slightly easier to explain if the comet first appeared in an area of the sky in which the astrologers focused their observations—namely, the zodiacal region. It is striking that the surviving Babylonian cometary records seem to reflect a heavy bias toward comets that appeared in the zodiacal band (they specifically mention comets in Scorpius [210 BC], Taurus and Sagittarius [164 BC], Libra [138 BC], and Aries [120 BC]).[68]

For these reasons, it seems very likely that the cometary Star of Bethlehem was narrowly inclined to the ecliptic, like the planets and most asteroids and Jupiter-family comets. Since we have already made the case for the comet having a long period, we can with some confidence identify it as one of a relatively small group of such comets that are narrowly inclined to the ecliptic.[69]

We can go further: the comet was probably also retrograde. This is because the only way a narrowly inclined comet can switch rapidly from the eastern to the western sky some weeks after its perihelion/heliacal rising is if it is at that point cutting through the Earth-Sun line. Only a retrograde comet can do this. That the Star proceeded to migrate to the southern evening sky confirms this.

Accordingly, the Star of Bethlehem was most likely a very large, intrinsically bright, narrowly inclined (and hence zodiacal), retrograde, long-period comet that appeared many months before its perihelion passage, rose heliacally in the eastern morning sky around the time of its perihelion, and then subsequently moved between the Sun and Earth, therefore switching to the western evening sky and then the southern evening sky.

POSITIVE INTERPRETATIONS OF COMETS IN THE ANCIENT WORLD

One of the most peculiar features of recent works on the Star of Bethlehem is their tendency to dismiss cavalierly the comet

[68] Moreover, we know that the 110 BC comet was first seen by the Chinese in the zodiacal constellation Gemini. In addition, orbital calculations of Halley's Comet's performance in 87 BC (built on the assumption the Chinese comet of 87 BC was Halley's Comet) suggest that during the month when the Babylonians first observed the comet it was beside Taurus's left horn, in Gemini, next to Cancer, and then in Leo and Virgo.

[69] A zodiacal comet would have set in the west to west-southwest, the direction of Judea from Babylon, in the middle of the night if the season was early summer or in the first half of the fall. Only in the fall, however, would the ecliptic have been near-vertical in the middle of the night.

hypothesis simply by lambasting the notion that the Star was Halley's Comet and/or by insisting that comets were always interpreted negatively in the ancient world. We have already dealt adequately with the former idea; because of how widely disseminated the latter claim is, we must give careful consideration to it now.

The charge that the comet hypothesis is implausible because it conflicts with the universally negative ancient interpretation of comets is made by many. Moore, for example, asserted that comets were always perceived to be unlucky and evil and therefore no magus would have interpreted one to be an omen of a royal birth.[70] Molnar likewise rejects the comet hypothesis on the basis that "The evidence . . . is strong that people of Roman times feared rather than welcomed comets. Long-haired stars were thought to be harbingers of disaster, usually the death of a king or an emperor."[71] Similarly, P. A. H. Seymour dismisses the idea that the Star of Bethlehem was a comet on the ground that comets were always omens of death and disaster and never of happy things.[72] Sumners and Allen echo this indictment of the comet view: comets were "considered harbingers of evil. Their appearances foretold war, pestilence, or the death of a ruler, not the birth of a king."[73]

Regardless of how much this charge is repeated, it is over-simplistic and indeed profoundly misleading.

First, it stands to reason that, if a comet's apparition was perceived to be bad news by and for the ruling elite, it would naturally have been interpreted as good news by and for those who were eager to see regime change. As Humphreys puts it, "although a comet was regarded as a bad omen for the king who was about to die or for the side that was going to lose a war, equally a comet was regarded as a good omen heralding a new king or a major victory for those on the winning side."[74] Tacitus wrote concerning AD 60 that "In the meantime a comet star blazed out. The opinion of the masses is that [a comet] portends revolution for kingdoms. Therefore, as if Nero had already been dethroned, [the masses] began to ask who should be chosen [as his replacement]. On the lips of everyone was the name Rubellius Plautus, who inherited his nobility from the Julian family through his mother" (Tacitus, *Ann.* 14.22).[75] This demonstrates that a comet's announcement of a change of regime was not necessarily perceived by all to be bad news. As we have already highlighted, the Star seen by the Magi in the east was interpreted negatively by Herod but positively by the Magi.

Second, there is some evidence that in the Greco-Roman world comets were sometimes regarded as being omens of good. Origen, *Contra Celsum* 1.59, refers to a Stoic called Chaeremon (fl. AD 30–65), who in a work on comets insisted that comets sometimes appeared "when good things were to occur" and proceeded to substantiate his claim with examples.[76]

It is worth giving a few Greco-Roman examples of comets being perceived to be good omens.

Diodorus Siculus (16.66.3) and Plutarch (*Timoleon* 8) refer to a bright light that at-

[70] Moore, *Star of Bethlehem*, 70.
[71] Michael R. Molnar, *The Star of Bethlehem: The Legacy of the Magi* (New Brunswick, NJ: Rutgers University Press, 1999), 17–18.
[72] Seymour, *Birth of Christ: Exploding the Myth*, 102–103.
[73] Carolyn Sumners and Carlton Allen, *Cosmic Pinball: The Science of Comets, Meteors, and Asteroids* (New York: McGraw-Hill, 2000), 31–32. Cf. Brown, *Birth of the Messiah*, 172; Boa, "Star of Bethlehem," 65.
[74] Humphreys, "Star of Bethlehem, a Comet in 5 B.C.," *Tyndale Bulletin*, 38–39.
[75] My translation. For an excellent modern translation of Tacitus's *Annals*, see Tacitus, *The Annals*, trans. A. J. Woodman (Indianapolis: Hackett, 2004).
[76] My translation. Hans-Rudolf Schwyzer, *Chairemon* (Leipzig: G. Harrassowitz, 1932), 62–63, suggests that Chaeremon's claim was rooted in a prior source. See also *Chaeremon: Egyptian Priest and Stoic Philosopher. The Fragments Collected and Translated with Explanatory Notes*, ed. Pieter Willem van der Horst (Leiden: Brill, 1984), 12–13.

tended Timoleon as he traveled to Sicily. If this was a comet,[77] it was a favorable one.

According to Justinus,[78] the birth and accession of Mithridates VI Eupator of Pontus, in 135 or 134 BC and 120 or 119 BC respectively, were announced by cometary apparitions: "The future excellency of this man was foretold by celestial signs. For both in the year in which he was born and in the one in which he started to reign, on each occasion, a comet star blazed for 70 days in such a way that the entire sky appeared to be on fire."[79] Interestingly, the Chinese[80] record comets in the years 135, 134 (possibly), 120, and 119 BC, and it is now widely accepted that Justinus's claims concerning comets at the time of Mithridates's birth and coronation are historically reliable.[81] Ramsey has argued convincingly that the 135 BC comet is the one that coincided with Mithridates's birth and that the 119 BC comet occurred at the time of his coronation.[82] In Justinus's account therefore we have excellent historical evidence of two cometary apparitions that were interpreted positively.[83]

Pliny the Elder, *Natural History* 2.23, records that in 44 BC, just under four decades before the Bethlehem Star, most probably late in July, a dramatic daytime comet occurred in the heavens in the time shortly after the death of Julius Caesar. Octavian, later called Caesar Augustus, regarded this comet as "auspicious." In his *Vita* he wrote,

> During the very time of these games of mine, a hairy star [comet] was seen during seven days, in the part of the heavens which is under the Great Bear. It rose in about the eleventh hour of the day, was very bright, and was conspicuous in all parts of the earth. The common people supposed the star to indicate that the soul of Caesar was admitted among the immortal gods; under which designation it was that the star was placed on the bust which was lately consecrated in the forum.[84]

Pliny then comments that "This is what [Octavian] proclaimed in public, but, in secret, he rejoiced at this auspicious omen, interpreting it as produced for himself; and, to confess the truth, it really proved a salutary omen for the world at large."[85]

Cassius Dio 45.7 states that "the majority . . . ascribed [the comet star] to Caesar, interpreting it to mean that he had become a god and had been included in the number of the stars" and that Octavian "took courage and set up in the temple of Venus a bronze statue of him with a star above his head."[86] Octavian's initial nervousness about the comet

[77] So Wilhelm Gundel, "Kometen," in *Paulys Realencyclopädie der Classischen Altertumswissenschaft* 11.1 (Stuttgart: Druckmüller, 1921), 1148–1149; A. A. Barrett, "Observations of Comets in Greek and Roman Sources before A.D. 410," *Journal of the Royal Astronomical Society of Canada* 72 (1978): 87–88.

[78] Justinus, *Epitome of the Philippic History of Pompeius Trogus* 37.2.1–3.

[79] My translation. See A. A. Barrett, "The Star of Bethlehem: A Postscript," *Journal of the Royal Astronomical Society of Canada* 78 (1984): L23.

[80] See Kronk, *Cometography* 1:15–16, on the comets of 120 BC and 119 BC. Kronk correlates Mithridates's coronation comet with the 120 BC comet. In his favor is the fact that the Babylonians cover the 120 BC comet's progress over at least 56 days, which is not far from the 70 days mentioned by Justinus. However, it is probably best to date the crowning of Mithridates to the spring of 119 BC (John T. Ramsey, "Mithridates, the Banner of Ch'ih-Yu, and the Comet Coin," *Harvard Studies in Classical Philology* 99 [1999]: 197–253; Adrienne Mayer, *The Poison King: The Life and Legend of Mithridates, Rome's Deadliest Enemy* [Princeton, NJ: Princeton University Press, 2010], 29) and hence to associate the 119 BC Chinese comet with the beginning of Mithridates's reign.

[81] Ramsey, "Mithridates," 198–199. Ramsey also points out that a coin from Mithridates's reign portraying a comet authenticates Justinus's claim with respect to one or both comets (213–253).

[82] Ibid., 202–253. J. K. Fotheringham, "The New Star of Hipparchus and the Date of the Birth and Accession of Mithridates," *Monthly Notices of the Royal Astronomical Society* 89 (1919): 162, regarded the first comet as appearing in 134 BC (cf. Yeomans, *Comets*, 365) and the second in 120 BC.

[83] Strikingly, Barrett, "Star of Bethlehem: A Postscript," L23, appeals to Justinus's quote in defense of his view that the Star of Bethlehem was a comet that marked a royal birth.

[84] Octavius's *Commentarii de vita sua* as cited by Pliny the Elder, *Natural History* 2.23. See *The Natural History of Pliny*, vol. 1, trans. John Bostock and Henry Thomas Riley (London: Henry G. Bohn, 1893), 58.

[85] Ibid.

[86] Cassius Dio 45.7. Translation from *Dio's Rome, Volume 3*, trans. Herbert Baldwin Foster (New York: Pafraets, 1906), 8–9.

makes plenty of sense—if the comet had been interpreted as a negative omen regarding his reign, it could have spelled serious trouble for him. However, the sources are clear that Octavian was delighted publicly and privately at the prodigious timing of the comet.

FIG. 6.11 The front and back of a silver denarius from about 19/18 BC displaying Caesar's Comet of 44 BC. The comet's tail is portrayed as oriented upwards. The other seven rays suggest that it was extremely bright. Image credit: www.forumancientcoins.com.

What Pliny and Cassius Dio give us here is a clear instance of a comet, indeed one of the most famous comets in history, having a happy association. Molnar, however, in his rejection of the idea that the Bethlehem Star was a comet, offers a different spin on what transpired: "Augustus Caesar, Julius Caesar's adopted son and political heir, knew that people would speculate that the new comet foretold his own death. . . . However, Augustus stemmed any thoughts about his demise by proclaiming that the comet was the wandering soul of Julius Caesar. Augustus proved to be one of history's greatest propagandists and spin-control artists: he commissioned coins and statues honoring the comet."[87]

However, this interpretation of the data by Molnar is overly cynical and contrary to the earliest and most reliable sources. That Octavian exploited the popular interpretation of the comet and used and developed it in his own propaganda is scarcely to be doubted (see fig. 6.11), but that is a long way from his inventing it. The evidence, such as we have, strongly suggests that this comet was almost universally embraced as a positive sign.

We conclude, then, that the great comet that appeared in the aftermath of Julius Caesar's death, less than four decades before the birth of Jesus, was perceived to be a resoundingly positive omen by most of the people of Rome and by Octavian himself.[88]

Moreover, Seneca, in his *Natural History* 7.17.2, speaks of a comet that appeared when Nero was emperor and "redeemed comets from their bad reputation,"[89] commenting that it was not the same one as had appeared in the aftermath of Julius Caesar's death in 44 BC. Seneca's statement reveals that there was a cometary apparition during the reign of Nero that, like the one in 44 BC, was broadly interpreted to be a decidedly positive omen, so manifestly good indeed that it seemed to transform the reputation of comets.

In addition, Pseudo-Hegesippus 5.44.1 claimed that a comet near the start of the First Jewish War was interpreted by Jews as a sign that they would succeed in their bid for freedom from the Romans.

Comets therefore could be interpreted positively in the Greco-Roman world.[90] The

[87] Molnar, *Star of Bethlehem*, 18.
[88] The 44 BC comet has been the focus of extensive debate and study. According to the ancient sources (for a useful collection of them, see John T. Ramsey and A. Lewis Licht, *The Comet of 44 B.C. and Caesar's Funeral Games* [Oxford: Oxford University Press, 1997], 155–177), while a minority may have interpreted the comet as a negative omen (Cassius Dio 45.6.6–7: "Some called it a comet and said that it signified the usual things [that comets signify]" [my translation]—this need not be interpreted to mean that the comet was regarded as a negative omen for Octavian in particular), the vast majority of people interpreted the appearance of the comet as marking the divinization of Julius Caesar. Octavian, while giving lip service to this popular notion (Pliny the Elder, *Natural History* 2.23; Servius on Virgil, *Aeneid* 8.681; and Servius on Virgil, *Eclogue* 9.47), privately regarded the comet as a positive omen for himself (Pliny the Elder, *Natural History* 2.23) and perhaps even as marking the inauguration of a Golden Age (Servius on Virgil, *Eclogue* 9.47). On the Golden Age, see Mary Francis Williams, "The *Sidus Iulium*, the Divinity of Men, and the Golden Age in Virgil's *Aeneid*," *Leeds International Classical Studies* 2 (2003): 1–29, esp. 20–25, contra Ramsey and Licht, *Comet of 44 B.C.*, 140–142, and 165n14, who claim that Octavian's propaganda twisted the truth, covering up intensely negative assessments of the cometary apparition. Pliny speaks for the consensus when he asserts that the comet was, "if we confess the truth," "a positive omen for the world" (*Natural History* 2.23, my translation).
[89] My translation.
[90] Ramsey and Licht, *Comet of 44 B.C.*, 135–136n5, also point to the "torch" of 204 BC, which may have been cometary and was interpreted by the Romans as heralding their victory over the Carthaginians in the Second Punic War (Livy 29.14.3).

widespread claim that they were always perceived to be negative omens must be laid to rest once and for all. Mercifully, some who advocate alternative hypotheses concerning the Star of Bethlehem, like Bulmer-Thomas, are willing to concede that, in view of the various examples of comets announcing good news in ancient literature, the argument that the Star of Bethlehem could not have been a comet because comets always had a negative significance has little force.[91]

In addition, it is important to remember that the Biblical tradition regarding comets has a markedly positive dimension. As we have already seen, the Hebrew Scriptures portray the birth of the Messiah in terms of a comet in Numbers 24:17 and Isaiah 9:2. The fact that the Magi interpret the cometary sign as that of "the King of the Jews" and embark on a pilgrimage to worship him (Matt. 2:2) indicates that they interpreted what they saw in the heavens with reference to the messianic tradition reflected in the Hebrew Bible. This direct or indirect exposure to Old Testament tradition concerning the Messiah evidently played a decisive role in their positive assessment of the cometary Bethlehem Star.

This naturally raises the question of why the Magi, trained in astrology, felt the need to turn to Jewish tradition to explain what they observed. Clearly there was something unusual and remarkable about the sight that demanded to be understood with reference to Jewish messianic tradition. But what was it? What did they see? What did the comet do in the heavens that convinced these astrologers that the Jewish Messiah had been born at that very moment?

COUNTERING TWO FINAL OBJECTIONS

Before we turn to consider those questions in the next chapter, we must respond to two final objections to the comet hypothesis.

"NO EXTANT RECORD OF A MATCHING COMET"

The first objection was raised by R. T. France: "Comets have long been held to herald the arrival of important figures on the world stage, and a comet visible in the western sky might well explain the journey of the magi, but unfortunately astronomers have not been able to identify a comet which would have been visible at about the right historical date."[92] Simo Parpola also argues against the comet view by stating that, aside from Halley's Comet in 12 BC, "No other suitable observations of comets are known from this period."[93]

However, as we have already shown, we lack anything like a comprehensive set of comet records from the period. Accepting that per century on average there are something like 87 comets, it is clear that the memory of over two-thirds of the comets from the period 50 BC to AD 50 has been lost to history. Moreover, the comets that we do know about were not recorded because they were astronomically impressive but because they occurred at auspicious moments and/or served to advance the agenda of a variety of ancient writers/historians. Therefore we have every reason to believe that many great cometary apparitions have long been forgotten.

In the case of the Greco-Roman comet data, we are dependent on a selection of scattered references to comets drawn from different writings.[94]

[91] Ivor Bulmer-Thomas, "The Star of Bethlehem—A New Explanation—Stationary Point of a Planet," *Quarterly Journal of the Royal Astronomical Society* 33 (1992): 368.
[92] R. T. France, *The Gospel of Matthew*, New International Commentary on the New Testament (Grand Rapids, MI: Eerdmans, 2007), 68. Cf. Henry C. King, *The Christmas Star* (Toronto: Royal Ontario Museum, 1970), 4.
[93] Parpola, "Magi and the Star," 15. Also Moyise, *Was the Birth of Jesus according to Scripture?*, 51n6, who rejects the comet hypothesis on the grounds that "the seventy-seven year orbit [*sic*] of Halley's comet would have been too early (c. 12–11 B.C.E.) and we know of no other comet that fits the bill."
[94] It should be borne in mind that, as Timothy D. Barnes, "The Triumphs of Augustus," *Journal of Roman Studies* 64 (1974): 21–26, and Ronald Syme, *The Crisis of 2 B.C.* (Munich: Verlag der Bayerischen Akademie der Wissenschaften, 1974), 3–34, point out, the decade beginning in 6 BC is one of the most obscure in the entire history of the Roman Empire, due to gaps in our historical sources.

As for the Far Eastern records, although some writers assume that the surviving Chinese records from the Former Han period are so comprehensive that a cometary Star of Bethlehem would certainly have been among them,[95] this is simply not true (see appendix 1). It is widely accepted by scholars in the field that most records were lost prior to the composition of the *Han shu*, and that the historians were very selective in those that they elected to integrate into their narrative, making their choices based on astrological and ideological considerations.

It should also be appreciated that Matthew is claiming to preserve observations of the Christ Star by astronomers who in all probability come from Babylon—specifically concerning the Star's first appearance and heliacal rising. Effectively, then, in Matthew 2 we have indirect access to *Babylonian* records of the Christ Comet. Matthew probably implies that the Magi had a written record regarding the comet's apparition from which they were able to draw in order to inform Herod regarding the date of the first appearance of the comet. Moreover, the reaction of Herod and the people of Jerusalem to the Magi's enquiry strongly implies that they themselves had been awed by the sight of the comet but had not been interpreting it messianically. Therefore there is good reason to believe that both the Babylonians and the Judeans had observed the comet.

In conclusion, the objection to the comet hypothesis based on its absence from the extant astronomical records should be firmly rejected. If we had comprehensive astronomical records, France and Parpola might have had a point. However, the records that we have are patchy at best.

"COMETS WERE NOT CONSIDERED 'STARS'"

The second of the two final objections to the comet hypothesis is that a comet is not, strictly speaking, a star. For example, Judith Weingarten insists that Matthew's choice of "star" (*astēr*) could not refer to a comet because the ancients knew their comets from their stars.[96] Likewise Raymond Brown presents as the first argument against the comet view that "a comet is not a star."[97] Then, later, he writes, "Matt[hew] says that the magi saw the star (not planets, not a comet) of the King of the Jews at its rising (or in the East)."[98]

However, to call comets "stars" is hardly a matter of confusing comets and stars! The Mesopotamians as far back as the first half of the first millennium BC could refer to comets as "stars."[99] In the rest of the ancient world as well, comets were often called "stars," as the following examples demonstrate.

Caesar Augustus, as cited by Pliny the Elder,[100] refers to the comet of 44 BC as "a hairy star."[101] Pliny himself regularly calls comets "stars."[102] For example, concerning comets, he asserts that "Some persons suppose that these stars are permanent."[103] And

[95] E.g., Frank Ramirez, *The Christmas Star* (Lima, OH: CSS, 2002), 34; "Bethlehem's Star," http://www.unmuseum.org/bstar.htm (accessed May 15, 2014).
[96] Judith Weingarten, "The Magi and Christmas," http://judithweingarten.blogspot.com/2007/12/magi-and-christmas.html (posted December 22, 2007).
[97] Brown, *Birth of the Messiah*, 172; the identical words are used in Marten Stol, *Birth in Babylonia and the Bible: Its Mediterranean Setting* (Groningen: Styx, 2000), 99–100. Brown, *Birth of the Messiah*, 172, does concede that Origen's statement concerning comets indicates that there was a common conception of comets as bearded stars and that this could "solve this difficulty." Unfortunately, his subsequent repetition of the argument on p. 612 suggests that he does not take on board the force of this concession.
[98] Brown, *Birth of the Messiah*, 612. Cf. Susan S. Carroll, "The Star of Bethlehem: An Astronomical and Historical Perspective," http://www.tccsa.tc/articles/star_susan_carroll.pdf (last modified February 22, 2010).
[99] See Hermann Hunger, F. Richard Stephenson, C. B. F. Walker, and K. K. C. Yau, *Halley's Comet in History* (London: British Museum, 1985), 17.
[100] Pliny the Elder, *Natural History* 2.23.
[101] Williams, "*Sidus Iulium*," 5, points out that Augustan literature tends to avoid using the word "comet," presumably because comets were generally bad omens. It prefers the word "*sidus*" ("star") for the great comet of 44 BC.
[102] Pliny the Elder, *Natural History* 2.22–23.
[103] *Natural History of Pliny*, vol. 1, 58.

again, "stars are suddenly formed in the heavens themselves; of these there are various kinds. The Greeks name these stars 'comets'; we name them Crinitiae [long-haired stars], as if shaggy with bloody locks, and surrounded with bristles like hair."[104] In a passage we have already had cause to cite, Pliny states concerning the comet of 44 BC that "The common people supposed the star to indicate that the soul of Caesar was admitted among the immortal gods."[105] In that same context Pliny also writes that a comet was generally regarded as a "terrifying star."[106]

Seneca quotes Artemidorus as saying that comets are "new stars."[107] Seneca himself makes reference to cometary apparitions as "the appearance of such stars."[108]

Cassius Dio calls them "comet stars."[109] He also speaks of the comet of 44 BC as "the star called comet."[110]

Josephus mentions "a star, resembling a sword, which stood over the city, and/namely a comet that continued for an entire year."[111]

Similarly, Origen clearly regards "comets" as a subset of "stars" and assumes that this was the common understanding.[112] He refers to different types of comets as "such stars." He also interprets Balaam's "star" in Numbers 24:17 as a comet. On that note, Numbers 24:17 itself is strong evidence that, within the Biblical tradition, a comet ("scepter") could be reckoned a "star."

It is clear, then, that the ancients frequently referred to comets as "stars." This is true even where there was a clear appreciation that comets were fundamentally different from fixed stars.[113]

CONCLUSION

In this chapter I have made a case for regarding the Star seen by the Magi in the east and then in Bethlehem as a narrowly inclined, retrograde, long-period comet that, around the time of its close perihelion, rose heliacally and thereafter crossed the Sun-Earth line to be on the western and eventually the southern side of Earth. Not only are the alternative theories fundamentally flawed (as we showed in chapter 4), but the comet hypothesis fits perfectly with the entire narrative of Matthew. The so-called Bethlehem Star was undoubtedly a "comet star."

Having concluded that the Star seen by the Magi was a comet, we shall in the following chapter turn to the task of identifying the celestial scene that greeted the Magi's eyes in the eastern sky and caused them to make their way to Judea. How did the comet reveal so much information to the Magi, namely, that someone had just been born, that he was divine in nature and hence worthy of worship, and that this person was the King of the Jews?

[104] Ibid., 55.
[105] Ibid., 58.
[106] Pliny the Elder, *Natural History* 2.23 (my translation).
[107] Seneca, *Natural Questions* 7.13.2 (my translation).
[108] Ibid., 7.14.4 (my translation); cf. 7.26.2.
[109] E.g., Cassius Dio 56.29; 60.35; 65.8; 66.17.
[110] Ibid., 54.29; translation from *Dio's Rome, Volume 4*, trans. Herbert Baldwin Foster (New York: Pafraets, 1905), 56.
[111] Josephus, *J.W.* 6.5.3 (§289) (my translation).
[112] Origen, *Contra Celsum* 1.58–59.
[113] It is also worth noting that sometimes, when comets first appear, only the bright central condensation of the coma is visible, making the comet look very like a star or nova. In addition, in cases where a comet becomes visible when it is far away from the Sun and Earth, its movement within the starry sky may not be detectable by the naked eye for days or weeks, as in the case of Sarabat's Comet (see Kronk, *Cometography*, 1:394). Therefore it is very possible that the Magi initially categorized the Christ Comet as a new star or nova.

7

"Yon Virgin Mother and Child"

The Celestial Wonder

INTRODUCTION

In the previous chapter I argued that the Star of Bethlehem could only have been a comet. Indeed I suggested that only an intrinsically bright, very large, narrowly inclined, retrograde, long-period comet could have done what Matthew recounts concerning the Bethlehem Star.

However, one key question remains: what in particular did the comet do to convince the Magi that, if they embarked on a long journey westward to Judea, they would be able to find a newborn king? We know from Matthew 2:2 that the part of the cometary apparition that played the decisive role in prompting the Magi to go to Judea related to the comet's heliacal rising. Needless to say, however, the mere presence of a comet appearing low in the eastern predawn sky would hardly have seemed of earthshaking significance to seasoned stargazers. Something extraordinary must have happened in connection with the comet's heliacal rising, something perceived by the Magi to be utterly astonishing and to communicate definitively that the divine Messiah was born at that time.[1]

We have already highlighted that no astronomical entity's heliacal rising is more surprising and dramatic than that of a bright comet that has just made a very close pass by the Sun and is reemerging over the eastern horizon in advance of the Sun, or appearing on the western horizon, in the wake of the Sun's setting. At that time (around perihelion) comets are at their most active, at the peak of their brightness, and at their most unpredictable. They may become visible during the daytime, and, in rare cases, shine with a brilliance surpassing that of the full Moon. Coma and tail growth go into overdrive in outer space and often within the dome of the sky.

In the case of the Christ Comet, there was unquestionably something remarkable about the brightness, size, shape, location, and/or movement of the comet at the time when it heliacally rose. But what? How did it spell out for the Magi that an important royal personage had been born, and persuade them to

[1] Sometimes, of course, the timing of a cometary appearance could be significant in determining its meaning. The great comet of 44 BC was interpreted as an "auspicious" omen, based on the fact that it had appeared during the Games of Venus Genetrix, at the start of Octavian's reign, in the aftermath of Julius Caesar's death (Pliny the Elder, *Natural History* 2.23). In this historical context people interpreted the comet to be the divinized soul of Caesar, and Octavian thought that it was a positive omen for himself.

undertake their 550-mile pilgrimage to pay homage to him?

This is an interesting question, the more so because the Christ Comet was no run-of-the-mill comet. The Magi's early and long observation of the Star indicates that the object was exceptional in intrinsic brightness and size, since only bright and large comets are capable of becoming and remaining visible to the naked eye when they are so far away. Of all the comets that have been detected in the modern era, the only ones that would compare to the Christ Comet in these respects are Sarabat's Comet of 1729,[2] Hale-Bopp,[3] and perhaps the Great Comet of 1811.[4] However, these comets did not venture close to the Sun or to Earth. Therefore, if the Christ Comet did indeed make a close pass by the Sun (within Mercury's orbit) and also by Earth, it puts us in unfamiliar, indeed to some extent theoretical, territory. Thankfully, however, we are not completely in the dark. Based on what we know about the behavior of Hale-Bopp and that of smaller comets that have made close approaches to the Sun and Earth, we have a good idea as to what could have happened around the time of the Christ Comet's perihelion. Almost certainly, it would have become as bright as the full Moon, its coma (head) would have become greatly enlarged, and its tail would have grown very long. But how bright, how large, and how long? We shall seek to shed some important light on these issues in this and subsequent chapters.

One important clue as to the form of the Christ Comet can be detected in what the Magi say to the people of Jerusalem, as reported in Matthew 2:2: "we saw his star at its rising." We have already observed that the Magi here are probably alluding to Balaam's oracle concerning the cometary scepter-star that would "rise" (Num. 24:17), implicitly claiming that they have witnessed a literal fulfillment of this prophecy. The most natural conclusion to draw from this is that around the time of its heliacal rising the comet as a whole looked like a scepter, that is, a long straight rod (see fig. 7.1). This would have been a magnificent and memorable phenomenon. In recent centuries only a select number of comets (most notably, the great comets of 1680 and 1843) have around the time of their heliacal rising had a length and general shape that would permit them to pass for a scepter.

As to the question of the comet's location as it heliacally rose, in the preceding chapter we proposed that the Christ Comet was narrowly inclined to the ecliptic plane and probably remained within the zodiacal band throughout its apparition. That obviously raises the tantalizing question: in which zodiacal constellation or sign did the comet heliacally rise? In light of that question, it is helpful to list the zodiacal constellations:

Aries (the ram)
Taurus (the bull)
Gemini (the twins)
Cancer (the crab)
Leo (the lion)
Virgo (the virgin)
Libra (the scales)
Scorpius (the scorpion)
Ophiuchus (the serpent-bearer)
Sagittarius (the archer)
Capricornus (the goat)
Aquarius (the water-bearer)
Pisces (the fishes)

The names of the zodiacal *signs* (the 30-degree segments into which the zodiacal band was divided by the Babylonians around

[2] Sarabat's Comet had an absolute magnitude of between -3 and -6 and a nucleus with a diameter of over 100 km, and it attained to naked-eye visibility when it was still over four times farther from the Sun than Earth is.
[3] Hale-Bopp had an absolute magnitude of -1 and a nucleus with a diameter of 40–70 km, and it attained to naked-eye visibility 10½ months before perihelion.
[4] The Great Comet of 1811 had an absolute magnitude of 0 and a nucleus with a diameter of 30–40 km, and it attained to naked-eye visibility 5½ months before perihelion.

the middle of the first millennium BC) are identical except for the exclusion of Ophiuchus. So in which constellation or sign was the comet when it rose over the eastern horizon? The answer to this question is very important, for there can be little doubt that at least part of the reason that the Christ Comet's behavior at the time of its heliacal rising was regarded as meaningful and significant was the celestial context in which it occurred.[5]

FIG. 7.1 Merodach-Baladan II, king of Babylon, and Sargon II, king of Assyria. Both kings are holding their royal scepters (the long, straight rods in their hands). Image credits: Wikimedia Commons (left); Wikimedia Commons/Jastrow (right).

In this chapter I propose that the story of what the Star did in connection with its heliacal rising is actually recorded in some detail in the New Testament, specifically in Revelation 12:1–5. That this text is the key to unlocking the mystery of what the Star did to mark the Messiah's birth has gone unappreciated for long enough.

Before we turn to Revelation 12, it is important to recognize that we are well positioned to evaluate the credentials of any specific proposals regarding what the Christ Comet did, because we have extrapolated from Matthew significant data about the comet—specifically concerning its profile (large, intrinsically bright, long-tailed), orbit (retrograde, long period, narrow inclination, small perihelion distance), and behavior (heliacally rising around perihelion time and then passing between Earth and the Sun). Only if what Revelation 12:1–5 records is fully consistent with Matthew 2:1–18 should it be accepted as holding the key to unlocking the mystery of the Bethlehem Star.

REVELATION 12:1–5

In Revelation 12:1–5 we read of a woman in heaven who is pregnant and brings forth a special child:

> And a great sign appeared in heaven: a woman clothed with the Sun, with the Moon under her feet, and on her head a crown of twelve stars. And she was pregnant; and she was crying out because of labor pains and the agony of giving

[5]The zodiac features in the apparently first-century BC book of *2 Enoch* (long recension) 21:6; 30:6. See Wilhelm Bousset, *Die Offenbarung Johannis* (Göttingen: Vandenhoeck und Ruprecht, 1904), 336, 355; R. H. Charles, *A Critical and Exegetical Commentary on the Revelation of St. John*, 2 vols., International Critical Commentary (Edinburgh: T. & T. Clark, 1920), 1:315–316; W. K. Hedrick, "The Sources and Use of the Imagery in Apocalypse 12" (unpublished ThD dissertation, Graduate Theological Union, Berkeley, CA, 1971), 44; David E. Aune, *Revelation*, 3 vols. (Nashville: Thomas Nelson, 1997–1998), 2:681. We also find zodiacs on Jewish synagogue mosaic floors from the early centuries AD, including the Galilean Beth Alpha synagogue. A tomb at Khirbet el-Qom, 11 km from Lachish, contains graffiti that may represent the zodiacal constellations Leo, Virgo, and Libra; it is apparently from the eighth century BC (Emile Puech, "Palestinian Funerary Inscriptions," in *The Anchor Bible Dictionary*, ed. D. N. Freedman, 6 vols. [New York: Doubleday, 1992], 5:127–128).

> birth. . . . And the dragon stood before the woman who was about to give birth, so that when she bore her child he might devour it. She gave birth to a male child, one who is to rule all the nations with an iron scepter, but her child was snatched away to God and to his throne . . . (Rev. 12:1–2, 4b–5).[6]

MESSIAH'S BIRTH

It seems clear that what is in view here is the story of the birth of the Messiah. Within Revelation 12–14, which is in many ways the center of the book of Revelation, there are several indications that this woman's male child is the Messiah Jesus. The twelve stars in the mother's crown (v. 1) reveal that the son is born to Israel, with its twelve tribes. Revelation 12:5b, in speaking of the woman's son being taken to God's throne, strongly alludes to the ascension of Jesus. Verse 17 explicitly mentions Jesus, and verses 10–11 refer to him as "Christ" and "the Lamb." Moreover, the strong allusion in verse 5a ("one who is to rule all the nations with a rod of iron"; ESV) to Psalm 2:7b–9 ("You are my Son; today I have begotten you. Ask of me, and I will make the nations your heritage, and the ends of the earth your possession. You shall break them with a rod of iron and dash them in pieces like a potter's vessel") suggests that Jesus is in view, for Revelation 19:15 ("he will rule them with a rod of iron") employs the same language from the same psalm to refer to the imposition of Jesus's sovereign authority over the nations at the end of the age. Therefore in Revelation 12:1–5 the author, John, begins the narrative of the great conflict between the Messiah and the dragon, Satan, by telling the story of the nativity of Jesus the Messiah.

THE CELESTIAL PLAY

What is so remarkable about Revelation 12:1–5a, of course, is that the narrative of Jesus's birth is told in what are clearly celestial terms. The wonder is explicitly located "in heaven" and involves the heavenly entities of the Sun, Moon, and stars, as well as a great celestial serpentine dragon who throws the stars to earth with its tail, and a stellar woman. Indeed everything in verses 1–5 takes place "in heaven" (vv. 1, 3; cf. v. 4), with the shift to the earth occurring only in verse 6.

The overwhelmingly celestial nature of verses 1–5 obviously begs the question of why. Why does John offer his readers an astronomical version of Jesus's birth narrative? As much as scholars of the Apocalypse have noticed how peculiar the celestial framing of these verses is, they have never been able to explain it. Nor have they been able to shed light on why verses 1 and 3 specify that the scenes of the drama recounted in verses 1–5 constitute "signs" ("a great sign appeared . . . another sign appeared").

Quite simply, the only plausible explanation of the celestial and portentous nature of the messianic birth scene in Revelation 12:1–5 is that John is consciously recalling the heavenly wonder that attended Jesus's nativity. In other words, what we read in these verses is an account of the marvel that coincided with the Messiah's birth and that prompted the Magi to travel to Judea to worship the newborn King of the Jews.[7] This astronomical marvel establishes the narrative framework for the whole chapter of which it is a part.

Accordingly, what we find in these verses is the key to unlocking the mystery of the Star of Bethlehem.

[6] My translation.

[7] Most of what John describes in Rev. 12:1–5 would have been visible in both Babylon and Judea. The strong allusions to pagan mythology and paradigms may favor the hypothesis that the report that underlies these verses stems from Babylon. However, the Jewish and Biblical perspective reflected in verses 1–5 is consistent with Judean observations. Nevertheless, there was a Jewish community in Babylon too, and we have good reason to believe that some within it influenced the Magi to adopt a Jewish and Biblical interpretation of the Star. Therefore we should leave the question of the provenance of the report underlying Rev. 12:1–5 open, while acknowledging that the case for Babylon, the center of astronomical record-keeping in the ancient Near East, may be slightly more compelling.

THE CELESTIAL WOMAN VIRGO

The Greek word translated "sign" (*sēmeion*) in verse 1 may also mean "constellation,"[8] as a number of scholars have pointed out.[9] In this context a double meaning seems very likely—the "sign" is an empirical phenomenon disclosing some theological truth and it also concerns a stellar constellation.[10]

It seems clear who the heavenly woman crowned with twelve stars is. Since the Sun and Moon traverse the heavens along the ecliptic, the fact that they are here respectively described as clothing the woman and as being under her feet makes it clear that the female is positioned along the ecliptic and is therefore one of the zodiacal constellation figures. The only zodiacal female is Virgo the Virgin, and hence it is unquestionably she who is in view here.[11] Just as Virgo was often portrayed with wings, so the woman in John's vision is given wings in verse 14. Moreover, just as Virgo was typically envisioned by the ancients as a virgin of childbearing age and indeed often as a mother, so also the celestial woman in Revelation 12 is a young maiden who gives birth to a child. In addition, the fact that the serpentine dragon is said to have "stood before" the woman (v. 4b) supports this identification. As we will see below, the multiheaded serpentine dragon is the constellation figure Hydra, which is located immediately to the south of Virgo and rises in the eastern sky on her left side (on the right side, from an observer's perspective). We concur with the claim of Stephen Benko regarding the woman of Revelation 12: "Any Greek or Roman reading such a description would have thought of the constellation Virgo (*parthenos*, virgin), . . . who was represented as a woman holding an ear of corn and having wings."[12]

Virgo was the largest zodiacal constellation and the third largest of all the constellations after Argo Navis and Hydra. The constellation consists mostly of rather faint stars spread over a wide area. The brightest star is Spica, a stunning first-magnitude star halfway down the constellation, close to the ecliptic. Spica is actually in the top twenty of the brightest stars of the night sky. The next brightest stars in the constellation are the third-magnitude Porrima (γ), Vindemiatrix (ε), and Auva (δ), and the fourth-magnitude Zavijava (β). (Modern readers who wish to get a good view of Virgo are advised to look at it shortly after dark in April–June.)

The constellation was called AB.SIN (the Furrow) by the Mesopotamians, Parthenos (Virgin) by the Greeks, Virgo (Virgin) by the Romans, and Bethulah (Virgin) by the Jews.

The Mesopotamians and Greeks depicted the constellation as a young maiden who carried an ear of grain in her left hand. We have a sketch of the Furrow (Virgo) from Seleucid[13] Babylonia, in which she is doing precisely this (see fig. 7.2).[14] The ear of grain was particularly closely associated with Spica and reflects the fact that historically the constellation, and

[8] Euripides, *Rhesus* 528–533; *Ion* 1146–1158, specifically 1157; Aratus, *Phaenomena* 10; see Henry George Liddell and Robert Scott, *A Greek-English Lexicon*, ed. Henry Stuart Jones (Oxford: Clarendon Press, 1940), 1593.

[9] So, for example, Aune, *Revelation*, 2:679; Craig S. Keener, *Revelation* (Downers Grove, IL: InterVarsity Press, 2009), 312.

[10] Cf. Jürgen Roloff, *The Revelation of John: A Continental Commentary*, trans. John F. Alsup (Minneapolis: Augsburg Fortress, 1993), 145.

[11] So, for example, Franz Boll, *Aus der Offenbarung Johannis: hellenistische Studien zum Weltbild der Apokalypse* (Leipzig and Berlin: Teubner, 1914), 98–124; A. A. Farrer, *The Revelation of St. John the Divine* (Oxford: Clarendon, 1964), 139–145; Bruce J. Malina and John J. Pilch, *Social-Science Commentary on the Book of Revelation* (Minneapolis: Augsburg Fortress, 2000), 155–157; Roloff, *Revelation*, 145; Tim Hegedus, "Some Astrological Motifs in the Book of Revelation," in *Religious Rivalries and the Struggle for Success in Sardis and Smyrna*, ed. Richard S. Ascough (Waterloo, ON: Wilfred Laurier University Press, 2005), 77, 83–84; *pace*, among others, Charles, *Revelation*, 1:319; George Bradford Caird, *The Revelation of Saint John* (Peabody, MA: Hendrickson, 1966), 148–149 (who, despite deducing that the woman and the dragon are constellations, states that they have been newly imagined and so cannot be found in any star atlas), and Keener, *Revelation*, 312.

[12] Stephen Benko, *The Virgin Goddess: Studies in the Pagan and Christian Roots of Mariology* (Leiden: Brill, 1993), 110–111.

[13] The Seleucid Empire lasted from 323 BC to 63 BC.

[14] H. van der Waerden, *Science Awakening II* (Leyden, Netherlands: Noordhoff, 1974), 288 and 81 plate 11c. This part of the cuneiform tablet (a copy of an older original) is found at the Department of Oriental Antiquities of the Louvre in Paris: item number AO 6448. Note that Ulla Koch-Westenholz, *Mesopotamian Astrology: An Introduction to Babylonian and Assyrian Celestial Divination*

Spica in particular, was linked to the start of the grain harvest. Often Virgo was believed to hold a palm branch in her right hand. Greeks and many in the ancient Near East imagined her to have wings.

The constellation was associated with a number of different virgin goddesses in the centuries surrounding the birth of Jesus, including Ishtar/Asherah, Athena, Demeter, Atargatis, Tyche, Dike, Justa, Astraea, Juno, and Isis.[15] As Dike, Justa, and Astraea, Virgo was presented as an innocent and pure virgin so exasperated with humanity that she left the earth for the starry heavens. Pseudo-Eratosthenes[16] states that Hesiod identified her as Dike, daughter of Zeus, who became so weary of human injustice in all its forms that she departed for the mountains and ultimately ascended into the heavens. Hyginus claims that the celestial virgin Astraea became the constellation Virgo.[17] According to Apuleius,[18] Psyche spoke of Juno as being worshiped "as a virgin who travels through the sky on the back of Leo." However, Isis (who called herself "The Great Virgin" in a hymn to Osiris) was the predominant identity of Virgo in our period[19] and it is as Isis that Virgo is portrayed on the Dendera Zodiac (a famous, probably mid-first-century BC, sky map carved on the ceiling of the Hathor temple at Dendera, Egypt).[20]

FIG. 7.2 The Furrow (Virgo) holding an ear of grain. Based on a Babylonian astrological cuneiform tablet from Uruk which is now preserved in the Louvre Museum (AO 6448). Image credit: Sirscha Nicholl.

Strikingly, Virgo was widely regarded as a virgin and yet, paradoxically, often also as a mother. Tim Hegedus points out that,

> Mother goddesses were not incompatible with Virgo in ancient Greco-Roman religion. According to Frances Yates, "The . . . virgin is . . . a complex character, fertile and barren at the same time." . . . For example, . . . the figure of Isis holding her son Horus was identified with Virgo. Virgo was also associated

(Copenhagen: Museum Tusculanum Press, 1995), 41–42n3, suggests that the ears of grain on certain kudurrus represent Virgo. (Kudurrus are second-millennium BC Kassite boundary stones that were sometimes used to record grants of land from the king.)

[15] Pseudo-Eratosthenes, *Catasterismi* 9; Benko, *Virgin Goddess*, 111; André Le Boeuffle, *Les Noms Latins d'astres et de constellations* (Paris: Les Belles Lettres, 1977), 212–215; Boll, *Offenbarung Johannis*, 109–111; Malina and Pilch, *Social-Science*, 163; Tim Hegedus, *Early Christianity and Ancient Astrology* (New York: Peter Lang, 2007), 236–227. For ancient references to Virgo, see Aratus, *Phaenomena* 93–138; Cicero, *De Natura Deorum* 2.42; Hyginus, *Poetica Astronomica* 2.25; Nonnos, *Dionysiaca* 2.655.

[16] Pseudo-Eratosthenes, *Catasterismi* 9.

[17] Hyginus, *Poetica Astronomica* 2.25; Mary Amelia Grant, *The Myths of Hyginus* (Lawrence: University of Kansas, 1960), http://www.theoi.com/Text/HyginusAstronomica2.html#25 (accessed March 26, 2014); Theony Condos, *Star Myths of the Greeks and Romans: A Sourcebook* (Grand Rapids, MI: Phanes, 1997), 206.

[18] Apuleius, *Metamorphoses* 6.4.

[19] For example, the fourth-century AD Avienus, *Phaenomena* 273–292. See Bruce J. Malina, *On the Genre and Message of Revelation: Star Visions and Sky Journeys* (Peabody, MA: Hendrickson, 1995), 159. On Isis worship throughout the ancient world, see R. E. Witts, *Isis in the Ancient World* (Baltimore: Johns Hopkins University Press, 1971).

[20] On Isis as Virgo, see Boll, *Offenbarung Johannis*, 108–110; Benko, *Virgin Goddess*, 111–112; M. Tsevat, "bᵉthûlâh," in *Theological Dictionary of the Old Testament*, vol. 2, ed. G. J. Botterweck, H. Ringgren, and H.-J. Fabry, trans. J. T. Willis (Grand Rapids, MI: Eerdmans, 1977), 338–343, esp. 338–339. On the Dendera Zodiac, see note 43, below.

> with various other mother goddesses in antiquity, such as Juno, Dea Caelestis, Ceres, Magna Mater, Atargatis, and even Ilithyia, the Greek goddess of childbirth. . . . As Boll concludes, ". . . *alles ist eins*" ["everything is the same"].[21]

Theony Condos comments that, in identifying the constellation as a maiden, the Greeks were probably indebted to the Babylonians, who associated it with the virgin aspect of the Great Mother Goddess.[22]

Those who deny that Virgo is in mind in Revelation 12 claim that the twelve stars on her crown (Rev. 12:1: "and on her head a crown of twelve stars") represent the twelve zodiacal signs or constellations (excluding Ophiuchus).[23] However, even if we accepted that Virgo was being presented as having the twelve zodiacal constellations/signs on her head, that could be explained in another way, namely, that she encapsulated and represented the entire zodiac.[24] At the same time, as we shall see shortly, a careful study of the stars in the uppermost region of Virgo reveals that there is a much better way to understand her twelve-star crown.

THE PLACEMENT OF THE FIGURE VIRGO WITHIN THE CONSTELLATION

How did the ancients in this general period conceive of Virgo relative to the stars of her constellation? We have evidence of at least four versions of Virgo.

The first version of Virgo was described by the second-century BC Greek astronomical writer Hipparchus. Beginning at the bottom of the constellation, he regarded her feet as corresponding to the stars μ (Mu) and λ (Lambda) Virginis, her shoulders (evidently the lower part of the shoulders) as γ (Gamma) and δ (Delta) Virginis, and the top star in her head as ξ (Xi) Virginis. This conceptualization of Virgo is unquestionably the most bizarre—it requires Virgo to have an extraordinarily long neck and/or massive head. Indeed, in Hipparchus's analysis, the distance from the top star in her head to the star in her right shoulder is the same as the distance from her left elbow to her left foot!

A revised version of Hipparchus's Virgo is found in the second-century AD *Almagest* of Ptolemy. Although Ptolemy largely followed Hipparchus, he perceived the need to make modifications. Ptolemy freely admitted that he was exercising some creative license in his portrayal of the great celestial woman:

> We do not employ the same figures of the constellations that our predecessors did, just as they did not employ the same figures as their predecessors. But in many cases we make use of different figures that more appropriately represent the forms for which they are drawn.[25]

Ptolemy went on to give Virgo as an example of his innovative reimaginings of the constellations: "For instance, those stars which Hipparchus places 'on the Virgin's shoulder' we place 'on her side,'[26] because their distance from the stars in her head seems too great for the distance from the head to the shoulder in his constellation of Virgo. And so, by making those stars to be on her sides, the figure

[21] Hegedus, "Some Astrological Motifs," 80–81. Cf. Malina, *Genre and Message*, 160: "in the mind of the first-century Mediterraneans, the female statuses and roles of virgin, mother, and queen could . . . readily reside in the same person."
[22] Condos, *Star Myths of the Greeks and Romans*, 207.
[23] E.g., George R. Beasley-Murray, *Revelation*, New Century Bible Commentary (Grand Rapids, MI: Eerdmans, 1981), 197; Brian K. Blount, *Revelation: A Commentary*, New Testament Library (Louisville, KY: Westminster John Knox, 2009), 225.
[24] It should also be remembered that the woman represents Israel/Jacob, and that the twelve stars represent the twelve tribes. Had there been only eleven stars on the woman's crown, the symbolism on that level would have been ruined.
[25] Ptolemy, *Almagest*, book 7, H37 (my translation). For an excellent recent translation of Ptolemy's *Almagest*, see *Ptolemy's Almagest*, trans G. J. Toomer (Princeton, NJ: Princeton University Press, 1998).
[26] In fact, in his description of the constellation, while Ptolemy does describe δ (Delta) Virginis as on Virgo's right side (under the girdle), he portrays γ (Gamma) Virginis as in the left wing. But he probably meant that the star was both on Virgo's wing and on her side.

will be agreeable and appropriate, which it would not be if those stars were drawn 'on her shoulders.'"[27]

However, in decreasing the size of Virgo's neck and head, Ptolemy introduced a new problem: Virgo's torso and arms became disproportionately long—the distance from the star on her left side (γ) to her left hand (α) is greater than the distance from her left hand (α) to her left foot (λ). Quite simply, maintaining Hipparchus's placement of Virgo's head could not produce a properly proportioned constellation figure—measuring from just above Virgo's right buttock (which in a normal human body would be roughly halfway between the top of the head and the bottom of the feet), in Ptolemy's Virgo it is 29 degrees to the top of her head and yet only 18 degrees to her right foot (and 17 degrees to her left foot).

A third version of Virgo existed in the ancient world, around the time of Jesus's birth, which did not entail her having an extraordinarily long neck or elongated upper body (from the waist upwards). The first-century BC work *Poetica Astronomica*, by Hyginus, and the first/second century AD work *Catasterismi*, by Pseudo-Eratosthenes, portrayed Virgo in rather more vague terms than Hipparchus or Ptolemy, but nevertheless in readily identifiable terms. These two authors described Virgo very similarly. According to Hyginus, she had 19 stars, and, according to Pseudo-Eratosthenes, she had 20 stars. The faint star that they associate with the Virgin's head is 16 Virginis (mag. +4.96). Pseudo-Eratosthenes, *Catasterismi* 9, stated that Virgo was regarded as headless. The headless version of Virgo is actually easy to explain astronomically, because, as Pseudo-Eratosthenes himself went on to point out, there is only one faint star in the region of sky where Virgo's head was perceived to be (16 Virginis).[28] The shoulders are δ (Auva) (mag. +3.37) and γ (mag. +3.43). The elbows are σ and ψ (both mag. +4.75). The hands are α (Spica) (mag. +0.96) and ζ (Heze) (mag. +3.34). Her feet are μ (mag. +3.84) and λ (mag. +4.5). With respect to the wings, ε (Vindemiatrix) (mag. +2.84) and ρ (mag. +4.87) are on the right wing, and β (Zavijava) (mag. +3.56) and η (Zaniah) (mag. +3.87) on the left wing. The six faint stars that make up her dress, that is, the hem of her dress, are υ (mag. +5.12), φ (mag. +4.78), ι (Syrma) (mag. +4.06), 106 (mag. +5.4), 95 (mag. +5.43), and κ (mag. +4.15), all of which are fourth- or fifth-magnitude stars—dim, but well within the range of naked-eye observation in good atmospheric conditions.[29] This solution incorporates most of the major stars in the relevant part of Virgo in a natural way and is very probably correct.

Therefore Hyginus and Pseudo-Eratosthenes represent a view of Virgo that is very similar to that of Hipparchus from the feet (μ and λ Virginis) to the shoulders (δ and γ Virginis), but has the head much lower down and a body with more reasonable proportions. According to their assessment, the stars ξ, ν, ο, and π were not a part of Virgo's body, but were above the head.

A fourth version of the constellation figure was also prevalent around the turn of the ages. Teukros (Teucer) of Babylon portrayed Virgo as a goddess sitting on a throne.[30]

[27] Ptolemy, *Almagest*, book 7, H37 (my translation). Ptolemy was referring to the star δ (Delta) Virginis. Hipparchus had located δ on Virgo's right shoulder. The third-century BC Aratus, *Phaenomena* 137–138, writing before Hipparchus, stated that Vindemiatrix was located above her shoulders, which reveals that the right shoulder was perceived to be around δ Virginis. Ptolemy argued against Hipparchus's interpretation of δ Virginis on the ground that the head stars would then be too far from δ. Ptolemy made an excellent point: in Hipparchus's constellation the proportions of Virgo's upper part were utterly absurd—her neck was giraffe-like!

[28] Pseudo-Eratosthenes, *Catasterismi* 9: see Condos, *Star Myths of the Greeks and Romans*, 205.

[29] See the translation in ibid., 205–206. Unfortunately, because Theony Condos is too influenced by Ptolemy's version of Virgo, she is unable to identify which stars Hyginus and Pseudo-Eratosthenes had in mind in most cases. Moreover, she acknowledges that ε Virginis was the wing by the shoulder, even though in her scheme it is nowhere near the shoulder, but rather at the mid-belly region.

[30] This astrologer lived in the late first century BC or early first century AD; see Boll, *Sphaera*, 8; and Otto Neugebauer, *The Exact Sciences in Antiquity* (Mineola, NY: Dover, 1969), 189, 175, 189 (between 100 BC and AD 50). T. Boiy, *Late Achaemenid and Hellenistic Babylon* (Leiden: Brill, 2005), 310, states that Teukros lived in the first century BC; and Bruce M. Metzger, "Ancient

Teukros is an important representative of Babylonian astrology at the time of Jesus's birth; indeed he is the only Babylonian astrologer of the time that we know anything about. Specifically, Teukros of Babylon describes the sign of Virgo as "a certain goddess seated on a throne and nursing a child. Some say that she is the goddess Isis in the Atrium nursing Horus."[31] Although restricting a large constellation figure like Virgo to a 30-degree zone (to get it to work as a zodiacal sign) naturally results in distortion, and so cannot accurately capture the underlying constellation figure, Teukros nevertheless gives us a good sense of how she was envisioned and a general idea of her proportions. In his description of the sign of Virgo, Teukros locates Spica two-thirds of the way down.[32] His Virgo, and that of the Egyptians, is oriented parallel to the ecliptic[33] and seems closest to that of Hyginus and Pseudo-Eratosthenes, except that Virgo is imagined as sitting

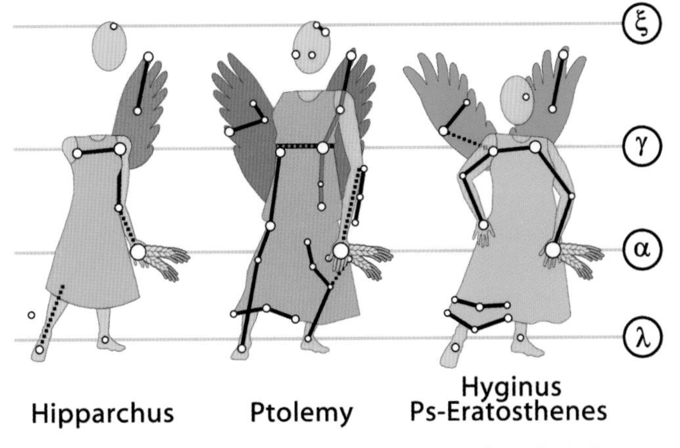

FIG. 7.3 Virgo as represented by Greek astronomers from the 2nd century BC to the 2nd century AD. Image credit: Sirscha Nicholl.

FIG. 7.4 An artistic re-creation of Ptolemy's representation of Virgo. Plate 21 in Richard Rouse Bloxam, *Urania's Mirror* (London: Samuel Leigh, 1825), a set of drawings accompanied by *A Familiar Treatise on Astronomy*, by Jehoshaphat Aspin. Image credit: oldbookart.com.

Astrological Geography and Acts 2:9–11," in *Apostolic History and the Gospel: Biblical and Historical Essays Presented to F. F. Bruce*, ed. W. Ward Gasque and Ralph P. Martin (Exeter: Paternoster, 1970), 130, said that Teukros "flourished about 10 B.C."

[31] My translation of the Greek text in *Catalogus Codicum Astrologorum Graecorum*, ed. and trans. Franz Boll, vol. 7 (Brussels: Lamertin, 1908), 202; cf. idem, *Sphaera*, 210; idem, *Offenbarung Johannis*, 109–110. The Arabian astrologer Abu Ma'shar, from the ninth century AD, in *The Great Introduction to Astrology*, book 6, chapter 1, wrote of Virgo that she was "a Virgin that Teukros called Isis; she is a pretty, pure virgin with long hair, with a beautiful face; she has two ears of corn in her hand and is seated on a throne on which lie cushions. She is looking after a little boy and gives him bread to eat, in a place called the Atrium; this boy is called by some peoples Isu (Jesus)" (translation influenced by Alfred Jeremias, *Handbuch der altorientalischen Geisteskultur* [Leipzig: J. C. Hinrichs, 1913], 172; cf. A. E. Thierens, *Astrology in Mesopotamian Culture: An Essay* [Leiden: Brill, 1935], 34). Abu Ma'shar was referring to a Persian translation of the book *Sphaera Barbarica* by Teukros of Babylon.

[32] Boll, *Catalogus*, 7:203. Teukros locates the beginning of the emergence of the head at 1–3 degrees, the nose at 4–6 degrees, the neck at 7–10 degrees, the arms at 11–13 degrees, the fingers at 14–18 degrees, Spica at 19–21 degrees, and (the upper section of) the lower half of the leg at 22–23 degrees.

[33] In earlier centuries, the Babylonians regarded Virgo as two distinct constellations consisting of the Furrow (who carried an ear of grain), who rose first, and the Frond (who carried a palm branch). While the Greco-Roman Virgo and the later Babylonian Virgo were parallel to the ecliptic, these constellations were oriented at close-to-90-degree angles to the ecliptic. It seems that at some stage around the middle of the first millennium BC, when the zodiacal band was divided into twelve equal signs, the Furrow and Frond were united into a "single unified figure" who carried an ear of grain in her left hand and a palm branch in her right hand (Gavin White, *Babylonian Star-Lore: An Illustrated Guide to the Star-Lore and Constellations of Ancient Babylonia*, 2nd ed. [London: Solaria, 2007], 115).

on a throne, presumably with her legs and feet parallel.[34]

Strikingly, a Jewish zodiac wheel from the sixth century AD, discovered at the Beth Alpha synagogue, portrayed Virgo as seated on a throne dressed in a long gown that reached down to her ankles, and wearing royal red shoes.[35]

Likewise Antiochus of Athens (from the first or, more likely, second century AD) envisioned Virgo as a woman holding a child, and hence probably as seated.[36]

So, it would seem, there were at least four portrayals of Virgo in the centuries around the birth of Jesus: three envisioned her standing and one imagined her sitting. Ptolemy, Hipparchus, and Pseudo-Eratosthenes and Hyginus represent versions of Virgo standing, and Teukros of Babylon and the Dendera Zodiac represent the version of Virgo sitting on a throne, holding an infant. Hipparchus has Virgo's head high, as does Ptolemy, but Pseudo-Eratosthenes, Hyginus, and Teukros (and Egyptian art) reflect a conception of Virgo in which her head is lower, at or near 16 Virginis rather than at ξ, ν, ο, and π Virginis. Happily, there was widespread agreement regarding the level of Virgo's groin and legs.

FIG. 7.5 Isis with Horus, her child. From the Late Egyptian period (7th–4th century BC). Image credit and copyright: The Walters Art Museum, Baltimore.

How does Revelation 12:1–5 compare with these visualizations of Virgo?

It is important to give due attention to the fact that Revelation 12:1 regards Virgo as wearing a "crown"[37] of twelve stars.[38] If we take on board that the generally acknowledged maximum naked-eye visibility in ideal

[34] A. Sachs, "A Late Babylonian Star Catalog," *Journal of Cuneiform Studies* 6 (1952): 146–150, and N. A. Roughton, J. M. Steele, and C. B. F. Walker, "A Late Babylonian Normal and *Ziqpu* Star Text," *Archives of the History of the Exact Sciences* 58 (2004): 566–567 (cf. van der Waerden, *Science Awakening II*, 99) point out that a pre-Seleucid-period fragmentary star catalog portrays Zavijava (β Virginis) as the rear foot of the Lion (Leo). Obviously, this imagining of the Lion, if it continued into the era when Virgo (consisting of the Furrow and the Frond) was conceived of as resting along the ecliptic, would not have been consistent with Virgo's head being where Hipparchus and Ptolemy located it, namely at ξ, ν, ο, and π. The only other catalogued Babylonian stars of Virgo, also from the pre-Seleucid period, are Spica (α Virginis) (called "the bright star of the Furrow") and Porrima (γ Virginis) (called "the single star in front of the Furrow" and "the root of the furrow").

[35] The Hebrew name for Virgo, *Bethulah*, was given alongside the image. For more on this fascinating image, see Eleazar Lipa Sukenik and Steven Fine, *The Ancient Synagogue of Beth Alpha* (London: Oxford University Press, 1932), 31 and plate XIII.2.

[36] See Boll, *Sphaera*, 58; cf. 210–211. In addition, Franz Boll described an ancient gem on which Isis was portrayed as Virgo holding Horus, who is carrying an ear of grain, with a star over Virgo's head and another ear of grain standing beside her (ibid., 211).

[37] Although many scholars claim that *stephanos* was not used of royal crowns, this is incorrect. Gregory M. Stevenson, "Conceptual Background to Golden Crown Imagery in the Apocalypse of John (4:4, 10; 14:14)," *Journal of Biblical Literature* 111.2 (1995): 260, points out that in Hellenistic Greek *stephanos* was used of royal crowns and that this usage was "more common among Jewish authors." The LXX translators and Josephus consistently preferred *stephanos* to *diadēma* when speaking of the crown worn by Israelite kings. This practice, he concludes, does not imply that these crowns were wreaths but rather "demonstrates that in the minds of some later Israelites *stephanos* was considered an acceptable term for describing a royal crown." Aside from Rev. 12:1, Revelation uses *stephanos* of royal crowns in 6:2 (of the rider of the white horse); 14:14 (of the Messiah); 4:4 (of the 24 elders); and 9:7 (of locusts).

[38] Ernest L. Martin, *The Star of Bethlehem: The Star That Astonished the World*, 2nd ed. (Portland, OR: Associates for Scriptural Knowledge, 1996), available at http://www.askelm.com/star/star008.htm (accessed March 26, 2014) stated that Prof. John Thorley, "When Was Jesus Born?," *Greece and Rome*, 2nd series, 28.1 (April 1981): 87–88, had successfully identified from a star atlas the twelve stars of Virgo's crown. However, Thorley's list included eight stars firmly associated in the ancient mind with the constellation Leo (σ, χ, ι, θ, 60, δ, 93, and β, all Leonis) and only four associated with Virgo (β, ν, π, and ο, all Virginis). Moreover, the resultant crown was absurdly large, 16 degrees wide and 20 degrees tall, extending up as far as Leo's hindquarters. We can be confident that no ancient ever imagined Virgo's crown where Thorley did!

conditions, such as those that would normally have prevailed in ancient Babylon and Jerusalem, is up to and including the sixth magnitude, that is, up to +6.5, it is remarkable that there are precisely twelve stars of up to +6.5 magnitude in the relevant part of Virgo. These 12 stars are 10 (mag. +5.93), 11 (mag. +5.71), 7 (mag. +5.34), HIP58809 (mag. +6.37), π (mag. +4.62), o (mag. +4.09), 6 (mag. +5.56), ν (mag. +4.03), 4 (mag. +5.28), ξ (mag. +4.81), ω (mag. +5.21), and HIP6756 (mag. +6.15). These stars form a conical shape that is reminiscent of a tiara, mitre, or tall royal crown (fig. 7.6). Strikingly, an image of Virgo next to Leo on a relief on the ceiling of the portico of the Temple of Khnum at Esna in Egypt portrays her with a tall crown of Egyptian style (fig. 7.7).[39]

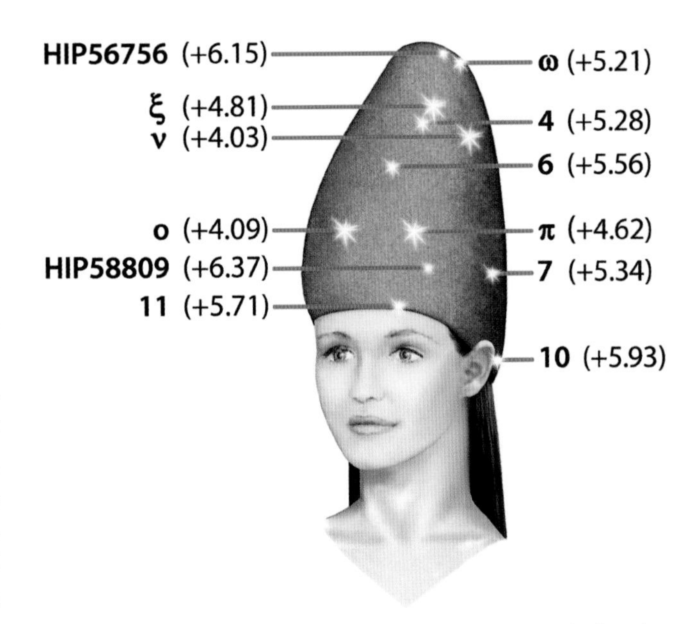

FIG. 7.6 Virgo's crown: the 12 stars form a tall cone shape, which makes for a very natural ancient tiara or tall crown. The name of each star is in bold, and its magnitude value is in parentheses. Image credit: Sirscha Nicholl.

The tall crown was the most common style of crown in the ancient Near East, worn by royalty in, among other places, Assyria, Babylonia, Egypt, and Parthia. It was used before and after the time of Jesus's birth. Indeed Musa, from 2 BC to AD 4 queen of the Parthian empire, in which Babylon was located and from which the Magi hailed, is pictured on coins wearing such a tall crown (fig. 7.8). Up to now, the 12 stars have not been correctly identified, probably because scholars looking for them have been presupposing the portrayal of Virgo found in Ptolemy.

With a crown on Virgo's head, Revelation 12:1's constellation figure evidently contrasts with that of Hipparchus and Ptolemy but is similar to that of Hyginus, Pseudo-Eratosthenes, Teukros, and Egypt.

In addition, both of Virgo's feet are envisioned in verse 1 as being above the Moon. The Virgo of Hipparchus, Hyginus, Pseudo-Eratosthenes, and Ptolemy had her left foot at λ Virginis and her right foot at μ Virginis. In certain years, depending on the angle of the Moon's orbit relative to the ecliptic plane of Earth's orbit, the Moon may pass through this area of sky. In the years 7 to 2 BC, at times when the Sun was in Virgo, the Moon ventured under λ Virginis in 7–5 BC (in the other years it was too far south of the ecliptic to be plausibly regarded as in any way under her feet). It seems therefore that Revelation 12:1 is portraying both of Virgo's feet as being in the vicinity of this star.

With the crown and feet identified on the star map, we know the boundaries within which we are to fit Virgo's body from her forehead to the end of her legs, that is, between the crown and the area around λ Virginis. Revelation 12:1 would seem to concur with the widespread view regarding the level

[39] A sketch of the Esna zodiac can be viewed at http://www.repertorium.net/rostau/secondary/esnae.html (last modified August 29, 2006).

within the constellation where her groin and legs were. With respect to the upper body of Virgo, verse 1 reveals that she is being viewed in terms very similar to the Virgo of Hyginus and Pseudo-Eratosthenes and the Virgo of Egypt and Babylon.

At the same time, since Virgo is wearing a crown[40] and is clothed with the radiance of the Sun, she is most naturally regarded as sitting on a throne. The fact that the Moon is stationed under her feet suggests that it was in subservience, paying homage to Virgo.[41] Indeed the Moon seems to be forming a footstool for her feet. Virgo, it would appear, is being exalted in glorious splendor. That she is seated is consistent with the fact that relatively little room is available for her legs (above λ Virginis). Just as many in the ancient world envisioned Virgo as Queen Isis seated on her throne, so Revelation 12:1 pictures Virgo as "Queen Israel" on her throne. It presents her in terms that anticipate her exaltation and sovereignty over the nations in the new age.

Virgo was usually envisioned as having wings,[42] as in Revelation 12:14. However, she was not always represented as having them, particularly when she was conceived of as Isis (as apparently, for example, on the Dendera Zodiac and at Esna), and indeed in Revelation 12 she lacks wings up until verse 14.

FIG. 7.7 Virgo as represented on the ceiling of the Esna Temple in Egypt. Image credit: Sirscha Nicholl, based on a photograph by Shaun Osborne (2012).

As for the precise location of Virgo's body relative to the stars of the constellation, the fact that in verse 1 the Sun is clothing Virgo and the Moon is under her feet gives us a sense of the celestial position of Virgo's throne in relation to the fixed stars and the ecliptic. Virgo *rises* with her crown up and her legs down, but she *sets* upside-down, with her crown (or, if she is envisioned as having wings, the tip of her left wing [the star Zavijava]) being the first part of her to disappear below the horizon. The Sun is perceived in this verse to be within Virgo, and hence her belly is envisioned as encompassing the ecliptic.

All in all, Revelation 12:1 gives us a rather clear idea how the constellation figure Virgo is being imagined with respect to the stars. John's Virgo is similar to that of Hyginus and Pseudo-Eratosthenes, but the portrayal of her as sitting on a throne is more reminiscent of Teukros of Babylon, the Dendera Zodiac, and the Jewish Virgo from the sixth century AD

[40]The crown functions in various ways in Rev. 12:1: to reveal the woman's identity, and to give a baseline for the proportions of Virgo as she is envisioned during this scene. At the same time, the crown contributes to the impression that this is an enthronement scene and that Virgo is destined to be greatly exalted.

[41]Cf. Gen. 37:9, where the Moon, together with the Sun and eleven stars, is envisioned as bowing down before Joseph.

[42]So Hipparchus, Hyginus, Pseudo-Eratosthenes, and Ptolemy. See Boll, *Offenbarung Johannis*, 113.

FIG. 7.8 Musa, queen of Parthia from 2 BC to AD 4 (the time of Jesus's childhood), as featured on the back side of a coin. Image credit: Classical Numismatic Group, Inc., http://www.cngcoins.com. Source: Wikimedia Commons.

FIG. 7.9 King Nebuchadnezzar of Babylon (7th–6th century BC) as portrayed on the damaged Tower of Babel stele (The Schøyen Collection, MS 2063). Note the tall crown. Image credit: Sirscha Nicholl, based on photographs by The Schøyen Collection (www.schoyencollection.com).

that was discovered at the Beth Alpha synagogue (fig. 7.10).

At the same time, it is important to appreciate that ancients seem to have had somewhat elastic conceptualizations of the constellation figures. For example, individual peoples were capable of imagining Virgo as both standing and sitting.[43] As Ptolemy made clear and an analysis of constellation descriptions and drawings from the ancient world confirms, there was some fluidity in how given cultures could conceive of Virgo in relation to her stars.[44] This is an important observation that may have a bearing on the interpretation of Revelation 12:1–5, where Virgo is envisioned as coming to life. While in verse 1 Virgo appears to be enthroned in majesty, in verse 2 she is portrayed as screaming in torment as she strives to deliver her child. It seems reasonable to allow that observers might have assigned her a slightly different posture for childbirth. Therefore we do well to imagine Virgo in Revelation 12:1–5 as alive and active within the general constraints of her constellation.

FIG. 7.10 A Jewish version of Virgo (Bethulah)—seated on a throne—from a zodiac wheel in a synagogue at Beth Alpha (early 6th century AD). Image credit: Wikimedia Commons.

[43] In a famous sculptured zodiac from a temple in Dendera, Egypt, Virgo seems to be presented as standing with an ear of grain in her hand, and right below her is a woman sitting down, with Horus standing by her right side, which some have suggested may possibly be a second representation of Virgo (see Boll, *Sphaera*, 243). For beautiful images of the Dendera Zodiac, go to the website of the Louvre Museum: http://cartelen.louvre.fr/cartelen/visite?srv=car_not_frame&idNotice=19044 (accessed April 4, 2014). While Teukros of Babylon portrayed Virgo as seated on a throne, a cuneiform tablet from Uruk in Seleucid Babylonia, also in the holdings of the Louvre (AO 6448), portrays the Furrow (an early conceptualization of Virgo) as standing.

[44] Ptolemy, *Almagest*, book 7, H37.

In fig. 7.11 we have sought to portray Virgo in a manner consistent with how the author of Revelation describes her.

THE DATE OF THE OPENING SCENE OF THE CELESTIAL DRAMA (REVELATION 12:1)

Remarkable as it may seem, if we accept that Revelation 12:1–5 is describing the celestial nativity drama that marked the Messiah's birth, the indication of the locations of the Sun and Moon with respect to Virgo in verse 1 enables us to pinpoint the year and, within it, the day when the opening scene took place.

In the scene, Virgo is "clothed with the Sun" (v. 1). This suggests that the Sun is located over her midriff,[45] which covers a roughly 10- to 11-degree zone from Virgo's chest (that is, just below the level of the stars δ–γ [Porrima]) to her groin (that is, the level on Virgo's body where 80 Virginis is).

At the same time, the Moon is under Virgo's feet. What is most remarkable about this is that it occurs when the Sun is clothing her. Because of how close the Moon is envisioned as being, we know that it is a waxing Moon and indeed a very young lunar crescent. To grasp what this means, it is important to pause and reflect briefly on the lunar cycle.

From the occurrence of a full Moon in the middle of a lunar month, the Moon enters a waning (shrinking in apparent size) phase. Eventually the Moon disappears from the sky for a few days at the end of the lunar month. Then it reemerges as a very thin crescent, barely visible over the western horizon in the aftermath of sunset. The Babylonians and Hebrews would therefore, during a short, 1–2 hour window of time between sunset and moonset, scan the western sky for the new crescent Moon descending in the Sun's wake.[46] At this time the Moon is moving away from the Sun, falling on average a further 12 degrees behind it every day. As it does so, it waxes (grows in apparent size) until, on the fifteenth day of its cycle, it again becomes a full Moon.

What is described in Revelation 12:1 is not an annual occurrence. When we review the years 7 BC to 4 BC, the time period during which Jesus was born, we see that the Moon was under Virgo's feet[47] at the point when the Sun was clothing her only in 6 BC—on September 15,[48] to be precise (see fig. 7.12 on why the Sun appeared to be clothing Virgo).[49] The

[45] Cf. Martin, *Star of Bethlehem* (http://www.askelm.com/star/star006.htm [accessed March 26, 2014]): the Sun was "mid-bodied to her, in the region where a pregnant woman carried a child." Martin chose September 12 in 3 BC as Jesus's day of birth based on Rev. 12:1 (ibid.). However, (1) v. 1 is not the moment of birth, but the prelude to the birth, the birth itself being related in the following verses; (2) dating Jesus's birth to 3 BC is based on a flawed chronology of Herod and the Herodian dynasty, as we have already seen.
[46] For a useful overview of new lunar crescent observation in the ancient Near East, see Leo Depuydt, "Why Greek Lunar Months Began a Day Later than Egyptian Lunar Months," in *Living the Lunar Calendar*, ed. J. Ben-Dov, W. Horowitz, and J. M. Steele (Oxford: Oxbow, 2012), 153–164; and John M. Steele, "Living with a Lunar Calendar in Mesopotamia and China," in ibid., 374–380. For treatments of ancient Jewish New Moon Observations, see Sacha Stern, "The Rabbinic New Moon Procedure: Context and Significance," in ibid., 211–230; and Lawrence H. Schiffman, "From Observation to Calculation: The Development of the Rabbinic Lunar Calendar," in ibid., 231–243.
[47] The description of Virgo in Ptolemy's *Almagest* indicates that, to the ancient mind, the feet of Virgo corresponded with μ Virginis and λ Virginis, unlike some modern representations of Virgo (such as in Starry Night® Pro 6.4.3) that portray her feet at 109 and μ Virginis.
[48] This is a Julian date.
[49] In 5 BC, the other major candidate for the year of Jesus's birth, the Sun was too high in Virgo, namely over her left shoulder, to be plausibly regarded as clothing her. Some revisionist chronologists date Herod's death to early 1 BC (Ormond Edwards, "Herodian Chronology," *Palestine Exploration Quarterly* 114 [1982]: 29–42; W. E. Filmer, "Chronology of the Reign of Herod the Great," *Journal of Theological Studies* 17 [1966]: 283–298; Ernest L. Martin, *The Birth of Christ Recalculated*, 2nd ed. [Pasadena, CA: Foundation for Biblical Research, 1980]; Andrew E. Steinmann, "When Did Herod the Great Reign?," *Novum Testamentum* 51 [2009]: 1–29). Doing so facilitates dating Jesus's birth in 3–2 BC. The Moon passed over 5 degrees to the south (left) of λ Virginis (which the Greeks regarded as her left foot), and was essentially level with it, while the Sun was over the upper abdominal region on September 11, 3 BC. That far from λ Virginis, the Moon could not plausibly have been construed as being under her feet. In 2 BC the Moon similarly passed over 5 degrees to the south (left) of λ Virginis and then only when the Sun was far from Virgo's torso. Moreover, the case for dating the death of Herod to 1 BC is weak. It cannot plausibly explain why Herod's sons Archelaus, Antipas, and Philip II all dated their reigns to 4 BC (see Timothy D. Barnes, "The Date of Herod's Death," *Journal of Theological Studies* 19 [1968]: 204–209). The idea that Herod appointed all three as coregents years before his death is uncompelling. Herod was, after all, having extraordinary problems, right up to his death, deciding who would succeed him—he wrote six or seven wills denominating who would succeed him, the final one five days before his death. Since

FIG. 7.11 An artistic representation of Virgo as envisioned in Revelation 12. The dots are stars, their different sizes representing their brightness in magnitudes (the larger the dot, the brighter the star). Image credit: Sirscha Nicholl.

Moon was technically under Virgo's feet (corresponding to λ Virginis) for that whole day, from the moment it rose in the east (when it was 24 degrees from the Sun) until Virgo set in the west (when it was 28–29 degrees from the Sun). Subject to favorable weather, the young crescent Moon would normally have become clearly visible to the naked eye shortly after the Sun set that evening.[50]

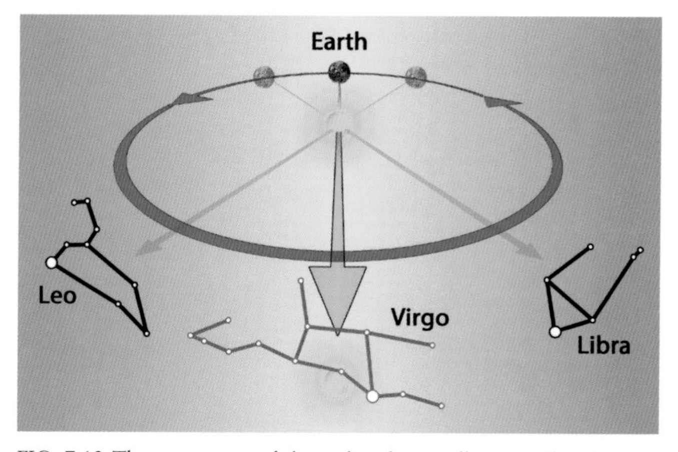

FIG. 7.12 The astronomy of the zodiacal constellations. The plane on which Earth orbits the Sun is called the ecliptic, and it is along this that the zodiacal constellations lie. In this image Earth (in grey and blue) is depicted revolving around the Sun. From the perspective of Earth on September 15, 6 BC, the Sun looks like it is located in Virgo's midriff. Image credit: Sirscha Nicholl.

Does Revelation 12:1 indicate that the crescent Moon was actually observed on September 15, 6 BC? On the one hand, since some astronomers at the time of Jesus's birth calculated the locations of the Sun, Moon, and stars in advance and knew where they were even when they could not see them, it cannot be assumed that the Moon was actually observed on the evening of September 15. On the other hand, since the astronomical focus of a scene such as is described in Revelation 12:1 would usually be the Moon, and since the sight is part of a great "sign," it seems most natural to conclude that it was observed. If observers did see the lunar crescent, they most likely did so between sunset and moonset.[51]

We conclude that Virgo, with her 12-star crown, clothed with the Sun, and with the Moon under her feet describes an astronomical phenomenon that can be dated to September 15, 6 BC.

THE COMETARY DRAMA OF REVELATION 12:1–2

The Actors

In Revelation 12:1–2, the heavenly entities are acting out key roles in the grand drama.

In the chapter as a whole, the woman symbolizes Israel—it alone can be reckoned to have given birth to the Messiah (v. 5) and to have fathered Christianity, the religious movement consisting of "those who keep the commandments of God and hold to the testimony of Jesus" (v. 17).[52] She consists of the twelve tribes represented by Virgo's twelve-starred crown. In the opening verses, however, Virgo represents Israel as embodied in the mother of Jesus, that is, so to speak, Israel with the face of Mary. The constellation figure is playing the part of Israel/Mary pregnant, in labor, and delivering her son, the Messiah. John may be

Archelaus and Philip II were introduced into Herod's will only in the immediate run-up to Herod's death, it is hardly likely that they were appointed coregents three years before it. The claim that the events that Josephus reports to have occurred between the eclipse and Passover cannot be accommodated between the lunar eclipse on March 13, 4 BC, and Passover, over 4 weeks later (e.g., Steinmann, "When Did Herod the Great Reign?," 15–16), is artificial. For instance, Steinmann's estimates (pp. 15–16) of the time necessary for travel and treatments are exaggerated, failing to take any account of the urgency of the moment or to reckon with overlaps. The suggestion that a partial lunar eclipse would not have been taken as a bad omen (pp. 16–17) is wrong—in the ancient Near East throughout the first millennium BC, partial eclipses could be regarded as terrible portents that augured death for kings or disaster for kingdoms.

[50] Actually, the new lunar crescent should technically have become visible on the evening of September 14, in ideal conditions.

[51] However, it is theoretically possible that it could have been seen before sunset by eyes shielded from the Sun.

[52] While Matthew presents the Messiah's mother as Mary, Rev. 12:1–5 identifies her as Israel. The two are compatible, for Mary is the representative of Israel when she gives birth to the Messiah (cf. Isa. 9:1a; Mic. 5:2–3).

deliberately recalling the portrayal of Israel as the Messiah's mother in Micah 5:2–3, the text to which the Jewish scholars turned in order to identify the location of the Messiah's birth for Herod the Great (Matt. 2:5–6).[53]

The role of the messianic baby that appeared in the virgin's womb could only have been played by a bright new celestial object. The fact that, according to Revelation 12:2–5, the pregnancy progressed, climaxing with an agonizing delivery, strongly suggests that this celestial entity grew in the manner of a human baby in its mother's womb and eventually descended down and out of her belly. Virgo's baby was unquestionably, therefore, a comet, more particularly a cometary coma. Only a comet can grow large and move relative to the fixed stars. Moreover, a cometary coma that is essentially oval (elliptical) in shape could readily pass for a baby when located in an area of sky regarded as a womb.[54] It is very normal for a large, long-period comet with a close perihelion distance to have this type of coma. Naturally enough, viewers would assume that a baby in its mother's womb was head-down. This assumption might have been reinforced if the coma had an area of condensed brightness around the nucleus on the sunward side of the coma—this might have looked like the shining face of the baby.

As regards the Sun in verse 1, it functions as the royal robe of Virgo as she is enthroned in splendor and is highly exalted. At the same time, the fact that in verses 2–5 Virgo is portrayed as heliacally rising with a cometary baby in her womb may inject the earlier scene on September 15, 6 BC, with additional significance. In and of itself, verse 1 does not explain why Virgo is portrayed in such exalted terms. However, verses 2–5 do explain why[55]—she becomes pregnant (literally, "having in her womb") and gives birth to a special baby representing the Messiah. The Messiah will conquer the forces of Chaos and establish his kingdom of justice and righteousness in the world, exercising dominion on behalf of Israel over all nations. In light of this, it is natural to interpret the opening scene in the celestial nativity play as a conception scene, with the Sun playing the part of God, the father of Virgo's baby. This is, of course, reminiscent of what Luke reports concerning Jesus's conception: Gabriel announced to Mary that "The Holy Spirit will come upon you, and the power of the Most High will overshadow you" (Luke 1:35).

The Moon in Revelation 12:1 seems to function as a footstool under Virgo's feet, emphasizing her great glory at or near the start of the new lunar month.[56] At the same time, the Moon was closely associated with menstruation and conception in the ancient world. In the ancient Greco-Roman world the close association of the Moon with conception was made explicit by a number of Greco-Roman medical writers and various natural philosophers.[57] Indeed, Aristotle and some Hippocratic authors claimed that conception was largely controlled by the Moon.[58] As for the Babylonian astrologers, we have evidence that they believed that omens relating to the Moon could speak of conception or childbirth. For example, they believed that if the Moon was surrounded by a halo within which was Scorpius, then high priestesses would conceive.[59] Consequently, in light of the fact that verses

[53] Cf. Isbon T. Beckwith, *The Apocalypse of John* (New York: Macmillan, 1919), 622.

[54] For example, the coma of the large Comet Hale-Bopp in 1996–1997, and the coma of the small but hyperactive Comet Hyakutake, which made a close pass by Earth (see the photographs of Hyakutake in Robert Burnham, *Great Comets* [Cambridge: Cambridge University Press, 2000], 7; and those of Hale-Bopp in ibid., chapter 4 [pp. 100–135]). It is also worth noting that a more fan-shaped parabolic coma can look like a glorious human or even possibly a baby draped in a blanket. Interestingly, Pliny referred to an extraordinarily bright comet that looked like a god in human form (*Natural History* 2.22).

[55] Note that, in her Magnificat (Luke 1:46–55), Mary's exaltation through her *conception* of the Messiah is highlighted.

[56] Farrer, *Revelation*, 141, suggested that v. 1 implied that Virgo would be especially exalted in the lunar month thus initiated.

[57] Tamsyn Barton, *Ancient Astrology* (London: Routledge, 1994), 103.

[58] Ibid.

[59] See Marten Stol, *Birth in Babylonia and the Bible: Its Mediterranean Setting* (Groningen: Styx, 2000), 105–106.

2–5 portray Virgo as having become pregnant, there is good reason to wonder if the Moon's presence in the celestial scene on September 15, 6 BC, detailed in verse 1, would have been interpreted as suggesting that Virgo had at that auspicious moment conceived her special child.

Having introduced the main actors in verses 1–2, we must now consider the narrative that unfolds in these verses.

The Course of the Celestial Drama in Revelation 12:1–2

The wonder of Revelation 12:1–2 has three essential elements: (1) The crowned Virgo is clothed with the Sun and has the Moon under her feet. (2) Virgo is pregnant. (3) Virgo is crying out because she is in labor and in torment due to giving birth (or in order to give birth). This last element may be divided into two distinct components: dilation and fetal expulsion.

Technically, these three or four elements constitute three or four different chronological moments,[60] but in verses 1–2 they are brought together to form a single wonder distinct from, though closely related to, verses 3–5.

By its very nature the celestial scene described in verse 1 can occur only around the start of a new lunar month, the beginning of either the first or the second (evening-to-evening) day, and in the western evening sky. The location of the Sun with respect to Virgo in verse 1 makes it clear that the area of sky associated with Virgo's womb was not visible at the time, because it was hosting the Sun and hence its stars were bleached out by intense sunlight. Obviously, therefore, the sighting of Virgo's baby did not occur then. It could only have occurred a few weeks later in the month inaugurated by the scene in verse 1, when, as Virgo heliacally rose over the eastern horizon, observers would have been able to catch a glimpse of the contents of her womb for the first time since September 15. Consequently, the pregnancy and labor mentioned in verse 2 follow the scene in verse 1, just as they are themselves followed by the successful delivery of the child in verse 5.

One implication of this is that the celestial activity in verse 2 is based on comet sightings in the eastern sky in the period leading up to dawn from a number of different observing sessions. In effect, then, in verse 2 we have a series of film frames that, when viewed as a moving picture, tell a grand narrative of a celestial pregnancy.

Each predawn observing session would have revealed more and more of Virgo as she heliacally rose over the eastern horizon before the stars were bleached from the sky by the rising Sun. As Virgo reemerged from her annual encounter with the Sun, steadily rising up over the eastern horizon in the last part of the night, she was seen to be with child—"she had [a baby] in her womb."[61] In other words, a bright comet had made its way into her womb to play the part of her baby in the celestial nativity play.

We may presume that the baby was relatively small when it first appeared. However, it grew in size over the following days even as it remained within the womb of Virgo.

As she begins labor, Virgo, playing the part of Israel/Mary, is described in the following terms: "[she] was screaming because of labor pain"[62] (v. 2b). That Virgo was scream-

[60] As we will highlight below in chapter 8, the influence of LXX Isa. 7:14 ("the virgin shall be with child and bear a son") on Rev. 12:2 ("And she was pregnant; and she was crying out because of labor pains and the agony of giving birth") is strong. Therefore, it is preferable to regard the clause "she was with child" in Rev. 12:2 as relating to a distinct phase of the vision and not merely as a redundant subset of the delivery scene (contra NIV, "she was pregnant and cried out in pain as she was about to give birth"). For this reason, the NASB is justified in placing a semicolon (rather than a comma) after "she was with child" and before "and she cried out" in Rev. 12:2: "she was with child; and she cried out, being in labor and in pain to give birth."

[61] In the Greek text a participial phrase (*en gastri echousa*, "having in the womb," "being pregnant") is used. The employment of the participle here is probably Semitic, it being used for a finite verb, as in Rev. 1:16 (*echōn*) and 19:12 (*echōn*) (so also Charles, *Revelation*, 1:316). Alternatively, it is possible that we have here a periphrastic participle, in which case *estin* is implied.

[62] My translation. It should be noted that there is a shift from the aorist tense in v. 1 to the present tense in v. 2. The choice of the present tense is no doubt for the sake of vividness.

ing due to her pain suggests that the cometary baby has descended within Virgo to the point that it appeared to be weighing down on her pelvic floor muscles. The pelvic floor muscles would most naturally be regarded as corresponding approximately to the level of the star 80 Virginis (see fig. 7.11).[63] Evidently the cometary coma had dropped down so that the baby's head seemed to be where it is when labor begins.

Verse 2 states that Virgo was crying out in pain not only because of the labor but also because of the torment she was experiencing in connection with the delivery of the baby. Unfortunately, the Greek text is unclear regarding whether Virgo's suffering here is essentially the same as her labor pain ("in torment *to* be delivered")[64] or is due to the actual delivery of the baby ("in torment *associated with* giving birth").[65] In the former case, John is simply underscoring dramatically the extraordinary suffering experienced by Virgo in connection with her labor. In the latter case (which is arguably more compelling[66]), he is portraying Virgo as suffering not just in the first stage of labor—active labor (dilation)—but also in the second stage—expulsion of the fetus. Virgo in dilation would have consisted of the cometary coma baby resting head-down on a point along Virgo roughly at the level of 80 Virginis. Virgo in the expulsion phase would have consisted of the baby passing out of the area associated with the womb. Of course, the birth would occur only when the area of sky corresponding to the level of Virgo's vaginal opening (80 Virginis) rose above the eastern horizon in advance of any part of the coma-baby, so that no part of the baby remained within her womb.

When ancient observers of the sky saw Virgo's baby located relative to Virgo where a fetus is relative to its mother at the point of dilation and during the stage of fetal expulsion, it was natural for them to attribute labor pain to the constellation figure. Of course, any growth in the baby's size as it descended within Virgo would have caused them to assign more distress to her, as would a long delivery.[67]

We conclude, then, that Revelation 12:2 essentially consists of a series of predawn comet observations in the eastern sky as Virgo was heliacally rising. When observers got their first glimpse into the womb of Virgo since September 15, 6 BC, they saw inside it a new occupant—a bright cometary coma that, as it heliacally rose over the eastern horizon, looked like a baby in Virgo's womb.[68] Over the following days, Virgo's special baby grew and grew. Eventually, the baby dropped down to the level on Virgo where 80 Virginis is, making it seem to observers that Virgo had begun labor. Then over the following observing sessions the coma-baby seemed to emerge from Virgo's womb, moving into the region of her legs.

We now briefly turn our focus to a discussion of the conflict theme of Revelation 12.

THE COMBAT MYTH IN THE GRECO-ROMAN AND EGYPTIAN WORLD AND REVELATION 12

In Revelation 12:3 and following it becomes clear that what is in view in chapters 12–14

[63]The star 80 Virginis is approximately at the level of Virgo's genitalia. This seems to be true in the Virgo of Hyginus, Pseudo-Eratosthenes, and Ptolemy, and probably also that of the Egyptians and Babylonians.

[64]My translation. Cf. NASB. This approach treats *tekein* as a verbal infinitive of purpose.

[65]My translation. Cf. ESV ("the agony of giving birth"). This approach treats *tekein* as an epexegetical infinitive (i.e., the infinitive explains the agony).

[66]In addition to the presence of *kai*, the fact that the usual indications of the purpose infinitive are absent here suggests this.

[67]It is interesting that in the fifth century AD Hephaistio of Thebes, *Apotelesmatics*, 1.24 (Hephaistio of Thebes, *Apotelesmatics, Book I*, trans. Robert Schmidt, ed. Robert Hand [Berkeley Springs, WV: Golden Hind Press, 1994], 58–60), refers to a "comet which is called Eileithyia," the name of the goddess of childbirth and one of the names of Virgo. It is possible that Hephaistio's assessment was based on an actual cometary apparition in Virgo, as Boll (*Offenbarung Johannis*, 105n1) proposed. This comet, according to Hephaistio, had the face of a virgin and golden hair and portended the dawn of a better era. See Hegedus, "Some Astrological Motifs," 82.

[68]Curiously, *Sibylline Oracles* 8:456–457, from the second or third century AD, states concerning the Messiah that "as a new light he rose from the womb of the Virgin Mary" (J. J. Collins, "Sibylline Oracles," in *The Old Testament Pseudepigrapha*, ed. James H. Charlesworth, 2 vols. [New York: Doubleday, 1983], 1:428).

is a great cosmic war between Virgo's son, the Messiah, and the great serpentine dragon Hydra. The Messiah is destined to conquer the dragon at the end of the age, but not before the dragon has persecuted the woman (Israel) and the woman's other offspring (the church). The climax of the dragon's reign comes at the end of the age, when he invests his royal authority in one man, the first beast, a megalomaniac who mounts an unprecedented offensive against the church (13:1–10). However, the Messiah will return and overthrow the dragon and the beast, establishing the kingdom of God on the earth (14:6–20).

A number of scholars, including Adela Yarbro Collins, have drawn attention to the employment of the Combat Myth in Revelation 12. This myth, which was widespread in the ancient world, took a variety of forms, most notably Greco-Roman and Egyptian.

Greco-Roman

The Greco-Roman version of the Combat Myth featured Leto as the pregnant goddess. According to one tradition (related by Hyginus), Python, the dragon, discovered that Leto's son, the child of Zeus, would slay him. Python pursued Leto and sought to kill her and her unborn infant. However, thanks to the north wind and Poseidon, Leto escaped to Delos, where she gave birth to Apollo and Diana. Sure enough, not long afterwards, four days in fact, Apollo slayed the dragon with arrows. For this reason he came to be called Pythian Apollo.[69]

Egyptian

In the Egyptian Combat Myth, which was Egypt's major national myth for some three millennia[70] and was well known internationally,[71] the combat was all about sovereignty. The red Chaos Monster, the serpent-dragon Seth-Typhon, killed Osiris (god of the underworld), but the dead king nevertheless managed to impregnate Isis. The child in her womb was Horus. The pregnancy was exceptionally long and her labor excruciatingly painful and difficult.[72] But she eventually bore the child. The dragon, discovering that Isis had delivered her child, determined to slay the boy, but Isis fled on a papyrus boat through the marshes to Chemnis in lower Egypt. Her son was destined to become king and to restore order to a world that had been dominated by Chaos since the dragon had slain Osiris. In due course Horus fought with Seth-Typhon and defeated him, imprisoning him and then ultimately destroying him with fire. Horus became king of the living and Osiris became king of the underworld.[73]

Revelation 12's Version

In Revelation 12 the woman is Virgo qua Israel and she is pregnant with the Messiah, and the dragon is Hydra, the massive neighboring constellation figure to her south (fig. 7.13).

The sea-dragon is presented in terms that highlight his royal authority—he has ten horns

[69] Hyginus, *Fabulae* 140. For more standard versions of the Apollo birth myth, see Robin Hard, *The Routledge Handbook of Greek Mythology* (London: Routledge, 2004), 188–190.

[70] Geraldine Pinch, *Egyptian Myth: A Very Short Introduction* (Oxford: Oxford University Press, 2004), 76.

[71] On the popularity of Isis within the ancient world, note that the first-century BC author Diodorus Siculus (1.25) stated that "virtually the whole populated world" testified that Isis was to be honored for revealing herself in the healing of her devotees from all kinds of diseases. In that same passage Diodorus states that Horus was widely regarded as a benefactor of the human race, healing and granting oracles to those who sought his aid. Witts, *Isis in the Ancient World*, 222–254, details the history of the Roman emperors' devotion to Isis—for example, he points out that Tiberius in AD 23 is represented as sacrificing to Horus and Isis. In the Greco-Roman environment Horus was often called Harpocrates. As Sharon Kelly Heyob (*The Cult of Isis among Women in the Graeco-Roman World* [Leiden: Brill, 1975], 76) has written, "To judge by the number of representations of Isis and Harpocrates it was in her role as mother that Isis achieved the greatest popularity in the Graeco-Roman world."

[72] Geraldine Pinch, *Egyptian Mythology: A Guide to the Gods, Goddesses, and Traditions of Ancient Egypt* (Oxford: Oxford University Press, 2004), 146.

[73] For a fuller account of the Horus birth myth, see Bousset, *Offenbarung*, 354–355 (for his exegesis of Revelation 12, see pp. 335–358) and Charles, *Revelation*, 1:313. Bousset, in particular, points out the closeness between the Egyptian myth and the account in Revelation 12. As Charles (*Revelation*, 1:313) highlights, there were different variants of the story—another version has Isis giving birth at the conclusion of her flight, at Chemnis.

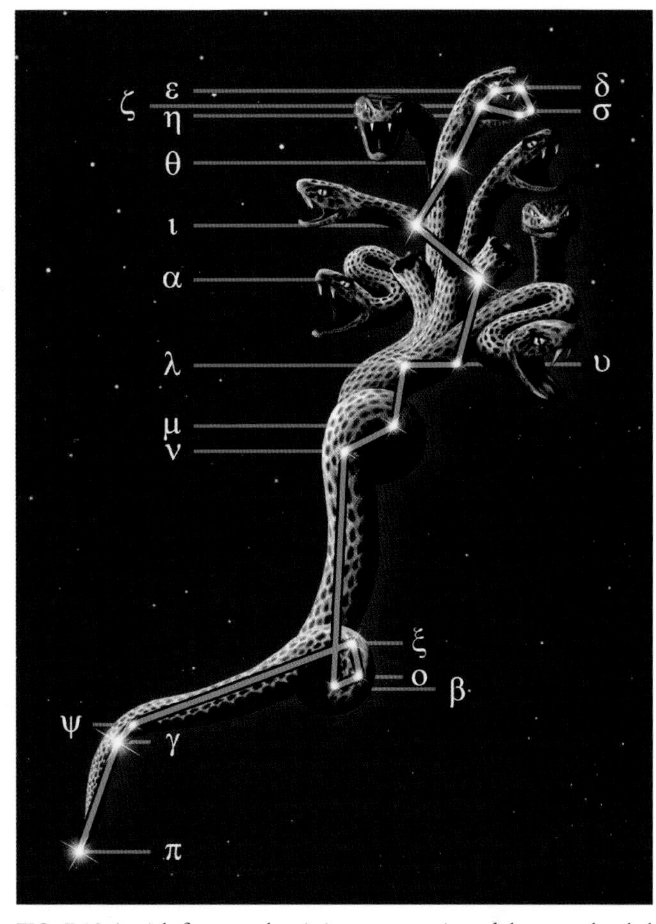

FIG. 7.13 A stick figure and artistic representation of the seven-headed constellation figure Hydra, along with the names of its major stars. Image credit: Sirscha Nicholl.

and seven crowns. When Virgo was giving birth to the messianic baby, the dragon dragged, or swept from their places, one-third of the stars, throwing them to the earth. Then he stood up aggressively, ready to kill the newborn (v. 4), no doubt because he realized that this baby posed an enormous threat to his dominion—destined to destroy him and inaugurate the kingdom of God in the world. Thereafter Virgo's son was quickly snatched away to God and his throne—the description of this in verse 5b strongly recalls Psalm 110:1, where the divine Messiah is invited by God to sit down at his right side until God has made his enemies a footstool for his feet. The dragon was thus thwarted in his bid to derail the divine plan and hence was doomed to see the evil world empire he sponsored decisively terminated. Consequently, the dragon mounted a vicious attack on the baby's mother, the woman. However, she was given wings with which to escape into the wilderness for 3½ years and there was aided by the earth (vv. 6, 13–16). Frustrated once more, the dragon then turned his attention to all-out persecution of the other offspring of the woman (v. 17).

The Egyptian myth is closer to Revelation 12[74]—in both, the pregnant woman gives birth to her divine son before the dragon chases her, and what is at stake in the conflict is royal authority.[75]

Of course, the fact that Virgo was often

[74] Contra many, including Adela Yarbro Collins, *The Combat Myth in the Book of Revelation* (Missoula, MT: Scholars Press, 1976), 61–71; Grant R. Osborne, *Revelation*, Baker Exegetical Commentary on the New Testament (Grand Rapids, MI: Baker, 2002), 454; Frederick J. Murphy, *Fallen Is Babylon* (Harrisburg, PA: Trinity Press International, 1998), 279–280; and Beasley-Murray, *Revelation*, 192, who find Revelation 12's narrative to be closest to the Greco-Roman form of the myth. According to Murphy (*Fallen Is Babylon*, 280), the Greco-Roman version of the myth and Revelation 12 have in common that a woman was pregnant with one who "will share in the divine rule of the universe," that a dragon attacked her, that she was rescued, that water features in the story, and that the son goes on to defeat the dragon. However, (1) Leto bears not only Apollo but also Artemis/Diana; (2) the predominant version of the myth has Apollo born after the flight from the dragon, not before it, contrary to Revelation 12; (3) Apollo slayed the dragon within just 4 days of his birth, whereas Virgo's son in Rev. 12:5 is caught up to God and his throne and does not vanquish the dragon until he returns at the end of the age; (4) the means of rescue and the role played by water in the Greco-Roman version are very different from the Biblical story; (5) the idea of "sharing in the divine rule of the universe" is hardly the way Revelation frames Jesus's destiny.

[75] However, it is important to remember that, in the general period we are considering, it was common to identify gods from different regions. In particular, Diodorus Siculus 1.25 informs us that Horus was popularly identified with Apollo in the Greco-Roman world—like Apollo, Horus was believed to benefit humanity by his oracles and healings.

regarded as Isis and that Revelation 12:1 is portraying Virgo in terms strongly reminiscent of Isis—enthroned and gloriously enrobed—strengthens the case for regarding the Isis-Horus myth as more important for Revelation 12 than the Greco-Roman myth concerning Apollo. Some have suggested that Seth-Typhon was identified with Hydra by Teukros of Babylon and others.[76] Regardless, the case for regarding Revelation 12 as picking up on contemporary associations between Virgo and Isis is formidable.

Isis was frequently portrayed as having the Moon on her crown, and Horus as the Sun God (fig. 7.14).

FIG. 7.14 An 8th-century-BC statue of Isis, with a crown consisting of the Moon within horns, preparing to nurse her son Horus. Image credit: the Walters Museum, Baltimore.

When, therefore, Revelation 12:1 describes Virgo as enthroned, with the Sun clothing her and the Moon under her feet, it naturally brings images of Isis to mind, although it challenges Isis theology as it does so. First, the locations of the celestial lights with respect to Virgo's body are important, making the point that Virgo did not give birth to the Sun, nor was she sovereign over the Moon, but the Sun could beautify her and the Moon could exalt her. Moreover, her son was not Horus, but rather the Messiah. Furthermore, it is very possible that the Sun was playing the part of the divine Father who begot the Messiah.

Of course, Teukros the Babylonian mentions that, in his day, Virgo was thought to be Isis enthroned and nursing Horus. If this indicates that the Greco-Egyptian identification of Virgo as Isis had taken root in Babylon, that would mean that the celestial narrative that is preserved for us in Revelation 12, and in which the comet plays the star role, would have spoken powerfully to Babylonian astronomers-astrologers at the time of Jesus's birth. Indeed the fact that Revelation 12 is interacting with pagan mythology regarding Isis and Horus may well hint at the kind of paradigms through which some pagans (perhaps especially those in Babylon) had sought to interpret the celestial drama. More important, since the celestial drama was essentially an adaptation of the internationally known combat myth, it offered pagan observers a ready paradigm by which to understand the importance of the Messiah's nativity in salvation history. It highlighted that Jesus was the fulfillment of the hope of all humans for deliverance from Chaos and Disorder.

[76] On identifications of Seth-Typhon in the stars, see Herman Te Velde, *Seth, God of Confusion: A Study of His Role in Egyptian Mythology and Religion* (Leiden: Brill, 1977), 86–87; Boll, *Offenbarung Johannis*, 98–124, esp. 108–109; W. Hadorn, *Die Offenbarung des Johannes* (Leipzig: Deichert, 1928), 131–132; H. Kraft, *Die Offenbarung des Johannes* (Tübingen: Mohr, 1974), 164; Jan Willem van Henten, "Dragon Myth and Imperial Ideology in Revelation 12–13," in *The Reality of Apocalypse: Rhetoric and Politics in the Book of Revelation*, ed. D. Barr (Atlanta: Society of Biblical Literature, 2006), 186–188.

THE DRAGON, THE METEOR STORM, AND THE BIRTH

The Relationship of Revelation 12:3–5 to verses 1–2

Revelation 12:3ff. is set apart from verses 1–2 by the new introduction ("And another sign . . ."), but it is clear from verse 4b that the story of Virgo's pregnancy and delivery of her child is still very much in view. While everything in verses 1–2 seems to relate to Virgo's pregnancy, verses 3–5 are united by their focus on the grand conflict between the dragon and Virgo's special son. However, although the spotlight in verses 3–4a is on a dragon rather than a woman, the dragon is introduced here because he is playing the role of antagonist in the nativity drama. It is important to appreciate the continuity between verses 1–2 and verses 3–5. For one thing, they are both astronomical in nature ("in heaven"; vv. 1, 3). For another, verses 3–5 continue the narrative that began in verses 1–2. The action of verses 3–5 clearly follows the events of verses 1–2 in time. The delivery of Virgo's baby began in verses 1–2 and is completed in verse 5. The action of verses 3–4 manifestly belongs chronologically between the action of verses 1–2 and that of verses 5ff. Accordingly, what is described in verses 3–4 occurs while Virgo is in labor and indeed on the verge of fully delivering her baby. As the baby is about to be fully delivered, the dragon is ready to devour him (v. 4b). With this in mind, let us examine verses 3–5. As we do so, we must remember that verses 2–5 record moments drawn from predawn astronomical observing sessions stretching over a few weeks (long enough to make room below Virgo's womb into which the comet could descend to be born). The cometary baby was slowly emerging from her belly, descending into the region of sky associated with her legs.

The Hydrid Meteor Storm

In verses 3–4 John writes, "And another sign appeared in heaven: behold, a great fire-colored[77] dragon, with seven heads and ten horns, and on his heads seven crowns.[78] His tail swept/dragged[79] a third of the stars of heaven and cast them to the earth. And the dragon stood before the woman who was about to give birth, so that when she bore her child he might devour it."

What is described in verses 3–4 is specifically placed on the eve of the birth of the child ("the dragon stood before the woman who *was about to* bring forth the child";[80] v. 4b). That is, the dragon's throwing of the stars to the earth and his aggressive standing in front of the woman must have been seen on the last predawn observing session before the coma-baby had descended to the point that it could be regarded as having been born.

The focus in verses 3–4 is not on Virgo but on a second constellation, one consisting of a great fire-colored dragon, with seven heads and ten horns. There can be little doubt that this constellation is the Greek Hydra,[81] known in Babylon as the Serpent, which was seen as a serpent-dragon "adorned with horns, wings and a pair of legs" and was identified with the Chaos Monster.[82] The serpentine dragon in Revelation 12 is, evidently, closely associated with the seven-headed Chaos Monster Tiamat in Babylonian tradition[83] and with the seven-headed Leviathan

[77] I have substituted "fire-colored" for the ESV's "red." The Greek word is *purros* and most naturally connotes the color of fire. See Johannes P. Louw and Eugene A. Nida, *Greek-English Lexicon of the New Testament: Based on Semantic Domains* (New York: United Bible Societies, 1988), §79.31.

[78] My translation.

[79] The Greek word *surō* most often means "drag" or "pull" (as in John 21:8; Acts 8:3; 14:19; 17:6), but here, with reference to the action of a dragon's tail, it could mean "sweep" (so most English versions).

[80] My translation.

[81] So Boll, *Offenbarung Johannis*, 101–102.

[82] White, *Babylonian Star-Lore*, 183.

[83] Charles, *Revelation*, 1:317–318. Aune, *Revelation*, 2:685, points out that some Mesopotamian cylinders portrayed divinities fighting a serpent with seven heads (James B. Pritchard, *Ancient Near East in Pictures Relating to the Old Testament with Supple-*

of Canaanite mythology.[84] The Greek Hydra/Babylonian Serpent is located in the area of the sky right beside the zodiacal constellation Virgo, just to its south. As a constellation, Hydra was "great" (v. 3) because it was the largest of all, aside from Argo Navis.

The manifestation of Hydra is astonishing and deeply disturbing. The fiery color, coupled with the self-assertion implied by the horns and crowns, suggests that the dragon is dangerous and angry, and enjoys great royal authority. When we get to chapter 13, we realize that this serpentine Chaos Monster is the power behind the first beast, who will rule over the whole world at the end of the age. But here, on the eve of the birth of the cometary baby, he is seething with rage because he realizes that this child is destined to destroy him. Consequently, in an act of fearful power, he casts a third of the stars to the earth.

Revelation 12:4 is very specific in saying that the means by which the stars are thrown toward the earth is Hydra's tail. Hydra's tail is a segment of the bottom of the constellation figure. Its lowest point (see fig. 7.13), the tip of the tail, is π (Pi) Hydrae. This was the opinion of Ptolemy in the second century AD, but it preceded him.[85] Pseudo-Eratosthenes (*Catasterismi* 41) and Hyginus (*Poetica Astronomica* 3.39) offer quite vague matchups of Hydra's stars to the constellation figure.[86] They speak of 9 dim stars in the section of Hydra from the tail to the fifth coil. These 9 stars may be identified as π, γ, ψ, β, ο, ξ, HIP56332, HIP56280, and HIP57613. Together those nine constitute the brightest stars in that long section of Hydra. Moreover, Hipparchus in the second century BC also regarded the tip of Hydra's tail as being π (Pi) Hydrae.[87] That the Babylonians in this general period thought of the constellation similarly is suggested by a drawing of the Serpent on a Babylonian astrological cuneiform tablet from Uruk in the Seleucid period, in which the small constellation known to the Babylonians as the Raven (and known to the Greeks as Corvus, or the Crow) is portrayed as a bird perched on the Serpent's tail (fig. 7.15). Accordingly, π (Pi) was almost certainly understood to be the tip of Hydra's tail at the time of the birth of Jesus.

The highest point of the tail segment of Hydra is not very clear, although Pseudo-Eratosthenes locates the Crow (Corvus) on the tail and reveals that what he calls the fifth coil of Hydra (evidently associated with β, ο, and ξ Hydrae) is not part of the tail.[88] The Babylonian representation of Hydra was very

ment, 2nd ed. [Princeton, NJ: Princeton University Press, 1969], 221 no. 691; cf. no. 671). On Tiamat sometimes having been portrayed as having seven heads, see Eberhard Schrader, *Die Keilinschriften und das Alte Testament*, 3rd ed., ed. H. Zimmern and H. Winkler (Berlin: Reuther und Reichard, 1903), 504, 512; Collins, *Combat Myth*, 77.

[84] On the Canaanite seven-headed Chaos Monster, see James B. Pritchard, *Ancient Near Eastern Texts Relating to the Old Testament with Supplement*, 3rd ed. (Princeton, NJ: Princeton University Press, 1969), 138; J. C. L. Gibson, *Canaanite Myths and Legends*, 2nd ed. (Edinburgh: T. & T. Clark, 1978), 50, 68–69 (*Ugaritic Baal Cycle* 1.3.5.1–3, 27–30); and Caird, *Revelation*, 150. Walter Burkert, *Structure and History in Greek Mythology and Ritual* (Berkeley: University of California Press, 1979), 80–83, discusses evidence that Hydra was sometimes considered to have seven heads.

[85] *Ptolemy's Almagest*, trans. Toomer, 393. Aratus, *Phaenomena* 443–448 (third century BC) states that Hydra's tail hangs over the Centaur, which is consistent with the tail extending to π (Pi) Hydrae. This is the predominant conceptualization of Hydra today: see, for example, Jim Kaler's discussion regarding π (Pi) Hydrae at http://stars.astro.illinois.edu/sow/pihya.html (last modified April 24, 2011).

[86] See Condos, *Star Myths*, 120, 122.

[87] See Hipparchus's *Commentary on the Phenomena of Aratus and Eudoxus*, in C. Manitius, ed. *Hipparchi in Arati et Endoxi Phaenomena Commentariorum Libri Tres* (Leipzig: Teubner, 1894), 219. I am grateful to Roger MacFarlane and Paul Mills for graciously giving me prepublication access to relevant parts of their forthcoming English translation (the first of its kind), *Hipparchus' Commentary on the Phaenomena of Aratus and Eudoxus*. On the influence of Babylonian astronomy on Hipparchus, see especially F. X. Kugler, *Die Babylonische Mondrechnung. Zwei Systeme der Chaldäer über den Lauf des Mondes und der Sonne* (Freiburg: Herder, 1900), 4–8, 50–53; and G. J. Toomer, "Hipparchus and Babylonian Astronomy," in *A Scientific Humanist: Studies in Memory of Abraham Sachs*, ed. Erle Leichty, Maria Ellis, and Pamel Gerardi (Philadelphia: Occasional Publications of the Samuel Noah Kramer Fund, 1988), 353–362.

[88] This assessment is consistent with Teukros of Babylon's description of that part of the sky (see Boll, *Catalogus*, 202; for an English translation, see James H. Holden, ed. and trans., *Rhetorius the Egyptian* [Tempe, AZ: American Federation of Astrologers, 2009], 176–178) and with Eudoxus's claim that Hydra's tail did not set until Pisces rose (in Manitius, *Hipparchi in Arati et Endoxi Phaenomena Commentariorum*, 170, 172).

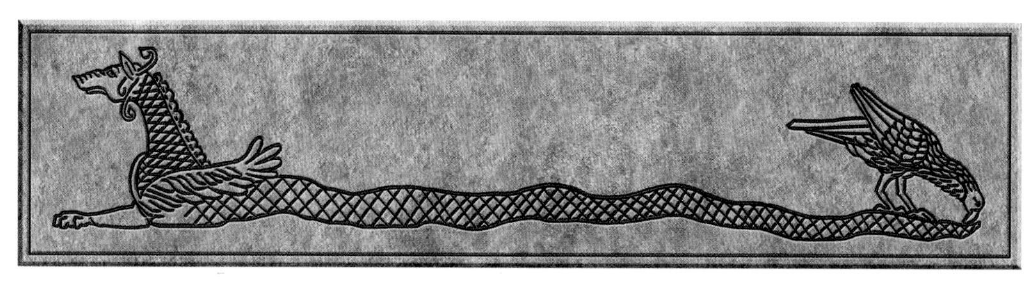

FIG. 7.15 The Babylonian constellation figure Serpent (equivalent of Hydra) as imagined in Seleucid-era Uruk—reconstructed from two pieces of a cuneiform tablet, one of which is in the Louvre in Paris (AO 6448) and the other in the Vorderasiatisches Museum in Berlin (VAT 7487) (photographs of which appeared in Ernst F. Weidner, "Eine Beschreibung des Sternenhimmels aus Assur," *Archiv für Orientforschung* 4 [1927]: 73–85). The constellation Raven is perched on Hydra's tail. Image credit: Sirscha Nicholl.

similar, except that the bird sitting on the tail is the Raven and is oriented toward the tip of the tail rather than, in the case of Corvus in Greco-Roman imagination, away from it.

From somewhere along this stretch of the tail, from π (Pi) Hydrae to where the feet of the Crow/Raven rested on Hydra, one-third of the stars of heaven seemed to streak toward the earth. Since the scene climaxes in verse 4b with Hydra standing, that is, with the tip of the tail level with the eastern horizon, the shooting stars must have seemed to stem from a higher point in the tail, probably between γ (Gamma) Hydrae (which is about 11 degrees basically straight above the tip) and a star like HIP59373 (about 25 degrees from π [Pi] Hydrae), under Corvus the Crow/the Raven. All or part of this section of Hydra's tail was above the eastern horizon when the stars seemed to be thrown toward the earth.

It seems clear that what is being described in Revelation 12:3–4 is a great meteor storm, when thousands, tens of thousands, or even hundreds of thousands of meteors per hour streak through the sky. Meteor storms occur when Earth moves through a dense collection of meteoroids in a meteoroid stream. As many who witnessed the Leonid meteor storms of the nineteenth century noticed, the meteors in a meteor storm all appear to streak away from one particular point in the sky, called a radiant. It is not that *all* the meteor streaks in a given shower begin at the radiant—they do not—but rather that if one draws lines from the meteor streaks backwards, those extended lines will converge at the radiant. In the case of the meteor storm in view in verses 3–4, the radiant is manifestly the tail of the serpentine dragon. We shall give more detailed consideration to verses 3–4a and the meteor storm in appendix 2.

According to verse 4b, the serpent-dragon is intent on killing the newborn child as soon as he is fully delivered. Specifically, Hydra the dragon "stood before the woman who was about to bring forth the child, so that when she brought forth her child he might devour it." The Greek verb for "stood" here is in the perfect tense, to make the Monster's "standing" more vivid.[89]

Of course, used of the serpent, the language is very forceful: snakes often "stand" before they strike out. The dragon's standing before Virgo is therefore aggressive—Hydra is intent on attacking the baby as soon as it emerges completely from Virgo's belly. He is desperate to kill and devour his ultimate opponent.

But what did Hydra's "standing" actually look like in astronomical terms?

Hydra only "stands" when the constellation is ascending over the eastern sky and

[89]It is an aoristic or dramatic perfect; see Daniel Wallace, *Greek Grammar: Beyond the Basics* (Grand Rapids, MI: Zondervan, 1996), 578.

the tip of the tail seems to be resting on the ground, that is, at the point of just emerging over the horizon. For Hydra to stand, π (Pi) Hydrae would simply have to be level with the eastern horizon (see fig. 7.13).[90]

Viewed from Babylon in the run-up to dawn in September/October, the lowest part of Virgo is almost level with π (Pi) Hydrae—λ (Lambda) Virginis, associated with Virgo's left foot by Hipparchus, Hyginus, Pseudo-Eratosthenes, and Ptolemy, is less than 4 degrees in altitude, and μ Virginis, regarded as her right foot by the Greco-Roman astronomers, is just over 2 degrees up. So someone in the Near East gazing at the eastern horizon as π (Pi) Hydrae was rising would have been looking at the whole of Hydra and the whole of Virgo. The image of Hydra standing beside Virgo as she is about to bring forth her baby is a powerful and deeply troubling one.[91]

The occurrence of a dramatic meteor storm radiating from Hydra's tail on the eve of the birth of Virgo's son, when little of the coma-baby remained inside her womb, would have been truly extraordinary.

The Birth

In verse 5 we return to the focus of verses 1–2—that is, Virgo's delivery of her child. According to verse 5a, she proceeded to give "birth to a son, a male child, who will rule all the nations with an iron scepter."[92]

Evidently, the cometary baby had now, finally, descended to the point where the whole of it rose after the rising of Virgo's vaginal opening. Her baby could therefore be regarded as born. Of course, that seemed to imply that the baby on the earth whom the cometary baby represented was born at that very time. In a special sense, then, the terrestrial newborn was Virgo's son, born in conjunction with her celestial baby.

It seems most natural to infer that the coma-baby at the time of the birth was approximately the size of a full-term baby relative to its mother—hence about 9–12 degrees long.

Needless to say, the sight of Virgo giving birth to her baby would have been a celestial wonder to behold!

The Iron Scepter

Verse 5's reference to a "scepter," the symbol of royal authority, at this point is striking. It recalls Psalm 2:9's prophecy that the Messiah will rule the nations with an iron scepter, and Numbers 24:17's oracle that the Messiah would be a "star" and "scepter."[93] That the scepter is "iron" suggests, as in Psalm 2:9, that it is a symbol not just of royal authority but also of overwhelming power.

The narrative of Revelation 12:1–5 has been about an astronomical nativity. Why, therefore, does John interrupt the flow of the celestial story to refer to the future reign of the Messiah? The answer is presumably that the remarkable play unfolding in the heavens above alluded to this future reign. Since the birth consisted of the cometary coma playing the part of a baby in Virgo's womb that grows and is born, the most obvious and natural explanation of this parenthetical note ("one

[90] Although in theory Hydra's "standing" could refer to its actual heliacal rising, that is most unlikely, because third-magnitude stars close to the ecliptic, like π (Pi) Hydrae, have to be a few degrees above the horizon in order to be visible at their heliacal rising.

[91] It is unclear from Rev. 12:3–4 whether the meteor storm lasted for a short time (with Earth quickly passing through the dense section of the meteoroid stream) or endured until sunrise (or beyond!). The fact that the scene climaxes with Hydra's standing may possibly suggest that the most intense part of the meteor storm was over by the point at which π (Pi) Hydrae rose.

[92] My translation.

[93] Note that the same Hebrew word (*shbt*) is used in Num. 24:17 ("*scepter*"); Ps. 2:9 ("*rod* of iron"); and Isa. 11:4 ("*rod* of his mouth"). Essentially, as Murphy, *Fallen Is Babylon*, 140, comments, "The scepter is a rod symbolizing kingship, but it can also be seen as a weapon. Kingly power carries destructive potential." Ps. 2:9 itself was probably picking up on Num. 24:17. Recall also the connection between Ps. 2:8–9 and Num. 24:17 in Rev. 2:26–28: the conquering believer's reward will be participation in the Messiah's iron-scepter reign (cf. Ps. 2:9) and reception of the "morning star" (cf. Num. 24:17). The linking of star and scepter in Rev. 2:26–28 clearly reflects Num. 24:17 ("a star shall rise . . . , a scepter . . ."). In addition, Rev. 22:16 ("I am the root and the descendant of David, the bright morning star") recalls Num. 24:17 when it refers to Jesus as the Star (the verse also alludes to Isa. 11:1–16 [particularly v. 4], a text that highlighted the messianic significance of the sign of the virgin's giving birth and revealed that Balaam's prophecy against Moab and Edom in Num. 24:17–19 still awaited future fulfillment in connection with the Messiah).

who is to rule all the nations with an iron scepter"; v. 5) is that the whole comet, including its long tail, at that very point took the form of an iron scepter. Scepters in the ancient world were typically straight, long sticks, often with some decoration or fancy design at the top. At the time that the baby Messiah was born, the cometary tail was apparently silvery-grey in color and was extremely long. Numbers 24:17 is very important here, because that prophecy by the Mesopotamian seer Balaam had suggested that, at the time of the Messiah's coming, a cometary scepter would "rise" or "stand." As we have already seen, what Balaam prophesied suggested that the Messiah's comet-star would look like a scepter at the time when it rose. Revelation 12:5 appears to be claiming that the Messiah's cometary scepter was a prominent celestial feature at the point when Virgo's baby was fully delivered.[94] The fact that Revelation specifies that the Messiah would rule "all the nations" with an iron scepter may well imply that the comet as a whole at that moment stretched all the way across the sky from low on the eastern horizon to the western horizon. Such a phenomenon would have made it seem that the scepter was resting on the ground in the west (namely, Israel). Balaam's scepter-star would literally have looked like it was rising up out of Israel to announce the coming of the Messiah, who would exercise dominion over the entire world (Num. 24:17).

Accordingly, it would appear from Revelation 12:5 that the newborn baby was not the only image being created by the comet at the point of birth. Even as the comet's coma played the part of Virgo's baby boy, the comet as a whole seems to have been forming an enormous iron scepter that symbolized his eschatological reign over the whole earth. Together, these two images powerfully revealed that the Messiah, future autocrat of the world, was being born at that time in Judea.[95]

In light of the fact that Hydra was presented in Revelation 12:3–4 in terms that highlighted his royal authority and indeed his determination to attack, kill, and devour the Messiah as soon as he was born, the mention of the iron scepter in verse 5 is very striking, for it makes the point that Hydra is destined to fail. Even Hydra's final worldwide kingdom at the end of the age would be overthrown and replaced by the Messiah's everlasting kingdom (Revelation 12–14). The Messiah's scepter would loom large over the whole world.

The Deliverance of the Child

In verse 5b we discover that, although the dragon Hydra was determined to devour the son of Virgo, he was unable to, because the infant child was snatched forcibly away and quickly taken to God and his throne. Scholars disagree regarding what is in view here. As far as the immediately preceding context is concerned, a reference to the deliverance of baby Jesus from Herod the Great would seem the most natural interpretation. However, as far as the rest of the chapter goes, the most plausible interpretation is that Jesus's resurrection-ascension (the two events being viewed as one) is in view. It is probably best to allow for both senses, with John deliberately conflating the two. In other words, baby Jesus's rescue from Herod was a foretaste of the ultimate deliverance he would come to know at his resurrection-ascension,

[94] Balaam's star and scepter are frequently linked to the Magi's Star in other early Christian literature—for example, Matt. 2:2 ("his [i.e., the Messiah's] star"); *Testament of Levi* 18:3; *Testament of Judah* 24:1; Justin, *Dialogue with Trypho* 106:5–6 (cf. 126:1) and *First Apology* 32:12–13; Ignatius, *To the Ephesians* 19:2–3; Irenaeus, *Adversus Haereses* 3.9.2–3, and *Demonstration of the Apostolic Preaching* 58; Origen, *Contra Celsum* 1:60, and *Homilies on Numbers* 13:7; Eusebius, *Demonstratio Evangelica* 9.1; also possibly LXX Zech. 6:12 (*Anatolē*). For more, see Jean Daniélou, *Primitive Christian Symbols*, trans. Donald Attwater (London: Burns & Oates, 1964), 102–123.

[95] On the tradition that Balaam was a father of the School of the Magi, see Tim Hegedus, "The Magi and the Star in the Gospel of Matthew and Early Christian Tradition," *Laval théologique et philosophique* 59 (2003): 87–89, who mentions that, among others, Ambrose of Milan, Origen, and "a certain history" used by Eusebius all held this view.

when he would take his seat at the Father's right hand (Acts 1:9; 2:32–36; cf. Ps. 110:1).

This somewhat awkward combination of ideas, as well as the skipping over of Jesus's ministry and death and the most peculiar conceptualization of the resurrection-ascension as a forcible snatching of Jesus away to God and his throne (Rev. 12:5b), calls for some explanation. How are we to explain these things?

The answer is most probably found in the realization that the action remains strictly astronomical until verse 6. Only there does the scene shift from the heavens to the earth, and the focus from that point onward is the woman rather than the male child. What therefore John sees in verse 5 is apparently still part of the heavenly nativity play marking Jesus's birth. Consequently, the reason that the male child's snatching to God's throne immediately follows the birth is presumably that this is how the drama marking the Messiah's birth unfolded in the heavens. In other words, the comet baby, having descended to the point where it was clearly below Virgo's belly and therefore was born, continued to move down toward the sunlight and the horizon, quickly disappearing from the predawn sky. If we are correct that the Sun had played the part of God on September 15, 6 BC, then it was very natural for the Sun to be envisioned as continuing to assume the role of God in the celestial nativity drama. Therefore, when the cometary baby appeared to move down toward the sunlight, it would have seemed to dedicated observers that the child was being taken into the presence of God. In the context of the conflict between the Messiah and Hydra, the speedy removal of the cometary baby would have seemed to refer to a divine rescue from Hydra.

REVELATION 12:1–5: A SUMMARY

In summary, Revelation 12:1–5 reveals the multifaceted celestial wonder that coincided with the birth of Jesus—the very sight that the Magi had seen in the eastern sky and that had prompted them to make a long journey west to Judea to worship the Messiah. In this astonishing celestial nativity drama, Virgo was playing the part of Israel/Mary, and the comet's coma was playing the role of the messianic baby. After rising heliacally in Virgo's womb, looking like a baby, the cometary coma remained there for many days, growing in size in the manner of a normal human baby in its mother's womb. While the comet rose in altitude, each passing day would have meant that it was observable earlier and in darker skies. Then, after descending within Virgo's belly, the coma would have moved down out of it, making it seem that the baby was being born. Eventually, the baby appeared to have completely vacated Virgo's womb and at this point it was regarded as having been born. At that moment the comet as a whole apparently formed an immense scepter that stretched from the eastern horizon all the way to the western horizon. Those attuned to what was happening and interpreting it messianically would have had no question but that the Messiah was born at that very time. Finally, the cometary baby speedily disappeared into the Sun's light (i.e., heliacally set), bringing an end to the wonder in the eastern sky.

We infer from Revelation 12:1–5 that the comet's coma became extraordinarily large, equivalent in size to a large full-term baby at the point of birth; that the comet as a whole took the form of a long iron scepter at the point of the child's birth; and that it must have been very bright. Further, Revelation suggests that, on the eve of the birth, there was a meteor storm radiating from the tail of Hydra.

What John writes enables us to narrow down when the celestial events took place—during the months of Ululu and Tishratu (Babylon) or Tishri and Heshvan (Judea), namely in September and October of 6 BC. Moreover, Revelation 12:1–5 enables us to narrow down the time of Jesus's birth to

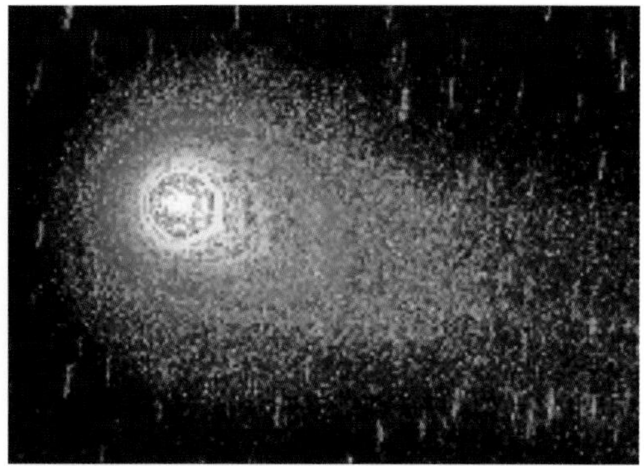

FIG. 7.16 Comet Hyakutake on March 21/22, 1996. The image to the right is a close-up of the coma that has been altered to highlight the pattern of the coma's condensation of brightness. Image credit: Herman Mikuz, Črni Vrh Observatory, Slovenia: http://www.observatorij.org/BrightComets/96b2c.html.

mid-October (early Tishratu in Babylon and early Heshvan in Judea)[96] of 6 BC. This is a plausible time of year for Jesus's birth—it was when the Romans tended to have their censuses[97] and when shepherds would certainly have been out in the fields (Luke 2:1–18).[98] The cometary baby would have heliacally risen on September 29 or 30 and remained in her belly for about two weeks before slowly descending out of it to be born.

Essentially, the wonder that marked Jesus's birth was an incredible full celestial nativity drama focused on Virgo and a great comet that seemed to bring her to life.

THE ASTRONOMICAL DIMENSION

As to how a comet could have done what Revelation 12:1–5 indicates that this 6 BC comet did, it is helpful to consider the following.

We have already seen that a cometary coma that is elliptical (oval) is ideally qualified to represent an upside-down baby.

For a comet to do (from the vantage point of an Earth-dwelling observer) what Revelation 12:2–5 describes requires very unusual comet-Earth-Sun geometry. It is possible only in the case of a narrowly inclined retrograde comet that has a close encounter with the Sun near the end of September. A few days later, the coma of such a comet would emerge over the eastern horizon in Virgo's womb just in advance of the Sun. Observers on Earth would be able to see the coma in Virgo's belly each day during a short window of time just before dawn, as they looked at the eastern horizon. As Earth continues on its orbit around the Sun, each day observers are able to see more and more of the stars of Virgo (from the top down) before the Sun rises. For the comet to remain in the same small area of sky, within Virgo's belly, its orbital course and velocity would have to work in synchronization with Earth's. The comet's relative "stability" within Virgo's belly would have had to last for a couple of weeks, to provide sufficient time for the lower part of Virgo to be far enough off the horizon to accommodate the cometary baby as it emerged from the womb to be born.

[96] Richard A. Parker and Waldo H. Dubberstein, *Babylonian Chronology 626 B.C.—A.D. 75* (Providence, RI: Brown University Press, 1956), 45, match the Babylonian luni-solar calendar with the Julian calendar, and their accommodation of the intercalary months at that time can be relied upon. We know this because the Babylonian almanac for 7/6 BC indicates that an intercalary month was added in the spring of 6 BC (Addaru II) (A. J. Sachs and C. B. F. Walker, "Kepler's View of the Star of Bethlehem and the Babylonian Almanac for 7/6 B.C.," *Iraq* 46 [1984]: 49). With respect to the Hebrew calendar at the time, like the Babylonians, the Jews evidently added 7 months in a 19-year cycle, but the method they used is not entirely clear. It seems that they attempted to make sure that the Passover fell shortly after the vernal equinox, and so declared Nisan 1 only on or close to the vernal equinox. According to this principle, September–October in 6–5 BC would have corresponded to the Hebrew Tishri–Heshvan.

[97] William Ramsay, *Was Christ Born at Bethlehem? A Study on the Credibility of St. Luke* (New York: G. P. Putnam's Sons, 1898), 193.

[98] Cf. Martin, *Star of Bethlehem* (http://www.askelm.com/star/star006.htm [accessed March 26, 2014]).

What might explain the growth of the baby?

At the moment of perihelion, cometary comas are smaller than they were in the run-up to it, or will be in its aftermath, since they are receiving their most intense blast of the Sun at close range. As they subsequently move away from the Sun, escaping the worst of its compressing effect, they grow bigger. Moreover, as productive comets come toward Earth, their comas naturally become larger and larger. These two factors obviously were working together to create the impression that Virgo's pregnancy was developing normally. The baby would have become larger and larger each day, just like a growing fetus in its mother's womb, albeit at a greatly exaggerated rate.

FIG. 7.17 Comet Hale-Bopp on April 4, 1997. Image credit: E. Kolmhofer and H. Raab, Johannes Kepler Observatory, Linz, Austria (http://www.sternwarte.at).

FIG. 7.18 Comet Hyakutake on April 13, 1996. Image credit: The High School Astronomy Class at Alssundgymnasiet Sønderborg, Denmark, www.astronomy-ags.dk.

At some stage, the cometary baby would have seemed to drop and descend within Virgo in the manner of a baby being born, because the retrograde comet's orbit was straightening out after the sharp U-turn around the Sun. This would have meant that the comet could no longer keep in sync with Earth, and hence the coma could no longer maintain its relatively stable position within Virgo's womb. The more the comet's orbit straightened out and the more the comet moved toward Earth, the faster the comet would have seemed to human onlookers to descend within Virgo toward the eastern horizon and the Sun.

At one point, as it slowly dropped within Virgo's belly, it would have seemed to be weighing down on her pelvic floor muscles, which naturally would have caused onlookers to envision the constellation figure as beginning labor. Educated observers could not help but attribute labor pain to Virgo. The delivery would have stretched over days, and during that time the coma would have continued to grow. Only when the whole coma had descended below the area regarded as Virgo's vaginal opening (approximately at the level of the star 80 Virginis; see fig. 7.11) would they have regarded the baby as having been born. By that time the comet would have been getting much closer to the eastern horizon. In space, the comet was actually preparing to cut through the Earth-Sun line, moving from the east side of it to the west.

The enormous length of the tail at the point of the birth would have been due to the fact that the comet was exceptionally large and productive and close to Earth.

All of this coheres perfectly with the data about the comet that we extrapolated from

FIG. 7.19 Comet Hyakutake on March 28, 1996. Image credit: Bojan Dintinjana and Herman Mikuz, Črni Vrh Observatory, Slovenia, http://www.observatorij.org.

Matthew's account, for in both Matthew and Revelation we discover a very large, intrinsically bright comet with a long tail, a retrograde, narrowly inclined orbit, and a small perihelion distance. Both Matthew and Revelation reveal that the comet heliacally rose (this is explicitly stated in Matt. 2:2, and implied in Rev. 12:1–2[99]) and that it was in connection with this event that it spoke most powerfully. Moreover, both accounts suggest that the comet moved from the western evening sky to the eastern morning sky and then returned to the western evening sky.[100] When we recognize that Revelation 12:1–5 records the celestial wonders that so impressed the Magi, suddenly their behavior—traveling hundreds of miles to worship the Messiah and securing gifts of gold, frankincense, and myrrh—appears eminently more reasonable.

It should also be noted that the massive size of the Christ Comet is curiously supported by the apocryphal and pseudonymous *Protevangelium of James* (*Gospel of James*) 21:2–3, which dates to around AD 150. According to that source, in response to Herod's question regarding what celestial sign the Magi saw that related to the newborn King, the Magi answered, "We saw an immense star [*astera pammegethē*] shining among these stars and causing them to become dim, so that they no longer shone; and we knew that a king had been born in Israel."[101]

"WE SAW HIS STAR AT ITS RISING": MATTHEW'S ACCOUNT OF THE COMETARY STAR

In light of what we have discovered in Revelation 12:1–5, it is helpful to reread Matthew 1–2. Is Matthew's account of the comet consistent with the testimony of Revelation 12 concerning the celestial sign that attended the Messiah's birth?

Matthew 1:18–25 discloses what happened in Judea at the very time when the Magi in the east were gazing in astonishment at the spectacle unfolding in the eastern sky. Then

[99] The astronomical scene of Rev. 12:1 occurred in the western evening sky, when Virgo's belly was hosting the Sun. The scene in v. 2 evidently took place in the eastern sky not long afterwards (since it is part of the same wonder), when Virgo's belly was beginning to reemerge after its encounter with the Sun. The narrowly inclined comet—as we have seen, only a narrowly inclined comet could have played the role of Virgo's fetus and newborn—must therefore have been in conjunction with the Sun before rising to play the part of Virgo's child (that is, it heliacally rose in the eastern sky).

[100] Matthew's Star rose over the eastern horizon after being in conjunction with the Sun, which implies that it had previously been present in the western evening sky. Since, after its performance in the eastern sky, the Star seems to have guided the Magi westward and, later, southward, it must have shifted quickly from the eastern to the western sky. As for the comet of Rev. 12:1–5, since it was narrowly inclined, it must have been present in the western sky before heliacally rising in Virgo's womb in the eastern sky. Then, after descending toward the eastern horizon and the sunlight to be born and to be delivered from the dragon, it must have returned to the western sky.

[101] My translation of the Greek text in Emile de Strycker, *La forme la plus ancienne du Protevangile de Jacques* (Brussels: Société des Bollandistes, 1961), 168–170.

2:1–12 relates the story of the Magi's pilgrimage to worship the newborn King of the Jews.

MATTHEW 1:18–25

With respect to the question of what happened in Judea at the point when the celestial sign took place in the eastern sky, Matthew 1:18–25 emphasizes that the terrestrial event was the birth of the Davidic Messiah to a virgin girl who had conceived by the miraculous intervention of the Holy Spirit, in fulfillment of Isaiah 7:14.

Matthew 1:18a highlights the subject of the paragraph: "Now the birth of Jesus Christ took place in this way." We discover that Mary, betrothed to Joseph, a descendant of Abraham and David (vv. 1–17), had not had sexual intercourse with him but nevertheless "was found to be with child through the Holy Spirit."

According to Matthew, Joseph, not realizing that a remarkable miracle had transpired as God was putting into effect the glorious salvation promised through the Prophets, concluded that Mary had been immoral with another man and that he had no choice but to divorce her. However, Joseph was gracious and kind and was fond of Mary, and so he resolved to carry out the divorce in as quiet a manner as possible, so that he might not add to her disgrace (v. 19).

Matthew then tells of how God let Joseph in on his secret plan, because it was to be through Joseph that the Messiah would have, as his legal father, a descendant of King David. An angel commanded Joseph not to be afraid to take Mary as his wife, explaining that the child "conceived in her is from the Holy Spirit" (v. 20b). It was, in other words, a virginal conception. The angel went on to say that "She will bear a son, and you shall call his name Jesus, for he will save his people from their sins" (v. 21). Mary would be a virgin mother and would bear for the line of David a son. Joseph would then name the son Jesus.

After waking up from his revelatory dream, Joseph "did as the angel of the Lord had commanded him": he married Mary but avoided having sexual intercourse with her until she had given birth to her son, and he named the son Jesus (vv. 24–25).

In the middle of the narrative, Matthew inserts a parenthetical comment to explain the Scriptural background to what was taking place (vv. 22–23): "All this took place to fulfill what the Lord had spoken by the prophet: 'Behold, the virgin shall be with child and bear a son, and they shall call his name Immanuel' (which means, God with us)." Matthew's claim is that Mary's virginal conception and birth of Jesus occurred in fulfillment of Isaiah 7:14.

The focus of Matthew 1:18–25 therefore is on the fact that Jesus was conceived in and born to a virgin but with a descendant of David as his legal father. Isaiah 7:14 was fulfilled when Joseph's betrothed, Mary, a virgin, became pregnant by the Holy Spirit and gave birth to Jesus.

According to Matthew, then, the terrestrial event that accompanied the celestial wonder in the eastern sky was the birth of the Davidic messianic King to a virgin who had conceived him by a miraculous work of God through the Holy Spirit. Whatever the celestial phenomenon was, it was highlighting that this was happening at that very time in the land of Judea, in fulfillment of Isaiah 7:14.

MATTHEW 2:1–12

Occasion of the Wonder

The opening two verses of Matthew 2:1–12 link the terrestrial event detailed in 1:18–25 to the Magi's journey from the east to Judea—and therefore to the natal sign the Magi saw when back in their homeland: "After Jesus was born in Bethlehem of Judea in the days of Herod the king, behold, magi from the east came to Jerusalem" (2:1). The Magi's visit to Jerusalem therefore is chronologically located in the aftermath of the birth of the Davidic Messiah Jesus to the Virgin Mary.

Despite being hundreds of miles away

from Judea, the Magi knew that the divine Messiah had been born, and so they set out to worship him (v. 2). What the comet had done in the heavens in connection with its heliacal rising had revealed to them that the Messiah's nativity had taken place.

Meaning and Nature of the Wonder

According to verse 2, the Magi asked the people of Jerusalem, "Where is he who has been born king of the Jews, for we saw his star at its rising and have come to worship him?" The logic of the Magi is striking: at or around the time of its recent heliacal rising, the Star, identified with the baby Messiah, had by its behavior in the heavens communicated to them that he had been born and could now be found in Judea. What the Magi saw therefore was a celestial nativity scene, an equivalent to the actual terrestrial nativity transpiring in Bethlehem. At the same time, in view of the probability that the Magi were alluding to Balaam's oracle (Num. 24:17) when they reported to the people of Jerusalem that they had seen "the star at its rising," it seems very likely that during this phase of its apparition the comet as a whole looked like a scepter.

Terrestrial House of Grain

The emphasis on the birth of the Messiah is continued in verses 3–4. King Herod, convinced by the Magi's deeds and words that the Messiah had indeed been born, inquired of the chief priests and scribes as to where the birth was to have taken place according to the Hebrew Scriptures (v. 4). The Jewish teachers informed him, based on Micah 5:2, that the Messiah was to be born in Bethlehem, which means "House of Grain"[102] (Matt. 2:5–6).

First Appearance of the Star

When Herod was told that the Prophets had predicted that the Messiah would be born in Bethlehem, he summoned the Magi and discovered from them the precise time when the Star had first appeared (v. 7). This information, Herod evidently assumed, disclosed the maximum possible age of the Messiah. We are informed in verse 16 that, at the point when Herod sent his troops to massacre the babies of Bethlehem and environs, he commanded them to kill all children in their first or second year. That means that the Star was first visible to the Magi at least 12 lunar months (if 7–6 BC) before the slaughter in Bethlehem and that the comet was a historically great one, visible to the naked eye for over a year in total.

When Herod issued his order to massacre the infants of Bethlehem, he was allowing that the Magi might have been wrong in their contention that the Messiah's birth coincided with the celestial wonder relating to the Star's heliacal rising. The Judean king evidently felt that it was only prudent to take seriously the possibility that the Messianic child might have been born at a stage of the Star's apparition prior to the sign in the eastern sky. In particular, Herod may have wondered if the Messiah's birth might have occurred at the point when the Star had first appeared. The first appearance, or "birth," of a "star" in the celestial realm could naturally have been regarded as the equivalent of the appearance on the terrestrial stage, or birth, of the one whom the "star" represented. It is conceivable that what the Magi reported to Herod concerning their first observation of the Star encouraged the king to consider this moment in particular a candidate for the Messiah's birthday.

Armed with the maximum age and geographical location of the Messiah, the tyrannical king of Judea felt confident that he could succeed in an audacious bid to kill the newborn Davidic King (vv. 7–8, 16).

[102] We shall explore, in a moment, the significance of "House of Grain." This meaning would have been evident to Hebrew/Aramaic readers of the Gospel of Matthew if the Gospel was originally written in Hebrew or Aramaic (as seems to be suggested by the early church father Papias, as cited by Eusebius, *Hist. Eccl.* 3.39).

Following Yonder Star . . . from Jerusalem to Bethlehem

When the Magi learned from Herod that the Scriptures had prophesied that the Messiah was to be born in Bethlehem, they assumed that the messianic baby would still be there and so made their way under the guidance of the Star to the town of David (v. 9). Perhaps the Magi assumed that the Messiah's family resided there permanently. If so, of course, they were wrong. The Messiah was to be found at Bethlehem even though Mary and Joseph ordinarily resided in Nazareth, because they had come down to Bethlehem for the census and found it more practical to remain there until after they had been purified (and performed the presentation of the child) at the Jerusalem temple on the fortieth day after the childbirth, in accordance with the Torah.[103]

As the Magi set out from Jerusalem to Bethlehem, the Sun set and, along with the stars and constellations, the comet reappeared in the southern sky in front of them (v. 9b). They recognized it as the celestial entity that they had been tracking for at least a year and that had revealed the Messiah's birth to them by what it did in connection with its rising—what they had referred to as the Messiah's Star when speaking to the people of Jerusalem.

Following Yonder Star . . . from Babylon to Judea

We have already seen that the fact that the Magi, on the road from Jerusalem to Bethlehem, recognized the Star as the one they had previously seen back in their homeland suggests that they had been following its course in the heavens through the intervening period. The comet was now in a completely different part of the sky than it had been in when it had risen. At its rising it had been low on the eastern pre-dawn horizon, but now, within the space of about a couple of months, it was appearing in the southern evening sky.

The comet's presence in the evening and night sky presumably encouraged and urged on the Magi as they traveled across the wilderness toward Judea. The comet, as it set over the western horizon each night, may well have seemed to the Magi to be traveling toward Judea ahead of them, urging them onward. Indeed the comet's behavior as it set may have urged the Magi to depart quickly and may have influenced their choice of route. However, it was not their main reason for electing to journey to Judea. Their primary reason for undertaking this long trip was because of what they had seen the Star do in the eastern sky in connection with its heliacal rising (v. 2).

When the comet appears as the Magi are making their way from Jerusalem to Bethlehem, Matthew refers to it as "the star that they had seen at its rising" (v. 9). This retrospective reference is akin to the kind of thing we see in the Babylonian astronomical diaries and monthly summaries, where each new report of a comet is introduced with a summary recollection of an earlier point in its apparition (these retrospective references in the Babylonian records do *not* imply that the comet was not also seen during the time subsequent to the referenced event). For example, Babylonian entries relating to the 138 BC and 120 BC comets retrospectively referenced earlier heliacal settings, while one entry concerning the 110 BC comet recalled its first appearance.[104] In Matthew, the reference back to the Star's rising functions to remind the reader of the most important stage of the 1+-year cometary apparition, namely, what it did in the eastern sky to reveal the Messiah's birth

[103] A mother who had just delivered a baby boy was permitted to enter the temple only after 40 days of purification, day 1 of which was the day of childbirth (Lev. 12:2, 4, 6–8). In Western exclusive counting, therefore, the mother went to the temple to be purified on the fortieth day.

[104] See F. Richard Stephenson, "The Ancient History of Halley's Comet," in *Standing on the Shoulders of Giants*, ed. Norman Thrower (Berkeley: University of California Press, 1990), 244.

and prompt the Magi to travel to Judea in search of the baby Messiah. This reminder highlights that the very same astronomical entity that launched them on their pilgrimage to Judea was now present at its culmination, to help them complete their journey.

Following Yonder Star . . . to the House Where the Virgin and Child Were Staying

The comet would have given the impression of forward movement in front of the Magi on a basically horizontal plane as they traveled southward from Jerusalem to Bethlehem. Then, having reoriented itself as it descended toward the western horizon, the comet pinpointed the particular house where the messianic child was. The comet did this by seeming to stand over the house as it was about to set (v. 9). Matthew's description mandates that the comet at that point had a long tail which, from the Magi's perspective at the time, projected upwards into the sky from a coma that was stationed over the visible horizon behind the house where Jesus was. The description suggests that the comet that night was probably at least 30 degrees long but no more than about 45 degrees long. According to Matthew, the Star was clearly of such brightness and size that it seemed from the Magi's perspective to be standing right over the house. For the comet to be regarded as doing so, it must have been angled at between approximately 70 and 110 degrees from the horizon.

The Magi felt great joy when they saw the comet standing up over the house. Incredibly, it had enabled them to complete their mission. The Messiah's Star had led them right to the Messiah. The comet that had represented the messianic baby in the great celestial wonder marking his birth was now pinpointing his precise location on the earth.

When the Magi went into the house, "they saw the child with Mary his mother" (Matt. 2:11a). The absence of Joseph is striking and pushes the reader to recall that Jesus was born to a virgin mother. The Magi at the climax of their journey therefore saw the virgin and her special child. They immediately realized that this was what they had come to Judea seeking. The Eastern visitors therefore fell down and worshiped the infant and presented him with their gifts of gold, frankincense, and myrrh (v. 11), tokens of their acknowledgement that he would, as the Hebrew Scriptures had prophesied, die, be buried, rise, and then reign over all nations.

MATTHEW'S EVIDENCE

We have every reason to believe that Matthew knew exactly what the Star had done in the eastern sky that prompted the Magi to journey westward to Judea. The question is, is Matthew's account of the nativity, the Magi, and the Star consistent with Revelation 12:1–5's portrayal of the natal sign? Our brief overview of Matthew 1:18–2:12 suggests that it is.

First, Matthew highlights that the Magi's observation of what the Star did in relation to its rising coincided with the terrestrial nativity of the Messiah in Bethlehem (2:1–2). Second, the Magi believed that the comet was the Messiah's "star" and interpreted its behavior in connection with its heliacal rising to signify not only that he was born, but that he was born at that time. In other words, what they saw in the eastern sky they interpreted as a nativity scene. Third, at the climax of the Magi's journey to Judea in search of the Messiah, the Star led them to the newborn King and his virgin mother (v. 11). The Magi instantly recognized that this is what the Star, during the eastern phase of its apparition, had commissioned them to find, and so they offered their gifts to the child.

Obviously, the most natural celestial context for a heavenly nativity scene announcing the birth of the Messiah to a virgin is the constellation Virgo, the sole zodiacal female. As to where within Virgo the comet would have risen, there is only one plausible suggestion: in her womb.

Moreover, in Matthew 2:2 the Magi seem to allude to Numbers 24:17's oracle concerning the rising scepter-star, implying that it was fulfilled by the Star. This probably implies that the comet looked like a scepter at the time when it rose (Matt. 2:2). This is consistent with our conclusion that Revelation 12:5 may well be revealing that the comet as a whole looked like a scepter at the point of the baby's birth.

Matthew's narrative is therefore very compatible with what Revelation 12:1–5 recounts regarding the nature of the wonder seen in the eastern sky in connection with the Messiah's birth.

When it is appreciated that the Magi witnessed a cometary coma create an unfolding nativity drama in the eastern sky even as the comet as a whole simultaneously formed a spectacular scepter, their wonderment and eccentric pilgrimage suddenly make sense. Moreover, when the Magi entered the house in Bethlehem, they were awestruck because they were seeing on the earth what they had earlier seen in the heavens—the divine baby with his virgin mother.

To those who spoke Aramaic (as did people in Mesopotamia and Judea) or Hebrew, the town of Bethlehem, meaning "House of Grain," may well have seemed a remarkably fitting place for the terrestrial representative of Virgo and her newborn child to be, for Virgo was strongly associated with grain. From at least the first part of the first millennium BC, the Mesopotamians identified the constellation associated with Spica as AB.SIN ("the Furrow"). The astronomical compilation MUL.APIN, from around 1000 BC,[105] stated that Spica was the goddess Shala's ear of grain.[106] The Babylonians portrayed this constellation as a virgin with a sprig of grain. Initially "the Furrow" referred only to half of the constellation we know as Virgo; the other half was called "the Frond," which rose in advance of "the Furrow." However, when these two constellations were combined into one (probably around the time when the zodiacal band was divided up into twelve equal segments), the new unified constellation seems to have taken on the identity of "the Furrow," including its close association with Spica and grain, although combining this with some of the traits of "the Frond." The Greeks regarded Spica as the ear of grain in Virgo's left hand.

Moreover, Babylonian astrologers referred to the zodiacal constellations as "houses."[107] Therefore the constellation Virgo was astronomically the "House of Grain," just as, terrestrially, Bethlehem was the "House of Grain."

We conclude, then, that Matthew's account of the comet is perfectly consistent with what Revelation 12:1–5 reveals concerning the celestial sign marking Jesus's birth: the cometary coma played the part of the baby Messiah in a celestial nativity play featuring Virgo as the Messiah's mother. Back in their homeland, the Magi had seen the heavenly Virgin with her divine baby in the celestial House of Grain. Then, at the climax of their journey west to Judea, they saw on the earth the Virgin Mary with baby Jesus in the terrestrial House of Grain. In addition, at this same time the Magi witnessed the comet as a whole looking like a scepter.

SUMMARY

We suggest, then, based on our study of Revelation 12:1–5 and our fresh analysis of Mat-

[105] Koch-Westenholz, *Mesopotamian Astrology*, 43n6, 78, following David Pingree, "Mesopotamian Astronomy and Astral Omens in Other Civilizations," in *Mesopotamien und seine Nachbarn: Politische und kulturelle Wechselbeziehungen im alten Vorderasien vom 4. bis 1. Jahrtausend v. Chr.*, ed. Hans Jorg Nissen (Berlin: Reimer, 1982), 613–631; Hermann Hunger and David Edwin Pingree, eds. *MUL.APIN: An Astronomical Compendium in Cuneiform* (Horn, Austria: Ferdinand Berger, 1989), 67–69, 144. J. Koch, *Neue Untersuchungen zur Topographie des babylonischen Fixsternhimmels* (Wiesbaden: Otto Harrassowitz, 1989), 34–52, thinks that it was composed around 700 BC.

[106] Van der Waerden, *Science Awakening II*, 288.

[107] Francesca Rochberg, *Babylonian Horoscopes* (Philadelphia: American Philosophical Society, 1998), 46–50; cf. Koch-Westenholz, *Mesopotamian Astrology*, 134–136.

thew 1:18–2:12, that while the Virgin Mary was giving birth to Jesus in Bethlehem, the zodiacal constellation figure Virgo was giving birth to a cometary baby.[108]

What we have preserved in Revelation 12:1–5 is a series of astronomical observations from 6 BC.

The heavenly birth was the climax of the year-plus cometary apparition. It was also the culmination of a pregnancy that had been apparent from the moment that a cometary baby was observed in Virgo's womb as she heliacally rose, emerging in the eastern predawn sky.[109] The cometary coma would initially have looked small in her belly, but over the following weeks, as the comet approached Earth, the "baby" would have become larger and larger, just like a fetus in its mother's womb. In due course, it descended within Virgo until it made it seem that she was in labor.[110] Then, when the coma-baby had fully emerged from its mother's womb, it was "born." Revelation implies that this celestial birth coincided with the birth of the Messiah to the terrestrial virgin, Mary. At that time the comet as a whole may well also have formed a massive celestial scepter that stretched from the eastern to the western horizon and seemed to rest on Israel in the west.

According to the New Testament, after the comet completed its time in the eastern sky and crossed to the west, it proceeded to

[108] Curiously, the Syriac *Cave of Treasures*, perhaps attributable to Ephrem in the fourth century, states, "Now, it was two years before Christ was born that the star appeared to the Magi. They saw the star in the firmament of heaven, and the brilliancy of its appearance was brighter than that of every other star. And within it was a maiden carrying a child, and a crown was set upon his head" (*The Book of the Cave of Treasures*, trans. E. A. Wallis Budge [London: Religious Tract Society, 1927], 203–204). The Ethiopic *Conflict of Adam and Eve with Satan* has a very similar section but lacks mention of the two years: at the time of Jesus's birth, "a star in the east made it known, and was seen by Magi. That star shone in heaven, amid all the other stars; it flashed and was like the face of a woman, a young virgin, sitting among the stars, flashing, as it were carrying a little child of a beautiful countenance. From the beauty of His looks, both heaven and earth shone, and were filled with His beauty and light above and below; and that child was on the virgin woman's arms . . ." (translation from S. C. Malan, *The Book of Adam and Eve, also Called the Conflict of Adam and Eve with Satan* [London: Charles Carrington, 1882], 204). It is conceivable that these late sources reflect a garbled awareness of the original sign, although more likely they simply reflect speculation.

[109] No doubt the appearance of the baby in Virgo's womb at the point of the heliacal rising proper deeply impressed the Magi and played an important part in disclosing to them that a divine figure was coming into the world, the offspring of a virgin's womb. In a sense, everything that the Star did thereafter within Virgo unpacked the meaning of that unforgettable opening scene. Moreover, since the comet as a whole probably looked like a scepter as it heliacally rose, this might well have seemed a powerful fulfillment of Balaam's oracle (Num. 24:17). Therefore the Magi could have been referring to the heliacal rising proper alone in Matt. 2:2 ("at its rising"). However, it was in the predawn observing sessions following the heliacal rising proper that the Magi would have come to appreciate that the sign consisted not just of a single snapshot but of a developing drama; not just of the first trimester of a virgin's pregnancy but of the virgin's whole pregnancy and delivery; not just of a fetus but also of a newborn baby; not just of a bright scepter but also of an extraordinarily long one that eventually stretched from one horizon to the other; and not just of a remarkable cometary rising but also of an extraordinary astronomical wonder. In the days following the rising, the allusions to key Old Testament oracles concerning the Messiah's birth would have become clearer and stronger. Moreover, only at the climax of the drama was the Messiah's birthday disclosed. So, while it is possible that when the Magi spoke of "his star at its rising" they were reflecting narrowly on the heliacal rising proper alone, it seems more likely that they were speaking generally (perhaps using synecdoche) of all that the Star did in the eastern sky at its rising proper and in the following days and weeks. Since ancient Near Eastern astronomers who observed comets took special note of key moments in an apparition, such as heliacal risings, it was natural for them to identify a particular stage of a comet's apparition with reference to such chronological astronomical landmarks.

[110] On this note, it is interesting that in Luke 1:78–79 Zechariah, at the birth of John the Baptist, prophesied that, on account of God's tender mercy, "the rising [star]" (not "sunrise," contra ESV and many English versions) (*anatolē*; cf. Matt. 2:2, 9) shall visit us from on high to give light to those who sit in darkness and in the shadow of death, to guide our feet into the way of peace." This striking prophecy concerning the coming of the Messiah draws heavily on astronomical imagery, particularly from Isa. 9:2. As I. Howard Marshall (*The Gospel of Luke: A Commentary on the Greek Text*, New International Greek Testament Commentary [Grand Rapids, MI: Eerdmans, 1984], 94–95) points out, although *anatolē* could be used by metonymy for "the rising Sun," it could equally refer to a rising celestial body other than the Sun. He observes that here, where its second intended meaning is "Shoot" or "Branch," alluding to Isaiah's oracle concerning the Messiah's coming (Isa. 11:1ff.; cf. LXX Jer. 23:5; Zech. 3:8; 6:12; *Testament of Judah* 24:1, 6), the rising entity is probably the rising "star" prophesied by Balaam (Num. 24:17). Astronomically, what Luke writes in Luke 1:78–79 seems most compatible with an extremely bright comet that rises and shines into the heart of darkness. Observe that the "rising" entity will "visit us from on high." Visiting from on high most naturally suggests descent. Stars that rise keep rising. However, like inferior planets (Mercury and Venus), many comets descend after rising, and certainly the comet that announced Jesus's birth did—it rose within Virgo and then descended to be born. On *anatolē* here being double entendre, see, for example, François Bovon, *Das Evangelium nach Lukas*, Evangelisch-Katholischer Kommentar zum Neuen Testament, 3 vols. (Zurich: Benziger, 1989–2001), 1:109; Darrell L. Bock, *Proclamation from Prophecy and Pattern: Lucan Old Testament Christology* (Sheffield: JSOT, 1987), 73; Luke Timothy Johnson, *The Gospel of Luke*, Sacra Pagina (Collegeville, MN: Liturgical Press, 1991), 47. It is also just possible to detect in *anatolē* a quiet allusion to the palm branch in Virgo's right hand. Incidentally, the Christ Comet would have been visible by this time, but its heliacal rising (and the birth of the Messiah) was still some months off (note the future tense: "the rising [star] shall visit").

guide the Magi to the place where the terrestrial virgin mother and her child were located. While the Messiah's Star at its rising had revealed to the Magi the fact, time, and manner of his birth, it subsequently turned into a massive celestial pointer, disclosing to them precisely where the baby Messiah was located. The comet that had played the part of Virgo's messianic baby in the celestial play eventually led the Magi right to the virgin and her special baby!

The Biblical account suggests that, as the Magi entered the house in Bethlehem, they finally saw on the earth what they had seen in the heavens less than 1½ months beforehand: the virgin with her newborn child.[111] Their divine mission was now complete. Heaven and earth were united.

THE INTERPRETATION OF THE CELESTIAL DRAMA

What interpretation were the Magi to give to the remarkable celestial phenomenon in the eastern sky? Those operating on the faulty assumption that comets were always interpreted as negative omens might judge that ancient astrologers would necessarily have concluded that a terrible curse was about to befall pregnant women in general or perhaps one particular pregnant woman or a royal dynasty.[112] However, while such an interpretation was theoretically possible, the image of Virgo being pregnant with a child at her heliacal rising and then going on to have a full pregnancy was more susceptible to a positive interpretation than most cometary apparitions. The presumably baby shape of the cometary coma, coupled with its growth and downward, birth-like movement would have confirmed that this particular sign should be interpreted more joyfully. The only plausible explanation was that the heavens were signaling that a wonderful natal event was transpiring somewhere on the earth around that time. Together with the scepter-like form of the comet as a whole, it suggested that the birth of someone destined to be a mighty king was being heralded.

We can perhaps go further. This astronomical wonder might naturally have been interpreted in light of the broader sequence of celestial events. Back on September 15, 6 BC, the Sun had been located in the region of Virgo's womb. Subsequently, as Virgo rose heliacally in the eastern sky and the presence of a gloriously bright cometary coma in her womb was detected, this may well have prompted onlookers to interpret the Sun's role on September 15 as being that of God begetting his divine Son within the celestial Virgin. That is, Virgo clothed with the Sun, and with the Moon under her feet, was liable to be regarded as the first scene in the celestial nativity drama relating to God's Son. After the cometary baby moved out of Virgo's womb and down through her birth canal and was born, it would then quickly have seemed to descend toward the horizon and the Sun, a phenomenon which, in context, was susceptible to the interpretation that this newborn king would be delivered from danger by God.[113]

Based on what they saw in the heavens, one can well understand why observers like the Magi would have concluded that an important birth was taking place at that very time, and indeed that the baby represented a great person

[111] In the light of our findings in this chapter, representations of Mary as Virgo in Christian history (on which see Wilhelm Gundel, "Parthenos," in *Paulys Realencyclopädie der Classischen Altertumswissenschaft* 18.4 [Stuttgart: Druckmüller, 1949], 1936–1957) do not seem inappropriate.

[112] Cf. Richard Hinckley Allen, *Star Names: Their Lore and Meaning* (New York: G. E. Stechert, 1899), 465, who claimed that Pliny "said that the appearance of a comet within [Virgo's] borders implied many grievous ills to the female portion of the population." To the best of my knowledge, Pliny said no such thing.

[113] In the Israelite tradition Yahweh is portrayed as "the sun of righteousness" with "healing in its wings" (Mal. 4:2; cf. Ps. 139:9). On Yahweh as the Sun, see Num. 6:24–26; Ps. 84:11; 1 Sam. 6:9, 12–21 (note Beth-shemesh, "House of the Sun"); and 2 Sam. 23:3–4. On the portrayal of Yahweh in terms of the Sun, see Karel van der Toorn, "Sun," in Freedman, *Anchor Bible Dictionary*, 6:237–239; M. S. Smith, "The Near Eastern Background of Solar Language for Yahweh," *Journal of Biblical Literature* 109 (1990): 29–39. In an astronomical drama, the Sun is the most natural candidate to play the role of God.

who was being born right then to a terrestrial equivalent of Virgo, most naturally herself a virgin. The astrologers would also probably have come to the conclusion that this person's father was the Most High God, and indeed that he himself was divine. Moreover, it is easy to see how the scepter-like form of the comet as a whole might have prompted them to believe that the baby would become a great ruler. Needless to say, observers would have been eager to know who the mysterious divine child was and where on the earth he was located.

Many pagans in the Greek, Roman, and ancient Near Eastern world in the first and second centuries BC thought of Virgo in terms of Isis, the consort of Osiris and mother of Horus. We recall that Teukros of Babylon reported that Virgo was understood by some of his contemporaries as Isis sitting on a throne, feeding her young son Horus. Moreover, the whole celestial story as it played out in September/October of 6 BC was reminiscent of the story of Isis's pregnancy and delivery of Horus in the face of Seth-Typhon's determined hostilities. However, the heavenly narrative was different at key points, and these differences challenged any attempt to read it in a thoroughgoing way through that paradigm. For one thing, it was not Osiris, god of the underworld and the dead, who was playing the role of the father of Virgo's son. Nor was the Sun the son. Rather, the Sun seemed to be playing the role of the father of Virgo's child. Nevertheless, this international combat myth did offer ancient Near Eastern observers a ready-made paradigm for interpreting the cometary drama in Virgo, one into which the story of the Messiah's birth could be fitted.

While many ancients would have sought to make sense of what was transpiring in Virgo by resorting to Babylonian, Egyptian, and Greco-Roman religious ideas, the Magi were thoroughly convinced that the comet was the Messiah's "star" and should be interpreted through the grid of the Hebrew Scriptures. They believed that it was the Messiah who would vanquish the forces of Chaos and would bring Order to the cosmos. Whether the Magi had some inkling of the Star's messianic significance prior to its heliacal rising we do not know, but they certainly interpreted the wonders in the eastern sky as announcing the Messiah's birth. We shall explore in the following chapter the particular prophecies of the Hebrew Scriptures that provided them with the keys to understanding the heavenly sign. For now it is important to note that, having watched the cometary show in the morning sky and having become convinced that it was disclosing the birth of the Jewish Messiah, the Magi set off urgently for Jerusalem on a quest to find Virgo's son and worship him. As they did so, they may well have felt concern that the great dragon, representing the forces of evil in the cosmos, was intent on attacking and killing the newborn king. Evidently Herod was a great actor, because the Magi had no idea that it was he who was playing the part of Hydra, determined enemy of Virgo and her son, until they were informed of his malign intentions in a dream.

CONCLUSION

From what they saw in the eastern sky the Magi could have deduced certain things about the newborn baby, Virgo's child *par excellence*: (1) His mother had conceived him through divine intervention without losing her virginity. (2) He had been born at the point when the cometary coma had in its entirety descended below Virgo's groin. (3) He was the son of God. (4) He was glorious. (5) He was divine. (6) He had a powerful enemy who was eager to kill him. (7) He was destined to reign over the whole world. However, the celestial wonders by themselves cannot explain why the pagan astrologers came to the conclusion that the one born to a virgin was the Messiah, the King of the Jews. It was the Hebrew Scriptures, mediated through one or more Jews in Babylon, that furnished them with the all-important messianic paradigm.

8

"With Royal Beauty Bright"

Messiah's Star

In the previous chapter we proposed that the celestial sight observed by the Magi consisted of a large cometary coma playing the part of Virgo's baby in a celestial birth scene even as the comet as a whole formed a scepter. The retrograde long-period comet's coma heliacally rose in Virgo's womb and proceeded to grow in the manner of a baby there before descending to be "born." The wonder occurred because of extraordinary comet-Earth-Sun geometry—the comet was approaching Earth after perihelion and was moving in sync with Earth. As they watched the magnificent celestial marvel unfolding before their eyes, the Magi became certain that the heavens were signaling the birth of an extraordinarily important, indeed divine, ruler to a virgin on the earth.

We must now ask how the Magi concluded that the heavenly phenomena pointed to the Messiah, the King of the Jews. Anyone living in Babylon, with its significant population of Jews, would have been familiar with the basic tenets of Judaism and probably would have known that the Jewish people expected a great future leader called the Messiah to come and ultimately reign over the world. Suetonius, *Vespasian* 4.5, confirms this: "An ancient superstition had spread throughout the east that out of Judea would come the rulers of the world. This prediction, which actually referred to a Roman emperor, as became clear after the event, the Jews interpreted to refer to themselves. Therefore they rebelled" (cf. Tacitus, *Ann.* 5.13).[1] However, one would be surprised if the average magus would have been aware of particular Biblical prophecies concerning the Messiah's birth. Almost certainly, as most scholars agree,[2] the Magi must have been aided in their interpretation of the cometary apparition by one or more Jewish exiles who knew their Scriptures, had a developed messianic expectation, and considered the cometary apparition to be the fulfillment of ancient prophecies recorded in the Hebrew Bible. Certainly by the time the Magi got to Judea, they were surprisingly well-versed in Jewish messianic traditions (although not Mic. 5:2).

But which Hebrew Scriptures played the decisive role in convincing the Magi that it was the Messiah who had been born at that

[1] My translation.
[2] For example, Donald A. Hagner, *Matthew*, 2 vols., Word Biblical Commentary (Dallas: Word, 1993–1995), 1:27.

time? In this chapter I will suggest that the key Biblical texts that gave the Magi their messianic paradigm were Numbers 24:17; Isaiah 7:14; and 9:2.

NUMBERS 24:17

One of the main Old Testament prophecies that almost certainly played a part in convincing the Magi to interpret the Star as a sign of the birth of the Messiah was in Numbers 24:17–19, in the Book of Moses.

NUMBERS 22–24

In Numbers 22–24, the people of Israel were encamped just east of the River Jordan, poised to enter the Land. Balak, king of Moab, was afraid of them because of how they had so overwhelmingly conquered the Amorites. But instead of engaging them in military conflict, Balak hatched the plan of employing an internationally renowned Mesopotamian prophet/diviner called Balaam[3] to curse the Israelites. Securing his services by flattery and a generous offer of remuneration, Balak called upon Balaam to pronounce curses on the people of Israel. However, each time Balaam opened his mouth to curse Israel, Yahweh obligated him to speak only words that were favorable to Israel. Balak was infuriated.

In Numbers 23–24 we find the four main oracles uttered by Balaam (23:7–10, 18b–24; 24:3b–9, 15b–19), followed by a cluster of short concluding oracles (24:20–24).

The fourth oracle, which is the focus of our concern, was given by Balaam without Balak's specific prompting. This prophecy, in context, highlighted the punishment due Moab for its unprovoked hostility toward Israel. In the oracle, Balaam revealed what would take place in the distant future. He foresaw that a great leader would emerge from Israel, who would conquer the Moabites and Edomites (24:17–19): "I see him, but not now; I behold him, but not near: a star shall come out of Jacob, and a scepter shall rise out of Israel; it shall crush the forehead of Moab and break down all the sons of Sheth. Edom shall be dispossessed; Seir also, his enemies, shall be dispossessed. Israel will do valiantly. And one from Jacob shall exercise dominion and destroy the survivors of cities!"

In the ancient Near East a monarch could be spoken of as a "star" (so, for example, in Isaiah 14:12). In addition, "scepter" sometimes represented (by metonymy) a king (as in Psalm 45:6: "The scepter of your kingdom is a scepter of uprightness") or royal authority (as in Genesis 49:10: "The scepter shall not depart from Judah, nor the ruler's staff from between his feet"). Here Balaam speaks of the coming of the future Israelite ruler in terms of a "star" and "scepter."

THE SCEPTER STAR

As we saw in chapter 6, the fact that the "scepter" is parallel to, and synonymous with, "star" makes a compelling case that the "scepter" in view is a straight, long-tailed comet. That a comet is in view is consistent with the fact that Balaam prophesied that the "star" would "come" or "move" (*drk*) and that the "scepter" would "rise" or "stand" (*qm*). In addition, the Babylonian Talmud tractate *Berakhot* 58b expressly refers to a comet as "a scepter star." Many scholars[4] and two recent Bible translations, the *New English Bible* and the *Revised English Bible*, have even rendered the Hebrew word used by Balaam "comet" rather than "scepter."[5] The peculiar conceptualization of this future

[3]The discovery in 1967 of the Deir 'Alla Inscription in Jordan from the eighth century BC confirmed that "Balaam son of Beor" was revered as a renowned seer in the Jordan Valley at that point and strengthened the case for regarding him as a historical figure. See especially P. Kyle McCarter Jr., "The Balaam Texts from Deir 'Alla: The First Combination," *Bulletin of the American Schools of Oriental Research* 239 (1980): 49–60.

[4]See chapter 6, note 44.

[5]So also the *Good News Bible: Today's English Version*, 2nd ed. (New York: American Bible Society, 1992), 187 ("Like a comet he will come from Israel"). Hans-Jürgen Zobel, "šēḇeṭ," in *Theological Dictionary of the Old Testament*, vol. 14, ed. G. J. Botterweck, H. Ringgren, and H.-J. Fabry (Grand Rapids, MI: Eerdmans, 2004), 305, comments: "Since the term šēḇeṭ parallels 'star'

ruler in terms of a cometary scepter strongly suggests that his birth would be attended by an extraordinary cometary apparition, anticipating his destiny as world monarch. Accordingly, it seems that Balaam was employing double entendre here—the scepter-star was both literal and metaphorical. The literal, astronomical scepter-star would announce the birth of the metaphorical scepter-star, the Messiah, who was destined to wield sovereign and military authority over Israel and its neighbors. The natal star would be a symbol of the Messiah.

As we mentioned in the last chapter, when Balaam speaks of the cometary "star" as a "scepter" that "shall rise," it is most natural to interpret him as suggesting that the comet would look like a scepter at the time when it rose to herald the Messiah's coming.

FULFILLMENT?

A glance at 2 Samuel 8 (especially vv. 2 and 13–14; cf. 3:18; 7:8–11; 1 Kings 11:15–16) shows that King David conquered Moab and Edom militarily. Was he regarded as the prophesied Scepter-Star? The answer is no. First, the historical books disclose that Moab and Edom did not remain subservient to the people of Israel on a permanent basis[6] (see 2 Kings 1:1; ch. 3; 8:20; 13:20; 2 Chron. 28:17). Moreover, Isaiah 11:14 ("They shall put out their hand against Edom and Moab") and 25:9–11 make it clear that the conquest foretold by Balaam was not regarded as entirely fulfilled in the centuries after David, but still awaited a future fulfillment in connection with the Messiah. Second, there is in the Hebrew Scriptures a notable lack of any claim that David's birth was attended by any special astronomical phenomenon.

ANCIENT JEWISH MESSIANIC INTERPRETATIONS OF NUMBERS 24:17

Jews around the time of Jesus's birth interpreted Balaam's oracle as referring to the coming of the Messiah. For example, the Qumran Community took it in this way.[7] Its Damascus Document unpacked the meaning of Numbers 24:17 in the following manner: ". . . it is written 'A star has journeyed out of Jacob and a scepter is risen out of Israel.' 'The scepter' is the Prince of the whole congregation, and at his coming 'he will break down all the sons of Seth.'"[8]

Philo summarized the oracle in the following terms: "A man shall emerge, says the oracle, leading his army to war—he shall conquer great and densely populated nations."[9]

The Septuagint rendered the verse, "A star [*astron*] will rise [*anatelei*] from Jacob, and a man [*anthrōpos*] will emerge [*anastēsetai*] from Israel," and another Greek translation[10] and a Syriac version (the Peshitta)[11] used "leader" in place of "scepter" or "man."

Some of the Targums, which are later but often preserve traditions from the time of Jesus, spelled out that the "star" in view is a human. For example, *Targum Neofiti* rendered the verse, "A king will arise from the house of Jacob and a redeemer and ruler from the house of Israel."[12] *Targum Onqelos* substituted "a king" for "a star" and "the

(kôḵāḇ) here, Berend Gemser offered the attractive suggestion that one understand šēḇeṭ as 'comet' (*BHS: stella crinata*) (Berend Gemser, "Der Stern aus Jacob [Num. 24.17]," *Zeitschrift für die Alttestamentliche Wissenschaft* 43 [1925]: 301–302)." While acknowledging that Num. 24:17 is referring to a comet, it seems preferable to me to maintain the formal translation "scepter."

[6] Timothy R. Ashley, *The Book of Numbers*, New International Commentary on the Old Testament (Grand Rapids, MI: Eerdmans, 1993), 503.

[7] 11QMelch 11:4–9; cf. 1QSb 5:27–28; 4QTest 9–13. Translation from Craig A. Evans, *Jesus and His Contemporaries: Comparative Studies* (Leiden: Brill, 1995), 71.

[8] Damascus Document (CD) 7:18–21. Translation from ibid., 87–88.

[9] *De praemiis et poenis* 16 §95 (my translation).

[10] As cited by Justin Martyr, *Dialogue with Trypho* 106.5–6.

[11] See A. P. Hayman, ed. *The Old Testament in Syriac, According to the Peshitta Version: Numbers* (Leiden: Brill, 1991), 78.

[12] Kevin J. Cathcart, "Numbers 24:17 in Ancient Translations and Interpretations," in *The Interpretation of the Bible: The International Symposium in Slovenia*, ed. J. Krašovec (Sheffield: Sheffield Academic Press, 1998), 512–513. Cf. B. Grossfield, *The Targum Onqelos to Leviticus and Numbers*, The Aramaic Bible (Wilmington, DE: Michael Glazier, 1988), 138–139.

Messiah" for "a scepter."[13] *Targum Pseudo-Jonathan* read, "a mighty king of the house of Jacob shall reign, and shall be anointed Messiah, wielding the mighty scepter of Israel."[14] The *Fragmentary Targum* stated, "A king is destined to arise from the house of Jacob, a redeemer and ruler from the house of Israel, who shall slay the mighty ones, . . . who shall destroy all that remains of the guilty city, which is Rome."[15]

Famously, the messianic claimant Simon Ben Kosiba in the second century AD was identified by Rabbi Aqiba as the Messiah and given the title Bar Kokhba ("Son of a Star"), under the influence of Balaam's prophecy. This title, as well as the images of a star over the Jerusalem Sanctuary on coins issued by him, may imply that Ben Kosiba was regarded by followers as having been authenticated by some celestial phenomenon, most likely a comet, either at his birth or at some key moment in his career, so that he was viewed as the fulfillment of Numbers 24:17. Certainly, his title implied that Numbers 24:17 had been fulfilled in the coming of Ben Kosiba, "a luminary who had come down to them from heaven" (Eusebius, *Hist. Eccl.* 4.6.2).[16]

Two passages from the *Testament of the Twelve Patriarchs* (which was completed by the second century AD) also took a strongly messianic interpretation of Balaam's oracle. *Testament of Judah* 24:1–6 drew heavily on Numbers 24:17 when it declared concerning the Messiah that "there shall arise for you a Star from Jacob in peace: And a man shall rise from my posterity like the Sun of righteousness. . . . This is the Shoot of God Most High. . . . Then he will illumine the scepter of my kingdom, and from your root will arise the Shoot, and through it shall grow a rod of righteousness for the nations, to judge and to save all that call on the Lord."[17] *Testament of Levi* (18:3) stated, concerning a messianic figure whom it calls a "new priest," that "his star shall rise in heaven like a king, kindling the light of knowledge. . . . And he shall be extolled by the whole inhabited world."[18]

It is striking that *Targum Pseudo-Jonathan* and the *Testament of Judah* and *Testament of Levi* strongly connect the star and scepter of Numbers 24:17 with Isaiah 11's great prophecy concerning the Messiah. Notably in both Numbers and Isaiah 11, the Messiah is associated with a "rod" or "scepter" (Isa. 11:4: "the rod of his mouth").[19]

There can therefore be no doubt that Balaam's oracle concerning the Star and Scepter was widely understood by Jews around the time of Jesus to relate to the coming of the Messiah.[20]

NUMBERS 24:17 AND THE MAGI'S STAR IN EARLY CHRISTIAN INTERPRETATION

The early church clearly believed that Balaam's oracle was fulfilled in connection with the coming of Jesus. Revelation 22:16 (cf. 2:26–28) refers to Jesus as "the bright morning star," which strongly alludes to Numbers

[13] Martin McNamara, *Targum Neofiti 1: Numbers*, The Aramaic Bible (Collegeville, MN: Liturgical Press, 1995), 140.
[14] Evans, *Jesus and His Contemporaries*, 71.
[15] Ibid., 71–72.
[16] For portrayals of Bar Kokhba in Jewish rabbinical sources, see Richard G. Marks, *The Image of Bar Kokhba in Traditional Jewish Literature: False Messiah and National Hero* (University Park: Pennsylvania State University Press, 1994), 14–18, 20–56. Some Jewish traditions claim that the name was inspired by the similarity of Simon's name (which was taken to be *Koziba*) and the Hebrew word for "star" (*kokhba*).
[17] H. C. Kee, "Testaments of the Twelve Patriarchs," in *The Old Testament Pseudepigrapha*, ed. James H. Charlesworth, 2 vols. (New York: Doubleday, 1983), 1:801. Many believe that this chapter reflects a Christian perspective (see, for example, J. J. Collins, *The Scepter and the Star: The Messiahs of the Dead Sea Scrolls and Other Ancient Literature*, 2nd ed. [Grand Rapids, MI: Eerdmans, 2010], 91; W. D. Davies and Dale C. Allison, *A Critical and Exegetical Commentary on the Gospel according to Saint Matthew*, 3 vols., International Critical Commentary [Edinburgh: T. & T. Clark, 1988–1997], 1:234).
[18] Kee, "Testaments of the Twelve Patriarchs," 794. On the allegedly Christian nature of this statement, see Cathcart, "Numbers 24:17," 516.
[19] For some rabbinical interpretations of Num. 24:17, see Marks, *Bar Kokhba*, 18–20.
[20] On the messianic interpretation of Num. 24:17, see Martin McNamara, "Early Exegesis in the Palestinian Targum (Neofiti 1) Numbers Chapter 24," *Proceedings of the Irish Biblical Association* 16 (1993): 57–79; idem, *Targum Neofiti 1*, 140; Cathcart, "Numbers 24:17," 511–519.

24:17. Similarly, 2 Peter 1:19 speaks of "the prophetic word . . . , to which you will do well to pay attention as to a lamp shining in a dark place, until the day dawns and the morning star rises in your hearts."[21] This passage too assumes that Balaam's prophecy was fulfilled in Jesus.

That Jesus was the fulfillment of the Mesopotamian seer's oracle was regarded by many early Christians as confirmed by the appearance of the Star at the time of his coming. In the early centuries of Christianity, Balaam's prophecy was widely believed to have been fulfilled literally by the Magi's Star.

With respect to Matthew himself, most scholars recognize that Numbers 24:17 is an important background text for the account of the Magi and the Star. More particularly, it is commonly accepted that when Matthew records that the Magi declared to the people of Jerusalem that "we saw his star at its rising" (Matt. 2:2), he was strongly alluding to Numbers 24:17 and expected his readers to recognize the allusion. Of course, if that was what Matthew was thinking, then it is more than likely also what the Magi themselves were thinking, since the words are attributed to them.

Further, we recall that Revelation 12:1–5, which preserves the memory of what the Star did to mark the birth of Jesus, makes reference to the iron scepter of Psalm 2:8–9 in connection with the birth of Virgo's son, ultimately recalling Balaam's oracle concerning the messianic scepter-star.[22] In a strongly astronomical context, particularly where a cometary apparition and the birth of the Messiah are in view, the reference to the iron scepter is most naturally interpreted as alluding to the literal fulfillment of Balaam's oracle concerning the cometary scepter. The implication of Revelation 12:5 is that the comet at its rising took the form of a gloriously bright scepter, in which case it would be little wonder that Balaam's prophetic word seemed a compelling interpretive key.

This New Testament evidence suggests that Numbers 24:17 played an important role in the earliest Christians' interpretation of the significance of the Christ Comet's appearance.

It is important to realize that the church fathers in the second and third centuries AD continued to regard Numbers 24:17 as having provided the Magi with the prophetic key to unlocking the meaning of the cometary apparition.

The second-century AD Christian apologist Justin Martyr wrote, "And that He should arise like a star from the seed of Abraham, Moses showed beforehand when he thus said, 'A star shall arise from Jacob, and a leader from Israel'. . . . Accordingly, when a star rose in heaven at the time of His birth, as is recorded in the memoirs of his apostles, the Magi . . . , recognising the sign by this, came and worshipped Him."[23]

Also in the second century, Irenaeus made this connection between Balaam's prophecy and the Magi's Star.[24]

So did Origen in the third century: "The Magi, seeing God's sign in the heavens, looked for its meaning. I think they knew the prophecies of Balaam that are recorded by Moses."[25] Elsewhere Origen asserted that, at the point when Jesus was born, the Magi did realize that the Star was the fulfillment of that oracle.[26]

[21] For more on early Christian interpretation of Num. 24:17, see Jean Daniélou, *Primitive Christian Symbols*, trans. Donald Attwater (London: Burns & Oates, 1964), 102–123. 2 Pet. 1:19 is also heavily dependent on Isa. 8:11–22.

[22] Cf. Rev. 2:26–28: vv. 26–27 draw on Ps. 2:9, and Rev. 2:28 draws on Num. 24:17.

[23] *Dialogue with Trypho* 106:5–6. Translation from *The Ante-Nicene Christian Library: Translations of the Writings of the Fathers Down to A.D. 325*, vol. 2, ed. Alexander Roberts and James Donaldson (Edinburgh: T. & T. Clark, 1868), 233. See also *Dialogue with Trypho* 126:1; and *First Apology* 32:12–13.

[24] *Adversus Haereses* 3.9.2–3; and *Demonstration of the Apostolic Preaching* 58.

[25] *Contra Celsum* 1:60 (my translation).

[26] *Homilies on Numbers* 13.7. Origen also reflected a new development in the tradition, namely the idea that Balaam was highly respected by the Mesopotamian Magi and indeed was a founding figure of the Magian community; Origen maintained that the Magi had the text of all of Balaam's oracles, including Num. 24:17 (*Homilies on Numbers* 13.7; cf. Eusebius, *Demonstratio Evangelica* 9.1).

Therefore the early Christians believed that the Balaam oracle was literal and metaphorical. Jesus, they were convinced, was the metaphorical Star and Scepter foreseen by Balaam. At the same time, the appearance of the Star at the time of Jesus's birth was a literal fulfillment of Balaam's prophecy that confirmed the messianic identity of Jesus.

NUMBERS 24:17 AND THE MAGI'S INTERPRETATION OF THE STAR

The case is strong for concluding that the Magi interpreted the cometary apparition in 6 BC in light of Numbers 24:17. No prophecy in the Hebrew Bible other than Balaam's oracle more clearly associated the Messiah's coming with the appearance of a celestial entity. Within Jewish thought in the Second Temple period, the prophecy was widely interpreted as referring to the coming of the Messiah. Further, the early Christians maintained not only that Jesus was the Star and Scepter of Numbers 24:17, but also that the new astronomical entity that had appeared to the Magi to signal his birth was a literal fulfillment of that ancient prediction and was regarded as such by the Magi. Moreover, most scholars agree that Matthew, our primary source on the Magi and the Star of Bethlehem, strongly intimates that the Star's appearance was a fulfillment of Numbers 24:17. Matthew also implies that this was the understanding of the Magi themselves. It stands to reason that the Magi may have become persuaded of the relevance of Balaam's oracle for the interpretation of the comet by the scepter-like form of the celestial body around the time of its rising.

In addition, Revelation 12:5 seems to intimate that, at the point when the messianic baby was born on the earth, the long, straight-tailed comet had the form of a massive scepter that was stretched out from the eastern horizon to the western horizon. Accordingly, the cometary scepter would have looked like it was resting on the earth in the west, where Judea was. Those witnessing such a phenomenon would have recalled that, according to Numbers 24:17, "a scepter shall rise out of Israel."

Accordingly, both Matthew and Revelation suggest that the comet had the credentials to qualify as an obvious bona fide literal fulfillment of Balaam's oracle.

It is very likely that the Magi were guided to this prophecy by one or more members of the Jewish exilic community in Babylon and found in it an important interpretive key to unlocking the mystery of the great cometary apparition.

We conclude, then, that Numbers 24:17 played an important part in helping the Magi deduce that the cometary apparition was marking the birth of the Jewish Messiah. However, it is probable that another passage from the Jewish Scriptures also strongly influenced the Magi's interpretation of the celestial phenomenon unfolding in the eastern sky, and prompted them to make a pilgrimage to Judea to find and worship the newborn Messiah. That passage is found in the book of Isaiah.

THE BOOK OF ISAIAH

THE INFLUENCE OF THE BOOK OF ISAIAH ON THE MAGI

The fact that the Magi brought gifts of gold and frankincense for the newborn Messiah probably reflects the influence of the book of Isaiah on their thinking regarding the comet. Gold and frankincense are specifically mentioned in Isaiah 60:1–6 as offerings that will be brought by pilgrims from the Arabian Peninsula to the Messiah during his eschatological reign over the nations. There, these two gifts are representative of the great wealth that will be presented to the Messiah by the nations. When therefore the Magi give the baby Messiah these same gifts, they are most likely intimating that they already fully acknowledge his royal authority over them and the world and look forward to the beginning of his reign.[27]

It is important to observe that this passage in Isaiah opens with a reference to the Messiah as Israel's extraordinarily bright celestial "light" that rises in a context of darkness:

> Arise, shine, for your light has come,
> and the glory of Yahweh has risen upon you.
> . . . Yahweh will arise upon you,
> and his glory will be seen upon you.
> And nations shall come to your light,
> and kings to the brightness of your rising.
> Lift up your eyes all around, and see;
> they all gather together, they come to you. . . .
> Then you shall see and be radiant;
> your heart shall thrill and exult,
> because the abundance of the sea shall be turned to you,
> the wealth of the nations shall come to you.
> A multitude of camels shall cover you,
> the young camels of Midian and Ephah;
> all those from Sheba shall come.
> They shall bring gold and frankincense,
> and shall bring good news,
> the praises of Yahweh.
> (Isa. 60:1, 2b–4a, 5–6)

Note the mention of gold and frankincense in verse 6. As for the Magi's gift of myrrh, as we suggested earlier, it probably reflects the influence of Isaiah 53, the mysterious Suffering Servant being regarded as the Messiah.

As important as these texts probably were for the Magi, the passage that contained the key for their interpretation of what the comet did with respect to Virgo in the eastern sky is almost certainly Isaiah 7–12. From these chapters they were able to deduce not only that the newborn was the Messiah but also that he was divine in nature and would be born of a virgin, and that his birth to a virgin mother would be attended by a celestial sign, a great light shining in the darkness.

ISAIAH 7:10–14

In Isaiah 7:1–25 the prophet was challenging the covenantally faithless king of Judah, Ahaz, to trust in Yahweh through the crisis precipitated by the Syro-Ephraimite invasion of Judah in 734/733 BC rather than turning for help to the regional superpower of his day, Assyria.[28] To encourage Ahaz to have faith in the God of David, God offered him an authenticating sign:

> Again Yahweh spoke to Ahaz, "Ask a sign of Yahweh your God; let it be deep as Sheol or high as heaven." But Ahaz said, "I will not ask, and I will not put Yahweh to the test." And [Isaiah] said, "Hear then, O house of David! Is it too little for you to weary men, that you weary my God also? Therefore the Lord himself will give you a sign. Behold, the virgin [*almah*] shall be with child and bear a son, and shall call his name Immanuel." (Isa. 7:10–14)

The Context of Isaiah 7:14

It is necessary to reflect on the historical and theological context and on the meaning of Isaiah's important oracle.[29]

[27] As Joachim Gnilka, *Das Matthäusevangelium*, 2 vols., Herders theologischer Kommentar zum Neuen Testament (Freiburg: Herder, 1986–1992), 1:41, comments, for Matthew the coming of the Magi to Jesus anticipates the eschatological reign of the Messiah over the world. In my opinion, the Magi themselves probably acted out of this same paradigm.

[28] Ahaz's subsequent appeal to Tiglath-pileser, king of Assyria, is recorded in 2 Kings 16:6–9.

[29] Most scholars accept that at least the core of Isaiah 7–8 is attributable to the prophet Isaiah during the Syro-Ephraimite Crisis. Many are willing to concede that all or most of it should be assigned to him. My own study suggests that the whole of Isaiah 7–12 is a unity and was composed by Isaiah son of Amoz during the Syro-Ephraimite Crisis. For a defense of Isaiah 7–12's unity, see John N. Oswalt, "The Significance of the 'Almah Prophecy in the Context of Isaiah 7–12," *Criswell Theological Review* 6.2 (1993): 223–235. According to John H. Hayes and Stuart Irvine (*Isaiah the Eighth-Century Prophet: His Times and His Preaching*

Isaiah's prophecy was delivered in the winter of 734/733 BC in Jerusalem, capital of the southern kingdom of Judah.[30] The eighth-century BC King Ahaz of Judah was terrified by the prospect of an imminent attack on Jerusalem by his enemies Syria and Israel. These northern kingdoms hated the resident regional superpower Assyria and loathed Ahaz, because he would not wholeheartedly join their anti-Assyrian alliance. Ahaz was frightened of Syria and Israel, because their kings Rezin and Pekah were relatively powerful and were dead set on ousting him from the kingship of Judah and replacing him with a puppet king more sympathetic to their anti-Assyrian agenda. Moreover, the Syrians and Israelites, along with their allies, the Edomites and Philistines, had in the immediate run-up to this crisis brought great destruction on Judah and killed many tens of thousands of Judahite men and taken captive countless women and children (2 Kings 16; 2 Chronicles 28). Ahaz felt that without foreign military assistance he stood no chance in the face of his enemies' determined advance to Jerusalem to oust him.

From the prophet Isaiah's perspective, what Ahaz was not sufficiently taking into account was that the Davidic dynasty and David's covenant with Yahweh were at stake. The future of Ahaz, Jerusalem, and Judah was determined by Yahweh, not Pekah and Rezin. In Isaiah's analysis, whether King Ahaz and the Davidic dynasty would remain in power in Judah would be determined in heaven and not on the earth. And God would make his decision regarding the future of Ahaz and the house of David on the basis of whether or not the king kept covenant with him, and on the basis of his own mercy and love. Consequently, as the prophet saw it, the crisis facing the king of Judah served to shine the spotlight on Ahaz's spiritual state and covenant performance and on the future of the Davidic covenant and dynasty. The spotlight exposed Ahaz's unbelief, for he was unprepared to trust his God to protect him. He was evidently determined to appeal to Assyria rather than Yahweh for help. In his heart, the Judahite king was not an admirer of Yahweh or of the religion of his fathers. He loved the gods of the nations and worshiped idols and celestial entities, and had even offered up his sons as sacrifices to a pagan deity (2 Kings 16:3; 2 Chron. 28:3).

The prophet Isaiah, representing Yahweh, went to meet Ahaz when the king was extremely worried about the vulnerability of Jerusalem's water supply, which was the city's Achilles' heel in times of siege. Isaiah sought to reassure Ahaz that the Syro-Ephraimite intervention in Judah would not succeed in its main objective of toppling the Davidic dynasty, because God had decreed that it would not. Moreover, the prophet challenged the king to put his faith in Yahweh.

According to Isaiah, although Ahaz would not be overthrown by Rezin and Pekah, his future and that of his dynasty were still very much at stake. This was so because Yahweh had decreed (Isa. 7:9) that, if at this moment of crisis Ahaz turned his back on his divine covenant partner, he would bring disaster upon himself: he would "not be firm at all" (v. 9b). Clearly, in the prophet's judgment, this was a momentous hour in the history of the House of David in Judah.

Probably at that time (or at least very shortly thereafter), Isaiah once again reached out to Ahaz (vv. 10–14), instructing him to request a sign from God to confirm his commitment to keeping his covenant with the Davidic dynasty in Judah and to preserving Ahaz through the Syro-Ephraimite invasion. The sign could be "deep as Sheol" (v. 11b, ESV) or "in the height above" (v. 11c, cf. KJV, ASV). "Deep as Sheol" presumably referred to some kind of seismic activity or resurrection from

[Nashville: Abingdon, 1987], 13), virtually all of Isaiah 1–39 "derives from the eighth-century B.C.E. prophet," and chapters 7–12 were written at the time of the Syro-Ephraimite Crisis.

[30] See Stuart A. Irvine, *Isaiah, Ahaz, and the Syro-Ephraimite Crisis* (Atlanta: Scholars Press, 1990), 107.

the dead. "In the height above" plainly referred to a celestial wonder such as an eclipse or some phenomenon against the backdrop of the stars and constellations. In effect, Ahaz was to tailor his very own special sign within the set parameters. The implicit deal was that, when Yahweh did the sign in Sheol or in the heavens, Ahaz would turn his heart back to Yahweh and trust him through the present crisis.

Ahaz, however, declined to choose a sign for Yahweh to do, obviously because he was privately resolute that he was not going to trust or obey the God of Israel. He had the audacity to try to cover up his lack of trust in Yahweh with a cloak of pseudo-piety: "I will not ask, and I will not put Yahweh to the test" (v. 12). His words are drawn from Deuteronomy 6:16: "You shall not put Yahweh your God to the test, as you tested him at Massah." Ironically, Ahaz, by his refusal to specify a sign, *was* in fact "putting Yahweh to the test." At Massah in the wilderness the Israelites were confronted with a trial—they lacked water to quench their thirst—and they refused to trust God to meet their need and rebelled against him and his servant Moses (Ex. 17:1–7). Ahaz was in his Massah, so to speak, and he was intent on rebelling against the word of Yahweh as represented by Isaiah. To Isaiah, the king's failure to request a sign was nothing short of covenant treachery.

By his refusal to stand firm in faith, Ahaz brought upon himself the fate decreed by Yahweh in verse 9: he ceased to "be firm at all." The House of David in Judah had been tested and found wanting. As it happened, subsequent history powerfully vindicated Isaiah. In the story of the Davidic dynasty and of Judah, this incident proved to be a decisive turning point. Ahaz needlessly sold Judah's independence to Assyria, and the southern kingdom quickly became a pathetic vassal state that struggled to bear the heavy financial burdens that the short-tempered superpower put on it (see Isa. 7:17–25; 2 Chron. 28:16–21; 2 Kings 16:17–18). Essentially, from this moment onward Judah was on a downhill slope to termination in 586 BC at the hands of the Babylonians.

In response to the hard-heartedness and rebellion of King Ahaz, Isaiah then declared to Ahaz (Isa. 7:13–14): "Hear then, O house of David! Is it too little for you to weary men, that you weary my God also? Therefore the Lord himself will give you a sign. Behold, the virgin shall be with child and bear a son, and shall call his name Immanuel." Ahaz had refused to specify which sign he wanted from Yahweh (v. 12)—whether a sign "deep as Sheol or in the height above" (v. 11)—and so Yahweh now chose "a sign" for him (v. 14). Disturbingly, Isaiah speaks of Yahweh as "my God," implying that he was no longer Ahaz's God.

The Immanuel Oracle: The Virgin and Her Child

One scholar has called Isaiah 7:14 "the most controversial passage in the Bible."[31] Debate has particularly focused on the identity of the prophesied child. Was it someone born in Isaiah's day? If so, was it a son of Isaiah,[32] or a son of Ahaz, either Hezekiah[33] or an anonymous younger sibling of Hezekiah?[34] Or was it someone born in the future, namely the Messiah.[35]

[31] Martin Buber, *Der Glaube der Propheten* (Zürich: Conzett & Huber, 1950), 201.

[32] It is generally reckoned to be his second, Maher-shalal-hash-baz—so, for example, Robert H. Gundry, *Matthew: A Commentary on His Handbook for a Mixed Church under Persecution*, 2nd ed. [Grand Rapids, MI: Eerdmans, 1994], 25; John N. Oswalt, *The Book of Isaiah*, 2 vols., New International Commentary on the Old Testament (Grand Rapids, MI: Eerdmans, 1986–1998), 1:207–213; Walter Mueller, "A Virgin Shall Conceive," *Evangelical Quarterly* 32 (1960): 206; and Herbert M. Wolf, "A Solution to the Immanuel Prophecy in Isaiah 7:14–8:22," *Journal of Biblical Literature* 91 (1972): 449–456.

[33] E.g., Walter C. Kaiser, "The Promise of Isaiah 7:14 and the Single-Meaning Hermeneutic," *Evangelical Journal* 6 (1988): 64.

[34] E.g., John H. Walton, "Isaiah 7:14: What's in a Name?," *Journal of the Evangelical Theological Society* 30 (1987): 289–297.

[35] E.g., John Calvin, *Calvin's Bible Commentaries: Matthew, Mark, and Luke, Part 1*, trans. William Pringle (Edinburgh: Calvin Translation Society, 1845), 99–102; idem, *Commentary on the Book of the Prophet Isaiah—Volume 1*, trans. William Pringle (Edinburgh: Calvin Translation Society, 1850), 244–249; Donald A. Carson, "Matthew," in *Expositor's Bible Commentary*, rev. ed., ed. Tremper Longman III and David E. Garland, vol. 9 (Grand Rapids, MI: Zondervan, 2010), 104–105.

Inextricably linked to this debate is the question of the identity of "the virgin" who gives birth to the child. Was she Isaiah's wife? Was she a member of Ahaz's royal harem? Or was she simply a woman who happened to be passing by when Isaiah was meeting with Ahaz? Or, was she the mother of the Messiah?

We must first ask a simple question: Does the oracle speak of the near future or of the distant future or of both? It is clear from 7:15–8:8 that it does speak to the near future.[36] At the same time, 8:8–10 and 9:2–7, as well as 11:1–16, reveal that the prophet is looking beyond the near future into the distant future. So it is best to conclude that Isaiah's oracle has a relevance to both the near future and the distant future. As we shall see below, this dual perspective comes to the surface in 9:1, where Isaiah speaks of "the former time," which brings distress and darkness, and "the latter time," which brings joy and light.

As to the identity of the newborn of 7:14 in Isaiah's near future, in light of what the prophet reports in 8:3–4, it is difficult to deny that "Immanuel" was Maher-shalal-hash-baz, Isaiah's second son: "And I went to the prophetess, and she *conceived* and *bore a son*. Then Yahweh said to me, '*Call his name* Maher-shalal-hash-baz, for before the boy knows how to cry 'My father' or 'My mother,' the wealth of Damascus and the spoil of Samaria will be carried away before the king of Assyria." Here not only do we have a clear allusion back to Isaiah 7:14's conception, childbirth, and naming, but we also have the birth being regarded as the beginning of a countdown to the fulfillment of Yahweh's promise to punish Syria and Israel, just as in 7:15–16 (where Immanuel's birth marks the start of a countdown to the desertion of Syria and Israel). A further indication that "Immanuel" in Isaiah's day was the prophet's second son is found in 8:18, where we read that Isaiah and his sons are "signs and portents in Israel from Yahweh of hosts." We recall that 7:14 spoke of the birth of Immanuel as a "sign." Of course, if Maher-shalal-hash-baz was "Immanuel," then the "virgin" (*almah*) of 7:14 was Isaiah's wife, the prophetess.

Compared to this view, other proposed identifications of the newborn baby boy in the late 730s BC come up short. In particular, the identification of Immanuel as Hezekiah fails on the grounds of chronology. Whichever chronology of the kings of Judah one adopts, Hezekiah was certainly born well before the Syro-Ephraimite crisis.

Moreover, it is hard to see why the naming of a young, non-succeeding son of Ahaz by a concubine in the royal household might constitute a sign signaling divine judgment on the king.

The idea that "the virgin" was simply a passerby can be safely rejected, since it is ridiculous to imagine Ahaz, who did not want the sign and had no intention of heeding it, proceeding to track an anonymous woman to her home and getting updates about her pregnancy, delivery, and naming of her child.

Although we judge that the "virgin" in Isaiah's day was his prophetess wife and the newborn was his second son, Maher-shalal-hash-baz, this interpretation is not without apparent difficulties. The main perceived problem centers on the particular word used to describe the woman: *almah*. Isaiah's wife, if she is the same woman who previously bore him Shear-jashub, does not seem a ready fit for the role of *almah*. That is because the word is not a natural one to use of a married woman[37] and certainly not one who has already had a child.[38] Indeed, although the term

[36] Of course, the fact that the "sign" is a response to Ahaz's rebellion (7:12–13) and is expressly directed at "you" (plural), namely, the Davidic house, favors a fulfillment in the near future.

[37] Brevard S. Childs, *Isaiah: A Commentary*, Old Testament Library (Louisville, KY: Westminster John Knox, 2000), 66; Edward J. Young, *The Book of Isaiah: A Commentary*, 3 vols. (Grand Rapids, MI: Eerdmans, 1965), 1:287.

[38] Walton, "Isaiah 7:14," 292; Paul Wegner, *An Examination of Kingship and Messianic Expectation in Isaiah 1–35* (Lewiston, NY: Edwin Mellen, 1992), 110.

does not refer to virginity as such, generally speaking an *almah* is sexually inexperienced.[39] It has been suggested that Isaiah's first wife, the mother of Shear-jashub, had died and that Isaiah had recently married a second wife, the prophetess, so that he is referring to a woman who at the point that he was speaking was a bona fide *almah*, in the sense that she had not previously borne a child.[40] That possibility should not be quickly dismissed, although it is probably unnecessary to resort to this.

With respect to the distant future, there can be no question but that the birth of the Messiah was in view.

The royal birth announcement in Isaiah 9:6–7 makes this abundantly clear: "For to us a child is born, to us a son is given; and the government shall be upon his shoulder, and his name shall be called Wonderful Counselor, Mighty God, Everlasting Father, Prince of Peace. Of the increase of his government and of peace there will be no end, on the throne of David and over his kingdom, to establish it and uphold it with justice and with righteousness from this time forth and forevermore." The child would be the ultimate son of David who fulfilled the Davidic covenant and was divine in nature.

That Isaiah 7:14 had in view the birth of the Messiah is also strongly supported by 11:1–3a, 4b–10. There, in the climax of chapters 7–12, Isaiah again referred to the arrival of the Messiah on the earthly scene and his ultimate destiny:

> There shall come forth a shoot
> from the stump of Jesse,
> and a branch[41] from his roots
> shall bear fruit.
> And the Spirit of Yahweh shall rest
> upon him,
> the Spirit of wisdom and
> understanding,
> the Spirit of counsel and might,
> the Spirit of knowledge and the
> fear of Yahweh.
> And his delight shall be in the fear
> of Yahweh. . . .
> and he shall strike the earth with
> the rod of his mouth,
> and with the breath of his lips
> he shall kill the wicked.
> Righteousness shall be the belt
> of his waist,
> and faithfulness the belt
> of his loins.
> The wolf shall dwell with
> the lamb,
> and the leopard shall lie down
> with the young goat,
> and the calf and the lion and the
> fattened calf together;
> and a little child shall lead
> them.
> The cow and the bear shall graze;
> their young shall lie down
> together;
> and the lion shall eat straw like
> the ox.
> The nursing child shall play over
> the hole of the cobra,
> and the weaned child shall put
> his hand on the adder's den.
> They shall not hurt or destroy
> in all my holy mountain;
> for the earth shall be full of the
> knowledge of Yahweh
> as the waters cover the sea.

[39] As R. Dick Wilson, "The Meaning of *'Almah* (A.V. 'Virgin') in Isaiah VII. 14," *Princeton Theological Review* 24 (1926): 316, put it, "the presumption in common law and usage was and is, that every *'almâ* is a virgin, until she is proven not to be."

[40] Wolf, "Solution," 450.

[41] The Hebrew word *netser* seems to be a wordplay on "Nazareth." This wordplay is recognized by Matthew in Matt. 2:23. It seems to presuppose that the village existed in the prophet's day and indeed was called Nazareth at that time. There is evidence of habitation there in the Bronze and Iron Ages and in the period running up to Tiglath-pileser's devastating invasion of Galilee in 733 BC (2 Kings 15:29), which transpired very shortly after Isaiah delivered his prophetic oracles of chapters 7–12 (for an imminent announcement of Galilee's doom, see Isa. 9:1). Thereafter the site seems to have been unoccupied until the third century BC (James F. Strange, "Nazareth," in *The Anchor Bible Dictionary*, ed. D. N. Freedman, 6 vols. [New York: Doubleday, 1992], 4:1051). Of course, it later became the hometown of Mary and Joseph and Jesus.

> In that day the root of Jesse, who shall stand as a signal for the peoples—of him shall the nations inquire, and his resting place shall be glorious.

In this passage Isaiah makes it explicit that the birth that was uppermost in his mind during the Syro-Ephraimite Crisis was, strikingly, that of the Messiah.

Accordingly, in the near future the *almah* in Isaiah 7:14 was Isaiah's wife, the prophetess, and in the distant future she was the Messiah's mother.

Whatever ambiguity there was concerning the word *almah*, the simple fact is that if the offspring was divine, as the name Immanuel ("God with us") implied (cf. 9:6), the only natural conclusion to reach was that the *almah* was a virgin, with the father of Immanuel being divine and the means of reproduction being nonsexual.

It is doubtful if Ahaz grasped the full meaning and significance of the sign. To him, the oracle may have sounded like an adaptation of a pagan myth concerning the birth of a child to a virgin mother goddess, like the ancient Egyptian myth of the "virgin" goddess Isis and her son Horus. The pagan king of Judah may well have assumed that the prophesied myth-like sign would be enacted in some kind of drama. Indeed drama was a common prophetic tool. Isaiah 20 records that Isaiah walked around "naked and barefoot" for three years "as a sign and portent against Egypt and Cush" (v. 3a), prophetically playing the role of an Egyptian or Cushite taken captive by the Assyrians (vv. 3b–4).

Isaiah 7:14 was indeed disclosing a drama, an extraordinary one. The dramatization of the virgin and Immanuel would be spread over ten months and involve a real pregnancy and birth. When Ahaz saw Isaiah's wife, the prophetess, pregnant, and then learned that she had given birth to Maher-shalal-hash-baz and (as Isaiah 7–8 implies) named him "Immanuel," he would have understood that the woman and her son were playing the parts of the virgin and her divine child.

For Isaiah, however, his wife was not playing the part of a virgin mother goddess in a mythical scene, but rather the part of the Messiah's mother as she became pregnant with and gave birth to the divine Son of God. It was the Messiah's mother who would be "*the* virgin." She alone could with full justification use the name "God with us" to describe her son.

That the drama unfolding in 733 BC was not the actual, full fulfillment of Isaiah's oracle was clear simply by virtue of the nature of the actors playing the key roles—Isaiah's wife, particularly if she was already a mother, seems to have lacked the qualifications to be a bona fide *almah*, and Isaiah's second son was certainly not divine (he was not, in truth, "God with us" [or "Mighty God"])—and by virtue of the participation of Isaiah in the drama, impregnating his wife.

However, the outworking of the "sign" in Isaiah's day was not devoid of impressive elements, because Isaiah's wife did conceive, did give birth to a boy, and did, evidently without Isaiah's personal intervention, prophetically name him Immanuel, all in accord with the prophetic word.

Maher-shalal-hash-baz was a "sign" pointing forward to the true son of the virgin, the divine Messiah. As Isaiah framed it, Maher-shalal-hash-baz's very existence was a rebuke to the Davidic dynasty—God would fulfill his promises to David not through the seed of Ahaz or one of his dynastic successors, but rather through an adopted son of David born of a virginal conception. After all, Yahweh's plan was to unite his own house with that of David (2 Samuel 7), causing a virgin girl to become pregnant by nonsexual means.

That the second son of Isaiah was a dramatic type, pointing forward to the Messiah, is clear in 8:8–10. In 8:5–8 Isaiah prophesied concerning the near future but in verse 8 he suddenly spoke of the land as belonging to

Immanuel. This makes it clear that Immanuel there was not Maher-shalal-hash-baz but the Messiah whom he represented. This conclusion is supported by the fact that verses 9–10 go on to present an eschatological word of judgment to the nations that scheme against Israel-Judah, "for God is with us." This final element intimates that Immanuel in the person of the Messiah would rescue his people from the international conspiracy at the end of the age. The allusion to the very ancient Davidic Psalm 2 is forceful—that psalm, widely regarded as a coronation psalm, warned a conspiracy of nations not to rebel against the Messiah but to submit to him in advance of his coming in wrath.

The sign had significance for Isaiah's own day too, of course. Positively, it demonstrated that God would fulfill his promise to neutralize Ahaz's enemies Syria and Israel (Isa. 7:15–16; 8:3–4). Negatively, since it was none other than God who was Judah's covenant partner, Ahaz's religious treachery could not go unpunished (7:17–25). The sign showed that Yahweh was turning away his favor from the Davidic dynasty that had ruled an independent Judah for 200 years. The dynasty of David, associated with Ahaz, had been exposed by Isaiah as unworthy. Judah would now lose its independence and become a vassal state, and embark on an inevitable path toward destruction and deportation.

The messianic drama acted out by Isaiah's wife and second son was a fitting sign both positively and negatively. Because the kings of Syria and Israel had shown contempt for the Davidic covenant, a sign that highlighted the ultimate fulfillment of God's promise to unite his house with David's was appropriate. In addition, because Ahaz had violated the Davidic covenant, a sign that emphasized that the future of the Davidic covenant lay with a virgin-born Messiah rather than the reigning Davidic dynasty had punch. To what extent the message conveyed by the sign was understood by Ahaz, however, is unclear—it is doubtful that he took much notice of Isaiah's word. Indeed 2 Chronicles 28:22–23 seems to imply that Ahaz by this time had started worshiping the gods of Syria in a desperate bid to secure their favor against his enemy, Rezin king of Syria.[42]

Isaiah 7:14 and the Birth of Jesus

How was Isaiah 7:14 interpreted in the period running up to the birth of Jesus? Regrettably, we lack Jewish texts from the pre-Christian period that comment on Isaiah 7:14's meaning. One tentative indication that the oracle may sometimes have been interpreted within Second Temple Judaism as referring to the Messiah may be the Septuagint's rendering of *almah* with *parthenos*. Usually *parthenos* implies sexual chastity. If the Septuagint (LXX) translator was using it in this sense, then he may have interpreted *almah* in the context of Isaiah 7:14 to imply virginity and therefore may have believed that the oracle revealed that the Messiah would be born by supernatural, nonsexual agency.[43]

[42] The king's decision to begin worshiping the gods of Syria is difficult to explain in the aftermath of the Assyrian campaign against Syria and Israel. If, however, he had been seeking to secure their aid against the Syrians during the Syro-Ephraimite Crisis, the Assyrian campaign against Syria would have confirmed his faith in the Syrian divinities rather than undermining it. I note that Josephus represented Ahaz as beginning his worship of the gods of Syria "even when he was at war with [the Syrians]" (*Ant.* 9.12.3 [§255]). The fact that Ahaz gave the instruction for a model of the altar of Ben-Hadad in Damascus to be constructed and put in the place of the altar of burnt offering of the Jerusalem temple at the time of his meeting with Tiglath-pileser in 732/731 BC (2 Kings 16:10–16) is consistent with this interpretation.

[43] See especially Martin Rösel, "Die Jungfrauengeburt des endzeitlichen Immanuel. Jesaja 7 in der Übersetzung der Septuaginta," *Jahrbuch für Biblische Theologie* 6 (1991): 135–151. Also W. F. Albright and C. S. Mann, *Matthew*, Anchor Bible (Garden City, NY: Doubleday, 1971), 8; Hagner, *Matthew*, 1:20; Craig Blomberg, "Matthew," in *Commentary on the New Testament Use of the Old Testament*, ed. G. K. Beale and D. A. Carson (Grand Rapids, MI: Baker Academic, 2007), 4. Even though Rodriga F. de Sousa, *Eschatology and Messianism in LXX Isaiah 1–12* (New York: T. & T. Clark, 2010), 70–102, rejects the idea that the translation *parthenos* reflected a high Christology on the part of the LXX translator, he does think that there are other indications in LXX Isa. 7:14–16 that are consistent with a messianic reading of this passage. For a general discussion of the history of interpretation of this controversial verse in the early centuries AD, see Adam Kamesar, "The Virgin of Isaiah 7:14: The Philological Argument from the Second to the Fifth Century," *Journal of Theological Studies* 41 (1990): 51–75.

Certainly, early Christians regarded Isaiah 7:14 as predicting Jesus's birth to a virgin. In no place is this clearer than in Matthew 1:18–25, where Matthew explicitly quotes Isaiah's oracle: "All this took place to fulfill what the Lord had said through the prophet: 'Behold, the virgin [*parthenos*] shall be with child [*in gastri hexei*] and shall give birth to a son [*texetai huion*], and they shall call him Immanuel' (which means, 'God with us')" (vv. 22–23).[44] The use of *parthenos* here probably indicates that Matthew was convinced that Isaiah was predicting that the Messiah would be conceived nonsexually in the body of a virgin.

Matthew 1:18's "she was found to be with child [*en gastri echousa*]" echoes LXX Isaiah 7:14: "The virgin shall be with child [*en gastri hexei*]," and Matthew 1:21's "She will bear a son [*texetai huion*]" and verse 25's "she [gave] birth to a son [*eteken ton huion*]" are clearly drawn from LXX Isaiah 7:14's "[the virgin] will bear a son [*texetai huion*]." Likewise, "he called his name [*ekalesen to onoma autou*]" in Matthew 1:25 echoes the wording of LXX Isaiah 7:14's "she will call his name [*kaleseis to onoma autou*]." It is evident that Matthew is inviting his readers to interpret the events relating to Jesus's nativity with reference to Isaiah's prophecy concerning the pregnancy and childbirth of "the virgin."

It should also be noted that when, at the very end of Matthew's Gospel, Jesus declares, "I am with you always, to the end of the age" (Matt. 28:20), he is picking up on Isaiah 7:14, claiming that he himself is the prophesied Immanuel ("God with us").[45] The fact that Isaiah 7:14 is so prominent in the first and last main parts of the Gospel, bracketing the whole, suggests that the theme of Jesus as the presence of God with humans was very important to Matthew.[46]

Luke 1:27 also calls Mary a "virgin" (*parthenos*), and verse 31 records Gabriel as saying that she would conceive "in [her] womb [*en gastri*]" and would "bear a son [*texē huion*]" and indeed would "call his name [*kaleseis to onoma autou*] Jesus." The allusion to Isaiah 7:14 is undeniable.[47] This strongly suggests that Luke, like Matthew, believed that Isaiah 7:14 was fulfilled in connection with the events of Jesus's nativity.

Moreover, it is striking that when, according to Luke's Gospel, Simeon met Joseph, Mary, and Jesus in the Jerusalem temple on the 40th day after the birth, he declared, "Behold, this child is appointed" to be "a sign that is opposed" (Luke 2:34). This statement strongly alludes to Isaiah's word to Ahaz in Isaiah 7:14: "The Lord himself will give you *a sign*. Behold, the virgin shall be with child and bear *a son*, and shall call his name Immanuel." That Maher-shalal-hash-baz was a "sign" is reiterated in 8:18, where Isaiah refers to himself and his children as "signs and portents." That Jesus as a newborn infant was declared to be "a sign" closely identified him with this son of Isaiah. Since, according to Luke, Simeon was speaking just 5½ weeks after the

[44] My translation.

[45] The importance of the bracketing of the Gospel by the Immanuel references of 1:23 and 28:20 has been appreciated by many scholars: e.g., W. C. van Unnik, "*Dominus Vobiscum*: The Background of a Liturgical Formula," in *New Testament Essays*, ed. A. J. B. Higgins (Manchester, England: Manchester University Press, 1959), 287, 293; and Gunther Bornkamm, "The Risen Lord and the Earthly Jesus: Matthew 28:16–20," in *The Future of Our Religious Past*, ed. J. Robinson (New York: Harper & Row, 1971), 203–229.

[46] See, for example, Ulrich Luz, *Matthew 1–7: A Continental Commentary*, trans. Wilhelm C. Linss (Minneapolis: Augsburg Fortress, 1989), 121–22 (who also states that "Allusions to God's being-with-us permeate the whole Gospel [17:17; 18:20; 26:29]"); and J. D. Kingsbury, *Matthew as Story*, 2nd ed. (Philadelphia: Fortress, 1986), 41–42, who writes,

> The key passages 1:23 and 28:20, which stand in a reciprocal relationship to each other, highlight this message. At 1:23, Matthew quotes Isaiah in saying of Jesus: in "Emmanuel . . . God [is] with us." And at 28:20 the risen Jesus himself declares to the disciples: "I am with you always, to the close of the age." Strategically located at the beginning and the end of Matthew's story, these two passages "enclose" it. In combination, they reveal the message of Matthew's story: *In the person of Jesus Messiah, his Son, God has drawn near to abide to the end of time with his people, the church, thus inaugurating the eschatological age of salvation.* (italics his)

[47] Especially as represented by the Septuagint. See I. Howard Marshall, *The Gospel of Luke: A Commentary on the Greek Text*, New International Greek Testament Commentary (Grand Rapids, MI: Eerdmans, 1984), 66.

ultimate fulfillment of Isaiah 7:14, the allusion was particularly powerful. It should be observed that this oracle about Jesus being a sign that would be opposed and would cause many to fall is directed by Simeon exclusively "to Mary his mother" (Luke 2:34a). Luke is probably implying that Simeon perceived that Mary was the virgin about whom Isaiah had been prophesying and whose part Isaiah's wife had played back in 733 BC. In other words, it seems likely that Simeon is being portrayed as conscious that both parties ultimately in view in Isaiah 7:14—the Messiah and his virgin mother—were standing before him at that moment.

Furthermore, according to Luke 2, the angel who announced the birth of Jesus during the night to the shepherds gave them "the sign" [*to sēmeion*]: "You will find a baby wrapped in swaddling cloths and lying in a manger" (v. 12). The reference to "the sign" arguably recalls Isaiah 7:14's "sign." The fact that the angel was proclaiming the birth of "a baby," indeed of the Messiah, would seem to confirm this. Moreover, the angel's phrase "unto you is born this day . . . a Savior, who is Christ the Lord" in Luke 2:11 alludes to Isaiah 9:6–7's great prophecy anticipating the Messiah's birth ("to us a child is born, to us a son is given"), which was inextricably linked to the "great light" that Isaiah proclaimed would shine in the darkness (Isa. 9:2).[48] What the angel said was therefore strongly influenced by Isaiah's prophecies, particularly Isaiah 7:1–9:7.

Like Matthew, therefore, Luke maintains that Isaiah 7:14 was fulfilled in connection with the birth of Jesus to Mary.

This Christian approach to Isaiah 7:14 continued into the post–New Testament period. For example, Justin Martyr famously disputed with Trypho the Jew concerning the meaning of this oracle and in particular the identity of Immanuel:

> And hear again how Isaiah in express words foretold that He should be born of a virgin; for he spoke thus: "Behold a virgin shall conceive, and bring forth a Son, and they shall say for his name, "God with us." . . . This, then, "Behold a virgin shall conceive," signifies that a virgin should conceive without intercourse. For if she had had intercourse with any one whatever, she was no longer a virgin; but the power of God having come upon the virgin, overshadowed her, and caused her while yet a virgin to conceive.[49]

In his *Dialogue with Trypho*, Justin strongly affirmed that Jesus "was born of a virgin, and that His birth of a virgin had been predicted by Isaiah."[50]

Shortly thereafter, he wrote to Trypho, his Jewish interlocutor:

> But since you and your teachers venture to affirm that in the prophecy of Isaiah it is not said, "Behold, the virgin shall conceive," but, "Behold the young woman shall conceive, and bear a son," and [since] you explain the prophecy as if [it referred] to Hezekiah, who was your king, I shall endeavor to discuss shortly this point in opposition to you, and to show that reference is made to Him who is acknowledged by us as Christ.[51]

[48] The relevance of Isaiah 1–12 for Luke 2:12 may also be demonstrated by the fact that the sign consists of a baby in a manger. This appears to be based on Isa. 1:3: "The ox knows its owner, and the donkey its master's crib, but Israel does not know, my people do not understand." The angel seems to be implying that the messianic identity of the newborn baby would be demonstrated to the shepherds by the facts that the bed of this newborn infant still wrapped in swaddling clothes was literally a feeding trough and that he shared a room with animals. Is the angel claiming that Isaiah had been prophesying about the circumstances of the newborn Messiah in Isa. 1:3?

[49] *Dialogue with Trypho* 33. Translation from *Ante-Nicene Christian Library*, 35–36.

[50] *Dialogue with Trypho* 66. Translation from *Ante-Nicene Christian Library*, 178–179.

[51] *Dialogue with Trypho* 43. Translation from *Ante-Nicene Christian Library*, 142.

So it is clear that the early Christians strongly believed that Isaiah 7:14 was an oracle concerning the birth of the Messiah and indeed that it had predicted a virginal conception.

Isaiah 7:14 as the Key to Understanding the Celestial Drama in Virgo

When the Magi watched the celestial Virgin, Virgo, emerge in the eastern sky pregnant and then saw her give birth to her cometary baby, no Biblical text would have seemed more obviously pertinent than Isaiah 7:14: "Behold, the virgin shall be with child and bear a son. . . ." Before their eyes, Virgo seemed to be acting out in the celestial realm the very drama envisioned by Isaiah. The fact that Isaiah declared that the baby would be divine in nature ("Immanuel" and "Mighty God") would have seemed very compatible with the celestial apparition. This Isaiah connection is strengthened by the fact that the Magi were inspired to bring their gifts of gold and frankincense by Isaiah 60:4–6, a passage with close links to Isaiah 7–9.

Revelation 12:1–5, which discloses the celestial wonder that was observed by the Magi unfolding in the eastern sky, strongly hints that Isaiah 7:14 was being fulfilled in heaven. It is important to take due note of the indications of Revelation 12:1–2's dependence on Isaiah's oracle.[52] The word *sēmeion* ("sign") in Revelation 12:1 is used in LXX Isaiah 7:14 of the sign consisting of the virgin being with child and giving birth. Furthermore, in both texts she is "with child" (*harah* in the Hebrew text; *en gastri hexei* in the Septuagint of Isa. 7:14; *en gastri echousa* in Rev. 12:2) and goes on to "give birth to a son" (*weyoledet ben* in the Hebrew; *texetai* [from the verb *tiktō*] *huion* in the Septuagint of Isaiah 7:14; *eteken* [from the verb *tiktō*] *huion* in Rev. 12:5). In portraying the heavenly scene in terms of Isaiah 7:14, John is, I would suggest, reflecting how the early Christians, and indeed the Magi, interpreted the cometary nativity drama that coincided with Jesus's birth.

This astronomical approach to the fulfillment of Isaiah 7:14 raises an interesting and important question: Was Isaiah's Immanuel oracle being interpreted in strictly terrestrial terms but applied to the heavenly phenomenon that attended Jesus's birth, or was the Immanuel oracle believed to have prophesied the celestial wonder that occurred in Virgo, low on the eastern horizon, in September and October, 6 BC?[53] In other words, did those who interpreted what the comet did in Virgo in terms of Isaiah 7:14 believe that "the virgin" who conceived and gave birth to the child was exclusively terrestrial, or did they believe that she was simultaneously celestial *and* terrestrial?

It must be conceded that there are a number of elements within Isaiah 7–12 that might have been regarded as favoring a celestial aspect to 7:14: the reference to a "sign . . . in the height above"[54] (v. 11); the prophecy of a "great light" in 9:2; the conceivable allusion to Virgo's "branch" (11:1); the fact that Virgo may have been present at the meeting between Isaiah and Ahaz (in December/January, 734/733 BC) (7:10–25); and that the

[52] Those who acknowledge Revelation's echo of Isa. 7:14 here include J. Massyngberde Ford, *Revelation: Introduction, Translation, and Commentary*, Anchor Bible (Garden City, NY: Doubleday, 1975), 195; Brian K. Blount, *Revelation: A Commentary*, New Testament Library (Louisville, KY: Westminster John Knox, 2009), 227; and G. K. Beale, *The Book of Revelation: A Commentary on the Greek Text*, New International Greek Testament Commentary (Grand Rapids, MI: Eerdmans, 1999), 631.

[53] Franz Boll, *Aus der Offenbarung Johannis: hellenistische Studien zum Weltbild der Apokalypse* (Leipzig and Berlin: Teubner, 1914), 121–123, stressed the importance of Isa. 7:14 (LXX) for early Christians' Christology and for the interpretation of Revelation 12. However, he drove a wedge between Matthew and Revelation, claiming that the first Gospel reflects a strictly terrestrial interpretation of Isa. 7:14, taking it to refer to the Virgin Mary and Jesus, in contrast to Revelation 12, which, he maintained, reflects a celestial interpretation, taking it (creatively) to refer to the celestial Virgin and her divine Child. Boll did not appreciate that Rev. 12:1–5 is describing an astronomical sign that attended the Messiah's birth and was perceived to disclose the nature and significance of the terrestrial moment and in particular to intimate that what was transpiring on the earth was bringing to fulfillment Isaiah's prophecy concerning the Messiah's virginal conception.

[54] Note how the Targum of Isaiah renders v. 11: "Request a sign from Yahweh your God, that a great wonder may be done for you on earth or that a sign may be shown to you in the heavens" (cf. Bruce Chilton, *The Isaiah Targum* [Wilmington, DE: Michael Glazier, 1987], 16; C. W. H. Pauli, trans., *The Chaldee Paraphrase on the Prophet Isaiah* [London: London Society's House, 1871], 23).

sign was a punishment for Ahaz, who had strong pagan tendencies (2 Chron. 28:1–4, 22–25) and probably worshiped astral deities (2 Kings 20:8–11 [the sundial of Ahaz]; 23:5 [priests of astral cults], 11–12 [horses and chariots dedicated to the Sun, and rooftop altars dedicated to the worship of astral deities]). These considerations might conceivably have prompted some Jews in Babylon to interpret the celestial wonder unfolding in the eastern sky in 6 BC as the fulfillment of Isaiah 7:14. The Babylonian Magi would have known that the zodiacal constellations, which originated in Babylon, were being studied in the eighth century BC, when Isaiah issued his oracle.[55] Moreover, they would have been very well aware that the conceptualization of the sixth zodiacal constellation (or at least part of it) as a young virgin was very ancient.[56]

However, it is much more natural to interpret Isaiah 7:14 in strictly terrestrial terms, with the "virgin" being Isaiah's wife and the Messiah's mother, and with the child being Maher-shalal-hash-baz and the Messiah. Indeed a celestial interpretation of 7:14 would have been difficult to sustain.

First and foremost, a constellation figure could not, of course, have named the terrestrial baby "Immanuel." Virgo could only do that through her terrestrial counterpart, the earthly virgin mother. However, introducing such a shift of referent would rupture the logical flow of 7:14. The text is more naturally interpreted as speaking exclusively of a terrestrial "virgin." Moreover, the name Immanuel ("God with us") reflects a particularly human and Judahite perspective rather than a celestial one.

Second, in light of the fact that Ahaz rejected a celestial sign in 7:11, it is unlikely that Yahweh granted him one. The logic of verses 11–14, particularly the "therefore" in verse 14, which refers back to verse 13, indicates that Yahweh withdrew his offer and issued the House of David a "sign" quite unlike the options he had just offered. In verse 11 God had offered to do a "sign" that would satisfy Ahaz's need for reassurance concerning Isaiah's oracle, calling on him to trust and not capitulate to his fear. However, in verse 14 the "sign" is not tailored to encourage Ahaz in faith but rather to confirm Isaiah's word in the face of Ahaz's unbelief and rebellion and also to express Yahweh's disfavor concerning the Davidic dynasty. Therefore it would actually be surprising if the "sign" of verse 14 was either "deep as Sheol" or "in the height above."

We suggest therefore that Revelation 12:1–5 does not reflect a celestial interpretation of Isaiah 7:14 but rather is simply claiming that the fulfillment of the Immanuel oracle on the earth was attended by a dramatization of it in the heavens. In the celestial drama, Virgo played the part of the virgin and the cometary coma played the part of Immanuel.

However, this does not mean that Isaiah did not anticipate the occurrence of a great comet to mark the fulfillment of the Immanuel oracle. He did anticipate this, and this can be seen in Isaiah 9:2. To demonstrate this, we must turn our focus to the latter half of Isaiah 7:1–9:7, namely 8:11–9:7.

ISAIAH 8:11–9:7: THE GREAT LIGHT SHINING IN THE DARKNESS

Any remaining thought that Isaiah was predicting that a celestial sign would occur in Ahaz's day is ruled out by 8:11–22.[57] There

[55] James Evans, *The History and Practice of Ancient Astronomy* (Oxford: Oxford University Press, 1998), 39.

[56] MUL.APIN, a Babylonian astronomical text from around 1000 BC, identifies the constellation Furrow (which, together with the Frond, occupied the part of the sky that became known as Virgo) with the virgin goddess Shala (meaning "maiden"). A Babylonian line drawing from the Seleucid-era Uruk portrays the Furrow as a virgin holding an ear of grain (Louvre Museum, item AO 6448; see fig. 7.2). See H. van der Waerden, *Science Awakening II* (Leyden, Netherlands: Noordhoff, 1974), 81 plate 11c, 125, 288; F. Thureau-Dangin, *Tablettes d'Uruk à l'usage des prêtres du Temple d'Anu au temps des Séleucides* (Paris: Paul Geuthner, 1922), no. 14.

[57] It is widely agreed that these verses belong to the period of the Syro-Ephraimite Crisis. The fact that Isaiah is here issuing advance warning that there would be no fresh revelation or signs during the period of Assyria's conquest of Syria and Israel and the regional superpower's subsequent oppression of Judah mandates that the section be dated to 733 BC.

the prophet declared that the people of Judah would long for some divine revelation during the Assyrian oppression that was about to descend on the kingdom. But they would be granted only the revelations given in oral and written form through Isaiah and the "signs and portents" consisting of the prophet himself and his sons (vv. 16–20). At that time the people of Judah would "turn their faces upward" and "look to the earth" (vv. 21–22). As Wildberger highlights in his comments on 8:21–22, this language recalls Yahweh's offer in 7:11 to do a sign "deep as Sheol or in the height above":

> One does not only utter curses, but also looks around for help: "He turns himself toward what is above and looks to the earth." In 7:11, Isaiah offered the king a sign "deep within the underworld" or "high above in the heights." Instead of Sheol, here he mentions the earth (but we must also remember that *'eretz* can also designate the underworld), whereas in both passages *lemā'lāh* ("toward what is above") is used when speaking of the realm above. In this particular passage, one probably ought to take it to mean something like the following: One seeks in the heavens above and the earth below for some sign of a coming change in fortune. It is possible that this means careful observations which seek omens (flight of birds, movements of or constellations of stars, or something similar . . .). However, no trustworthy sign is to be discovered, for there is nothing but oppression and darkness.[58]

Isaiah's point is that there would be no celestial wonder in 733–727 BC. Indeed the people of Judah would be so frustrated that they had no revelation other than the prophetic word (8:16–20) and the "signs and portents" that were Isaiah and his sons (v. 18) that they would desperately look for some astronomical or (sub-)terrestrial sign of the kind Ahaz had rejected. There may well be an implication that they would do so with bitterness in their hearts against their king for spurning God's offer, and against God himself for refusing to give that kind of "sign." God would not give them the celestial sign, the light, for which they longed. No, he would only give them "distress and darkness, the gloom of anguish" and cast them into "thick darkness" (v. 22).[59] This provides the context in which 9:1–2 must be understood.

Isaiah 9:1–7 is a birth announcement, like 7:14–16 (Immanuel) and 8:1–4 (Maher-shalal-hash-baz). In 9:1–7 the one whose birth is being announced and celebrated is obviously the Messiah.[60] What Isaiah is declaring here is the ultimate fulfillment of the Immanuel oracle of 7:14.[61] This is clear in 9:6–7:

> For to us a child is born,
> to us a son is given;
> and the government shall be upon his shoulder,
> and his name shall be called
> Wonderful Counselor, Mighty God
> Everlasting Father, Prince of Peace.

[58] Hans Wildberger, *Isaiah 1–12*, trans. Thomas H. Trapp, Continental Commentary (Minneapolis: Augsburg Fortress, 1991), 380–381.

[59] Young, *Book of Isaiah*, 1:322, rightly comments with respect to this verse, "Only light can dispel the gloom of despair and desperation, but that light is not to be seen."

[60] That the Messiah's birth is in view is patently obvious. Commenting on this section, Childs (*Isaiah*, 81) wrote the following: "The description of his reign makes it absolutely clear that his role is messianic. There is no end to his rule upon the throne of David, and he will reign with justice and righteousness forever. Moreover, it is the ardour of the Lord of Hosts who will bring this eschatological purpose to fulfillment. The language is not just that of a wishful thinking for a better time, but the confession of Israel's belief in a divine ruler who will replace once and for all the unfaithful reign of kings like Ahaz."

[61] Since this subsection is very closely related to the preceding verses and is focused on the fate of Galilee in 733/732 BC, there can be no serious doubt that 9:1–7 stems from the prophet Isaiah during the Syro-Ephraimite Crisis (so, for example, Wildberger, *Isaiah 1–12*, 393; cf. M. E. W. Thompson, "Israel's Ideal King," *Journal for the Study of the Old Testament* 7 (1982): 79–88, although, while dating 9:1 to 732 BC, he assigns 9:2–7 to 731–723 BC).

Of the increase of his government and
of peace
there will be no end,
on the throne of David and over his
kingdom,
to establish it and to uphold it
with justice and with righteousness
from this time forth and
forevermore.

The birth announcement in 8:1–4 was evidently intended by Isaiah to refer to the coming of the first "Immanuel," Isaiah's son, while the birth announcement in 9:1–7 was manifestly intended to refer to the coming of the second, greater "Immanuel," the Messiah. Whereas the birth of Immanuel ("God with us") is simply anticipated in 7:14, in 9:1–7 it is portrayed as having already occurred:[62] Isaiah, speaking from the perspective of one present at the time of the Messiah's birth, declares that the prophesied Messiah, "Mighty God,"[63] has now been born (9:6–7).

This ultimate fulfillment of Isaiah 7:14 is associated in 9:1–2 with the coming of a great light:

> But there will be no gloom for her who was in anguish. In the former time he brought into contempt the land of Zebulun and the land of Naphtali, but in the latter time he has made glorious the way of the sea, the land beyond the Jordan, Galilee of the nations.
>
> The people who walked in darkness
> have seen a great light;
> those who dwelt in a land of deep
> darkness,
> on them has light shone.

These verses can be understood only in the context of chapters 7–8, particularly 7:14 and 8:11–22. The opening sentence ("But there will be no gloom for her who was in anguish"; 9:1a) simultaneously recalls the virgin who gave birth to Immanuel in 7:14 and the people who looked for a celestial sign but knew only "distress and darkness, the gloom of anguish" (8:21–22). That the virgin mother of Immanuel was in anguish was, of course, due to the pains of childbirth—such a reference is especially fitting in a birth announcement. That the people of Israel were in anguish was due to the way that Yahweh had punished them in the eighth century BC at the time of Maher-shalal-hash-baz's birth. It seems that the virgin mother has now come to represent Israel as a whole (cf. 26:17–27:1).

The focus in 9:1 is specifically on Galilee. Galilee is spoken of in terms of its divisions before Tiglath-pileser III's invasion of 733/732 BC—the tribal territories of Zebulun and Naphtali—and after it the Hebrew equivalents of the three Assyrian provinces of Way of the Sea, Transjordan, and Galilee of the Nations. The deep darkness that encompassed Judah according to 8:20–22 was also enshrouding Galilee.

Now, however, there is a major contrast between "the former time" and "the latter time." The fact that the first of the two moments is associated with the Israelite tribal names reveals that "the former time," when the land came into contempt, would begin with the invasion of Tiglath-pileser late in 733 BC. We know from 7:14–8:8 that this Assyrian conquest of Galilee started on the heels of the birth of Maher-shalal-hash-baz/Immanuel in September/October of 733 BC. Indeed it is almost certainly the Immanuel oracle of 7:14 that establishes the chronological schema of "the former time" and "the latter time." The "former" moment relates to the birth of the first fulfillment of that oracle in 733 BC. "The latter time" is the time of the Messiah's birth, the ultimate fulfillment of

[62] Wildberger, *Isaiah 1–12*, 392.
[63] That "Mighty God" signifies divinity is clear from 10:21, where it is used of Yahweh.

FIG. 8.1 Tiglath-pileser III, king of Assyria. A stone panel from the Central Palace in Nimrud now located at the British Museum (WA 118900). Image credit and copyright: Marie-Lan Nguyen/Wikimedia Commons.

Isaiah 7:14.[64] This approach makes excellent sense—9:1–7 is, after all, the Messiah's birth announcement.[65]

According to Isaiah, "the latter time," associated with the Messiah's birth, brings an end to the gloom and the anguish. It augurs glorious days for Galilee. More specifically, it brings the shining of celestial light in the midst of the deep darkness (9:2). Whereas in 8:21–22 we read of those who longed for a celestial sign but were merely thrust into deep darkness, now we learn that "The people who walked in darkness have seen a great light; those who dwelt in a land of deep darkness, on them has light shone." The people of Israel are no longer abandoned to the gloom of deep darkness at the point of the second fulfillment of Isaiah 7:14, in connection with the Messiah's birth. They are finally permitted to see a magnificent celestial light.

Those walking and dwelling in a land of deep darkness, as Wildberger has pointed out, are those living in Sheol—"Whoever has to wander in 'darkness' is, for all intents and purposes, already in the realm of the underworld. . . . The OT does not only speak of human beings who are close to death finding themselves already in Sheol, but can say the same thing about those who are harshly oppressed by their enemies."[66]

However, now, finally, at the time of the Messiah's birth, those living in Sheol get a heavenly sign in addition to the prophesied earthly one. As we have already highlighted, there can be little doubt that Isaiah's "great light" was an extraordinarily bright comet. The description of the light recalls Genesis's description of the Sun and Moon as the two "great lights" (Gen. 1:16). They were most likely called this because of their large size and brightness.[67] So brilliant would this comet be that it would dispel "gloom" and "darkness." It would be a beautiful picture of the Messiah's presence in Galilee.

The momentous celestial phenomenon which Ahaz had spurned and for which the people of Judah and Israel would long without satisfaction during "the former time," namely the Syro-Ephraimite crisis and the Assyrian reign of terror in Ahaz's latter years, would be granted only in "the latter time." The "latter" time is the second and ultimate fulfillment of Isaiah 7:14's oracle about the virgin conceiving, giving birth to a child, and naming him Immanuel—that is, the time of the Messiah's nativity.

The celestial sign that would mark the ul-

[64] The "latter time" of the northern tribes' glorification was certainly not the period after the Assyrian termination of the northern kingdom in 722–721 BC, when the land was desolate and in ruins.

[65] Interestingly, M. E. W. Thompson, *Situation and Theology: Old Testament Interpretations of the Syro-Ephraimite War* (Sheffield: Almond Press, 1982), 14, came to a similar conclusion: "The 'latter time' spoken of in 8.23 [9:1 in English translations] . . . is a reference to the new situation to be brought about through the advent of the ruler spoken of in 9.1–6 [9:2–7]. . . . 8.23 [9:1] [is] the connecting verse used to apply the Isaianic oracle of 9.1–6 [i.e., 9:2–7], concerning the ideal Davidic king, to the situation in the lands of the northern kingdom after they had fallen to Tiglath-pileser III."

[66] Wildberger, *Isaiah 1–12*, 395.

[67] Umberto Cassuto, *A Commentary on the Book of Genesis (Part 1)* (Jerusalem: Magnes, 1989), 45 ("those that seem the biggest to us").

timate fulfillment of Isaiah's Immanuel oracle would be a magnificent cometary apparition. The comet would be uniquely large and bright. Its light would penetrate into the deep darkness.[68]

Of course, the great light shining in the darkness is not merely literal; it is also metaphorical, encapsulating the nature and effect of the Messiah's presence and ministry in Galilee.[69]

Isaiah 9:2 is, then, reminiscent of Numbers 24:17. It predicts that a literal cometary phenomenon would mark the birth of the Messiah, the metaphorical great light.

Notably, Isaiah is not explicit concerning what the comet would do within the celestial dome to mark the fulfillment of Isaiah 7:14 in the terrestrial sphere. However, the Magi and some Jewish observers in Babylon came to the conclusion in 6 BC that the prophesied comet had been destined to play the role of the messianic baby in a heavenly drama, broadcasting in the heavens what was happening somewhere on the earth.[70]

THE FULFILLMENT OF ISAIAH 9:2

With respect to how Second Temple Jews interpreted Isaiah 9:1–2, Richard Beaton summarizes that it was "a well-known passage in Judaism that was thought to foretell the coming Messiah."[71]

The New Testament cites and alludes to Isaiah 9:1–2, regarding it as having come to fulfillment in connection with the coming of Jesus.

Most important, Matthew 4:13–16 explicitly quotes the verses and claims that they were fulfilled when Jesus began doing ministry from his base in Capernaum in Galilee:

> And leaving Nazareth [Jesus] went and lived in Capernaum by the sea, in the territory of Zebulun and Naphtali, so that what was spoken by the prophet Isaiah might be fulfilled:
>
> "The land of Zebulun and the land
> of Naphtali,
> the way of the sea, beyond
> the Jordan, Galilee of the
> Gentiles—
> the people dwelling in darkness
> have seen a great light,
> and for those dwelling in the region
> and shadow of death,
> on them a light has risen."[72]

Matthew's extension of his quotation to Isaiah 9:2 suggests he perceived that the verse had prophesied that the Messiah himself would be "a great light" in Galilee as he did his ministry there. This interpretation of Isaiah 9:2 is in accord with the originally intended meaning of Isaiah—not only would a literal great light

[68] Note that the great light associated with the Messiah's coming shines during the night, when deep darkness holds sway. It cannot therefore be the Sun.

[69] Marvin A. Sweeney, *Isaiah 1–39, with an Introduction to Prophetic Literature*, The Forms of the Old Testament Literature (Grand Rapids, MI: Eerdmans, 1996), 183, states concerning 8:19–9:1 that "the contrasting images of light and darkness in this passage have defied adequate explanation." When, however, the imagery is interpreted in light of Yahweh's offer to do a celestial sign in 7:10–14, it can be readily explained.

[70] A probably second- or third-century BC passage in the *Sibylline Oracles* refers to the Star in connection with the fulfillment of Isa. 7:14 by the virgin birth. According to *Sib. Or.* 1:323–324 (J. L. Lightfoot, *The Sibylline Oracles* [Oxford: Oxford University Press, 2007], 311),

> When the heifer God the Highest's word shall bear, (323a)
> The manless maid the Logos give a name. (323b)
> Then from the east a star in fullest day (323c)
> That brightly shines shall from the heavens beam (323d)
> Announcing a great sign for mortal men. (323e)
> Then God's great son will come to humankind (324)

Note the allusions to Isa. 7:14 ("shall bear," "give a name," and "sign"). In fact, the reference to the Star (lines c and d) is sandwiched between the strong allusions to Isa. 7:14 (lines a and b and line e).

[71] Richard Beaton, "Isaiah in Matthew's Gospel," in *Isaiah in the New Testament*, ed. S. Moyise and M. J. J. Menken (London: Continuum, 2005), 67.

[72] My translation of the Greek verb *anatellō*, which was used in 2:2 and 9 of the Star's "rising."

shine in the darkness to signal the coming of the Messiah, but, metaphorically, the Messiah himself would shine his light in Galilee. Matthew has, in his second chapter, already spoken of the literal light; in Matthew 4:13–16 only the metaphorical meaning is relevant. At the same time, Matthew's subtle change from the Septuagint's "has shone" to "has risen" [*aneteilen*] seems to recall the literal fulfillment by the Star at the time of Jesus's birth, implying, as Gundry puts it, that "the dawn of Jesus' ministry fulfill[s] the promise contained in the rising of the messianic star (2:2, 9)."[73] In this way Matthew quietly links the literal and metaphorical fulfillments of Isaiah 9:2.

Similarly, the words spoken by the priest Zechariah, John the Baptist's father, at the prophet's birth, according to Luke 1:78b–79, reflect the expectation that Isaiah's prophecy concerning the strange new celestial light was about to be fulfilled: "the rising [star] [*anatolē*] shall visit us from on high to give light to those who sit in darkness and in the shadow of death, and to guide our feet into the way of peace." Although *anatolē* is sometimes used of "the Branch" (see Jer. 23:5; Zech. 3:8; 6:12; Isa. 11:1) and this meaning is probably secondarily present here, the context favors the conclusion that "the rising [star]" first and foremost is in view here, recalling Numbers 24:17. The priest, drawing especially on Isaiah 9:2, prophesies vividly concerning the Messiah in terms of an extraordinarily bright comet that rises and descends and dispels the darkness and illuminates the earth's residences and roads. The focus is on the metaphorical fulfillment of Isaiah's prophetic word, but it is clear that Zechariah is strongly alluding to a literal fulfillment of the prophecy.

The Gospel of John frequently speaks of Jesus as the "light" prophesied by Isaiah (John 1:4–5, 7–9, 14; 3:19–21; 8:12; 9:5; 11:9–10; 12:35–36, 46). This is especially clear in John 8:12, which is set in the Court of the Women in the Jerusalem temple (8:20) during the Feast of Tabernacles. Each year during the Feast the four golden candelabras in the Court of the Women were ritually lit, beaming bright light to the entire city, a practice probably inspired by Isaiah 9:2. In saying, "I am the light of the world," Jesus "is literally claiming to be the fulfillment of an Isaian text that explicitly promises light from and on Galilee, Isaiah 9:1–2."[74] Essentially, as David Ball puts it, "Jesus takes the idea of light from Isaiah 9 and applies it to himself. Thus he claims to be the light that was to arise in Galilee of the Gentiles."[75] Coming on the heels of the chief priests and Pharisees' statement, "Search and see that no prophet arises from Galilee" (John 7:52), Jesus's allusion to the fulfillment of Isaiah 9:2 in and through his person and ministry is powerfully ironic.[76]

In John 12:35 Jesus seems to recall Isaiah 9:2 once again: "The light is among you for a little while longer. Walk while you have the light, lest darkness overtake you."

In the prologue of John's Gospel Jesus is described as "the light of men" (1:4). John then states that "The light shines in the darkness, and the darkness has not overcome it" (v. 5). John the Baptist, we read, "came to bear witness about the light" (v. 8). "The true light, which gives light to everyone, was coming into the world" (v. 9).[77] The language here strongly

[73] Gundry, *Matthew*, 60.

[74] Frederick Dale Bruner, *The Gospel of John: A Commentary* (Grand Rapids: Eerdmans, 2012), 516–517.

[75] David Mark Ball, *"I Am" in John's Gospel: Literary Function, Background, and Theological Implications* (Sheffield: Sheffield Academic Press, 1996), 221.

[76] Bruner, *Gospel of John*, 516–517.

[77] Marcus J. Borg and John Dominic Crossan, *The First Christmas: What the Gospels Really Teach about Jesus's Birth* (New York: HarperCollins, 2007), draw out how Matthew's Nativity account is in some respects similar to John's prologue: in Matthew's account, "the uses of light and darkness . . . are thus many and rich. Jesus's birth is the coming of light into the darkness. But the darkness seeks to extinguish the light (Herod's plot to kill Jesus). Drawn to the light, wise men from the nations pay homage to Jesus. Jesus is the light of the nations. Thus Matthew's story makes the point made in only slightly different language in John: 'Jesus is the light of the world'" (184).

recalls Genesis 1:3 and Isaiah 9:2, making the point that Jesus is the light anticipated by the Hebrew Scriptures. The emphasis on Jesus as the prophesied light in John's prologue is all the more striking when we remember that, as Borg and Crossan have helpfully observed, the prologue takes the place of a nativity story in John's Gospel: "Recall that John does not have a birth story. But this passage virtually functions as its equivalent: the coming of Jesus, the incarnation, is the coming of 'the *true light*, which *enlightens* everyone.'"[78] To the extent that this is true, John is essentially replacing a focus on the fulfillment of the literal dimension of Isaiah 9:2 with a focus on the fulfillment of the metaphorical dimension of this oracle, that is, on Jesus himself as the prophesied light.

It would seem, therefore, that the Gospel of John regards Isaiah 9:2 as having been fulfilled in and through the ministry of Jesus. Isaiah 9:2, like Numbers 24:17, had in mind both a literal and a metaphorical light in connection with the coming of Jesus. John does not refer to a literal light shining in the darkness, but only to a metaphorical light. He emphasizes that the coming of Jesus to the world, and more specifically his people, is the coming of the prophesied light.[79]

Likewise Paul, in 2 Corinthians 4:4, speaks of "the god of this world" preventing unbelievers from "seeing the light of the gospel of the glory of the Christ." Then, in verse 6, he declares that "God, who said, 'Let light shine out of darkness,' has shone in our hearts to give the light of the knowledge of the glory of God in the face of Jesus Christ." Here the apostle is drawing not only on Genesis 1:3–4 ("And God said, 'Let there be light'"), but also on Isaiah's prophecy concerning the great light that would shine in the darkness to mark the inauguration of God's plan to bring salvation with the birth of the Messiah (LXX Isa. 9:1: "O people walking in darkness, behold a great light! You who live in the country and in the shadow of death, a light shall shine upon you").[80] Paul assumes that Isa. 9:2 has come to fulfillment metaphorically in the person of Jesus.

The New Testament authors, therefore, clearly regard Isaiah 9:2 as having been fulfilled in connection with the coming of Jesus. They are particularly concerned to highlight that the oracle of the great light was fulfilled metaphorically through Jesus's ministry. But Matthew and Luke also hint at the oracle's literal, astronomical fulfillment at the time of Jesus's birth.

We have already noted that the precise way in which the great light would confirm the terrestrial events associated with the ultimate fulfillment of Isaiah 7:14 was not made explicit in Isaiah. When, however, a great comet appeared that brought the heavens to life, creating a celestial nativity scene in which Virgo played the part of the virgin and the comet played the part of Immanuel, those with eyes to see (see Isa. 6:8–10) would have quickly realized that this had to be the celestial sign predicted in 9:2. The only logical deduction one could reach was that the Messiah's birth was occurring at that time.

CONCLUSION

All in all, it would seem safe to say that the Magi probably came to the conclusion that

[78] Ibid., 180 (italics theirs). Richard Bauckham, "The Qumran Community and the Gospel of John," in *The Dead Sea Scrolls Fifty Years after Their Discovery: Proceedings of the Jerusalem Congress, July 20–25, 1997*, ed. L. H. Schiffman, E. Tov, and J. C. Vanderkam (Jerusalem: Israel Exploration Society, 2000), 113, writes that "John's image of Christ as the light of the world is also, more directly, a form of messianic exegesis of prophecies in Isaiah. It reflects Isa. 9:1[2] (. . . cf. John 1:5; 8:12; 12:35)," in addition to Isa. 42:6–7 and 49:6 (cf. John 9); and Isa. 60:1–3. According to Bauckham, together "These passages . . . readily supply the central Johannine image of the great light shining in the darkness of the world to give light to people, as well as the christological-soteriological significance which this image bears in the Fourth Gospel" (113).

[79] It is interesting that John speaks of the light as "coming into the world" (1:9; compare what Jesus claims in 3:19), referring to Jesus's incarnation and birth.

[80] So Murray J. Harris, *The Second Epistle to the Corinthians*, New International Commentary on the New Testament (Grand Rapids, MI: Eerdmans, 2005), 334; Craig S. Keener, *1–2 Corinthians*, New Cambridge Bible Commentary (Cambridge: Cambridge University Press, 2005), 174. In favor of Isa. 9:1–2 being in mind is Paul's use of the same Greek words as the Septuagint—"light," "darkness," and "shine"—and, in fact, the same Greek phrase (*phōs lampsei*).

the great leader whose birth was being so dramatically announced in the heavens was the Messiah based on a number of key prophecies in the Hebrew Bible—particularly, Numbers 24:17; Isaiah 7:14; and 9:2. Together these texts disclosed the identity, nature, destiny, and general location of the newborn.

What we find, then, is that Matthew 1:18–25 documents the fulfillment of Isaiah 7:14, while Matthew 2:1–12 details the fulfillment of Isaiah 9:2 and Numbers 24:17. The virgin had given birth to the Messiah on the earth and, simultaneously, the prophesied great light had announced the momentous event in a celestial IMAX drama. If we consider the two Gospel nativity accounts together, in Luke 2:6–20 the shepherds testified to the birth of the Messiah as prophesied in Isaiah 7:14 (and Mic. 5:2), while in Matthew 2:1–12 the Magi confirmed the fulfillment of Micah's prophecy that the Messiah would be born in Bethlehem (Mic. 5:2) and the appearance of the extraordinarily bright comet to mark the occasion in fulfillment of Isaiah 9:2 and Numbers 24:17.

When Revelation 12:1–2 described the Star that signaled the Messiah's birth in terms that drew on the oracle of Isaiah 7:14, it did not do so because it interpreted the prophecy celestially. Rather, it did so because the celestial sign marking the ultimate fulfillment of the Immanuel oracle in accordance with Isaiah 9:2 mirrored the terrestrial "sign." That is, the celestial sign of the great light in the deep darkness partly consisted of a heavenly drama in which the virgin's pregnancy and delivery of the messianic child were acted out, with Virgo playing the part of the virgin and the comet playing the part of the messianic child.[81] From the moment that the cometary coma rose in Virgo's belly, Isaiah 7:14 and 9:2 offered the Magi a compelling paradigm to understand the meaning and significance of what they were observing. Viewing this magnificent celestial play through the grid of these ancient Hebrew prophecies, the Magi naturally concluded that the cometary "birth" narrative was disclosing the time when the Immanuel oracle was being fulfilled on the earth. At the same time, the comet as a whole formed a great scepter which demanded to be interpreted in light of Balaam's oracle concerning the scepter-star in Numbers 24:17.

It was only natural for the Magi to assume that the birth of the Messiah would take place in Judea. Indeed Isaiah 7:14 seemed to presuppose this, and Numbers 24:17 made it clear. As the capital of Judea, Jerusalem would have been the obvious first port of call for the Magi as they sought to track down the Messiah. They would probably have presumed that he would be in Jerusalem but reckoned that, even if he was not, someone there would surely know where he was. Moreover, since they had no doubt that the God of the Jews was launching them on their quest to find and worship the king of the Jews, they would have felt confident that God would ensure that they successfully located the baby Messiah within Judea.

SUMMARY

In chapter 7 we argued that the heavenly sign observed by the Magi in the eastern sky was caused by a great comet emerging from a close perihelion and heading toward Earth.

For one thing, the comet's coma appeared, looking like a baby, in Virgo's womb at her heliacal rising (that is, as the constellation was emerging from a conjunction with the Sun). Just a few weeks before, on September 15, 6 BC, the Sun had been over her womb when the Moon was under her feet. Now, in the very place where the Sun had been, there emerged a cometary coma. The impression that the cometary baby was conceived as the result of the Sun's life-giving presence would have been quite compelling. Thereafter the

[81] When the Magi read Isa. 7:14 in light of the celestial sign, they may well have been more inclined to interpret Isaiah's oracle as referring to the nonsexual conception of a virgin girl, since it was widely thought that Virgo was a pure virgin.

baby appeared to grow while remaining in Virgo's belly because (1) the retrograde comet, after perihelion, was moving away from the Sun and toward Earth, and (2) the comet's movement on its orbit countered the effect of Earth's movement on its orbit, causing the coma to appear to Earth-dwellers to remain in approximately the same location in the sky. Then, since the comet's course on its orbit was straightening out and taking it toward Earth, the comet fell out of sync with Earth and hence appeared to human observers to move down toward the Sun. This meant that the coma appeared to descend within Virgo in the manner of a baby moving through the birth canal. When the coma-baby descended to such an extent that it was clearly about to be fully born the next morning, there was a major meteor storm radiating from the tail of Hydra. After being "born," the coma appeared to descend rapidly toward the horizon and into the Sun's light (i.e., it heliacally set), a phenomenon that in context may well have seemed to some observers to indicate that the child would be rescued by God, whom the Sun may have represented in the celestial drama.

At the same time as the drama was unfolding in Virgo, it seems that the Magi were also impressed by another celestial show that the comet as a whole was putting on: it had become a majestic scepter that dominated the sky, eventually stretching right across the whole sky from eastern to western horizon.

As they watched these phenomena unfolding before their eyes, observers could have understood much: a divine dignitary destined to exercise dominion over the whole world was being born on the earth to a terrestrial surrogate of Virgo. However, the wonders did not directly reveal who or where the child was.

In this chapter we have suggested that one or more members of the Jewish community in Babylon probably introduced the Magi to Numbers 24:17 and Isaiah 7–12, enabling them to grasp the meaning and significance of what the comet was doing to mark the Jewish Messiah's birth. It was the Messiah—called Immanuel and "mighty God"—who had been conceived by a virgin, who would be born at the point that the cometary baby completely exited Virgo's belly, and who was destined to reign as King over the whole earth. Therefore at the conclusion of the wonders in the eastern sky the Magi set off to Judea in a bid to find the terrestrial virgin and her divine son, and to worship and honor him (Matt. 2:2, 11).

Needless to say, what the comet did during its time in the eastern sky left the Magi in no doubt that the Messiah had indeed been born. It is little wonder that the people of Jerusalem and Herod were deeply shaken by the Magi's appearance and report. If Eastern Magi were so confident that the great comet that had been gracing their skies in recent months was signaling the fulfillment of key oracles in the Hebrew Bible concerning the Messiah's birth, that they traveled hundreds of miles to Jerusalem to worship him, this constituted powerful evidence that the Messiah had indeed been born.

Of course, that the prophesied sign marked the birth of Jesus was a remarkable attestation of Jesus's messianic status. It made it rather clear that the time of fulfillment had come and the Messiah was now finally on the earthly stage.

Having defined the celestial sign precisely, and having highlighted the Biblical keys to its interpretation, we shall in the next chapter develop an astronomical profile of the Bethlehem Star Comet and determine its orbit.

9

"Lo, the Star Appeareth"

Profiling the Comet

Having identified what the Star was, what it did in connection with its rising to impress the Magi, how they understood it, and why they interpreted it as they did, we shall now turn our focus to the Christ Comet as a historical astronomical object. In particular, we shall attempt to develop a profile of the Star, determining approximately its orbit and its intrinsic brightness. This is possible because of the wealth of information concerning the Christ Comet in the Biblical text.

The claim of the Bethlehem Star to be historical is strong, grounded not only in multiple independent sources but also in the sheer implausibility that anyone in the ancient world could have fabricated a cometary apparition so complex and unusual.

Determining the comet's orbit and its intrinsic brightness will enable us to recreate the whole cometary apparition in 8/7–6 BC. In particular, it will put us in the position of being able to figure out where the Star was when it was first observed by the Magi, how close to the Sun and Earth it came, how bright it became at its rising, and how it guided the Magi from their homeland to Jerusalem, ushered them to Bethlehem, and enabled the Magi to find the baby Messiah. It will also give us the opportunity to work out roughly what the comet looked like at the different stages of the long apparition. In the next chapter we will overview the comet's career from its first appearance to the point when it stood over the house in Bethlehem where Jesus was.[1]

A CHRONOLOGY OF THE MAGI'S JOURNEY

Before going any further, we must pause to consider the chronology of the nativity story, asking how the Magi's journey fits into the story of Mary, Joseph, and Jesus. In particular, we need to give special attention to the matter of how much time separated the end of the comet's performance in the eastern sky and its appearance in the southern sky to escort the Magi to the baby Jesus.

We shall begin with a consideration of Mary and Joseph's schedule. According to Luke 2:1–7, when Mary was already in the advanced stages of her pregnancy, she and Joseph traveled from Galilee to Bethlehem to

[1] Note that this is not where Jesus was born, but where he was some weeks after his birth.

be registered for tax purposes. But when did they plan to leave Bethlehem?

A long journey back home to Nazareth with a newborn baby would have been impractical. Moreover, Mary had to be in Jerusalem, just five or six miles from Bethlehem, on the 40th day after the birth to perform her ceremonial obligations in the temple (Lev. 12:2, 4–8): a woman who had given birth to a baby boy had to avoid contact with anything holy for 40 days after the delivery (including the day of delivery) and then go to the temple and make the prescribed offerings so that she could be ritually clean again. During that visit Mary and Joseph could also present the child to God at the temple, in accordance with Leviticus 12 and Exodus 13. Quite simply, it would have made no sense for Mary and Joseph to trek all the way north to Nazareth only to turn around shortly thereafter and head back to the very area they had just left.

At the same time, however, it is difficult to imagine that the holy family would have wished to remain in Bethlehem any longer than was necessary before heading back home to Galilee.

It seems safe therefore to conclude that Mary and Joseph had planned a 6-week stay in Bethlehem, intending to depart within a few days of their visit to the Jerusalem temple. As it happened, they did not immediately return to Galilee, but fled to Egypt instead, because Joseph had a dream alerting him to the imminent threat to the baby's life posed by Herod.

Where does the Magi's visit fit into this chronology? Matthew 2:12–14 makes it clear that the holy family's flight to Egypt occurred on the heels of the Magi's departure from Bethlehem:

> And being warned in a dream not to return to Herod, [the Magi] departed to their own country by another way. Now when they had departed, behold, an angel of the Lord appeared to Joseph in a dream and said, "Rise, take the child and his mother, and flee to Egypt, and remain there until I tell you, for Herod is about to search for the child, to destroy him." And he rose and took the child and his mother by night and departed to Egypt.

It was the failure of the Magi to return to Jerusalem to report to Herod concerning the location of the newborn Messiah that prompted Herod to adopt Plan B of his scheme to kill the Messiah, namely, to send in his troops to massacre all the male babies in Bethlehem.

So did the Magi's visit occur before or after Mary and Joseph's visit to the Jerusalem temple? This is a complex matter.

What we can say with some confidence is that the Magi would have felt great urgency about leaving their eastern homeland on their mission to worship the newborn Messiah. In light of this, the Magi would probably have departed very shortly after the conclusion of the celestial drama connected with the Star's heliacal rising.

As regards their journey, the fact that their homeland is said to be in "the east" may well imply that the Magi traveled due west across the inhospitable desert from Babylon to Jerusalem (in the Biblical tradition, those coming from Babylon by the normal trade routes were normally portrayed as coming from the north, because they would have approached Jerusalem by way of Syria in the north).[2] Camel caravans traveled at the speed of humans walking (about 2–3 miles per hour), since caravaneers would typically walk with the camels or go ahead of them.[3] A study of journeys in the ancient Near East re-

[2] If the Magi journeyed across the desert from Judea back to their homeland, as seems to be suggested by Matt. 2:12, this would mandate that they had traveled across the desert by camel caravan from their homeland to Jerusalem. Otherwise they would have been ill-equipped to make the return journey through the desolate, inhospitable, and dangerous desert.

[3] James K. Hoffmeier, *Ancient Israel in Sinai: The Evidence for the Authenticity of the Wilderness Tradition* (Oxford: Oxford University Press, 2005), 120.

veals that an average day's journey by camel caravan would have consisted of between 15 and 20 miles.[4] For those taking a straight route from Babylon to Jerusalem, the journey is about 550 miles long.[5] An average rate of progress toward Jerusalem of 15 miles per day would have got the Magi to the capital of Judea on the 37th day of their journey. Progress of 20 miles a day on average would have resulted in the Magi arriving on the 28th day of their trip.[6]

For how long did the Magi remain in Bethlehem?

According to Roland de Vaux,[7] the cultural expectation among Arabs and ancient Israelites would have been that visitors should remain for three days and nights. This could be extended if the host was positively inclined toward his guests. The host would furnish food and beverages for his guests and provide necessary staples for their animals.[8] We can therefore assume that the Magi, after their long, arduous journey and with so much to discuss, spent about 3 days in Bethlehem with Joseph, Mary, and baby Jesus. It might have been cut short on the 3rd day or extended by a day or 2—so we should allow for the possibility that the Magi were in Bethlehem for 2–5 full days in total (i.e., they left on the 3rd–6th day).

If the Magi's journey started shortly after the end of the celestial wonders in the eastern sky, and lasted approximately 28–37 days, it is possible that they arrived in Bethlehem either before or very shortly after Mary and Joseph's temple visit on the 40th day after Jesus's birth. In view of the fact that Herod's massacre of the innocents occurred on the heels of the Magi's visit, and that Joseph, Mary, and Jesus left Bethlehem for Egypt immediately in advance of it, the most natural time for the Magi's Bethlehem stay would be in the week prior to the temple trip or in the week following it.

Obviously, if the Easterners' Bethlehem visit occurred before the holy family went to the Jerusalem temple, the Magi would have traveled more quickly to Judea, and Mary and Joseph's trip to the temple would have been rather tense. After all, it is very likely that, immediately prior to departing Bethlehem, the Magi informed Joseph about their dream (and perhaps explained to him that Herod was the personification of the serpentine dragon Hydra, preparing to kill the baby Messiah).

Alternatively, the Magi may have arrived in Bethlehem in the immediate aftermath of Mary and Joseph's return from Jerusalem. As we have noted, Mary and Joseph were almost certainly only intending to remain in Bethlehem until they had fulfilled their religious obligations at the temple on the 40th day. Had the Magi arrived any more than a few days after that, the holy family would presumably have been on their way back up north to Galilee.

Unfortunately, we are forced to leave the question of whether the Magi arrived in Beth-

[4] "Based on comparative travel distances derived from ancient texts," Hoffmeier, *Ancient Israel in Sinai*, 119–120, 143–144, concludes that ancient travelers generally traveled at 15–20 miles per day. The numbers suggested by scholars on this topic are all rather similar, although some are more conservative—H. Clay Trumbull, *Kadesh-Barnea: Its Importance and Probable Site* (Philadelphia: J. D. Wattles, 1895), 71–74, suggested 15–18 miles per day—and some are more liberal—Graham Davies, "The Significance of Deuteronomy 1:2 for the location of Mount Horeb," *Palestine Exploration Quarterly* 111 (1979): 96, opts for 16–23 miles a day, while Barry Beitzel, "Travel and Communication (OT World)," in *The Anchor Bible Dictionary*, ed. D. N. Freedman, 6 vols. (New York: Doubleday, 1992), 6:646, suggests 17–23 miles per day.

[5] My calculations suggest that the two cities are precisely 543 miles apart as the crow flies.

[6] A camel caravan traveling 550 miles that was 15–18 miles closer to their destination at the end of an average day would have taken about 31–37 days, one that made average progress of 16–23 miles per day would have taken 24–35 days, and one advancing an average of 17–23 miles per day would have taken 24–33 days. Cf. Mark R. Kidger, *The Star of Bethlehem: An Astronomer's View* (Princeton, NJ: Princeton University Press, 1999), 260.

[7] Roland de Vaux, *Ancient Israel: Its Life and Institutions* (Grand Rapids, MI: Eerdmans; Livonia, MI: Dove, 1997), 10.

[8] One gets a glimpse of ancient rules of hospitality in Judg. 19:1–9: when prevailed upon to stay with his concubine's father, the Levite "remained with him three days. So they ate and drank and spent the night there. And on the fourth day they arose early in the morning, and he prepared to go" (vv. 4–5). However, the girl's father persuaded him to remain his guest for a fourth day and night and then for a fifth day and night before the Levite finally declined to accept any further hospitality and so left on the sixth day.

lehem before or after the temple visit open, allowing for both possibilities.

Herod the Great eventually realized that his crafty plot to assassinate the Messiah had been scuppered (Matt. 2:16). The text does not make it explicit what the catalyst for his realization was, but it is most natural to believe that the time Herod had allotted for the Magi's finding the baby Messiah and completing their visit to worship him had expired. Herod would probably have been willing to wait the customary 3 days and nights and maybe, impatiently, a few days beyond that. Certainly it is difficult to imagine him waiting more than 6 full days in total before concluding that he had been double-crossed by the Magi.

WHAT WE HAVE LEARNED ABOUT THE STAR

In order to tell the full story of the Christ Comet's apparition, it is important to gather up all the key bits of information that we have gleaned in the course of our investigation.

1. *Long-period comet.* It was undoubtedly a long-period comet. This is demonstrated by the duration of the apparition and particularly how long prior to perihelion the comet first became visible to the naked eye, by the great size of the cometary coma shortly after perihelion when it was in the eastern sky, and by the length of its tail at that time and some weeks later when it stood over the house where Jesus was.

2. *Narrow inclination.* The comet was narrowly inclined to the ecliptic and probably orbited within the zodiacal band, rendering it of more astrological significance to the Magi. Among other things, that the comet set with its tail stretching upwards, when the Magi were in Bethlehem, is suggestive of narrow inclination.

3. *Retrograde.* The comet was unquestionably retrograde (moving in a clockwise direction, from the vantage point of the north pole). This is indicated by the fact that it increased in size within Virgo and, having left the eastern sky, quickly progressed to the southern evening sky.

4. *First appearance.* The comet first appeared at least 12 lunar months (assuming a Jewish [and not Babylonian] system of intercalation) before Herod the Great issued the order for the infants in Bethlehem and environs to be slaughtered. This information reveals that the comet was intrinsically very bright (i.e., it had an extraordinary absolute magnitude) and was very large and productive. Herod regarded the date of the Star's first appearance as the earliest possible birthday of the Messiah. He may even have wondered if it actually coincided with the Messiah's birth.

5. *The exaltation of Virgo.* The key celestial drama began on September 15 of 6 BC, when the Moon was under Virgo's feet and the Sun was clothing her (that is, was over her belly region). With respect to how the constellation figure Virgo is being imagined relative to the stars, the proportional constraints indicated by Revelation 12:1, together with the facts that she is wearing a crown and glorious robes and has the Moon under her feet, suggest that Virgo is being thought of as seated on a throne. This is consistent with how Teukros of Babylon (first century BC and/or beginning of first century AD) and others viewed Virgo. Since the Sun on September 15, 6 BC, is said to have been clothing Virgo, the author of Revelation is evidently imagining the constellation figure's midriff in particular to be wide enough to accommodate the ecliptic at that point. Since the Babylonians established the beginnings of months by calculation in advance, we can be confident that the sighting of the new crescent Moon after sunset occurred at the start of the second day of the Babylonian month of Ululu (since Ululu 1 fell on September 14/15, sunset to sunset). According to the observational Hebrew calendar, it would have been the beginning of Tishri 1 or 2, depending on whether the new Moon had been spotted on the evening of September 14.

6. *Heliacal rising*. The Magi saw the cometary coma rise heliacally in the eastern sky, meaning that it had been in conjunction with the Sun prior to its reappearance. The retrograde comet was emerging from the Sun around the time of perihelion (when a comet's brightness peaks). During this time the coma was an especially bright morning star and was observable in the northern hemisphere (specifically in the Near East). The comet as a whole probably looked like a scepter at this time.

7. *In Virgo's womb*. Since the comet was very bright, its coma was visible in Virgo's belly before sunrise each day from about September 29/30.[9] At that time Virgo was emerging in the eastern sky after its annual encounter with the Sun. We can be confident concerning where approximately Virgo's belly was thought to be: the area from a little above 80 Virginis up to a little below the level of δ.

8. *Cometary baby*. The comet's coma appeared to play the part of a baby in Virgo's womb. That it was viewed as a baby was not only because it was located in a womb, but probably also because it had the general shape of a baby in the womb. Elliptical (oval) comas may look like upside-down babies in the womb.

9. *Growing baby*. When the baby heliacally rose in Virgo's womb shortly after perihelion, it was relatively small. This makes sense because, when a comet makes a very close pass by the Sun, its coma is generally relatively small in the run-up to perihelion, due to its receiving the most intense blast of the Sun at that time. Then the baby seemed to increase in size within the womb area over the course of Virgo's pregnancy; comas tend to grow immediately after perihelion as they are increasingly liberated from the Sun's intense compression. More important, the comet was moving toward Earth and so, from the vantage point of humans, the coma appeared to become larger and larger.

10. *Stable in womb*. The comet's coma remained relatively fixed in Virgo's womb area. The cometary baby was able to stay within Virgo's belly because of unusual comet-Sun-Earth geometry. There was a remarkable synchronism between the movements of Earth and the comet in their respective orbits. The relative stability of the baby in Virgo's womb before its delivery may also have been partly due to the growth of the coma.

11. *Notable apparent magnitude*. Since dim comets would not be visible low on the horizon at sunrise or remain observable when very large, and since this comet is regarded as a great light shining into the deep darkness, the Christ Comet must have had extraordinary apparent magnitude values. It was probably a daylight object around perihelion. The coma's pseudonucleus (the area immediately surrounding the nucleus) may have had a condensed brightness, making it seem that the baby's face was shining.

12. *Duration of pregnancy*. The cometary coma's growth gave the impression that Virgo's pregnancy was developing normally. Together with the fact that the baby, during the delivery, was observed to descend into the area of Virgo's legs, and thereafter was seen completely below her womb (at the birth) and then at least once thereafter heading toward the Sun, this suggests that the coma remained wholly or partly within the confines of Virgo's womb until about October 19.

13. *The delivery*. Dilation would have been regarded as coinciding with the cometary coma seeming to weigh down on Virgo's pelvic floor (approximately the level on Virgo where 80 Virginis is), and delivery would have entailed the baby moving down over that level. This downward movement within Virgo was due to the comet's movement relative to Earth—essentially the comet's orbit

[9] By September 29/30, Virgo's lower womb area would have been about 8–9 degrees from the Sun. A very bright coma would have been detectable at that distance from the Sun.

straightened out after its sharp perihelion U-turn and hence it could no longer stay in sync with Earth.

14. *Hydrid meteor storm.* While Virgo was actively giving birth to the baby, but had not quite fully delivered it, observers in the Near East saw a predawn meteor storm (Rev. 12:3–4a). The storm was attributed to the action of Hydra's tail, almost certainly because it radiated from there, probably from the upper section of the tail, between γ (Gamma) Hydrae and the part of the tail near where Corvus (i.e., the Raven, in Babylonian thought) perched. Evidently the meteor storm occurred when the upper section of Hydra's tail was above the horizon and when the sky was dark enough for the dramatic meteor display to be visible to human observers. This meteor storm happened before Hydra "stood." This standing refers to the point when π (Pi) Hydrae, the tip of the tail, was level with the horizon, so that all of Hydra was above the horizon.

15. *The birth.* The night after the meteor storm, for the first time since the baby had appeared in Virgo's womb, no part of the coma rose in advance of Virgo's vaginal opening. The whole baby could be seen below the level where 80 Virginis was but above the visible horizon. To those interpreting what they saw through the grid of messianic prophecy, it seemed that the Messiah was born at this very time.[10] The birth was probably on or around October 20, 6 BC, which equates to Tishratu 6 in Babylon and Heshvan 5 or 6 in Judea (depending on whether the new crescent Moon was observed in Judea on the evening of October 14).[11] Both Matthew and Revelation imply that the heavenly birth coincided with Jesus's birth on the earth. It is most natural to conclude that the coma-baby appeared to be approximately the size of a newborn baby at the point of birth.

16. *Iron scepter.* At the point when the cometary baby was born, the comet's tail was apparently extraordinarily long and possibly silvery-gray. The whole comet seems to have formed an "iron scepter" that stretched right across the sky to the western horizon.

17. *Disappearance into the Sun's light.* In terms of the celestial narrative, the messianic child needed to be delivered from the grave danger posed by Hydra. This deliverance was communicated to Earth-dwelling observers when the cometary coma, after birth, quickly disappeared into the light of the Sun and below the horizon (i.e., it heliacally set).

18. *From Babylon to Jerusalem.* Having witnessed the entire celestial nativity drama, the Magi no doubt soon began their journey to Judea. The comet, having switched quickly from the morning to the evening/night sky, probably remained visible to them throughout their journey. It is likely that, every night, the comet moved toward and over the western horizon, seeming to urge them onward to Jerusalem.

19. *South-southwest to Bethlehem.* On the evening when the Magi traveled from Jerusalem to Bethlehem, approximately 30–40 days after the conclusion of the sign in the eastern morning sky, the Star reappeared, probably at sunset, in the southern sky. The Star seemed to go before the Magi to Bethlehem. Assuming a journey of 2 hours, the comet would have appeared in the sky to the SSE and moved to the SSW, the direction of Bethlehem from Jerusalem.

20. *Dropping in altitude.* After guiding

[10] The predawn celestial announcement of Jesus's birth is consistent with Luke's record of the angelic birth announcement to the shepherds who were keeping watch over their flock "by night" (Luke 2:8). The angel made it clear that the Messiah had been born at that very time and would be lying in a manger when they found him in Bethlehem (v. 11).

[11] Richard A. Parker and Waldo H. Dubberstein, *Babylonian Chronology 626 B.C.—A.D. 75* (Providence, RI: Brown University Press, 1956), 45, whose formula is similar to the one employed by the Babylonians in their calculations, start the Babylonian month Tishratu on the evening of October 14, so that Tishratu 1 = October 15 in 6 BC. See the helpful work of Rita Gautschy, "Last and First Sightings of the Lunar Crescent," at http://www.gautschy.ch/~rita/archast/mond/mondeng.html (last modified February 15, 2013), on first sightings of the new Moon in history, and especially her application of the "Yallop method" for different sites across the ancient Near East.

them to Bethlehem, the comet, having passed its culmination (its highest point, on the meridian in the south), then dropped in altitude ("coming"; Matt. 2:9) until it seemed to "stand."

21. *Setting on the western horizon.* The Star guided the Magi to one particular house in Bethlehem. We can deduce that the coma was close to the visible horizon at that stage and so was about to set.[12] The house where Mary and Jesus were was evidently, from the Magi's perspective, located along the skyline and was distinct from other structures. Since the Star "stood over" one particular house, it is most likely that at that time the comet had a near-vertical tail, 30–45 degrees long.

22. *Subsequent behavior.* There is no indication of the comet's behavior thereafter. However, the fact that it was observable for so long prior to perihelion suggests that it could have remained visible for many months afterwards.

DETERMINING THE ORBIT

We have already seen that astronomers have determined the orbits of many historical comets. In fact, we are in the fortunate position of knowing the orbits of comets as far back as the fourth and third centuries BC.[13] Astronomers are able to calculate a comet's orbit when they can derive sufficient positional information from the surviving records of observers. Remarkably, just three observations are adequate to determine the six orbital elements that fully describe any orbit. In the case of the Christ Comet, Revelation 12:1–5 provides more than enough observational data to make a reasonably precise determination of its orbital elements at that time. It is important to realize that only a comet with a very particular orbit can do what Revelation 12:1–5 describes. Everything has to be just right: the placement of Earth on its orbit around the Sun, the time and place of the comet's perihelion, the comet's inclination and the direction of its motion, and the whole orientation of its orbit. Accordingly, we are able to deduce the six orbital elements that describe the Christ Comet's orbit (table 9.1; see fig. 9.1 for a portrayal of the comet's orbit).[14]

Eccentricity	**1.0**
Perihelion Distance (AU)	**0.119**
Inclination	**178.3**
Argument of Perihelion	**9.47**
Longitude of Ascending Node	**200.08**
Perihelion Date (Julian)	**1719500.7/ September 27, 6 BC**

TABLE 9.1 The orbital elements of the Christ Comet.

Of course, the big question is whether this orbit is compatible with what Matthew records concerning the comet's behavior on the final stage of the Magi's journey (Matt. 2:9–11). Astonishingly, it is. Shortly after the cometary baby's birth in the eastern predawn sky, the comet shifts to the western evening sky and quickly migrates to the southern sky. 30 to 40 days after leaving the eastern sky, the comet is appearing at sunset in the south-southeast and, over the following couple of hours, is moving to the south-southwest, the direction of Bethlehem from Jerusalem.

[12] Ancient Bethlehem was located on a narrow ridge at a high altitude (over 2,500 feet above sea level) surrounded by valleys on all sides, stretching from east to west.

[13] For a useful collection of calculated orbits of historical comets up to the eighteenth century AD, see the first volume of Gary W. Kronk, *Cometography: A Catalog of Comets*, 6 vols. (Cambridge: Cambridge University Press, 1999–).

[14] Astronomers assume an eccentricity of 1.0 when observational data is limited or imprecise. We follow this standard practice. For our purposes, whether the eccentricity is 1.0 or a smidgeon above or below it has little bearing on the comet's course around perihelion time. Note too that when a solar system object's eccentricity is high and its inclination is close to 180 degrees, subtracting the argument of perihelion from the longitude of the ascending node gives us what can legitimately be reckoned the ecliptic "longitude of perihelion" (David Asher, personal email message to the author, May 8, 2013). The longitude of perihelion is essentially the angle between the First Point of Aries (the vernal equinox) and the point of perihelion (although, strangely, it encompasses measurements on two different planes). Different pairs of values for the argument of perihelion and longitude of the ascending node will produce essentially the same longitude of perihelion.

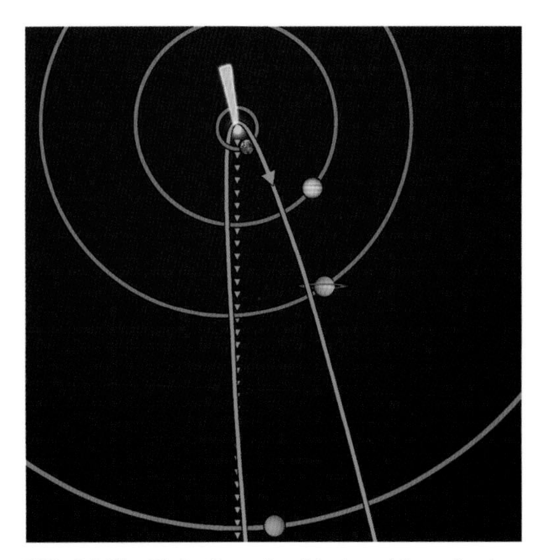

FIG. 9.1 The Christ Comet's orbit viewed from the time of perihelion on September 27, 6 BC. The outermost circle is Uranus's orbit. The green line is the comet's eccentric orbit, and the orange arrows are directed to the First Point of Aries, a base line for astronomical measurements. The comet is under the ecliptic plane on its way toward the Sun and over it on its way away from the Sun. However, because the comet is so narrowly inclined, it is above the orbital paths of the major planets (other than Earth) on its way toward perihelion. On its way out, the comet is above the orbital paths of all the major planets except for Mercury (because of the significant tilt in Mercury's orbit). Image credit: Sirscha Nicholl.

Then some 5 hours later it sets near-vertically over the western horizon. This remarkable compatibility is overwhelmingly strong evidence that our approximate orbit is correct, that Revelation 12:1–5 is indeed the key that unlocks the mystery of the Bethlehem Star, and that Matthew 2's account of the Star is grounded in history.

Since cometary orbits are so susceptible to change due to gravitational and nongravitational effects, it cannot be assumed that a calculated orbit is applicable a long time before or after the observations which were the basis of the orbit determination. In the case of the Bethlehem Star, our orbit suggests that the comet had a relatively close encounter with Saturn on the way toward the inner solar system and so the orbital elements before that would have been different.

CHRONOLOGICAL CONSIDERATIONS

Before we consider the comet and its orbit in more detail, it is important to pause to fill in some of the chronological gaps in our knowledge based on our orbit. Our orbit suggests that the predawn celestial show in the east ended a couple of days after the birth, namely on October 22, and that this would have become evident to observers by October 23. We should therefore reckon on the Magi departing their homeland at some point between October 23 and 25. A quick departure is strongly supported by the observation that at that time the comet, having shifted to the western sky, becomes capable of functioning as a westward prompt or guide for the Magi.

In light of this and our earlier conclusions regarding the duration of the Magi's journey (approximately 28–37 days) and their stay (2–5 full days, leaving on the 3rd–6th day), we are in a position to set out two scenarios regarding the chronology of the Magi's visit to Bethlehem: either (A) it occurred in the week running up to the holy family's trip to the temple (on November 29)—that is, the week of November 21–28;[15] or (B) it took place in the week immediately following it—that is, sometime in the week following Mary and Joseph's return from Jerusalem on the afternoon of November 29—that is, November 29–December 6.[16] In Scenario A, assuming a departure date between October 23 and 25, the Magi's camel caravan advanced approximately 17–18 miles per day toward Jerusalem and arrived in Bethlehem between November

[15] Calvin (*Matthew, Mark, and Luke, Part 1*, http://www.ccel.org/ccel/calvin/calcom31.html [accessed April 5, 2013]), believed that the Magi arrived in Bethlehem in advance of Mary and Joseph's temple visit.

[16] The view that the Magi arrived in Judea after Mary and Joseph's temple visit is held by many, including Robert H. Stein, *Jesus the Messiah: A Survey of the Life of Christ* (Downers Grove, IL: InterVarsity Press, 1996), 53; Darrell L. Bock, *Luke 1:1–9:50*, Baker Exegetical Commentary on the New Testament (Grand Rapids, MI: Baker, 1994), 235; idem, *Jesus according to Scripture: Restoring the Portrait from the Gospels* (Grand Rapids, MI: Baker, 2002), 68–71.

23 and 25, and they departed on or before the 28th. In Scenario B, having left Babylon on October 24 or 25, the camel caravan got 15 miles closer to Jerusalem each day, and the Magi arrived in Bethlehem late on November 29 or on the 30th, in the immediate aftermath of Mary and Joseph's return from Jerusalem, and departed for Babylon between December 2 and 6.

In our analysis we shall consider two representative dates for the arrival of the Magi in Bethlehem: November 23 (= an average journey speed of about 18 miles per day if they left Babylon on October 24); and November 30 (= an average journey speed of about 15 miles per day if they left Babylon on October 25), 6 BC.[17]

According to our reckoning, Herod's decree to massacre the infants was issued on (or around) either November 30–December 2 or December 3–7, 6 BC. The Star's initial appearance (1–2 Jewish luni-solar years before Herod issued his order) was therefore either between November 21–23, 8 BC, and December 10–12, 7 BC; or between November 24–28, 8 BC, and December 13–17, 7 BC.[18]

We propose, then, the following approximate chronology:

> Between November 21–28, 8 BC, and December 10–17, 7 BC: The Star first appears.
>
> September 15, 6 BC: The Moon is observed under Virgo's feet.
>
> September 29/30, 6 BC: The cometary coma rises heliacally in Virgo's womb.
>
> October 15, 6 BC: Virgo appears to begin active labor.
>
> October 19, 6 BC: A meteor storm radiates from Hydra's tail.
>
> October 20, 6 BC: The cometary baby, having descended, has completely emerged from Virgo's womb and so is regarded as having been born.
>
> Between October 23 and 25, 6 BC: The Magi leave Babylon on their mission to worship the Messiah in Judea.

SCENARIO A

> Between November 23 and 25, 6 BC: The Magi arrive in Jerusalem and, that same evening, are ushered by the Star to Bethlehem. Later that night the comet stands as it sets, pinpointing the house where Mary and Jesus are.
>
> Between November 26 and 28, 6 BC: The Magi depart Bethlehem to return home to Babylon.
>
> November 29, 6 BC: Mary, Joseph, and baby Jesus visit the Jerusalem temple on the 40th day to fulfill their religious obligations, and return to Bethlehem.
>
> November 30–December 2, 6 BC: Herod the Great orders the Massacre of the Innocents in the vicinity of Bethlehem.

SCENARIO B

> November 29, 6 BC: Mary, Joseph, and baby Jesus visit the Jerusalem temple on

[17] The fact that during the second half of their journey from Babylon to Jerusalem through the Syrian Desert they would have passed through the Harra, a basalt boulder-strewn region, could help explain a slower rate of progress. Lady Anne Blunt and her husband traveled through part of the Harra in December 1878 and recalled that it was slow going. For example, she mentioned that at the end of one day their camel caravan ended up "barely twelve miles from where we began" (Anne Blunt, *A Pilgrimage to Nejd, the Cradle of the Arab Race*, 2nd ed., 2 vols. [London: John Murray, 1881], 1:69). Later she stated that "That black wilderness had become like a nightmare with its horrible boulders and little tortuous paths, which prevented the camels from doing more than about two miles an hour" (75). By contrast, the first half of the Magi's journey would have been through the Hamad, an area of open desert plains through which faster progress might be expected.

[18] Assuming that the Judeans inserted an intercalary month in the late winter/early spring of 7 BC to ensure that the Passover fell shortly after the vernal equinox. The Babylonians inserted their intercalary month in the spring of 6 BC.

> the 40th day to fulfill their religious obligations and return to Bethlehem.
>
> November 29–30, 6 BC: The Magi arrive in Jerusalem and that same evening are ushered by the Star to Bethlehem. Then, later that night, the Star leads them to the house where Mary and Jesus are.
>
> Between December 2 and 6, 6 BC: The Magi depart Bethlehem to return home to Babylon.
>
> Between December 3 and 7, 6 BC: Herod the Great orders the Massacre of the Innocents in the vicinity of Bethlehem.

We must now turn to make some observations about the comet based on our orbital elements.

LONG-PERIOD NEARLY ISOTROPIC COMET

Within the category of long-period comets, the Christ Comet is a member of the broad family of "Long-period nearly isotropic" (NI) comets.[19] Basically, this is all the long-period comets that are not part of the sungrazer class—sungrazers are reckoned to total about one-third of all comets (if one assumes that recent centuries are representative).[20] The sungrazers have in common that they all approach within approximately 0.01 AU or 0.05 AU of the Sun. Astronomers also use a category of "sunskirters," which includes all comets that have a perihelion distance that is less than 0.1 AU but greater than that of the sungrazers.[21] According to our orbit, the Magi's Comet came as close as 0.119 AU to the Sun, which is very close, but not quite as near as the sunskirters.

NARROWLY INCLINED NI COMET

NI comets have a wide variety of inclinations across the full spectrum of 0° to 180°. Strikingly, however, most NI comets whose orbits have been calculated have inclinations between 40° and 160°. Of those between 0° and 40° and between 160° and 180°, the smallest proportion fall within the 170–180° range.[22] Therefore what we have in the Christ Comet may be relatively rare for comets—an essentially ecliptic, retrograde long-period comet. In this respect the Christ Comet, with its 178.3-degree inclination, is like Comet Lulin, the brightest comet in 2009,[23] which had an inclination of 178.4 degrees. It is also reminiscent of Comet Tempel of 1864 (C/1864 N1), which had an inclination of 178.13 degrees and came to within 0.1 AU of Earth on August 8, 1864, moving from the eastern morning sky to the western evening sky at that time.[24] Narrowly inclined comets sport relatively straight dust tails, from Earth's perspective, as opposed to the more curved dust tails of steeply inclined comets. Of course, the tails of narrowly inclined comets are curved in outer space, but the curvature is not apparent to those on Earth because it occurs on the plane on which Earth orbits the Sun.

REMINISCENT OF COMET HALE-BOPP

The fact that the Christ Comet was seen by the naked eye so long before perihelion is reminiscent of C/1995 O1 (Hale-Bopp). Hale-Bopp was first seen 10½ months before perihelion, when it was still 4.37 AU from the Sun. It has been calculated that Hale-Bopp started form-

[19] Richard Schmude, *Comets and How to Observe Them* (New York: Springer, 2010), 16–17.
[20] Ibid., 7 table 1.2.
[21] David Seargent, *Sungrazing Comets: Snowballs in the Furnace* (Kindle Digital book, Amazon Media, 2012).
[22] See Schmude, *Comets and How to Observe Them*, 20–21, esp. fig. 1.25; also 28–29, including figs. 1.32 and 1.33; cf. John C. Brandt and Robert D. Chapman, *Introduction to Comets*, 2nd ed. (Cambridge: Cambridge University Press, 2004), 75 fig. 2.7.
[23] Martin Mobberley, *Hunting and Imaging Comets* (Berlin: Springer, 2011), 167–168.
[24] Kronk, *Cometography*, 2:326.

ing its coma roughly 18 AU from the Sun.[25] The Bethlehem Star comet was first observed on or before December 10–17, 7 BC, about 9½ months before perihelion, when it was 4.71–4.63 AU from the Sun. Its coma probably began to form out beyond the orbit of Uranus (20 AU from the Sun).

Comets such as Hale-Bopp and the Magi's Star begin degassing so long before perihelion because they contain relatively high volumes of extremely volatile materials such as nitrogen, carbon monoxide, and methane.[26] These materials begin reacting to the Sun at long distances from the Sun, whereas water-ice begins to degas only when within 3 AU.[27]

BRIGHTNESS

Having determined an orbit, we must now turn to the matter of the Christ Comet's brightness. A comet's intrinsic brightness is called its absolute magnitude. Its brightness as it appears to observers from Earth at any given point in time is called its apparent magnitude.

The starting point of any investigation of a comet's brightness is to determine the apparent magnitude at first observation. Investigations of ancient cometary apparitions have determined that a tailless comet must attain to an apparent magnitude of about +3.4 to be observed with the naked eye.[28] It seems wisest therefore to assume that the Bethlehem Star comet had a +3.4 apparent magnitude when it was discovered.[29]

Next, the pattern of development of the comet's brightness must be estimated. Comets usually brighten exponentially as they approach the Sun. This increase in brightness is directly related to the increase in cometary degassing as the comet nears the Sun. The steepness of the rise in brightness as it nears the Sun and of the decline in brightness as it moves away from it is called the brightness slope, which is expressed as the value of n. The higher this value, the steeper the brightness slope is.

In the absence of adequate data to refine the pattern of cometary brightness development, scientists tend to assume that n=4.[30] This assumption is based on the idea that this is the average value for comets.[31] According to recent analysis by Schmude, long-period comets can have values of n from -2 to over 11, but most are grouped between 2 and 6, with the bulk of them concentrated between about 3 and 5, and the approximate average being 4.[32] For example, while Bennett's Comet of 1970 had a brightness slope of about 5, Hale-Bopp's was about 3. Accordingly, anyone assuming that Hale-Bopp's display in the inner solar system would abide by the average brightness slope n=4 would have been slightly (but not greatly!) disappointed.

In the tables accompanying my discussion of the Christ Comet's apparition, I shall offer brightness estimates based on the following values of n: 3, 4, and 5. It is likely that the bright-

[25] Schmude, *Comets and How to Observe Them*, 39–40; Robert Burnham, *Great Comets* (Cambridge: Cambridge University Press, 2000), 103.

[26] Schmude, *Comets and How to Observe Them*, 39–40, including table 1.10.

[27] Nick James and Gerald North, *Observing Comets* (London: Springer, 2003), 23.

[28] Yau et al., "Past and Future Motion," 314, maintain that Halley's Comet was first discovered at magnitude +4, but Swift-Tuttle at +3.4, and propose that the difference was due to Halley's Comet having a tail at the point of discovery (making it easier to spot). Donald K. Yeomans suggests that generally a tailless comet must attain to +3.4 to become visible to the naked eye ("Great Comets in History," http://ssd.jpl.nasa.gov/?great_comets [posted April 2007]). Since the Christ Comet was spotted extraordinarily early, it undoubtedly lacked a visible tail at the point of discovery—therefore it seems appropriate to assume a first sighting at +3.4. In his calculations of the brightness of the Bethlehem Star comet (personal email correspondence on September 28, 2012), Gary W. Kronk assumed discovery at +4 apparent magnitude. A first sighting of the Christ Comet at +4 magnitude would mean that it was slightly less intrinsically bright than the values given in our tables below.

[29] However, David W. Hughes, "Early Long-Period Comets: Their Discovery and Flux," *Monthly Notices of the Royal Astronomical Society* 339 (2003): 1103–1110, suggests that comets were often observed only when they reached an apparent magnitude of +2 (remember that, in the scale of astronomical magnitudes, a lower value means greater brightness).

[30] John C. Brandt and Robert D. Chapman, *Rendezvous in Space: The Science of Comets* (New York: W. H. Freeman, 1992), 221; idem, *Introduction to Comets*, 105; Fred Schaaf, *Comet of the Century* (New York: Springer, 1997), 348; Mobberley, *Hunting and Imaging Comets*, 17; David Seargent, *Comets: Vagabonds of Space* (Garden City, NY: Doubleday, 1982), 39.

[31] According to the analysis of Schmude, *Comets and How to Observe Them*, 32, and 35 table 1.7, the average value of n is approximately 4 for long-period comets, but closer to 6 for short-period comets.

[32] Ibid., 35 fig. 1.40. Cf. Brandt and Chapman, *Rendezvous in Space*, 221.

ness slope of the long-period Bethlehem Star comet comes within this range of values.

In establishing the correct value of n, we have more to go on than merely estimates based on the comet's probable apparent magnitude at first observation—we have data regarding the comet's performance in the inner solar system. Just as the astronomical community modifies the value of n in response to further observations of the comet after discovery, so also we are in a position to narrow down the brightness slope of the Bethlehem Star comet based on what we know of its behavior around the time of its closest encounter with the Sun.

Magnitude Slope (value of n)	Absolute Magnitude if it was first observed on November 21–28, 8 BC	Absolute Magnitude if it was first observed on December 10–17, 7 BC*
3	-8.1	-5.2
4	-10.4	-6.8
5	-12.7	-8.5

TABLE 9.2 The range of the Christ Comet's possible absolute magnitude values.

NOTE: *The absolute magnitude values given here (and the apparent magnitudes based on them) are for December 17, 7 BC. Had the comet first been spotted on December 10, 7 BC, its absolute magnitude would have been -5.2 (n=3), -6.9 (n=4) or -8.6 (n=5).

Magnitude Slope (value of n)	Apparent Magnitude on Sept. 30, 6 BC if it was first observed on November 21–28, 8 BC	Apparent Magnitude on Sept. 30, 6 BC if it was first observed on December 10–17, 7 BC
3	-13.6	-10.7
4	-17.7	-14.1
5	-21.8	-17.6

TABLE 9.3 The range of possible apparent magnitude values of the Christ Comet on September 30, 6 BC.

With our range of possible values of n, the apparent magnitude at first observation, and the orbit, we are in a position to determine the comet's absolute magnitude, that is, the brightness of the coma as a whole if it were precisely 1 AU from both the Sun and Earth. Since we know the range of dates within which the Bethlehem Star was first observed, we can work out a range of possible absolute magnitude values for each brightness slope (i.e., n=3, n=4, and n=5; see table 9.2).

If n=3, the absolute magnitude was between -5.2 and -8.1. If n=4, the absolute magnitude was between -6.8 and -10.4. If n=5, the absolute magnitude was between -8.5 and -12.7. Those familiar with comets will immediately grasp how astonishing these values are. The intrinsically brightest comets observed in recent centuries are Sarabat's Comet of 1729, with its absolute magnitude of between -3 and -6, and Hale-Bopp, which had an absolute magnitude of -2.7 in the early stages of its apparition. Even if we adopt a brightness slope (n) of just 3 and assume that the 6 BC comet was first observed at the latest possible time (December 10–17, 7 BC), the Bethlehem Star comet finds itself in an exclusive league with Sarabat's Comet and the progenitor of the Kreutz Sungrazers. If n=4, then the Christ Comet is distinguished as the intrinsically brightest comet in recorded history.

With an estimated value of absolute magnitude, we are in a position to calculate how bright a comet should become over the course of its visit to the inner solar system—that is, what its apparent magnitude will be. With respect to the Christ Comet, if we consider September 30, 6 BC, just a few days after perihelion, the comet's apparent magnitude would have been dramatic (table 9.3).

If n=3, then the comet would have been between -10.7 and -13.6 when it was seen in Virgo's belly before dawn on September 30, 6 BC. If n=4, the comet would have attained to between -14.1 and -17.7. If n=5, it would have been between -17.6 and -21.8.

If we remember that the apparent magnitude of the full Moon is -12.6 and that of the Sun -26.7, we can get some idea of how remarkable the comet's brightness would have been according to these statistics.

One major caveat should be mentioned: comets do not always develop in a very neat and orderly manner. With respect to the brightness curve and hence absolute and apparent magnitude values, they can vary within a single apparition. Comets may shift from one pattern of brightness development to another at a certain distance from the Sun or have one pattern before perihelion and quite another one afterwards.[33] Comets with perihelion distances of less than 1 AU, like the Bethlehem Star comet, often display a very different pattern of activity and brightness after perihelion than before. Moreover, sometimes comets have outburst events that dramatically increase their brightness and size. Therefore, as Schaaf points out, "Absolute magnitude and the brightening factor can only be regarded as useful, not perfect, guides for helping to predict and characterize a comet's brightness and brightness behavior during an apparition, or during part of one."[34]

At this point it is important to pull together some of the historical data that speaks to the Bethlehem Star's brightness. Ignatius, in his letter *To the Ephesians*, wrote of the Star in terms that suggested it was especially brilliant:

> A star shone in heaven [with a brightness] beyond all the stars; its light was indescribable, and its newness caused astonishment. And all the rest of the stars, together with the Sun and the Moon, formed a chorus to the star, yet its light far exceeded them all. And there was perplexity regarding from where this new entity came, so unlike anything else [in the heavens] was it.

According to Ignatius's authoritative-sounding statement (which was apparently rooted in first-century tradition), the Star was far brighter than all the stars, by which he is evidently including the planets such as Venus, which has a maximum apparent magnitude of -4.89.

Is Ignatius claiming that the Star was brighter than the Moon and even the Sun? That is certainly what commentators believe. If he is claiming this, we must allow for hyperbole but should probably take it to mean that the Star had an apparent magnitude more impressive than that of the full Moon (-12.6).

According to the less reliable *Protevangelium of James* (21:2–3), the Magi reported, "We saw an immense star shining among these stars and causing them to become dim, so that they no longer shone; and we knew that a king had been born in Israel."[35] This description suggests that the comet's brightness was greater than the Moon's. Moreover, this document portrays the Star as exceptionally large. For a very large comet to be extraordinarily bright means that the apparent magnitude must be most remarkable.[36]

[33] Joseph N. Marcus, "Forward-Scattering Enhancement of Comet Brightness. II. The Light Curve of C/2006 P1 (McNaught)," *International Comet Quarterly* 29 (2007): 119, 124.

[34] Schaaf, *Comet of the Century*, 349.

[35] My translation of the Greek text in Emile de Strycker, *La forme la plus ancienne du Protevangile de Jacques* (Brussels: Société des Bollandistes, 1961), 168–170.

[36] Passages in the *Sibylline Oracles* from the second or third centuries AD (which we have already cited) claim that the Star of Bethlehem was so bright that it was seen during the daytime. *Sib. Or.* 12:30–33 speaks of it as a celestial entity so extraordinarily bright that it shone forth in midday: "Whenever a bright star most like the Sun shines forth from heaven in midday, then indeed the secret word of the Most High will come wearing flesh like mortals" (J. J. Collins, "Sibylline Oracles," in Charlesworth, *Old Testament Pseudepigrapha*, 1:445). *Sib. Or.* 1:323c–324 describes the Star as brightly shining in the broad daylight (J. L. Lightfoot, *The Sibylline Oracles* [Oxford: Oxford University Press, 2007], 311):

> Then from the east a star in fullest day (323c)
> That brightly shines shall from the heavens beam (323d)
> Announcing a great sign for mortal men. (323e)
> Then God's great son will come to humankind (324)

In addition, Maximus the Confessor (early seventh century) maintained that Matthew's Star could be seen during the daytime (*Philokalia* 2:92; see *The Philokalia: The Complete Text*, vol. 2, trans. and ed. G. E. H. Palmer, P. Sherrard, and K. Ware [London: Faber & Faber, 1981], 166–167).

Apparent magnitude compares the brightness of *complete* entities to the star Vega (the second brightest star in the northern celestial hemisphere, after Arcturus). It does not compare the entity based on the brightness of a *set portion* of its area, that is, its "surface brightness." For a large object to be bright enough to bleach out the light of the stars, its apparent magnitude has to be extraordinary, because its overall brightness is distributed over a wider area, which means it is diluted. It is much like the brightness of a beam of light on a wall cast by a flashlight. When the beam is small, the brightness is more concentrated. When it is large, the brightness is more diffuse. If you compare a set area of the beam when small to a set area of it when large, the brightness of the set area of the more compact beam would, of course, be more intense than that of the more extended beam. That, in a nutshell, is surface brightness. On the other hand, do not forget that the same amount of light is being distributed—think of that as the apparent magnitude. The difference between apparent magnitude and surface brightness is the difference between the overall brightness of your whole computer display screen and the brightness of the average pixel on it. In the case of a large comet, the brightness of the whole coma (the apparent magnitude) needs to be high for the brightness of each small section of it (the surface brightness) to be really intense.

We have mentioned Ignatius and the *Protevangelium of James*, but what did the first generations of Christians claim about the brightness of the comet?

First, Revelation 12:2, 5 would seem to imply that the coma of the comet, in playing the role of the baby that grew in Virgo's womb and then seemed to cause her intense agony as it was "born," grew very large. It may possibly be inferred from this account that the coma became as large relative to Virgo as a newborn baby is in comparison with its mother: something like 9–12 degrees long (major axis), on October 20, 6 BC. That would be astonishing, because only a coma that had a strong apparent magnitude value would have a sufficiently great surface brightness at that size to be clearly visible to the naked eye. At apparent magnitude -9 to -11 (n=3, assuming that the comet was first seen between May and December, 7 BC), the large coma would have been easily detectable but it would not have been stunningly bright. At apparent magnitude -11 to -13 (n=4) it would have been more striking, with a surface brightness like that of Neptune (seen through a telescope).[37] Magnitude -13 to -16 (n=5) would have made for a stunning sight, with a "surface brightness" like that of Saturn.[38]

Second, we may perhaps tentatively glean indirect clues about the comet's brightness from texts that speak of Jesus in terms of the comet that heralded his birth. In particular, it is possible to detect some information about the comet's brightness from how New Testament texts draw upon Isaiah 9:2's great oracle concerning the coming of the Messiah and the great natal star: "The people who walked in darkness have seen a great light; those who dwelt in a land of deep darkness, on them has light shone." Luke 1:78–79 and Matthew 4:16 speak of the Messiah's coming in terms drawn from Isaiah's oracle, as does the Gospel of John.

Luke speaks of a rising star that "shall visit us from on high to give light to those who sit in darkness and in the shadow of death, and to guide our feet into the way of peace" (Luke 1:78b–79). The language describes a celestial

[37] It should be noted that larger celestial objects are easier for the eye to detect (Roger Nelson Clark, *Visual Astronomy of the Deep Sky* [Cambridge: Cambridge University Press, 1990], 12 fig. 2.5).

[38] In our estimations of apparent magnitude here we have not taken allowance of the brightness boost due to the forward-scattering effect. As we shall see in the following chapter, the comet would have been something like 3.5 magnitudes brighter due to this effect at this point in time. When this is taken into account, the comet's surface brightness, assuming n=4, would have been more like that of Saturn.

entity that will rise and become so bright that it will powerfully overcome the darkness, enabling travelers to walk safely by its light during the night hours. The point made by Zechariah (John the Baptist's father) and by Luke is that the Messiah's ministry will be like the great comet that marked his nativity, powerfully dispelling the darkness of night.

Matthew 4:16 claims that, through Jesus's ministry in Galilee, Isaiah's prophecy concerning a great light shining in the deep darkness came to fulfillment: "The people dwelling in darkness have seen a great light," and "a light has risen" on "those dwelling in the region and shadow of death." We remember that in Genesis 1:16–17 there are only two "great lights," so denominated because of their large size and more intense brightness. Matthew, following Isaiah, is claiming that there was a third great light. It powerfully vanquished the darkness, penetrating even into dark shady places. This great light is Jesus. As we observed earlier, Matthew's rewording of Isaiah 9:1–2 hints that what the Star at the time of Jesus's birth did literally in fulfillment of that oracle ("has risen"), Jesus did morally and spiritually in fulfillment of it.

Luke and Matthew therefore describe Jesus in terms drawn from Isaiah's oracle about the great light that would attend the Messiah's birth. These Gospel writers seem to assume that this great light had a brightness at least equal to that of the full Moon—probably greater.

The Gospel of John may also offer some indirect clues as to the brightness of the great light that announced Jesus's birth. In this Gospel Jesus is said to be "the light of men" (John 1:4b) and "the true light, which gives light to everyone" (v. 9), and he calls himself "the light of the world" (8:12). Moreover, John describes this light as shining in the darkness but not overcome by the darkness (1:5), and Jesus claims that the light is sufficiently bright to dispel darkness (8:12). No light other than an astronomical entity can shine very bright light on all of humanity in the whole world. Jesus is probably being portrayed in terms of the celestial luminary that announced his birth (cf. Num. 24:17; Isa. 9:2). The light is more like the Moon than the Sun in that it shines during the hours of darkness, but in its intense brightness it seems brighter than the Moon, the planets, and the stars.

From Luke, Matthew, and John therefore we may possibly derive indirect clues regarding the brightness of the great light that marked the occasion of Jesus's birth. What we find is consistent with the other, more explicit evidence that we have discovered regarding the comet's brightness: the comet marking Jesus's birth was astonishingly bright, and this great brightness was a literal picture of the moral and spiritual effect of the Messiah's ministry.

With regard to the brightness of the Bethlehem Star's tail, we have reason to believe that it was at the very least clearly visible around the time of its heliacal rising and specifically on October 20, 6 BC, when, together with the coma, it seems to have formed a scepter stretching across the whole sky. Also, on a night between November 23/24 and November 30/December 1, 6 BC, the comet tail stood up on the horizon and was bright enough to be seen when the Moon was present in the sky.

As a side note, it should also be borne in mind that any comet with a very low inclination (where tail curvature is essentially unobservable, because it occurs on the ecliptic plane) would have a brighter tail than more highly inclined comets, because "there is more material [i.e., dust] along the line of sight."[39] Moreover, a comet narrowly inclined to the ecliptic has a single, combined gas-dust tail, which is brighter by virtue of this union. In addition, the phenomenon of the forward-

[39] This was kindly highlighted to me by Andreas Kammerer, personal email message to the author, November 26, 2012. Andreas is a programmer of the Project Pluto Guide 9.0 software and a highly respected German amateur cometary astronomer.

scattering of light would have intensified the brightness of the comet when it was in the general area between Earth and the Sun in the period of Virgo's pregnancy and delivery.[40] Consequently, the Bethlehem Star comet, with its close perihelion pass, would have a dust tail that was strongly curved in outer space but that seemed straight and especially bright, brighter even than its apparent magnitude values would suggest. Furthermore, comets are subject to outbursts, small and large, which boost their brightness.[41]

In light of this evidence regarding the brightness of the Christ Comet, we suggest that, while the value of n could have been 3, it is more likely that it was closer to 4 (or even 5). We shall generally assume n=4 when we overview the comet's apparition in 7–6 BC in the following chapter.

THE SHAPE OF THE COMA

The Christ Comet's coma was probably elliptical (oval), with the nucleus relatively close to the top of the sunward side, in the month after perihelion. The oval shape is strongly suggested by the fact that it was judged to be playing the part of Virgo's baby during that period. Many comets, especially large, productive comets (such as Hale-Bopp) and comets that pass within the orbit of Mercury (such as the Comet of 1677 [C/1677 H1] and Comet West during part of its apparition in 1975–1976), have this shape of coma. What its form was in the preceding and succeeding period we do not know for sure. Cometary comas are capable of changing shape to a considerable extent during the course of a single apparition. For example, the Great September Comet of 1882 was round until perihelion, when it became notably oval.[42] However, in view of how large and productive the Christ Comet was, it is very possible that it remained oval-shaped throughout its time in the inner solar system.

NAMING THE COMET

The technical name of the Christ Comet, if we assume the latest possible date of first observation, between December 10 and 17, 7 BC, should perhaps be C/-6 X1 (Magi) or C/-6 Y1 (Magi). "-6" is 7 BC. Comets are assigned a prefix P, C, or D, depending on whether they have been observed at two or more passages around the Sun or judged to have a period of less than 200 years (P),[43] or not (C), or have disappeared or are considered extinct (D). In the case of long-period comets, after the prefix the year of the comet's discovery is given, followed by the part of the year when this occurred (in the form of a letter[44]) and a number indicating the order of the comet's discovery relative to others during that period ("1" for the first one discovered, "2" for the second, etc.). If the Christ Comet was spotted earlier in 7 BC than December 1, then the "X" or "Y" would be replaced by one of the

[40] Seargent (*Sungrazing Comets*) makes the point that, historically, the largest group of daylight comets consists of intrinsically bright comets with perihelion distances of 0.1–1.0 AU that have favorable forward-scattering geometry, such as Comet Skjellerup-Maristany in 1927.

[41] The Biblical data may put a couple of constraints on the coma's brightness: (1) Observers could tell that the meteor storm (Rev. 12:3–4), which we have dated to shortly before dawn on October 19, 6 BC, the eve of the celestial birth, was radiating from the tail of Virgo's southern neighbor, the serpentine Hydra. Depending on the precise location of the radiant within the tail of Hydra and on the time of the meteor storm, some or all of the coma may have been above the horizon. If any of the coma was visible at the time, it was clearly not so intensely bright that it prevented observers from seeing thousands of meteors streaking across that third of the sky. However, even if the coma was present, its large size would have greatly diluted its surface brightness, with the result that it would not have bleached out the meteors. (2) On the day that the Magi made their way from Jerusalem to Bethlehem (between November 23 and 30, 6 BC), the comet was no longer visible during the daytime ("behold" in Matt. 2:9 implies that it had not been present during the day in advance of the Magi's meeting with Herod). This absence could be explained by atmospheric conditions. However, it is also explicable with reference to the comet's surface brightness being inadequate to render it visible during the daytime. Our analysis of the comet's brightness and size at this point confirms that this was indeed the case.

[42] Gary W. Kronk, "C/1882 R1 (Great September Comet)," http://cometography.com/lcomets/1882r1.html (last modified October 3, 2006).

[43] At the time of writing, the sole exception is the comet discovered in 2002 by Kaoru Ikeya of Japan and Daqing Zhang of China, 153P/Ikeya-Zhang, which had been seen 341 years previously.

[44] January 1–15 is "A," January 16–31 is "B," February 1–15 is "C," February 16–29 is "D," etc., with December 16–31 being "Y."

FIG. 9.2 Halley's Comet (1P/Halley) (NASA/W. Liller). A photograph taken on March 8, 1986, by W. Liller, Easter Island, as part of the International Halley Watch Large Scale Phenomena Network. Image credit: W. Liller/NASA.

previous letters of the alphabet (A–W, excluding I), depending on the half-month during which it was first seen. For example, if the comet was first observed on September 30, 7 BC, its proper designation would be C/-6 S1 (Magi). If the comet was first seen in 8 BC, "-6" would be replaced with "-7," and the letter of the alphabet would be W–Y.

Over the last few centuries comets have generally been named after their discoverer(s) or the one(s) who determined their orbit. An example of the former is D/1993 F2 (Shoemaker-Levy 9), which is named after Gene and Carolyn Shoemaker and David Levy. An example of the latter is 1P/Halley, which is named after Edmond Halley (fig. 9.2). Since we do not know the name of the particular Magus who first observed the Christ Comet, we have no choice but to attribute the cometary discovery to the whole group. Alternatively, since many historically great comets are designated "Great Comet" (e.g., C/1811 F1 [Great Comet]), "Great January Comet" (C/1910 A1 [Great January Comet]), "Great Southern Comet" (C/1880 C1 [Great Southern Comet]), or the like, it may be preferable to name it "Great Christ Comet" (hence C/-6 Y1 [Great Christ Comet], assuming the latest possible date of first observation).

CONCLUSION

In this chapter we have sought to develop a profile of the comet, in particular nailing down its orbit and establishing the parameters of its brightness. In the following chapter we shall overview the cometary apparition on the basis of our orbital elements. This will help us get a vision of what the Star of Bethlehem did.

10

"Following Yonder Star"

Tracking the Comet

In this chapter we will put together the story of the apparition of the Christ Comet, tracking its progress from the point of its first appearance until it stood over the house in Bethlehem. We can do this based on our orbital elements and brightness calculations, the Biblical data, observations of comets over the last few centuries and especially recent decades, and cutting-edge cometary research. It is important to remember that a celestial object's orbital elements fully describe its orbit and enable us to plot the body's location in space and in the sky at any given point in time (planetarium software makes this easy). Our earlier work on the Christ Comet's absolute magnitude, in conjunction with the latest research on cometary brightness, enables us to figure out the comet's apparent magnitude (within a range) at any given point in its orbit.[1] Thanks to the Biblical text, extensive records of historical comets, and modern studies of the behavior of comets such as Hale-Bopp, we are in an excellent position to say something about the Christ Comet's appearance and behavior throughout its apparition.[2]

At the same time, we should bear in mind that comets are somewhat individual and are capable of springing surprises on observers, for example, when they have minor or major outburst events, when they inexplicably fade or brighten,[3] when their tails become discon-

[1] As noted in chapter 9, we shall accommodate, in the tables, apparent magnitude values based on first observation of the comet at magnitude +3.4 within the period November 21–28, 8 BC, to December 10–17, 7 BC, and according to a brightness slope (n) of 3, 4, and 5. I have used Guide 9.0 to calculate the estimated absolute and apparent magnitude values. Unfortunately, Starry Night® Pro 6.4.3 assumes that n=3, which is lower than the average for both short-period and long-period comets.

[2] Tail lengths in this chapter are derived using Project Pluto's Guide 9.0. The tail length formula used by Guide 9.0 (and Starry Night® Pro 6.4.3) is based on "the average comet." Andreas Kammerer, who developed the formula for Project Pluto, reckons that the software is accurate to within 30% in 80% of cases (http://www.projectpluto.com/update7b.htm#comet_tail [accessed March 26, 2013]). We have increased Guide's tail length estimations by approximately 12.5%, since Guide estimates that the Christ Comet's length on October 20 was about 160 degrees, but Rev. 12:5 would appear to indicate that at that stage the comet was approximately 180 degrees long. With respect to coma size, we have also based our estimations on Guide 9.0, doubling the coma diameter values to get an approximate length (major axis) of the elliptical coma. This takes account of the fact that the Christ Comet was not a small comet but rather a large one similar to Hale-Bopp. Kammerer notes that the formula he developed to calculate coma diameter for Guide comes to within 30% of the correct coma size for many comets, but does not work for large comets like Hale-Bopp, considerably underestimating their coma size (ibid.). In personal email correspondence (October 30, 2012), Kammerer estimates that the major axis (length) of Hale-Bopp was approximately double what Guide 9.0 estimates. Since the Christ Comet was large and productive, like Hale-Bopp, I have adopted this simple approach in estimations of the size of the Christ Comet's coma at different points of the apparition. This would suggest that the coma was about 11 degrees long on October 20, which, fittingly, is roughly the size of a newborn infant relative to its mother.

[3] Nick James and Gerald North, *Observing Comets* (London: Springer, 2003), 135. Allowance must be made for the possibility that a comet's brightness slope may change during the course of its apparition.

nected, and/or when they fragment. Nevertheless, it makes most sense for us to assume that the Christ Comet's apparition developed in a reasonably orderly fashion.

If the reader wishes to get a vision of how the cometary coma grew like a baby in Virgo's belly after perihelion, Starry Night® Pro software proves helpful.[4] Input the orbital elements along with a large nucleus diameter, the relevant absolute magnitude value, and the terrestrial location, and you will get a good general idea of the kind of thing that the Magi saw.[5] With respect to the growth of the comet tail in length during that period, Project Pluto's Guide 9.0 is the most useful software.

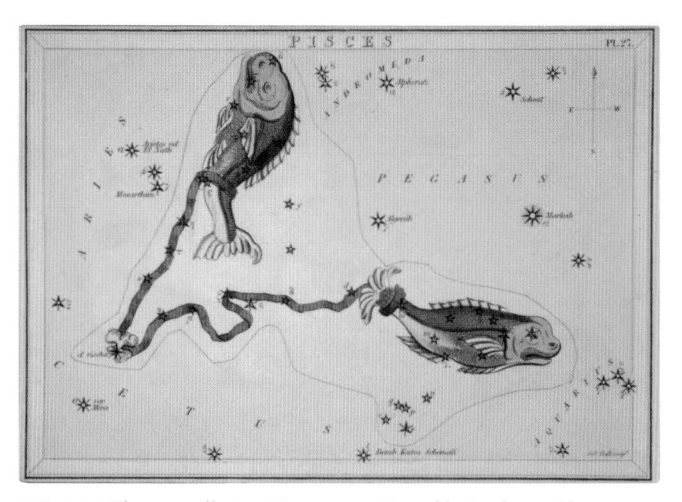

FIG. 10.1 The constellation Pisces as envisioned by Ptolemy. The western or advance fish is at the bottom right of the illustration. From Richard Rouse Bloxam, *Urania's Mirror* (London: Samuel Leigh, 1825). Image credit: oldbookart.com.

Overviewing the course of the comet during its 8/7–6 BC apparition serves a number of functions: (1) It demonstrates that a long-period, narrowly inclined, retrograde comet is capable of satisfying all the criteria to qualify as the Star of Bethlehem. (2) It authenticates Matthew's account of the Star. In view of how ignorant the ancients were concerning comets, there is, quite frankly, no way that anyone could have *invented* such a complex cometary apparition that is perfectly consistent with modern astronomical knowledge. (3) It enables us to understand why the Magi were so awed by the Star and why they responded to it by making a long pilgrimage across the wilderness in a bid to find and worship the baby Messiah. (4) It reveals why the Bethlehem Star deserves its reputation as the greatest astronomical entity in human history.

It should be noted that sightings of the comet on any particular day/night from any particular geographical location was dependent on favorable atmospheric conditions.

THE COMET'S HOME

The Christ Comet had spent the preceding centuries hidden in the region of the sky associated with the constellation Pisces ("The Tails," in Babylonian parlance[6]), more particularly under Pisces's western/advance fish (fig. 10.1). This constellation was the comet's home (from the perspective of Earth), where it long remained in darkness, far from curious human eyes. Surpassing even Hale-Bopp, the Christ Comet probably developed a coma more than 5 years before its discovery, farther away from the Sun than Uranus's orbit (i.e., 20 AU or beyond).

According to our orbit, on May 28, 8 BC,

[4] In this chapter I have used this software for determining the movements of the comet across the sky and the magnitude of stars, planets, and the Moon.

[5] However, the Virgo illustration used by Starry Night® Pro 6.4.3 does not match any ancient conceptualization of Virgo.

[6] The Babylonians called η (Eta) Piscium "the Ribbon of the Fishes" (N. A. Roughton, J. M. Steele, and C. B. F. Walker, "A Late Babylonian Normal and *Ziqpu* Star Text," *Archives of the History of the Exact Sciences* 58 [2004]: 566). They conceived of the constellation in the same way as the Greeks—namely, as two fish connected by a long v-shaped ribbon or fishing line. See Gavin White, *Babylonian Star-Lore: An Illustrated Guide to the Star-Lore and Constellations of Ancient Babylonia*, 2nd ed. (London: Solaria, 2007), 216–217.

the comet, on its way to the Sun, passed about 0.55 AU, that is, about half the distance between Earth and the Sun, from Saturn. This encounter had some effect on the comet's orbit, lengthening it.[7]

THE FIRST OBSERVATION

In our last chapter we suggested that the comet was first spotted between November 21–28, 8 BC, and December 10–17, 7 BC. Even the latest possible date of first observation would put the Christ Comet in a super-league of cometary greatness. At this point the comet was as far away from the Sun as Jupiter—it was 4.71–4.63 AU from the Sun (by comparison, Hale-Bopp was first observed when it was 4.37 AU from the Sun) and a little over 5 AU from Earth (see fig. 10.2). Back on November 21–28, 8 BC, the earliest possible date for the first observation of the comet, it was not far short of Saturn's closest distance to the Sun—between 8.34 and 8.28 AU from the Sun (and Earth).

FIG. 10.2 Hale-Bopp four days after it became visible to the naked eye (May 24, 1996). The false-color image highlights the comet's activity and structure at this early stage of its apparition. Images credit: David Hanon, Ringgold, GA.

A burst of brightness due to a cometary outburst could explain the early sighting. It is not uncommon for comets first to come to the attention of humans when they experience a sharp brightness outburst that causes them to cross the threshold of visibility.[8] However, if that was so, the fact that the Magi were able to keep track of the Star through the following months suggests that the comet was like Hale-Bopp in having copious outbursts, "puffing them out one after another like a locomotive."[9]

When a long-period comet is observed far from the inner solar system, its tail is typically not visible. Because the only observable brightness is concentrated around the nucleus, the comet can appear to be a new star and be initially classified as such. Such a first impression may have been difficult for the Magi to dispel, because the comet's movement against the backdrop of the stars was so slow and slight (due to its great distance from Earth) that it may have been detectable only after weeks of observation.[10]

In what circumstances did the Star first appear? Did it have an auspicious beginning? We recall that Herod's preoccupation with the Star's first appearance could conceivably have reflected a concern that the Messiah might have been born at that time. It is important to survey the comet's orbital course during the year between November 8 BC and December 7 BC to see where the comet would have been within the starry sky when the Magi discovered it.

On November 21–28, 8 BC, the comet would have been in the border region between the constellations of Aquarius the Water Bearer and Pisces the Fishes. Over the course of the following 6 months the comet would have slowly moved toward and under the western fish (the "Circlet") of Pisces.

According to our orbit, one striking as-

[7] Assuming our orbital elements in the aftermath of the comet's encounter with Saturn, prior to being exposed to the planet's gravitational pull, the comet's eccentricity would have been 0.9997, and its perihelion distance 0.127 AU, inclination 179.15 degrees, argument of perihelion 34.5 degrees, and longitude of ascending node 225.5 degrees. The backtracking was done by David Asher of the Armagh Observatory (personal email messages to the author, April 16, 19, and August 6, 2013).

[8] See David Seargent, *The Greatest Comets in History: Broom Stars and Celestial Scimitars* (Berlin: Springer, 2009), 21–22.

[9] Fred Schaaf, *Comet of the Century* (New York: Springer, 1997), 284.

[10] In the case of Sarabat's Comet of 1729, Father Nicolas Sarabat discovered it on August 1, 1729, but was persuaded that it was a comet rather than a nebulous star (namely, a cluster of hazy stars or a star in a haze) only after a couple of weeks of close observation with the naked eye (see Gary W. Kronk, *Cometography: A Catalog of Comets*, 6 vols. (Cambridge: Cambridge University Press, 1999–), 1:394).

pect of the comet's movement within the dome of the sky during the year when it first appeared was a conjunction with Jupiter under the belly of the western fish (equated with the stars κ and λ Piscium) in the constellation of Pisces (with respect to zodiacal sign,[11] it was crossing from Aquarius to Pisces).[12] On the western horizon shortly after sunset between January 30 and February 10, 7 BC, the comet would have been within 1 degree of the King Planet.[13] Our orbit suggests that the comet came to about half a degree from Jupiter on February 5. By February 10, the date when Jupiter made its last appearance in the western sky, the comet was about a degree from it. On the 5th, Jupiter and the comet would have been in the middle of a line of celestial entities consisting of the Sun and Venus on one side and Mercury and Saturn on the other, all within a 24-degree zone; by the 10th, the zone would have narrowed to 20 degrees.[14]

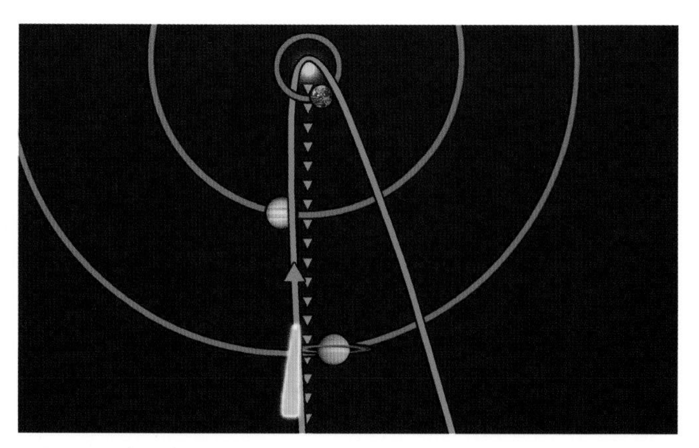

FIG. 10.3 The Christ Comet on November 21, 8 BC—it has passed Saturn on its way toward Jupiter and the Sun. (The comet icon is generic and is not intended to convey anything about the comet's size in outer space.) Image credit: Sirscha Nicholl.

Could the comet have come to visibility at this time? If it did, one could well imagine that it might have seemed to Babylonians that Jupiter, the "King," had begotten a new star. Striking and significant, such a sight would have ensured that the Magi paid close attention to the comet over the following months and years.

However, because the comet was getting very close to the Sun at twilight, it certainly would not have become visible for the first time then unless it experienced a sudden and dramatic surge in brightness, rendering it as bright as Jupiter. That is not impossible, but seems very improbable. For one thing, a first appearance 19½ months before perihelion seems implausibly early.

In the unlikely event that the Magi did observe the comet for the first time in early February of 7 BC, they would have found themselves frustrated because the comet and Jupiter would have disappeared below the western horizon very shortly thereafter. Having disappeared below the horizon, Jupiter would not have reemerged in the night sky until the middle of March. By the time of Jupiter's heliacal rising around March 17, 7 BC, it would have been more than 5 degrees from the comet, but both were still located in the constellation of Pisces, below the tail of the western fish. In terms of zodiacal sign, they were firmly in Pisces.

When Jupiter and Saturn had the first of

[11] The Babylonians divided up the zodiacal band from Aries to Pisces into 30-degree increments, each of which was given the name of one of the zodiacal constellations. The signs therefore corresponded approximately to the zodiacal constellations. The Babylonians do not seem to have been aware of precession of the equinoxes (the slight wobble of Earth on its axis). See Francesca Rochberg, *Babylonian Horoscopes* (Philadelphia: American Philosophical Society, 1998), 128, 131–133.

[12] The astronomical references in the 7/6 BC Babylonian almanac, although inexact and inconsistent, seem to assume a zodiacal starting point of about 356.5 to 357 degrees. On February 5, 7 BC, therefore, Jupiter was crossing the boundary from the sign of Aquarius into that of Pisces.

[13] The comet would have been within 3 degrees of Jupiter from January 15 until almost the end of February in 7 BC. Jupiter was 4.97 AU from the Sun on February 5, with an apparent magnitude of -2.03. The comet would have been about 7.7 AU from the Sun on February 5, 7 BC.

[14] All this time, Mars was around the location of Virgo's genitalia.

their three conjunctions in Pisces of 7 BC, coming to within 1 degree of each other on May 27–31, the comet, if visible, would have been less than 15 degrees away from each of them, still below the tail of Pisces's western fish (and within the zodiacal sign of Pisces).[15]

The comet remained below the tail of the western/advance fish of Pisces until mid-July, 7 BC, and then below its belly until the end of August.

FIG. 10.4 The constellation figure Aquarius, with his water jar. From Bloxam, *Urania's Mirror*. Image credit: oldbookart.com.

With respect to the comet as it traveled in outer space, it was hurtling toward Jupiter, the gas giant. According to our orbit, it came to about 2.07 AU from Jupiter on July 1, 7 BC.

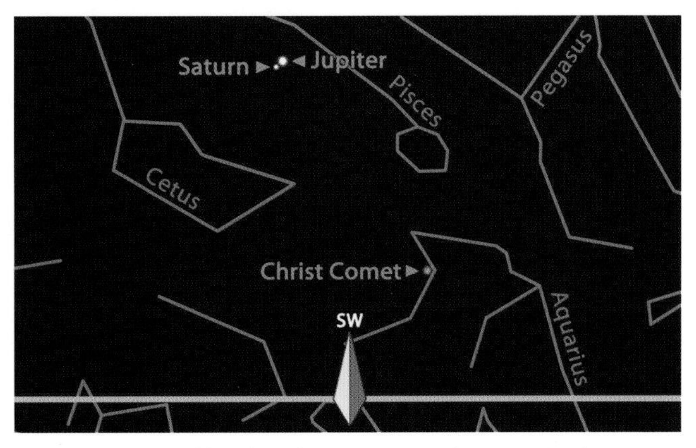

FIG. 10.5 The possible celestial scene on December 5, 7 BC, when Jupiter and Saturn were in conjunction in Pisces for the third and final time in 7 BC, if the comet was visible at the time. Image credit: Sirscha Nicholl.

In August/September, 7 BC, the comet drifted slowly out of the constellation Pisces into the neighboring Aquarius ("The Great One" in Babylon) (fig. 10.4), reaching the stream of water that flowed out of the water-bearer's water-jar[16] in mid-September, 7 BC.

The comet was present in Aquarius's water (which was, in Babylonian zodiacal reckoning, within the sign of Aquarius) for the second of the three Jupiter-Saturn conjunctions in Pisces—the two planets came to within 1 degree of each other from September 26 to October 5, 7 BC. If visible, the comet would have been just under 30 degrees from the planets at this time.

The comet remained in Aquarius's flowing water (and within the Babylonian zodiacal sign of Aquarius) for a long time. Strikingly, according to our orbit, the comet was within 1½ degrees of the star λ Aquarii all the way from October 22 through to December 20, 7 BC. It was here through the third of the Jupiter-Saturn conjunctions in Pisces, which climaxed on December 1–8 in 7 BC (fig. 10.5).

The comet was definitely visible to the Magi by December 10–17, 7 BC (fig. 10.6).

[15] If the comet appeared for the first time on May 29, 7 BC, at an apparent magnitude of +3.4, its absolute magnitude would have been -6.9 (n=3) or -9.0 (n=4), or -11.0 (if n=5).

[16] In Greek tradition the water is equated with, among other stars, κ, λ, φ, χ, ψ, and ω Aquarii—see *Ptolemy's Almagest*, trans G. J. Toomer (Princeton, NJ: Princeton University Press, 1998), 377–378.

Our overview of November 21, 8 BC, to December 17, 7 BC (see fig. 10.7), provides a few potentially auspicious celestial contexts for the comet's first appearance, all relating to the triple conjunction of Jupiter and Saturn in Pisces in 7 BC: (1) during the first conjunction in May–June, when the planets came to within 1 degree of each other (May 27–31) and the comet was within 15 degrees of them in the sign and constellation of Pisces; (2) during the second (and closest) conjunction in September–October, when the planets were within 1 degree of each other (September 26–October 5) and the comet was less than 30 degrees away, now in the sign and constellation of Aquarius, specifically in Aquarius's water; or (3) during the third and final conjunction in early December, when the planets came to within 1 degree and 4 arcminutes of each other (December 1–8) and the comet was about 30 degrees from them, still in Aquarius's water.

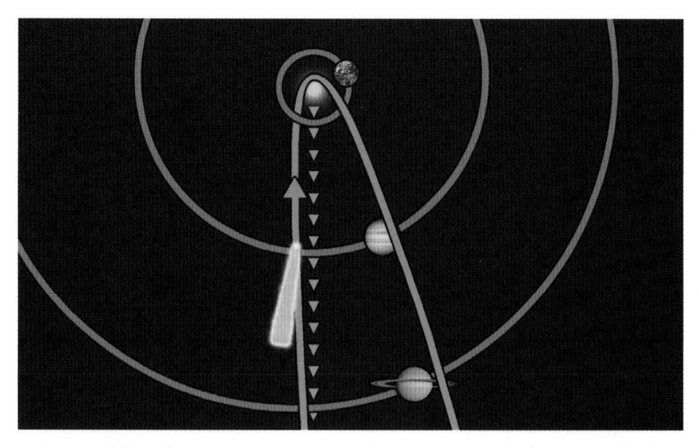

FIG. 10.6 The Christ Comet on December 17, 7 BC, the last possible date of first observation. The comet is near the orbit of Jupiter. Image credit: Sirscha Nicholl.

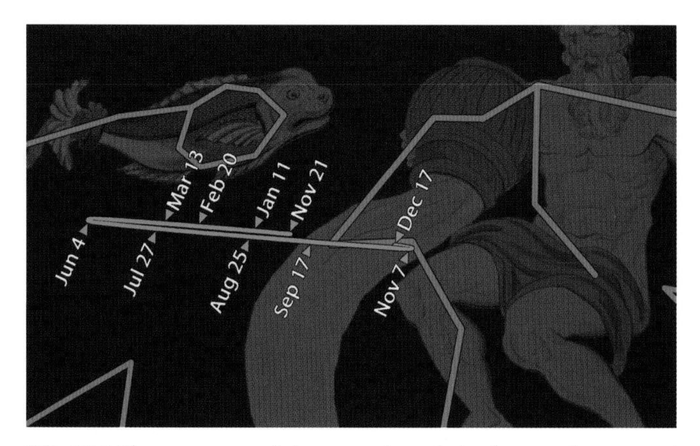

FIG. 10.7 The movements of the comet through the sky over the course of the year (from November 21, 8 BC, to December 17, 7 BC) when it was first seen. To the left is the western/advance fish of Pisces and to the right Aquarius with his water jar. Image credit: Sirscha Nicholl, using modified versions of the constellation illustrations from Bloxam, *Urania's Mirror* (images credit: oldbookart.com).

It is astonishing that John Nolland, in his commentary on Matthew's Gospel, proposed that the Star of Bethlehem was a "new star" that was first observed during the triple conjunction:

> One might speculate that a new star appearing close to a near conjunction of Jupiter and Saturn (it has been popular to identify [one of three] near conjunctions of these planets in 7/6 BC with the star of the Magi) might be taken to herald a key royal birth among the Jews, on the basis that Jupiter was the royal planet and that Saturn, as the star of Saturday, was at times associated with the Jews because of their sabbath observance.[17]

The Magi would certainly have been more liable to notice a new star in Pisces or Aquarius while they were viewing the conjunction be-

[17] John Nolland, *The Gospel of Matthew*, New International Greek Testament Commentary (Grand Rapids, MI: Eerdmans, 2005), 110–111. However, Nolland goes on to speak of "Matthew's miraculous star" (111).

tween Jupiter, the planet so closely associated with the Most High God, and Saturn. If the Star first appeared during one of the conjunctions, it might conceivably have transformed the scene into an investiture ceremony—the conjunction may have made it seem that Jupiter, representing God, was granting royal authority to Saturn, and that this authority would be exercised on behalf of Saturn by a ruler represented by the comet. Since oracles and omens marking the births of great kings tended to highlight their royal destiny, a comet appearing at such a propitious moment might have been initially perceived by the Magi to have a natal function (although the comet's dramatic heliacal rising in Virgo in September/October of 6 BC certainly convinced them that the birth occurred then).[18]

The planet Saturn was sometimes associated with Israel, and hence the conjunction could have been interpreted to mean that God would bestow sovereignty to Israel. The sudden appearance of a new "star" in that same region of the sky at this point might have seemed to symbolize the investiture of the Messiah, who would rule and reign on behalf of Israel. Whether this would have been the Magi's initial interpretation is questionable, but, as the allusions to Israel and the Messiah became clearer to them over the rest of the cometary apparition, it might well have become their preferred interpretation. A first appearance in Aquarius might (in retrospect) have been perceived to be significant in that it was thought to be ruled by Saturn, the planet of Israel, and Aquarius was sometimes considered the zodiacal sign of Israel.

If indeed there was something auspicious about the first observation of the comet and/or something susceptible to being interpreted as having natal significance, the comet most likely came to the Magi's attention in connection with one of the three Jupiter-Saturn conjunctions. Cometary precedent would favor the second conjunction, in September-October in 7 BC, or the third conjunction, in December.[19]

THE COMET'S COURSE FROM DECEMBER 17, 7 BC, TO SEPTEMBER 14, 6 BC

According to our orbit, the comet actually remained under the waters of Aquarius until the point at which it disappeared below the western horizon, in January of 6 BC.[20]

FROM PISCES TO VIRGO

The comet was absent from the night sky completely until late March or early April of 6 BC.[21] During this time Jupiter and Mars came very close to each other, when Saturn was in the vicinity, in what is called a massing or grouping of planets. This occurred in Pisces. Even though the Magi would not have been able to see the comet, they would have known that, if the comet was maintaining its pattern of behavior, it was within 30 degrees of Saturn under the western/advance fish of Pisces and would remain there until the spring. However, the astronomers may well have wondered if the comet might become visible in the dome of heaven sooner than this. After all, they would have been well aware of the fact that at some stage of their apparition many great comets suddenly start moving rapidly across the constellations and become brighter. However, this comet was highly exceptional; it was still far away from Earth and the Sun.

By the time that the comet did reemerge

[18] As we noted above, Herod's preoccupation with the Star's first appearance might reflect a concern that the Messianic baby could have been born at that time.

[19] New comets are generally discovered when the Moon is absent from the sky. Accordingly, if the comet was spotted in connection with the third and final conjunction, it would have been closer to the middle of December than its beginning.

[20] From December 31, 7 BC, to January 9, 6 BC, the Christ Comet, according to our orbit, was within 3 degrees of Mars, coming to about one-third of a degree from it on January 4.

[21] The precise moments when the comet disappeared and reappeared would have depended on its magnitude values as well as the visibility conditions.

in the eastern morning sky,[22] it had indeed returned to its home in Pisces, under its western fish.[23] During its period of absence it had moved only about 10–15 degrees from where it had been, and this movement had actually delayed its reemergence. The comet had continued to increase in brightness and size. It would have been something like 3–5 degrees long from the top of its coma to the end of its tail.[24]

By mid-April the comet would have been about 5½ degrees long.

The comet remained below Pisces's western fish and the part of the fishing line connected to it through April, May, June, and even July. Of course, during that whole time it would have been growing brighter, larger, and longer. Each night, it appeared higher in the sky and remained visible for a more extended period before dawn's light bleached it from view.

John the Baptist was born 5 to 6 months before Jesus—this would have occurred in late April or in the early or middle part of May. That is right around the time when the comet would have been becoming a prominent feature in the night sky. It was on the occasion of John the Baptist's birth that, according to Luke 1:78–79, Zechariah, John the Baptist's father, prophesied concerning the imminent coming of the Messiah in terms of a celestial body rising.[25]

By mid-May the cometary coma, its apparent length now greater than the Moon's diameter, was appearing more than 3 hours before sunrise and would likely have had a magnitude (-0.1) greater than Arcturus, the fourth brightest star in the night sky.[26] The comet may have been about 11 degrees in length. At this time the comet was "swimming" around the border region between the constellations Pisces and Cetus the Sea-Monster.

The comet's extremely slow motion through the sky for such an extended period, even as it brightened and its tail and coma grew, must have struck the Magi as very peculiar. They would not have known that this unusual behavior was because the comet had become visible much farther from Earth than other comets that appeared in their skies.

By mid-June the comet's brightness (-1.8) would have surpassed that of Sirius, the brightest star, approaching that of Jupiter. It would have been brighter than Hale-Bopp was during its entire 1996/1997 apparition.[27] The comet as a whole may have been about 18 degrees long. The coma may have been about the size of Hale-Bopp's at its largest. In length the coma may have been greater than the Moon's diameter. It was now appearing almost 5 hours before dawn. The comet was in the midst of an apparent sharp celestial U-turn in Pisces. This marked the beginning of its long journey across the sky to the Sun, although this would not become obvious until July.

By mid-July the comet would have been spectacular by any standard. With a mag-

[22] This was a heliacal rising. Because the comet was visible for so long and moving against the backdrop of the stars, it had at least two heliacal risings in the eastern sky. Babylonian astronomers would have made records of the comet on each of these occasions. However, only the heliacal rising around the time of perihelion was particularly striking and profoundly meaningful.

[23] According to our orbit, the comet was within 3 degrees of Mercury in the constellation Pisces from March 12 to 15, 6 BC, coming to within 1¾ degrees of it on March 14.

[24] On March 28–April 3, 6 BC, it came to within 3 degrees of Venus, climaxing at about ½ degree from it on March 31. If it had become visible again by this point, the comet would have made for a beautiful celestial partner for the morning star.

[25] Much-later Mandean tradition connects John's nativity with the appearance of "a star that came and stood over Jerusalem" (*Mandaean Book of John*, chapter 18). See James F. McGrath's translation, http://rogueleaf.com/book-of-john/2011/06/04/18-portents-of-the-birth-of-john-the-baptist (posted June 4, 2011). In this connection, it is interesting to note that the prologue of John's Gospel appears to imply that some people identified John as "the light" (see John 1:4–9). In verses 7–8 the author emphatically denies that John the Baptist himself was "the light" and three times declares that John came as a witness to "the light." These verses come between verse 5's reference to the light shining in the darkness and verse 9's reference to "the true light, which gives light to everyone, . . . coming into the world." It is conceivable that one of the factors behind the apparent notion that John the Baptist was "the light" was the coming of the Star to the attention of the public around the time of his birth. This may have caused some to conclude that John was the fulfillment of the oracles of Balaam in Num. 24:17 and of Isaiah in Isa. 9:2.

[26] If we assume n=4. If n=5, it would have been at least as great as the second brightest star, Canopus (-0.7).

[27] If we assume n=4.

nitude like that of Venus and a large coma (perhaps over 2 degrees long), the comet, now almost as long as the Big Dipper, must have been very eye-catching. It was now becoming visible seven hours before sunrise. Since it was appearing each night within a few hours of sunset, it inevitably would have become a talking point within the general population all across the northern hemisphere. In outer space, the comet was about as far from Earth as Earth is from the Sun, and was fast approaching the orbit of Mars.

The comet's procession to the Sun was truly majestic. No comet in recorded history ever put on a display like this. From the western fish of Pisces, the comet passed through the constellations Aquarius (July 26–August 8), Capricornus (August 8–13), and Sagittarius (August 14–21), past the right foot of Ophiuchus (August 21–24), and traversed Scorpius (August 25–September 2) and Libra (September 3–21) before ending up at the star λ Virginis (see fig. 10.8). This was the location of Virgo's left foot, according to the Greeks, but of both of her feet in the conceptualization of Revelation 12:1. The Sun was a couple of degrees from Virgo's brightest star, Spica, at the time.

During its procession, the comet's motion through the heavens was reminiscent of a bobsleigh from the point that it is approaching its greatest speed until it is almost stationary. All the while, the comet was growing in apparent magnitude. By the end of the journey it had the magnitude of the full Moon! During its procession the comet's coma appeared large, larger than any coma in the last few hundred years. As for the tail, although it briefly decreased in apparent length at the end of July and the start of August, thereafter it grew very long very quickly. By mid-September the whole comet would have stretched halfway across the sky. We shall now take a closer look at a few key stages of the comet's majestic procession.

WATER FROM AQUARIUS'S WATER JAR

We consider first the early stages of the comet's journey across the heavens. Ever brightening and growing on its way, the comet revisited the water flowing from Aquarius's water jug. Indeed, our orbit would suggest that, from July 25 to August 8, the appearance of the comet in the sky was radically reoriented, its direction swinging like a pendulum from left to right. It was reminiscent of a shuttlecock or birdie that is hit on the cork side and turns mid-air to fly cork-first. Initially the head was down to the left and the tail pointing up to the right, but by the end of the maneuver the head was down to the right and the tail up to the left.[28] As it reoriented itself in this way, the tail was very short. It was shortest on August 4–5, when the comet was close to upright (relative to the constellation figure). As for the coma, it was probably very large at this point (its longest side perhaps something like 5–6 degrees in diameter).

What was happening in outer space was that Earth was crossing the "interchange," or "crossroads," between its orbit and that of the comet (see fig. 10.9). As a result, from the perspective of Earth the comet's tail seemed to shorten, widen, and brighten dramatically as it shifted from one side to the other. The tail's intensified brightness was due to the fact that more tail debris was concentrated into a small area of the sky in Earth-dwellers' line of sight. Moreover, the geometry of the comet-Earth-Sun at this time would have boosted the brightness by something like 1 magnitude. Essentially, the comet was almost behind Earth from the perspective of the Sun (the phase angle of the nucleus peaked at under 3 degrees on August 4–5!) and the backscattering effect meant that the dust in the coma and tail became slightly brighter. The great size of the coma was because the comet was very close to Earth. Occurring in the context of Aquarius the Water-Bearer (in Babylonian

[28] On July 31 a 16-day, waning-gibbous Moon would have risen over the western horizon just 10 minutes after the last part of the comet (the head) had risen.

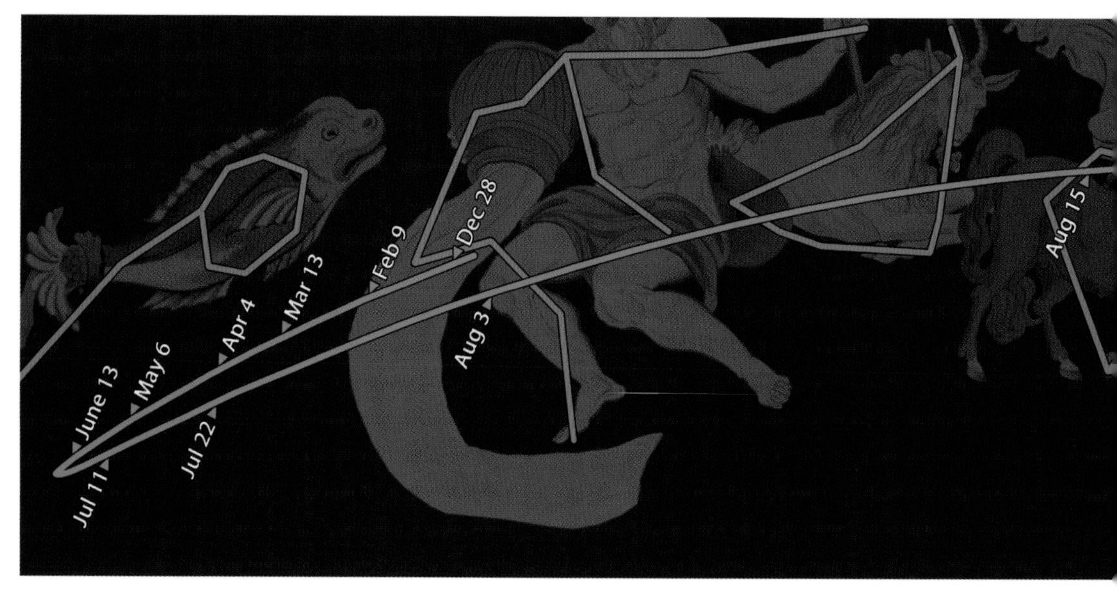

FIG. 10.8 The course of the Christ Comet across the sky from December 17, 7 BC, to September 29, 6 BC. Image credit: Sirscha Nicholl, using modified versions of the constellation illustrations from Bloxam, *Urania's Mirror* (images credit: oldbookart.com).

thought, the Great One), the comet's large, bright, and probably oval coma may possibly have appeared to be a water-jar and the short tail to be a stream of water issuing from it, on August 2 to 7. Over those nights, the angle of the coma and tail changed, making it look like the water jar was being swung by the constellation figure from his right to his left, rather like a pendulum.[29] The constellation figure may well have seemed to observers to have come to life.

On the nights of August 2/3 to 7/8, the comet was between 0.485 AU and 0.384 AU (approximately the same distance as Mercury is from the Sun) from Earth and heading straight toward the crossroads and the Sun, so that during this time humans would have been able to look directly at the sunward and active side of the large, oncoming comet. Had they been able to see what was happening in the coma, they would have caught a glimpse of magnificent fountains of dust jetting out from the nucleus (cf. figs. 10.10–11).[30]

The comet's apparent magnitude would have been at least -7.1[31] at the height of the display on August 5, 6 BC. Of course, the fact that so much dust was now in Earth's line of sight—and the backscattering effect—would have meant that the comet's brightness was greater than the apparent magnitude value would suggest. On the other hand, the large size of the coma would have diluted the intensity of the brightness.

Jews might have been prompted to reflect on Isaiah 12:3—"With joy you will draw water from the wells of salvation" (cf. Isa. 32:1–2; 33:17–21; 35:1, 6; 44:2–3; 51:3)—and Numbers 24:7, where Balaam prophesied concerning Israel that "Water shall flow [or overflow] from his buckets, and his seed shall be in many waters." In addition, the Feast of Taber-

[29] Babylonian conceptualizations of the Great One could envision a water-jar next to each of the constellation figure's feet and/or over his belly (see White, *Babylonian Star-Lore*, 122 fig. 71, and 123 figs. 72 and 73).
[30] For pictures of jets of dust erupting from the Sun side of the comet, see Carl Sagan and Ann Druyan, *Comet* (New York: Pocket Books, 1986), 174–182; and Jürgen Rahe, Bertram Donn, and Karl Wurm, *Atlas of Cometary Forms: Structures Near the Nucleus* (Washington, DC: NASA, 1969), passim.
[31] Assuming n=4. This does not take into account the effect of backscattering.

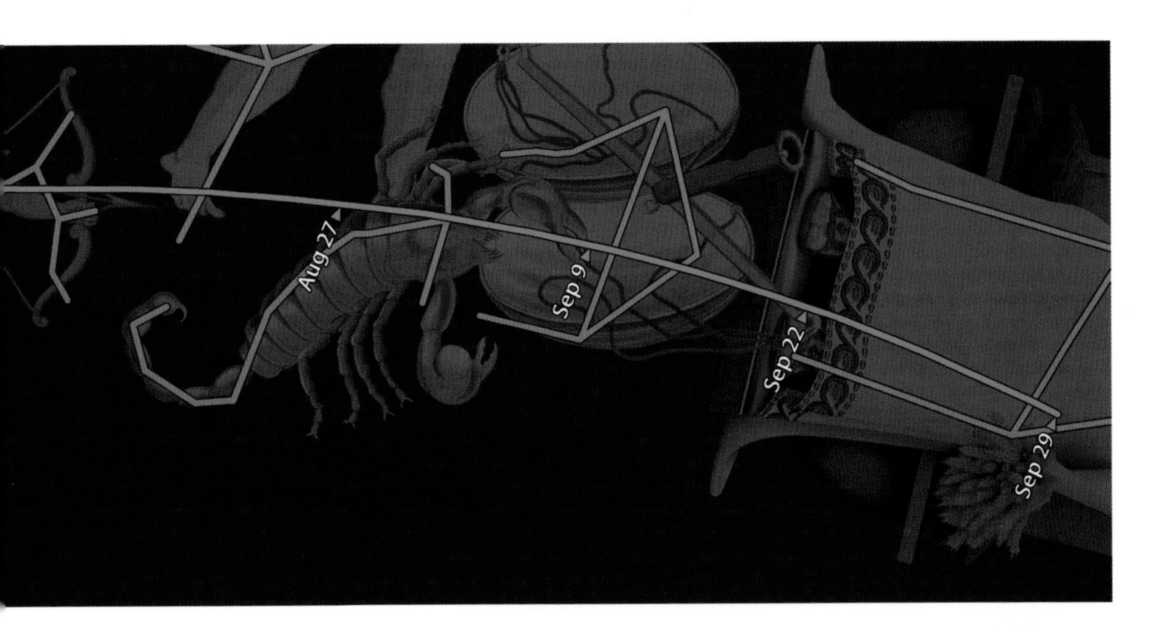

nacles in the fall featured a rite in which water from the Pool of Siloam was poured out at the base of the altar of burnt offering in the Jerusalem temple.

In Babylonian tradition, the waters of the Great One could be associated with judgment and danger in the form of severe rainstorms and floods, but they were more commonly associated with the coming of the rains and hence with fertility, prosperity, refreshment, purification, and cleansing.[32] Occurring in the summer, this celestial phenomenon would more naturally have been interpreted as having a positive meaning.

From early August onward, the comet appeared around sunset and remained visible until it set in the southwest. That meant that everyone could see the comet in all of its glory. It had become a third great light in the heavens.

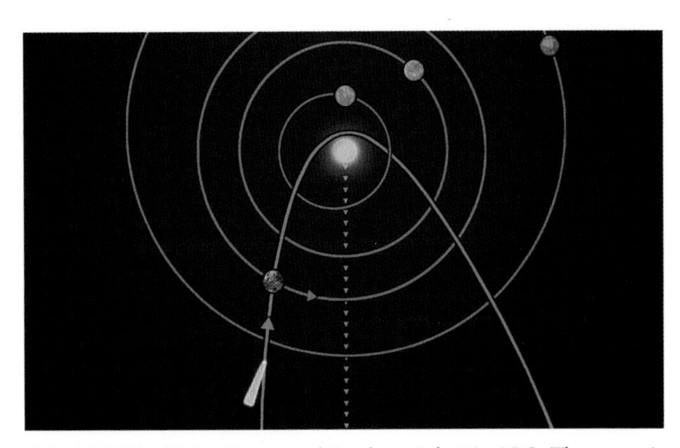

FIG. 10.9 The Christ Comet and Earth on July 30, 6 BC. The comet is just beyond the orbit of Mars as Earth crosses the comet's orbital line. Image credit: Sirscha Nicholl.

THE ARCHER'S ARROW

As the comet raced through Capricornus, its apparent magnitude would have become increasingly impressive, since the comet was not only getting closer to the Sun but was also coming closer to Earth. At the same time, this was counterbalanced by the fact that the brightness that the comet did have was being spread over a wider surface area.

By the time the comet entered Sagittarius

[32] On the Babylonian constellation "The Great One," see White, *Babylonian Star-Lore*, 121–123.

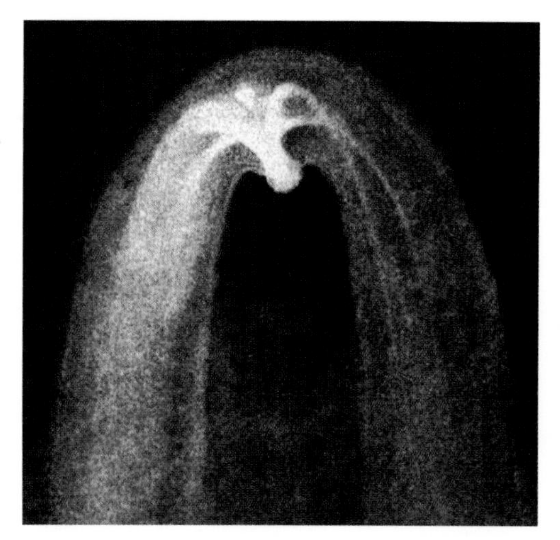

FIG. 10.10 A drawing of the jets of Comet Daniel's coma on August 5, 1907. Image credit: Max Wolf, Heidelberg Observatory/J. F. Julius Schmidt, in J. Rahe, B. Donn, and K. Wurm, *Atlas of Cometary Forms* (Washington, DC: NASA, 1969), 40.

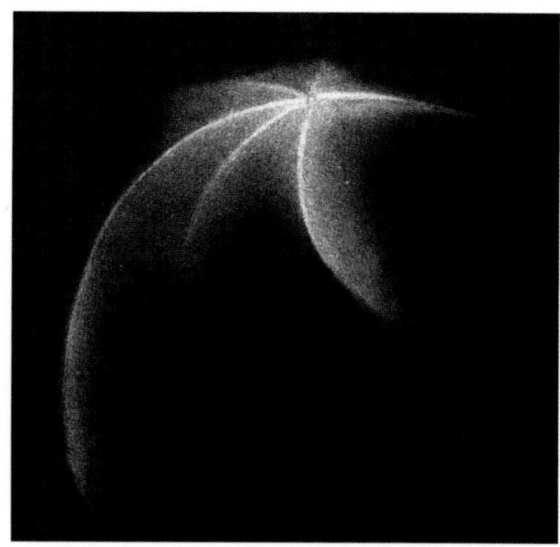

FIG. 10.11 A drawing of Comet Tebbutt's jets. Image credit: J. F. Julius Schmidt, in *Atlas of Cometary Forms*, 14.

the Archer, on August 14, 6 BC (according to our orbit), its apparent magnitude would have been at least -8.2,[33] and it was only 0.334 AU from Earth (1.26 AU from the Sun), the closest it would come to Earth before making its dramatic U-turn around the Sun (fig. 10.12). That, incidentally, is closer than most of the historically great comets have ever come to our planet! The comet may have been about 40 degrees long, the coma itself something like 7 degrees in diameter (major axis). Earth now found itself at a perfect vantage point from which to watch the comet as it hurtled toward perihelion at breakneck speed.

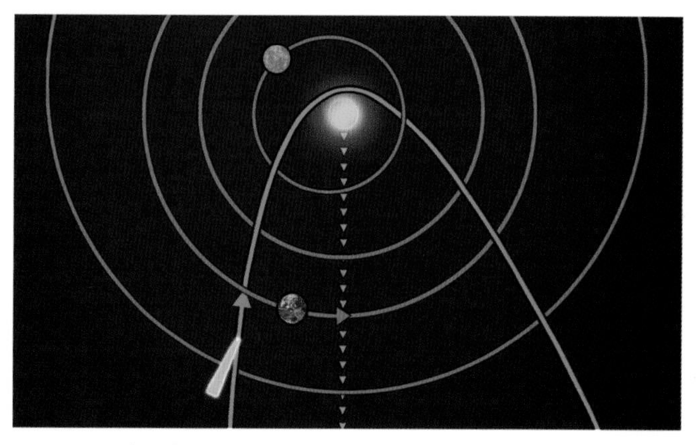

FIG. 10.12 The Christ Comet and Earth on August 14, 6 BC. The comet is between the orbit of Mars and that of Earth. Image credit: Sirscha Nicholl.

Over the next month or so the comet would have become an increasingly bright light shining in the deep darkness. Even as the coma shrank in apparent size as the comet moved away from Earth, its apparent magnitude was steadily becoming greater as it neared the Sun—which meant more brightness concentrated in a smaller area. In addition, more and more of the comet's tail would have been becoming visible to human observers each night. This was because the comet was continuing to race toward the interchange and was degassing more as it neared the Sun, while Earth, after crossing the underpass, was assuming a much better angle from which to see the whole comet. In fact, from the per-

[33] Assuming n=4.

FIG. 10.13 The Archer (Sagittarius/Pabilsag) fires his great cometary arrow and it strikes the Scorpion's heart and forehead in August 6 BC. Image credit: Sirscha Nicholl, using modified constellation images from Bloxam, *Urania's Mirror* (images credit: oldbookart.com).

spective of Earth, you would almost have thought that the tail was simply stretching as the comet raced across the constellations, because the end of the tail remained firmly fixed in Aquarius, back at the pivot point of the comet's radical pendulum-swing reorientation. When the comet head departed from Aquarius, the whole comet would have been just over 14 degrees in length, but by the time the head reached the far side of Sagittarius, it would have been over 70 degrees long.

The comet must have seemed to observers to have transformed itself into a dramatic celestial arrow! The coma was the arrowhead and the tail was the shaft and fletching. It had become a flaming arrow fired from the Archer's bow! (See fig. 10.13.)

To the Babylonians and Greeks, the Archer Pabilsag, or Sagittarius, was a winged half-horse and half-human creature known for his great bow and the arrow that is famously aimed right at the heart of the Scorpion (the star Antares). To the Babylonians, at least, Scorpius encapsulated the forces of wickedness, darkness, and death.[34] Over the nights of August 17–20, it looked like the Archer was sliding the awesome arrow into his bow and then firing it.[35] Even by the time the comet set on the 20th, the arrow could already be seen "in flight" toward Scorpius, such was the rapidity of the comet's apparent movement across the starry heavens at that time. The cometary arrow's movement would have been detectable in single observing sessions over the following 1½ weeks.

On August 25 the comet crossed Earth's orbit. At that point Earth was about half the distance from the comet that both Earth and the comet were from the Sun. The comet, both its coma and its tail, would have been brilliant and large. Even as the giant arrow flew through the heavens, it would have grown in length. Sure enough, the cometary arrow, having grown to well over 80 degrees long, and having an apparent magnitude of at least -8.4,[36] struck the Scorpion's heart (it

[34] See Robert Brown, *Researches into the Origin of the Primitive Constellations of the Greeks, Phoenicians, and Babylonians*, 2 vols. (Oxford: Williams & Norgate, 1899), 1:67–76.

[35] According to our orbit, on August 17, at sunset, the nucleus was less than a degree from σ Sagitarii (the Archer's left shoulder). August 19 brought the nucleus midway between δ Sagittarii (Kaus Meridianalis) and λ Sagittarii, on the bow. It was where the Greeks and Babylonians imagined the Archer's arrow.

[36] Assuming n=4.

came very close to Antares on August 27–28). The cometary arrow continued on its celestial flight path until it hit the middle of the Scorpion's forehead (δ Scorpii) on August 31.[37] (See fig. 10.13.) By that time the comet would have extended more than halfway across the sky.

This was a truly astonishing drama, extending from one constellation to another. It was the stuff of myth! The Magi, along with everyone else who knew their constellations, must have been glued to the heavens each night as they watched the unfolding nightly drama.

While the cometary arrow flew through the sky from bow to forehead, the coma would have steadily shrunk in length from something like 7 degrees to about 4 degrees. Together with the building apparent magnitude as the comet approached the Sun, this shrinking of the coma would have intensified the brightness.

The celestial display had already been remarkable. Having watched what it did, the Magi may have concluded that the Star was signaling the demise of the forces of Chaos and Darkness in the cosmos. These forces were destined to be decisively and overwhelmingly conquered. Order would once again prevail in the cosmos. It is doubtful at this stage, however, whether the Magi associated the Star with the Jewish Messiah. Only later would they conclude that this was all one awesome celestial show to put the forthcoming nativity in cosmological and eschatological perspective. The Magi must have wondered what the Star would do next. Little did they know that all the things that had happened up to this point were just the prelude to the marvels about to occur.

THE WONDER ON SEPTEMBER 15, 6 BC

On September 15, 6 BC, the comet was fast approaching perihelion—it was just 0.47 AU from the Sun and 0.93 AU from Earth (for comparison, Hale-Bopp never came closer to Earth than 1.32 AU). The increasing distance from Earth was because Earth's course had taken it away from the comet even as the comet had crossed the interchange, heading toward the Sun (fig. 10.14).

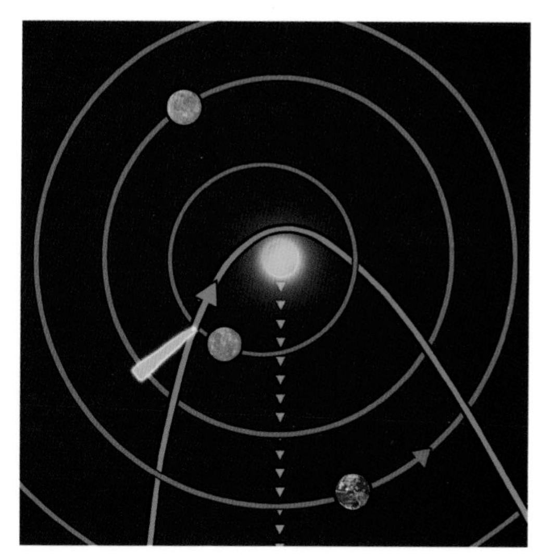

FIG. 10.14 September 15, 6 BC—the comet is making its way past Mercury on its way to perihelion. Because Mercury is inclined to the ecliptic at an approximately 7-degree angle, the comet passes over Mercury's orbital path before perihelion (even though the comet is below the ecliptic at that point) and under its orbital path after perihelion (even though the comet is then above the ecliptic). Image credit: Sirscha Nicholl.

On September 15 the Sun, making its way through Virgo, was located over her womb, while the Moon was under Virgo's feet. The occasion is memorialized in Revelation 12:1.

The comet's range of apparent magnitude values at that time is impressive (table 10.1).

If the comet was first discovered from May 29 to December 17 in 7 BC at an apparent magnitude of +3.4 and had an average slope value (i.e., n=4), then its magnitude on September 15 would have been between -10.3 and -12.5. If n=5, then the comet would have been between -12.8 and -15.3 on September 15. These numbers underscore the comet's brilliance. Even at the minimum of

[37] In Babylonian thought, β and δ Scorpii were two of three stars associated with the Scorpion's head (Roughton et al., "Star Text," 569)—evidently the third star was π Scorpii. The Greeks imagined the Scorpion's forehead in the same location.

Magnitude Slope (value of n)	Apparent Magnitude on September 15, 6 BC, if first observed on November 21–28, 8 BC	Apparent Magnitude on September 15, 6 BC, if first observed on February 5, 7 BC	Apparent Magnitude on September 15, 6 BC, if first observed on May 29, 7 BC	Apparent Magnitude on September 15, 6 BC, if first observed on August 17, 7 BC	Apparent Magnitude on September 15, 6 BC, if first observed on Septem-ber 30, 7 BC	Apparent Magnitude on September 15, 6 BC, if first observed on December 10–17, 7 BC
3	-10.7	-10.5	-9.5	-8.4	-8.1	-7.8
4	-13.9	-13.7	-12.5	-11.3	-10.9	-10.3
5	-17.0	-16.7	-15.3	-14.0	-13.5	-12.8

TABLE 10.1 The Christ Comet's apparent magnitude on September 15, 6 BC.

these magnitude values, a comet would normally have been visible during the daytime.[38] However, because the Bethlehem Star comet was large, the issue is not quite so simple. The coma could have been something like 2 degrees long. Accordingly, only a small part of the coma, namely the region of condensed brightness around the nucleus, would have been detectable during the daytime.

In normal circumstances that evening, observers in both Babylon and Judea, when they scanned the western sky under λ Virginis (= Virgo's feet, in Rev. 12:1's conception of the constellation figure; Virgo's left foot, in the imagination of most Greeks) in the wake of sunset, would have seen only the slim crescent of the new Moon. However, this evening, right beside the Moon, they would have seen the comet's coma, with its massive tail sweeping up to the left at a roughly 40-degree angle (relative to the ground). It is possible that the edge of the coma was slightly backlighting part of the edge (limb) of the Moon.

If these observations of the young crescent Moon were the first sightings of the new Moon in Babylon and/or Jerusalem, the comet's presence may have been perceived to be enduing the occasion with special importance.

However, even if this was the first sighting of the new Moon in Babylon, the Babylonian method of determining the start of new months by advance calculation would probably have resulted in the new month of Ululu starting on the evening of September 14, so that Ululu 1 was the evening-to-evening day September 14/15.[39] Accordingly, what happened after sunset on September 15 would have taken place at the start of Ululu 2.

The Judeans, however, had an observational luni-solar calendar. As a result, if the new crescent Moon went unobserved on September 14, this sighting of the crescent Moon after sunset on September 15 would have coincided with the formal beginning of the new month, Tishri. In the Jewish calendar, Tishri was no ordinary month.[40] It was the first month of the civil year. It was the month of months. And the first day of Tishri was no ordinary day—it was the Feast of Trumpets, Rosh Hashanah. It was a holy, solemn, and joyful holiday marked by the

[38] Gary W. Kronk: "The comet would have been seen in daylight from about 9 a. m. through the rest of the day" (personal email message to the author, September 26, 2012).

[39] The Babylonian almanacs, predictions of astronomical events in advance, stated when each new month would begin (see, for example, the translation of the 7/6 BC almanac in A. J. Sachs and C. B. F. Walker, "Kepler's View of the Star of Bethlehem and the Babylonian Almanac for 7/6 B.C.," *Iraq* 46 [1984]: 47–49).

[40] The Judeans' employment of an observational calendar meant that the determination of when a new month began partly depended on the sighting of the new crescent Moon. Months were generally either 29 or 30 days. If there was a new crescent Moon on the 30th day and it was observed, the 30th day became the first day of the new month. If there was no sighting of the new crescent Moon after sunset on that day, whether because the Moon was too close to the Sun to be detectable or because atmospheric conditions meant that it was unobservable, the new month began on the following day. With respect to Rev. 12:1, the Moon's location under the feet of Virgo indicates that September 15 in 6 BC is in view. If the new crescent Moon was not spotted on the evening of September 14, the new month, Tishri, would have been destined to begin the following evening regardless of whether the Moon was actually observed.

FIG. 10.15 A section of the frieze from the Arch of Titus in Rome commemorating the Roman conquest of Jerusalem. Note the trumpets to the right. Image credit: Yonidebest and Steerpike, Wikimedia Commons.

blowing of trumpets. Every Jewish month was announced by the blowing of trumpets, but Tishri 1 was set apart in that it was a day of trumpet blasts, presumably in that trumpets were to be blown all day long (Lev. 23:24; Num. 29:1–2a). The first sighting of the young crescent Moon in September was immediately followed by the blowing of the trumpets announcing the Feast of Trumpets and the beginning of the civil New Year. (See fig. 10.15.)

The comet at this time would have been around 90 degrees long, stretching across half the sky. Having played the part of Sagittarius's arrow that slayed the Scorpion at the end of August, the gloriously bright comet may have seemed to assume a new role to mark the beginning of the new Jewish civil year and the Feast of Trumpets. The long-tailed comet may well have looked to observers like an awesome celestial trumpet, with the coma being the mouthpiece (figs. 10.16–20).[41] During the whole time from sunset to moonset, the Moon may have appeared to observers to be playing the part of the trumpeter blowing into the trumpet's mouthpiece, proclaiming Rosh Hashanah and the start of the civil New Year. It may have appeared to be drawing the world's attention to the exaltation of Virgo—enthroned, wearing her crown, enrobed with the Sun, and with the Moon as her footstool. This might go some way toward explaining why the scene of Virgo's exaltation on September 15, 6 BC, is highlighted in the narrative of Revelation 12:1–5.[42] So spectacular would the celestial sight have been that some might have felt constrained to conclude that the marvel was announcing not just the start of a special month or the inauguration of a new civil year, but also the beginning

[41] For some examples of cometary trumpets, see figs. 5.27–28, 31; 10.16–20; also Roberta J. M. Olson, *Fire and Ice: A History of Comets in Art* (New York: Walker, 1985), 45 fig. 42; 53 fig. 49; 60 fig. 55; 69 fig. 62; 90 fig. 81. Byzantine texts report that the Comet of 467 was thought to look like a trumpet—according to the *Chronicon Paschale* (AD 628) and the *Chronographia* of Theophanes the Confessor (AD 813)—see Kronk, *Cometography*, 1:84.

[42] Rev. 12:1's description enables us to identify the mother as Virgo and see how she is being imagined against the backdrop of the stars, to appreciate that the drama of vv. 1–5 is one that concerns her exaltation, and to nail down the time of the astronomical events.

of a whole new era in world history.[43]

In view of the large Jewish population in Babylon, the Magi could have become aware of the coincidence of the celestial phenomenon and the Jewish calendar date. This may even have played a role in prompting them to appreciate that the cometary apparition had a particularly Jewish significance.

FIG. 10.16 A wood engraving of the trumpet-like Great Comet of 1843 on March 17, as observed from Blackheath, Kent. At the time, this comet was in a position relative to the Sun and Earth—0.7 AU from the Sun and 1 AU from Earth—similar to that of the Christ Comet on September 15, 6 BC—0.47 AU from the Sun and 0.93 AU from Earth. At the peak of the 1843 comet's performance, on approximately March 9, 1843, it was 0.5 AU from the Sun and 0.86 AU from Earth. From George F. Chambers, *The Story of the Comets* (Oxford: Clarendon, 1909), fig. 48 plate 14 (opposite page 132).

PERIHELION

Since the comet was now coming close to the Sun and hence was more exposed to the solar wind and radiation pressure, its coma shrank in size.[44] The visual effect of this in the sky was exaggerated because the comet was getting farther away from Earth. At the same time, the comet's brightness was building—not only was the apparent magnitude becoming progressively more impressive, but the smaller surface area of the coma was intensifying the brightness. In addition, as the comet began its slingshot around the Sun, the tail shortened as more and more of it moved behind the Sun, from Earth's perspective. This also intensified the brightness of the tail. All of this meant that the comet would have been an increasingly magnificent "evening star" in the days following September 15, although the fact that it was getting closer and closer to the Sun meant that it became increasingly difficult for humans to see the coma.

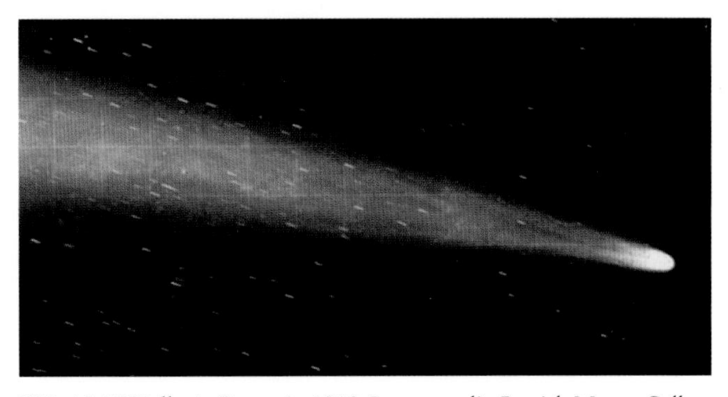

FIG. 10.17 Halley's Comet in 1910. Image credit: Patrick Moore Collection, www.patrickmoorecollection.com.

On September 23, as the comet was 0.2 AU from the Sun and beginning its hairpin turn around the solar disk, it found itself between Mercury and Venus. Venus, the comet, Mercury, and Earth were essentially in a straight line in that order. At that time the comet's 50+-degree tail would have been a striking feature in the evening sky, its last

[43] It is worth remembering that the comet's appearance (with respect to both brightness and form) would have been essentially the same on the evening of September 14. Therefore, whether the Feast of Trumpets began on the 14th or 15th of September, a cometary trumpet could have been present to mark the occasion.

[44] See George F. Chambers, *The Story of the Comets* (London: Clarendon, 1909), 8; Richard Schmude, *Comets and How to Observe Them* (New York: Springer, 2010), 41.

part only setting about 3 hours after sunset.

Between September 25 and 27, the nucleus was extremely close to the Sun from the perspective of Earth—within 4 degrees. Moreover, because the comet was essentially behind the Sun, the apparent length of the tail would have been relatively short. On the 25th, the end of it would have appeared for a brief time after sunset. However, no part of the comet would have been visible on the 26th.

FIG. 10.18 A photographic drawing of Halley's Comet on May 15, 1910. Image credit: Patrick Moore Collection, www.patrickmoorecollection.com.

FIG. 10.19 The Great Comet of 1680 over Nuremberg Observatory, Germany. From Simon Bornmeister, *Cometen Betrachtung* (Ulm: Scheurer, 1681). Image credit: Royal Astronomical Society.

According to our orbit, it was on the 26th, just before sunset in Babylon, that, from Earth's perspective, the comet shifted from the west side of the Sun to its east side. Because of this, when it reappeared, it would be a feature of the eastern morning sky. It was around this time that the comet crossed its ascending node, that is, passed from south of the ecliptic plane, on which Earth orbits the Sun, to north of it.[45]

The comet had its perihelion on September 27.[46] As with all comets, its speed on its orbit was greatest at perihelion, when it was midway through its astonishing U-turn around the Sun (fig. 10.21). Because the comet was coming so close to the Sun, just 0.119 AU away from it, the comet's activity may have increased dramatically from this point on.

The end of the comet's approximately 14-degree-long, intensely bright tail briefly reemerged over the eastern horizon shortly before sunrise on that day. The comet as a whole (not the coma) had started to rise heliacally; the scepter was beginning to emerge over the eastern horizon. The nucleus was at this point just 2 degrees from the Sun and so the coma and the sunward side of the tail were probably not detectable.

The comet was about 40 degrees long in the run-up to dawn on September 28. The comet's apparent magnitude was at least -15.5! (Remember that the full Moon's magnitude

[45] According to our orbit, about 2 hours after sunset in Babylon.
[46] According to our orbit, 1½ hours before dawn in Jerusalem and 53 minutes before dawn in Babylon.

is -12.6.) The end of the very bright tail, which stretched back as far as Mars in the hind part of Leo, would have begun rising 3½ hours before dawn. It would have been a striking sight. Part of the gloriously brilliant coma might theoretically have been glimpsed in advance of the rising Sun, although the comet nucleus was still very close to the Sun—about 5½ degrees from it.

Based on Revelation 12 and Matthew 2, however, we suggest that the coma was probably not spotted in Babylon or Judea at this point. This, of course, may be readily explained by the coma's proximity to the Sun and/or disadvantageous atmospheric conditions, whether because of cloud cover[47] or sand and dust in the atmosphere.[48]

Nevertheless, such was the intense brightness of the area around the nucleus that those under clear skies who shielded their eyes from the Sun might have been able to see it at some stage through the day.

FIG. 10.20 The Great Comet of 1618 as portrayed on the cover of Jacob Cats, *Aenmerckinghe op de tegenwoordige steert-sterre* (Middelburg, The Netherlands: n.p., 1619). Image credit: Google Books. Image enhanced and colorized by Sirscha Nicholl.

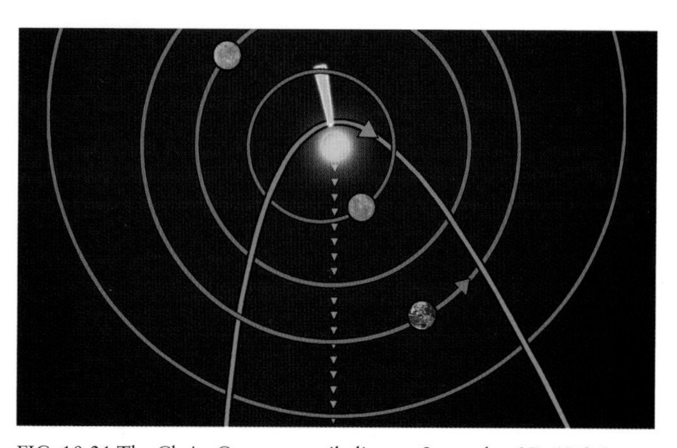

FIG. 10.21 The Christ Comet at perihelion on September 27, 6 BC. Image credit: Sirscha Nicholl.

VIRGO'S BABY

What occurred from September 29/30 to October 22 in 6 BC was surely one of the most amazing astronomical phenomena in history: each morning, in the time leading up to sunrise, as Virgo's womb rose over the eastern horizon, the cometary coma played the part of Virgo's baby in a magnificent celestial birth scene to mark the nativity of the Messiah.

THE BABY IN VIRGO'S WOMB

In the immediate run-up to dawn on September 29, when the nucleus was 0.15 AU from the Sun in outer space and just over 8 degrees from the Sun in the visible sky, and the comet had an apparent magnitude of at least -14.8 (see table 10.2),[49] the coma, or at least part of it, could have been spotted by observers in

[47] "The Average Weather in September for Iraq," http://weatherspark.com/averages/31352/9/Iraq-Babil (accessed May 13, 2014).
[48] "Al Hillah: Full Year Climatology," http://www.nrlmry.navy.mil/CLIMATOLOGY/pages/arabian_sea_region/iraq/pdf/alhillah_iraq.pdf (last modified April 24, 2013).
[49] Assuming n=4.

Magnitude Slope (value of n)	Apparent Magnitude on September 29, 6 BC, if first observed on November 21–28, 8 BC	Apparent Magnitude on September 29, 6 BC, if first observed on February 5, 7 BC	Apparent Magnitude on September 29, 6 BC, if first observed on May 29, 7 BC	Apparent Magnitude on September 29, 6 BC, if first observed on August 17, 7 BC	Apparent Magnitude on September 29, 6 BC, if first observed on September 30, 7 BC	Apparent Magnitude on September 29, 6 BC, if first observed on December 10–17, 7 BC
3	-14.1	-13.9	-12.9	-11.8	-11.5	-11.2
4	-18.4	-18.2	-17.0	-15.8	-15.4	-14.8
5	-22.8	-22.5	-21.1	-19.8	-19.3	-18.6

TABLE 10.2 The Christ Comet's apparent magnitude on September 29, 6 BC.

clear skies. The coma was very low in Virgo's womb and edging higher. The comet, with its long tail sticking up near-vertically, would have been about 65 degrees in length, making for a truly grand spectacle. It looked like a magnificent scepter (see Num. 24:17: "a scepter shall rise").

To those able to see it through the daytime, the condensed brightness of the part of the coma around the nucleus may well have been a stunning sight. Not only was the comet's apparent magnitude surpassing that of the full Moon, but its surface brightness would probably have been greater than that of the full Moon. The coma in length was possibly a little less than the Moon's diameter. Any glimpse that the Magi might have got of a very bright occupant in Virgo's womb during the day would certainly have grabbed their attention and whetted their appetite for what was to follow. But observers would have struggled to get a good look at it, because it was still so close to the Sun.

FIG. 10.22 The location of the Christ Comet's coma (the small white oval) in Virgo's womb on September 30, 6 BC. Image credit: Sirscha Nicholl.

When the cometary coma rose over the eastern horizon on September 30, some observers like the Magi were evidently astonished that it looked like a glorious and perfect baby within the womb of Virgo (fig. 10.22).[50] The comet had moved farther from the Sun, with the result that the coma was appearing in a darker sky—it was now 10 degrees in altitude at sunrise. The oval coma at its longest would have been about the Moon's diameter.

The baby would have seemed to be head-down. The cometary coma had transformed Virgo, making it look like she was well into a pregnancy. It would have been natural to reflect on recent celestial history for a narra-

[50] Because ancient conceptualizations of Virgo in the Greco-Roman and Near Eastern worlds were similar, observers in each of these places were capable of regarding the cometary coma as playing the part of a baby in Virgo's womb. It should also be remembered that the coma, as it played the role of a growing fetus in Virgo, was strongly shaping the precise way observers imagined the constellation figure relative to the fixed stars over the course of the celestial drama.

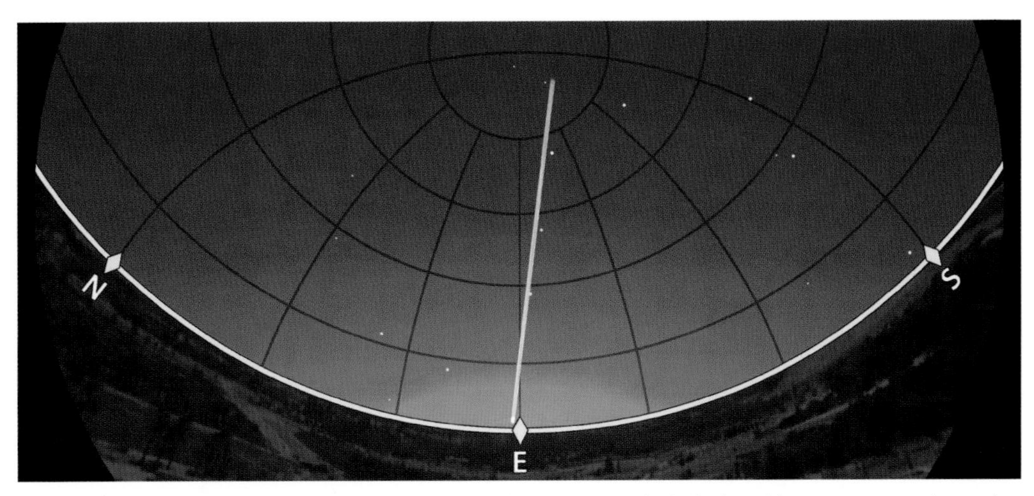

FIG. 10.23 The Christ Comet on September 30, 6 BC. The comet as a whole, looking like a scepter or rod, stretches up almost to the zenith (as shown by the altitude-azimuth grid in the background). Image credit: Sirscha Nicholl.

tive explanation of the conception; the recent clothing of Virgo by the Sun may have seemed to provide the obvious cause.[51]

The coma was probably regarded as having heliacally risen (i.e., it was first spotted by the Magi on the eastern horizon, rising in advance of the rising Sun) on September 29 or 30 (see Matt. 2:2 and Rev. 12:1–2). In view of the intensity of the brightness and the large size of the comet's coma and tail (in historical terms), and the shape and location of the coma, what a heliacal rising it was! And, to top it all off, at that moment, for the first time the entire cometary scepter was visible standing up over the eastern horizon (fig. 10.23).

It is astonishing that the heliacal rising of the cometary coma in Virgo's womb, the first view of the entire (heliacally risen) cometary scepter, and the visible peak of the whole comet's brightness all occurred in conjunction with the start of the great Hebrew Feast of Tabernacles. The Feast of Tabernacles was arguably the most joyful and important annual festival in the Jewish calendar and was celebrated on Tishri 15–22 (counting the Eighth Day as part of it). In 6 BC, if the new Moon was spotted on September 14, Tishri 15 coincided with September 29, but if the first sighting of the new Moon occurred the following evening (September 15), Tishri 15 was September 30.

The Feast of Tabernacles was an occasion of thanksgiving for the harvest and of prayer for the autumn rains to fertilize the land; a time to remember the exodus and wilderness wandering; and a week to look forward to the coming of salvation and the conversion of the nations in connection with the coming of the Messiah. In addition to making special offerings and hymns, the priests made a procession around the altar of burnt offering once daily for six days and then seven times on the seventh day. As they processed, they called out for Yahweh to bring salvation (Ps. 118:25).

During the festival, priests would draw water from the Pool of Siloam and then pour this water into a container at the base of the altar. It was a symbolic gesture, calling out for Yahweh to send the waters to enliven the land. Isaiah 12:3 may allude to this rite, implying that it had a spiritual significance, looking forward to the water that would at the time of the end be drawn "from the wells of salvation." According to John 7:37–38, Jesus claimed to be the giver of true spiritual water during the Feast of Tabernacles.

[51] For a very helpful and interesting overview of ancient Babylonian and Jewish ideas about conception, pregnancy, and delivery, see Marten Stol, *Birth in Babylonia and the Bible: Its Mediterranean Setting* (Groningen: Styx, 2000), 1–26, 109–145.

More important for our purposes, during this festival, indeed at the close of the first day of it, four golden candelabras were lit in the Court of the Women, giving light to all of Jerusalem. The rationale for this was presumably eschatological—Isaiah had prophesied that "the people who walked in darkness" would see "a great light" (Isa. 9:2), referring to the great cometary light that would announce the birth of the true Immanuel, the Messiah (7:14).[52] That the cometary coma rose heliacally, appearing at its stunning brightest, at this very time is astonishing. Of course, John records that, during the Feast of Tabernacles, Jesus, obviously alluding to the lighting ritual and indeed Isaiah 9:2, made the claim, "I am the light of the world. Whoever follows me will not walk in darkness, but will have the light of life" (John 8:12). In view of what had happened during the Feast of Tabernacles in 6 BC, Jesus's words would have been especially forceful.[53]

The Magi may have discovered the correspondence between the Star's heliacal rising and the Feast of Tabernacles from the Babylonian Jewish community.

As magnificent as the scene of the heliacally risen comet obviously was, the marvel was, in truth, only starting. For, as large as the coma (facing the full blast of the solar wind and the intensity of the Sun's radiation pressure) and the tail were at that point, they were actually small compared to what they would become.

Each day, the comet was moving farther away from the Sun both in outer space and in the sky, with the result that the coma was seen earlier and against a darker sky.

Every night, as more and more of Virgo rose over the horizon before dawn, the comet scepter appeared larger and longer. The comet tail's steady increase in length meant that it was appearing over the eastern horizon earlier and earlier in the night. On the evening of the 29th, the comet was about 78 degrees long and began rising about 5 hours after sunset. The following morning it was already 84 degrees long. By the evening of the 30th, the comet was about 94 degrees long and began to rise less than 4 hours after sunset. When predawn observers around that time followed the tail upwards to its end with their eyes, they were basically looking up straight above their heads. With the majestic scepter extended over them, the Magi may well have felt that the king represented by this comet was destined to exercise sovereign authority even over their own land (cf. Ps. 110:1–2).

As for the comet's coma, because it was located in Virgo's womb, its growth made it seem that the pregnancy was developing normally.

The growth in the size of the comet was because it was at its most productive and was emerging from its hairpin turn around the Sun, exposing more of its coma and tail to Earth-bound observers. It was also because the comet was steadily moving toward Earth, amplifying the apparent size of the coma and tail. Moreover, the coma was increasingly being liberated from the constricting effect of the Sun, causing it to become larger.

It is interesting to look at the range of possible apparent magnitude values for September 30 and October 3, 6 BC (tables 10.3–4).

From September 30 to October 3, the brightness of the comet would have been extraordinary.[54] If we assume an average brightness slope (n=4), the coma would have been at least -12.5 in apparent magnitude before

[52] At the same time, recall Isa. 42:6; 49:6; and 60:1–3.

[53] It is also interesting that the palm branch, which was widely associated with Virgo (she held one in her right hand), was a prominent feature of the Feast of Tabernacles. Worshipers were to carry palm branches in their right hands. Strikingly, during the Triumphal Entry the people used palm branches to welcome Jesus into the city, even as they blessed the Messiah as their Savior (John 12:12–13; Matt. 21:8–9).

[54] According to our orbit, the nucleus was located midway between α (Spica) and θ Virginis on September 30–October 1, 6 BC, after which it moved up closer to θ Virginis, being within 2 degrees of it from October 2 to 12.

Magnitude Slope (value of n)	Apparent Magnitude on September 30, 6 BC, if first observed on November 21–28, 8 BC	Apparent Magnitude on September 30, 6 BC, if first observed on February 5, 7 BC	Apparent Magnitude on September 30, 6 BC, if first observed on May 29, 7 BC	Apparent Magnitude on September 30, 6 BC, if first observed on August 17, 7 BC	Apparent Magnitude on September 30, 6 BC, if first observed on Septem-ber 30, 7 BC	Apparent Magnitude on September 30, 6 BC, if first observed on December 10–17, 7 BC
3	-13.6	-13.4	-12.4	-11.3	-11.0	-10.7
4	-17.7	-17.5	-16.3	-15.1	-14.7	-14.1
5	-21.8	-21.5	-20.1	-18.8	-18.3	-17.6

TABLE 10.3 The Christ Comet's apparent magnitude on September 30, 6 BC.

Magnitude Slope (value of n)	Apparent Magnitude on October 3, 6 BC, if first observed on November 21–28, 8 BC	Apparent Magnitude on October 3, 6 BC, if first observed on February 5, 7 BC	Apparent Magnitude on October 3, 6 BC, if first observed on May 29, 7 BC	Apparent Magnitude on October 3, 6 BC, if first observed on August 17, 7 BC	Apparent Magnitude on October 3, 6 BC, if first observed on September 30, 7 BC	Apparent Magnitude on October 3, 6 BC, if first observed on December 10–17, 7 BC
3	-12.5	-12.3	-11.3	-10.2	-9.9	-9.6
4	-16.1	-15.9	-14.7	-13.5	-13.1	-12.5
5	-19.7	-19.4	-18.0	-16.7	-16.2	-15.5

TABLE 10.4 The Christ Comet's apparent magnitude on October 3, 6 BC.

dawn on October 3.[55] In length it would have grown from about the Moon's diameter to over 2½ times that. The concentrated brightness around the nucleus would probably have been especially stunning, and indeed visible during the daytime. It may have given the impression that the face of the celestial Virgin's baby was especially glorious. This baby in Virgo's womb was truly magnificent! The comet had quickly gone from being a glorious evening star to being an astonishingly "bright morning star" (Rev. 22:16; cf. 2 Pet. 1:19) and indeed an afternoon star! Over these days, the comet was increasingly moving away from the Sun—for example, from September 30 to October 3 it moved more than 5 degrees further away from the solar disk. The Christ Comet, like the 373–372 BC comet[56] and Comet Tebbutt of 1861, would unquestionably have cast shadows.

It must soon have become apparent to the Magi that the cometary coma in Virgo's belly was playing the part of the Jewish Messiah and was signaling the fulfillment of the oracles of Isaiah 7:14 and 9:1–7. In addition, since the comet as a whole apparently looked like a scepter at this stage of its apparition, the conclusion that Numbers 24:17 was being fulfilled would quickly have become obvious. From the moment that the Magi came to the conclusion that the comet was at this time announcing the Messiah's coming, they would have anticipated that the celestial pregnancy would culminate in a birth scene that would coincide with the Messiah's nativity. Moreover, they would have been hungry to learn all they could concerning the Messiah as prophesied in the Hebrew Scriptures. In addition, around this time the Magi may have begun planning their trip to Judea to worship the Messiah.

[55] Gary W. Kronk comments: "On the 30th, the comet would have been visible in morning twilight shortly before sunrise and would have been visible throughout the day, setting about an hour before the sun. There would have been little change by the 3rd" (personal email message to the author, September 26, 2012).

[56] On the 373–372 BC comet, see Ephorus as cited by Diodorus Siculus 15.50.2–3.

GROWING BABY

According to our orbit, the baby (i.e., coma) seemed to move slightly upwards within Virgo's womb until October 6/7,[57] from which point it began to descend slowly (see fig. 10.24).[58] However, right up until October 14 the comet remained inside a set area within Virgo's belly.[59] The stability of the cometary baby within Virgo's womb was possible only because of a remarkable synchronism between the movement of Earth on its orbit and that of the comet on its orbit. Each day the comet could be seen in a darker sky, because more of the comet and Virgo were rising above the horizon in advance of the Sun. From October 5, the whole baby could be seen in a completely dark sky (i.e., when the Sun was 18 or more degrees below the horizon).

As the coma grew in size each day (from about 1 degree in length on October 2 to almost 6 degrees on October 15), the brightness would have been distributed over a larger area, diluting it. Of course, the darkening of the background sky counterbalanced this effect. More important, the unusual geometry of the comet relative to the Sun and Earth was intensifying the apparent magnitude beyond what calculations based on the absolute magnitude would suggest. Every day the comet came closer and closer to the Earth-Sun line (i.e., its "phase angle" increased), boosting the brightness of the comet due to the phenomenon of "forward-scattering." On September 30 the "phase angle" of the nucleus was 84 degrees; on October 11 it was 133 degrees, which gave the comet a 1-magnitude boost.[60] Nevertheless, the increase in the comet's size may well have caused it to fade from daytime visibility more quickly

FIG. 10.24 The Christ Comet nucleus's slow progress through Virgo from the perspective of Earth from September 30 to October 20, 6 BC. The slight rise coincided with the growth of the coma, so that the coma looked to be relatively stable within the womb. The drop (relative to the rising fixed stars of Virgo) constituted the baby's descent and birth. Image credit: Sirscha Nicholl.

than its apparent magnitude values and phase angle would suggest.

Until October 11/12 the comet's apparent brightness was dropping (the drop was in the order of about 2 or 3 magnitudes from September 30 to October 11/12). Thereafter the apparent magnitude began to build again as the comet neared Earth.[61]

As the cometary coma rose on October 11, it may have been backlighting part of the edge of the 27.5-day-old waning crescent Moon (magnitude -9.41).

[57] October 6 was either Tishri 21, the last of the seven main days of the Feast of Tabernacles, or Tishri 22, the eighth day.

[58] Since the nucleus is probably to be located at the focal point of the ellipsis (or parabola) (Andreas Kammerer, personal email message to the author, November 5, 2012), approximately 10% of the coma would have been on the sunward side.

[59] We recall that Babylonian astronomers were in the practice of making entries into their astronomical records not just when comets heliacally rose but also when they ceased moving relative to the fixed stars (i.e., became stationary; F. Richard Stephenson, "The Ancient History of Halley's Comet," in *Standing on the Shoulders of Giants*, ed. Norman Thrower [Berkeley: University of California Press, 1990], 244). Therefore what was happening in Virgo's womb during this period would have been duly noted.

[60] Joseph N. Marcus, "Forward-Scattering Enhancement of Comet Brightness. I. Background and Model," *International Comet Quarterly* 29 (2007): 57 fig. 15.

[61] Assuming n=4.

The remarkable growth of the coma-baby in Virgo's womb is reminiscent of the dramatic growth of Comet Holmes as a result of its outburst late in 2007 (fig. 10.25).[62] However, while a major outburst event cannot be ruled out in the case of the Christ Comet, there is probably no need to make recourse to such to explain the spectacular enlargement of the coma. The fact that the massive, productive comet had just made its closest pass by the Sun and was rapidly approaching Earth is probably sufficient to explain the stunning expansion.

FIG. 10.25 Comet Holmes on November 2, 2007. Image credit: Ginger Mayfield/Wikimedia Commons.

The whole time the coma was growing, so also was the tail. And, of course, the longer it grew, the earlier it rose in the east each night. On October 2 the cometary scepter was something like 110 degrees long. On October 14 it was over 155 degrees in length. By that time the tail was starting to rise around sunset and so was visible for the whole night. When the coma rose toward the end of the night, the massive comet must have been an awesome sight to behold. To Near Eastern astrologers probably attuned to perceive significance in the direction to which the tail pointed, it must have seemed remarkable that this extraordinarily long and ever-lengthening tail was pointing right to the western horizon.

THE BABY DROPS

From October 7 to 14 the cometary baby dropped slightly within Virgo's womb, even as it grew. This downward movement was due to the fact that the comet's orbit was straightening out after its hairpin turn around the Sun at perihelion and the fact that the comet was increasingly heading toward Earth. The more the orbit straightened out, the faster the comet seemed to descend within Virgo. The momentum of the descent appeared to build with every passing day, but in the first few days there was relatively little movement. Indeed, while the coma was descending within the context of the rising fixed stars of Virgo, the nucleus itself was technically still rising in altitude (i.e., increasing in its apparent distance from the Sun) up to October 14, when it reached its highest point in the predawn eastern sky (it was about 23½ degrees above the horizon at sunrise). By October 14 the comet coma's sunward side had descended (relative to Virgo's stars) to the level at which it had been on September 29.

Any intensification of the apparent magnitude of the comet as it came within about 0.5 AU of Earth (not to mention the steadily increasing brightness boost due to the forward-scattering effect as the phase angle grew) would have been offset by the continued growth of the coma, which caused the brightness of the comet to be more broadly distributed and hence less stunning. By October 11 the coma was probably something like 4 degrees long and at least -10.8 in apparent magnitude.[63] At this magnitude such a coma would have stood out in sharp contrast from the ancient night sky.[64] The baby's face may have seemed especially bright.

[62] Strikingly, Mark Bailey, Director of the Armagh Observatory, commented regarding Comet Holmes's outburst, "Astronomy is always full of surprises. These events make you think. Could the Star of Bethlehem have been a comet displaying a brightness outburst like this?" (http://star.arm.ac.uk/press/2007/cometholmes [last modified October 10, 2012]).

[63] Assuming n=4 and not taking into account forward-scattering. If the comet was spotted earlier than December of 7 BC, or if n=5, then the magnitude and the intensity of the brightness would, of course, have been greater.

[64] The surface brightness of the coma at this stage would therefore have been at least +8.8 magnitudes per square arcsecond (if n=4; +6.6 if n=5) (without taking into account the 1-magnitude brightness boost due to forward-scattering—Marcus, "Background

Magnitude Slope (value of n)	Apparent Magnitude on October 15, 6 BC, if first observed on November 21–28, 8 BC	Apparent Magnitude on October 15, 6 BC, if first observed on February 5, 7 BC	Apparent Magnitude on October 15, 6 BC, if first observed on May 29, 7 BC	Apparent Magnitude on October 15, 6 BC, if first observed on August 17, 7 BC	Apparent Magnitude on October 15, 6 BC, if first observed on September 30, 7 BC	Apparent Magnitude on October 15, 6 BC, if first observed on December 10–17, 7 BC
3	-11.5	-11.3	-10.3	-9.2	-8.9	-8.6
4	-14.3	-14.1	-12.9	-11.7	-11.3	-10.7
5	-17.1	-16.8	-15.4	-14.1	-13.6	-12.9

TABLE 10.5 The Christ Comet's apparent magnitude on October 15, 6 BC.

VIRGO'S LABOR BEGINS

While we cannot know for sure exactly where the nucleus was within the coma, it seems safe to conclude that about 10% of the coma was on the sunward side of the nucleus. This is important because it determines when the coma would have been perceived to be pushing down on Virgo's cervix/pelvic floor. Accordingly, October 15 emerges as the most likely date for the beginning of Virgo's "active labor." At that point the coma's major axis (i.e., the length of the baby) was something like 6 degrees. Since there was general agreement throughout the Greco-Roman world and the Near East (Egypt and Babylon) at the turn of the ages regarding the level of Virgo's groin relative to the stars of the constellation, the interpretation of the nativity scene transpiring in Virgo as described by Revelation 12:1–5 would have been widely shared.

At that time the comet's apparent magnitude was probably somewhere between -10.7 and -14.3 (table 10.5).[65] A 6-degree-long elliptical (oval) coma of this apparent magnitude would have been a striking feature of the night sky in the ancient Near East.[66] With such a large cometary baby at that location within Virgo, those watching the celestial drama in the eastern sky naturally assigned great pain to Virgo.[67] The slow descent of the cometary baby over the next few days, coupled with its still-increasing size, would have reinforced the perception that the labor was painful. October 17 and 18 were probably perceived to be the climax of Virgo's pain, since it was at that point that the widest part of the baby was passing through her vaginal opening.

Remarkably, the throne of Virgo had suddenly been transformed into a birthing chair! (See fig. 10.26.)

This extraordinary size of the coma was due to the largeness of the comet nucleus, the productive nature of the comet, its intrinsic brightness, the small perihelion distance, and the fact that the nucleus on October 15–17 was only about 0.4–0.3 AU from Earth (roughly the distance of Mercury from the Sun). All of these factors had to come together perfectly for the extraordinary phenomenon in Virgo to occur as it did.

and Model," 57 fig. 15). Especially given its size, the coma would have been a very striking sight. All estimates of "surface brightness" are based on the "Surface Brightness Calculator" by David Benn, http://www.users.on.net/~dbenn/ECMAScript/surface_brightness.html (last modified June 16, 2005). The width (minor axis) is reckoned to be 40% of the length (major axis) and the shape of the coma is assumed to be elliptical.

[65] These values do not take into account forward-scattering—the comet nucleus now had a phase angle of 140 degrees, which would have given the sunward side of the coma a brightness boost of about 2 magnitudes beyond what our apparent magnitude values based on absolute magnitude and slope values would suggest (Marcus, "Background and Model," 57 fig. 15).

[66] Even at n=3, a 6-degree coma would have had an apparent magnitude of -8.6 to -11.5 and a surface brightness of approximately +11.8 to +8.9 magnitudes per square arcsecond. More likely, n was 4 or 5. Furthermore, the forward-scattering effect and the fact that the comet was narrowly inclined to the ecliptic were boosting its brightness. The overall and surface brightness of the coma may have been greater (i.e., lower in negative numbers) by 5 magnitudes or more.

[67] It should perhaps also be observed that Jer. 4:31 associates the bringing forth of a first child with greater anguish (as noted in Stol, *Birth in Babylonia and the Bible*, 139). This may be relevant because Virgo was imagined to be a virgin and hence this was her first delivery.

THE METEOR STORM

From October 15 to 19 Virgo was in active labor, giving birth to the child. Relative to the stars of Virgo, the coma was clearly descending. From the 16th to the 19th, more and more of the cometary baby was rising above the horizon after 80 Virginis each morning. However, until the whole baby was completely below the level of her vaginal opening, the baby had not been born. The birth occurred on October 20. During these days the baby was growing dramatically; each morning it was 12–17% larger than the previous morning. The morning of October 15, when labor began, corresponded with Tishratu 1 in Babylon (the month had begun the previous evening). The first day of the Hebrew month of Heshvan probably also coincided with October 15 (although it is possible that it was delayed until the 16th).

FIG. 10.26 Virgo begins active labor on October 15, 6 BC. Image credit: Sirscha Nicholl.

On October 19 there was a dramatic new development in the grand heavenly play: radiating from the upper section of the tail of the serpentine constellation figure Hydra (the longest constellation in the sky), next to Virgo's left leg, was a fantastic meteor storm.[68] A full one-third of the stars of the sky seemed to be ejected from their places (Rev. 12:4). Because the radiant of the meteor storm was relatively low on the eastern horizon, the stars looked like they were being thrown toward the earth. Then the star in the tip of Hydra's tail (π Hydrae) rose to the point that it was level with the horizon. It was an unforgettable moment—there was the serpentine dragon, having just displayed his rage and power, looking like he was standing up on the earth.

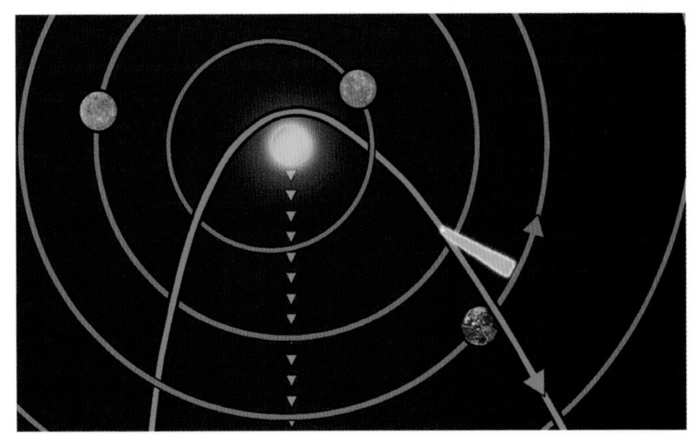

FIG. 10.27 October 15, 6 BC: What was happening in outer space at the point when Virgo commenced her labor. Image credit: Sirscha Nicholl.

The fact that this great 6 BC Hydrid meteor storm occurred when Virgo's cometary baby had almost fully emerged from her womb seemed to endow the events in that part of the sky with special significance. The association between the two was hard to avoid because much of Hydra (in particular the lower part of its tail) was located next

[68] For more on this meteor storm, see appendix 2.

Magnitude Slope (value of n)	Apparent Magnitude on October 19, 6 BC, if first observed on November 21–28, 8 BC	Apparent Magnitude on October 19, 6 BC, if first observed on February 5, 7 BC	Apparent Magnitude on October 19, 6 BC, if first observed on May 29, 7 BC	Apparent Magnitude on October 19, 6 BC, if first observed on August 17, 7 BC	Apparent Magnitude on October 19, 6 BC, if first observed on September 30, 7 BC	Apparent Magnitude on October 19, 6 BC, if first observed on December 10–17, 7 BC
3	-12.1	-11.9	-10.9	-9.8	-9.5	-9.2
4	-14.6	-14.4	-13.2	-12.0	-11.6	-11.0
5	-17.2	-16.9	-15.5	-14.2	-13.7	-13.0

TABLE 10.6 The Christ Comet's apparent magnitude on October 19, 6 BC.

to Virgo, to her south. The apparent throwing of stars from one-third of the sky to the earth by Hydra's tail seemed to add a new dimension to the grand narrative centered around the cometary coma. In this context, what Hydra was doing seemed ominous—the Chaos Monster was evidently furious and extremely dangerous. He was about to pounce on the unsuspecting baby and try to seize and eat it as soon as it had fully emerged from its mother's womb. It was a disturbing scene on the eve of the messianic child's birth.

The comet's apparent magnitude at the time was at least -9.2 (if n=3), -11.0 (if n=4), or -13.0 (if n=5) (see table 10.6).[69] At this stage the comet was very close to the Earth-Sun line (the phase angle of the nucleus was 155 degrees) and this would have boosted the brightness of the sunward side of the coma by 3 magnitudes beyond what our figures suggest.[70] Unquestionably, the cometary coma would have been clearly visible in the sky of ancient Babylon. Nevertheless, the massive size of the coma (perhaps approaching 10 degrees in length) meant that the brightness was not quite as striking as the resultant apparent magnitude and forward-scattering brightness boost calculations might suggest. If the radiant of the meteor storm on the eve of the birth of the cometary baby was lower on Hydra's tail—that is, closer to γ (Gamma) Hydrae than Corvus/the Raven—that would suggest that the cometary light at that stage was not sufficiently bright to bleach out the glory of the meteors in that third of the sky. Of course, if the radiant was high on Hydra's tail, that is, close to Corvus/the Raven, for example, at HIP59373, then the cometary coma might have been below the horizon when the radiant was above it.

At this point the comet's tail was extraordinarily long, stretching across most of the dome of the sky.[71] The previous evening, the 18th, the tail was already well above the eastern horizon when the Sun set. The cometary scepter must have been an awesome sight to behold.

The Magi and others who were paying attention to celestial events must have felt the tension of the scene on October 19. The momentum of the coma's descent was such that it must have seemed obvious that the baby would be born the next day (although, as seasoned astronomers, they would have been well aware that comets do not always do what is expected!). Now that Hydra was ready to pounce, what would happen when the baby emerged from Virgo's womb? The skies did not get any more gripping than this!

When the Sun set and the tail could be seen, it would have been apparent to the Magi that the tail would stretch all the way across the sky when the coma rose at the end of the night.

[69] Not taking account of the forward-scattering effect.
[70] Marcus, "Background and Model," 57 fig. 15.
[71] At the time the cometary coma was right beside Mercury (according to our orbit, the nucleus was about ¾ degree from it).

Magnitude Slope (value of n)	Apparent Magnitude on October 20, 6 BC, if first observed on November 21–28, 8 BC	Apparent Magnitude on October 20, 6 BC, if first observed on February 5, 7 BC	Apparent Magnitude on October 20, 6 BC, if first observed on May 29, 7 BC	Apparent Magnitude on October 20, 6 BC, if first observed on August 17, 7 BC	Apparent Magnitude on October 20, 6 BC, if first observed on September 30, 7 BC	Apparent Magnitude on October 20, 6 BC, if first observed on December 10–17, 7 BC
3	-12.3	-12.1	-11.1	-10.0	-9.7	-9.4
4	-14.8	-14.6	-13.4	-12.2	-11.8	-11.2
5	-17.4	-17.1	-15.7	-14.2	-13.9	-13.2

TABLE 10.7 The Christ Comet's apparent magnitude on October 20, 6 BC.

THE BIRTH SCENE

When Virgo rose on the morning of October 20, it was clear that the momentous occasion had come: the cometary baby had finally been fully delivered. The comet had jumped over 4 degrees in one day, and now the side of the coma nearest the horizon was located just over 2 degrees from λ Virginis, the star that the Greeks associated with Virgo's left foot but that Revelation 12:1 seems to associate with both of her feet. In the Babylonian and Hebrew luni-solar calendars, with sunset-to-sunset days, the date was Tashritu 6 and Heshvan 6 (or 5) respectively.

The coma was now possibly about 11 degrees long,[72] occupying the entire region from Virgo's feet to a few degrees below 80 Virginis. The whole coma was now well below Virgo's vaginal opening. Indeed the baby did not begin to rise above the horizon until after Spica. Judging by the widespread conceptualization of Virgo's lower half at that time, the baby was now born (fig. 10.28). The minimum brightness of the comet was apparent magnitude -11.2 (table 10.7)[73] and the actual figure may have been up to 4 or so magnitudes stronger.[74] Moreover, the phenomenon of forward-scattering (the nucleus at this stage had a "phase angle" of 160 degrees) would have boosted the brightness in the sunward side of the coma by about 3.5 magnitudes.[75]

At this climactic moment in the cometary apparition, the comet was apparently 180 degrees long, its silvery-grey tail streaming straight up to the zenith and all the way across the sky to the western horizon. The whole comet, including the coma and tail,

[72] From the point of the beginning of labor (October 15, when the comet was 0.38 AU from Earth) until the birth (October 20, when the comet was 0.21 AU from Earth) the coma essentially doubled in diameter (from 5.82 to 11.38 degrees). It is sometimes claimed that, as a comet makes a close approach to Earth (within about 0.4 AU), the outer edges of the growing coma may go undetected by the human eye for various reasons. This is called the Delta Effect, on which see Joseph N. Marcus, "The Need for Cometary Photometry," *International Amateur-Professional Photoelectric Photometry Communication* 8 (1982): 29; L. Kamel, "The Comet Light Curve Catalogue/Atlas," *Astronomy and Astrophysics Supplement Series* 92 (1992): 85–149; idem, "The Delta-Effect in the Light Curves of 13 Periodic Comets," *Icarus* 128 (1997): 145–159; David W. Hughes, N. McBride, J. Boswell, and P. Jalowiczor, "On the Variation of Cometary Coma Brightness with Comet-Earth Distance (the Delta Effect)," *Monthly Notices of the Royal Astronomical Society* 263 (1993): 247–255. However, the Delta Effect is highly controversial and it may simply reflect local observing conditions, the experience and eyesight of those observing, and the different altitude within the dome of the sky at which the object is observed.

[73] Assuming n=4. When the Sun is 18 degrees below the horizon, the sky is essentially as dark as it gets. When it is 15 degrees below the horizon, astronomers can see almost all the stars but struggle to observe faint and diffuse objects such as galaxies. When the cometary baby was born on October 20, 6 BC, the entire tail and most of the coma would have risen above the eastern horizon in a completely dark sky (when the Sun was more than 18 degrees below the horizon) and hence would have been easily seen. The very lowest segment of the coma would have risen when astronomical twilight was just beginning, with the whole coma and tail visible when the Sun was 15 degrees below the horizon. Since the comet was very bright, it would have been clearly visible to the Magi in Babylon at that point. Incidentally, from October 5 to 19 the whole coma appeared above the horizon when the Sun was more than 18 degrees below the horizon.

[74] If n=5 and the comet was first spotted in the latter half of 7 BC.

[75] Marcus, "Background and Model," 57 fig. 15. Earth was in a great location from which to see any sunward spike, consisting of the concentration of larger particles close to the cometary head, particles too large for the Sun to force behind the coma. Indeed, since the Christ Comet was narrowly inclined to the ecliptic plane, it, like Comet Lulin, may have had a continual sunward spike. However, the extent to which the Christ Comet's antitail would have been visible to the naked eye is impossible to determine.

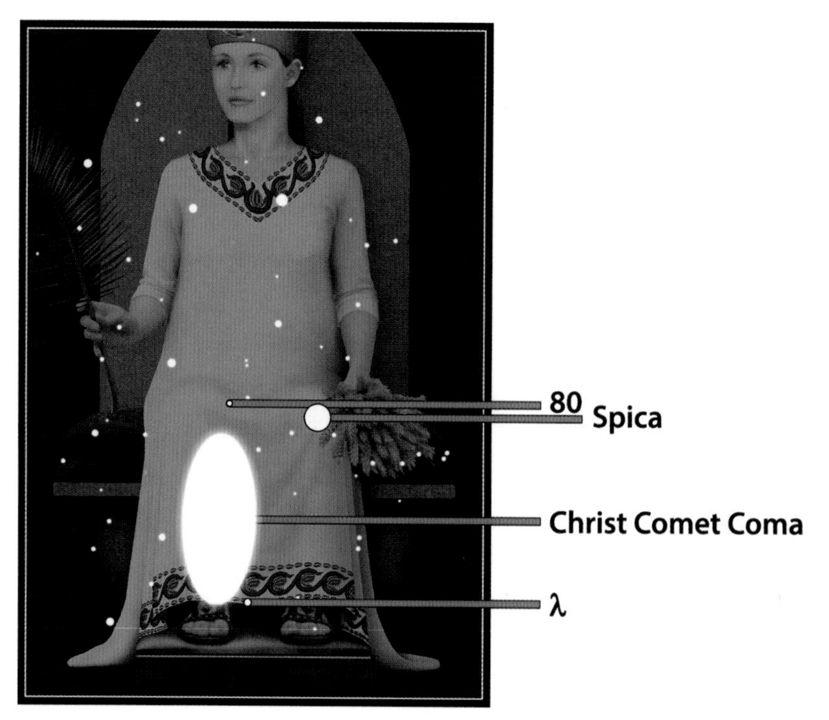

FIG. 10.28 October 20, 6 BC: Virgo's baby is fully delivered. Image credit: Sirscha Nicholl.

seems to have looked like an iron scepter with a beautiful orbed top. Indeed, as the Magi turned and gazed toward the western horizon at the tail jutting up from it, and followed its length all the way up over their heads and back behind them to the east, they would have got the impression that the majestic scepter was resting on the ground in the west. Since by this time the Magi understood that the comet represented the Messiah, they would naturally have regarded the scepter as touching the earth in Judea.[76] Those interpreting the celestial narrative in Biblical terms, including some within the Jewish community and the Magi, would have had no doubt that this symbolized the iron scepter of the messianic Son of God who was, according to Psalm 2, going to reign over the nations. The messianic prophecy of Numbers 24:17 would have strongly come to mind: "A star shall come out of Jacob; and a scepter shall rise out of Israel." The moment of fulfillment had come. Somewhere in the west, more specifically Judea, the Messiah was being born.

On October 20, Earth was located at what could be described as another crossroads (or "interchange") between its orbit and the comet's orbit, even as the comet was seemingly headed straight for it. The comet would get there only 9 days later. It was effectively staring Earth in the face. The comet was only 0.21 AU from Earth when it played the part of the baby newly born, and it was still coming closer (fig. 10.29).[77] Its long tail was extending well past Earth and close to it.

[76] At the time of the cometary baby's full emergence from the womb, the Magi might well have noted that Mercury was present in lower Virgo, next to κ Virginis (the nucleus was just under 2.5 degrees from Mercury). At the same time, Mars was located just to the south of Virgo's crown.

[77] The Magi would presumably have taken note of the date and the time (the watch of dawn), and the fact that the Moon was in the constellation Aquarius, the Sun in the constellation Libra, Jupiter in Taurus, Saturn in Pisces, and Venus in Scorpius, even while the comet, Mars, and Mercury were all in Virgo. At the same time, because in astrology, when a birth was in view, locations within the preestablished zodiacal signs (30-degree segments of the zodiacal region) were more important, astrologers would normally have classified the locations of the celestial bodies in those terms: with respect to the zodiacal signs, the Moon had just moved into

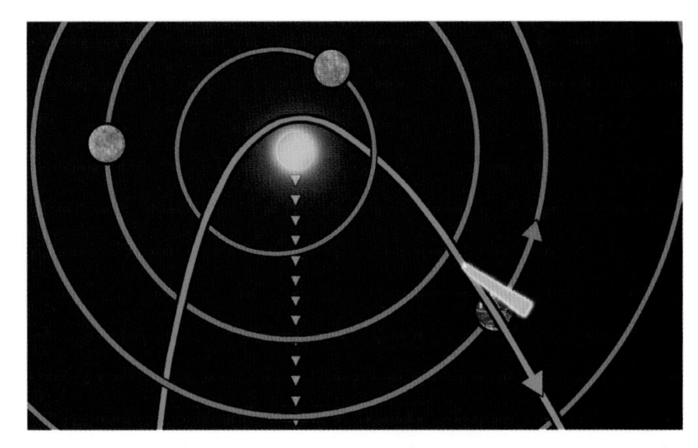

FIG. 10.29 October 20, 6 BC: What was happening in outer space. The comet was fast approaching Earth even as Earth was crossing the line of the comet's orbit. It should be remembered that the tail would have been much longer than this illustration shows. Also, the tail would have been sharply curved backwards, although the curvature would not have been apparent to Earth-dwellers. Image credit: Sirscha Nicholl.

DESCENT INTO THE LIGHT OF THE SUN

As thrilling as the cometary baby's birth was, onlookers following the narrative of the celestial drama would still have been concerned that Hydra might yet intervene to devour the newborn. Naturally the Magi and others enthralled by the unfolding story would have got up early on October 21 to see what would happen next. As they looked at the predawn scene, they would have seen that the cometary coma had jumped down a further 6 degrees (into the constellation Libra). In diameter the coma was something like 18% larger than the day before, although the increasing sunlight may well have meant that this was not apparent. The magnitude was now -11.4, with the coma benefiting from a remarkable brightness boost of about 5 magnitudes (due to its large phase angle—166 degrees). The coma appeared to be heading toward the horizon and into the intense light of the Sun; it was heliacally setting. It was clear that the coma-baby would soon completely disappear from the eastern horizon and that the predawn drama would come to an end. Since the Sun was playing the role of God in the celestial play, the movement of the coma-baby down toward the horizon and into the sunlight was most naturally interpreted to mean that the messianic baby would be rescued from Hydra by God.

By the morning of October 22, the nucleus had descended by more than 8 degrees, with the result that it was about 3 degrees in altitude at sunrise. However, the tail would have been streaming out diagonally across the whole sky. At that time the coma had an apparent magnitude of, at the very least, -11.7.[78] When we take into account that the comet was so close to the Earth-Sun line (the "phase angle" of the nucleus was 174 degrees!), we have to boost the brightness of the sunward side of the coma by 7.5 magnitudes (to an effective apparent magnitude of -19.2)![79] (Again, recall

Aquarius, the Sun was in Libra, Jupiter in Aries or Taurus (the Babylonians were not always consistent regarding the divisions of the zodiacal signs and tended to start the zodiacal system a few degrees before the First Point of Aries [as noted by Hermann Hunger, F. Richard Stephenson, C. B. F. Walker, and K. K. C. Yau, *Halley's Comet in History* (London: British Museum, 1985), 15]), Venus in Scorpius, Mercury in Libra, Saturn in Aries, and Mars in Virgo. However, in the case of the birth of Jesus, the wonders in the heavens were so explicit, dramatic, and awesome that even hardnosed Magi would have realized that horoscopy was inappropriate (not to mention unnecessary) on that occasion. In Jerusalem, the Magi emphasized that the omen announcing the Messiah's birth related to the Star at or around the time of its rising (Matt. 2:2). This suggests that the omen that impressed them was not the horoscope but a non-horoscopic nativity omen. Moreover, the comet was clearly interacting with the constellation figures as a whole, rendering the zodiacal constellations much more important for their analysis than the zodiacal signs.

In none of the extant horoscopes from Babylon do we have an interpretation of horoscopic data (Rochberg, *Babylonian Horoscopes*, 10; see Stol, *Birth in Babylonia and the Bible*, 95–98). This is not necessarily a great loss. J. D. North, *Horoscopes and History* (London: The Warburg Institute, University of London, 1986), xi, makes the point that historians frequently ask how ancient astrologers would have interpreted the horoscope of a particular historical figure, failing to appreciate that astrological interpretations typically consisted of whatever the astrologer thought the subject wanted to hear.

[78] Assuming n=4.

[79] What Zechariah prophesies in Luke 1:78–79 ("the rising [star]/Branch shall visit us from on high to give light to those who sit in darkness and in the shadow of death, and to guide our feet into the way of peace") may imply that the Star would become especially

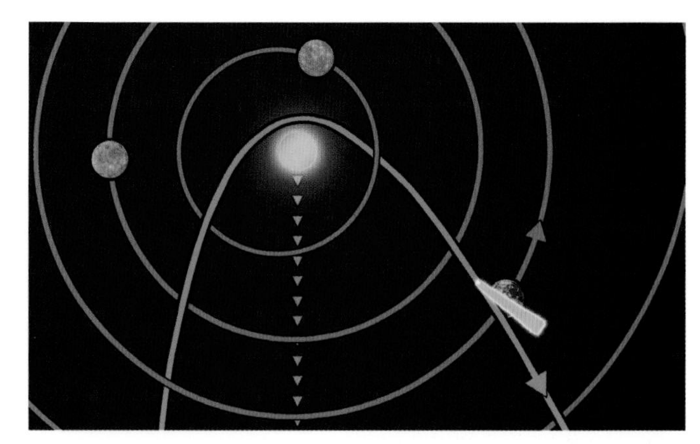

FIG. 10.30 October 23, 6 BC: the Christ Comet is heading toward the "overpass" just after Earth has gone under it. Note that the comet is switching from one side of Earth to the other (cf. fig. 10.28); this is why the focus of the celestial show shifts from the eastern morning sky to the western evening sky. Image credit: Sirscha Nicholl.

that the full Moon's apparent magnitude is -12.6 and that of the Sun is -26.7.) At least part of the back (upper) section of the large coma would have been visible before sunrise, as would the tail. The baby was, as it were, being rescued by the Sun. It was obvious that this was the last chapter of the nativity narrative in the predawn eastern sky. The coma was heliacally setting. The serpentine dragon had failed to devour the newborn child.

During the 22nd, the comet moved from the east side of the Sun to the west.

That evening at sunset, the nucleus was at 4 degrees altitude. The comet was now heliacally rising in the western sky. Part of the large coma could be seen over the western horizon for a short time after sunset, as could the tail, which streamed up at a slight angle (a slight right tilt), through the constellation Draco, and all the way to and beyond the far horizon. It must have looked like a gigantic neon sign suspended in the sky, directing the Magi to Judea to worship the baby Messiah. The Sun rose before the last section of the tail had set.

On the morning of October 23, the coma (the nucleus having descended 12 degrees in 24 hours to the point that it was the other side of the Sun) certainly did not appear in the eastern sky before dawn. The comet had moved 26 degrees in the space of 3 days! It was now between Libra and Ophiuchus. After impressing observers with its stability within Virgo's belly from September 30 to October 14, the coma had amazed them with its large leaps from October 20 to 23. This was due to the fact that the comet was getting very close to Earth[80] as it prepared to move in between the Sun and Earth. The Earth had already gone under the overpass 3 days before October 23, continuing on its counterclockwise orbit; in 6 days, the comet itself would zip across the overpass on its way out of the inner solar system (fig. 10.30).

THE MAGI'S DEPARTURE

Only when the Magi were exposed to the messianic traditions of the Hebrew Bible were they able to decode fully the multifaceted cometary apparition, most particularly the drama that unfolded in Virgo. This amazing celestial entity was the Messiah's Star and it was announcing his coming to the earth. The moment of birth had been marked by the full emergence of the coma-baby from Virgo's belly and also by the awesome sight of the comet as a whole, looking like an iron scepter, stretching from eastern to western horizon. It was clear to the Magi that the birth of the one destined to reign over the world from Jerusalem had taken place.

The Magi had followed this Star from their first observation of it. They had no doubt

bright at the point when, after rising, it seemed to "visit" the earth. If so, what Rev. 12:5 views as a divine snatching away to safety, Zechariah and Luke would be paradigming as a stellar visitation to the earth.

[80] On October 20, 6 BC, it was 0.21 AU from Earth; on the morning of October 23 it was only 0.125 AU away.

marveled as it made the constellations come alive. Now it had played the part of the messianic baby in an awe-inspiring drama to mark the occasion of his nativity. They must have felt a strong bond with the comet. They were firmly convinced that ancient oracles concerning the Messiah were now being fulfilled. The comet, they believed, was a divine messenger revealing the divine plan regarding the Jewish Messiah. And they clearly perceived that the comet was in a special way speaking to them and even commissioning them to participate in the grand narrative that was unfolding right then. They were not content merely to know that the divine Messiah had been born to Virgo's terrestrial counterpart some 550 miles away; they were determined to see the King and worship him.

According to our calculations, the baby was born before dawn on October 20, 6 BC. The celestial nativity show in the eastern sky was over on October 22, 6 BC. This conclusion would have been confirmed on the morning of October 23. One may presume that, in this period, the Magi made their final preparations to travel to Judea by camel caravan in order to worship the newborn messianic King.[81] Having made their travel preparations, the Magi must have got on their way to Judea quickly, probably departing Babylon sometime during October 23–25. That is when the comet began to behave in a way that could have been interpreted by the Magi as a celestial prompt or usher.

By the time the Magi arrived in Judea, they had secured gifts for the Messiah—gold, frankincense, and myrrh. Whether they purchased them in Babylon or en route, these gifts clearly show the influence of the book of Isaiah, which is hardly surprising, since Isaiah 7–12 had been the main key that unlocked the meaning of Virgo's pregnancy and delivery. In choosing gold and frankincense, they were consciously bringing to the Messiah at his birth what the prophet had prophesied the Gentiles would bring him during his reign (Isaiah 60). In selecting myrrh, they may well have been acknowledging that the Messiah had been born to suffer and die and be buried, bearing the sins of many so that he might win a people for himself who might reign with him (Isaiah 53).

WESTWARD LEADING

The comet was very close to Earth at this time, and getting closer. Because Earth had crossed the orbital interchange prior to the comet, the focus of the cometary drama had, for human observers, shifted to the western sky. This glorious apparition in the western sky, and later the southern and western sky, continued for the whole of the Magi's journey.

With surprising haste, the comet appeared to advance across Ophiuchus, upper Sagittarius, Capricornus's head, and Aquarius before making it back to its home under the western fish in Pisces by November 3.[82] Then over the next weeks it made its way to under

[81] When might the Magi have deduced that what was taking place in Virgo was signaling the Messiah's birth? The sight of a bright baby in Virgo's womb at the comet's heliacal rising proper (at the start of the Feast of Tabernacles) was a major clue; it was highly suggestive of Isa. 7:14's oracle concerning the Messiah's birth: "Behold, the virgin shall conceive." The fact that the comet as a whole at that time evidently looked like a scepter (in accordance with Num. 24:17) would have been a powerful clue as to the identity of the baby. After a few more predawn observing sessions, it would have been natural for the Magi to conclude that what was unfolding in the eastern sky was a heavenly dramatization of an earthly virgin's conception, pregnancy, and delivery of a baby, that this baby was the Messiah, and that the celestial birth would coincide with the terrestrial birth. In addition, the length of the scepter would have grown larger with every passing day, underlining that Num. 24:17 was being fulfilled. During the rest of the comet's time in the eastern sky the Magi would not only have been following developments closely but also presumably would have been eagerly learning more about the prophesied Messiah. In addition, they may have been making preparations to travel to Judea as soon as the astronomical drama in the eastern sky was over.

[82] Between October 23 (evening), when the cometary coma found itself at Ophiuchus's lower left leg (φ Ophiuchi), and the 24th (evening), the comet jumped almost 18 degrees to the space between his right thigh and the top of Sagittarius's bow. Its movement through the heavens was evident within a single observing session. On October 25 the comet passed over Sagittarius. On October 26 it was located over Capricornus's head and horns. The size of its jumps within the starry heavens from night to night was decreasing. Then on October 27 it traveled along the left arm of Aquarius (as envisioned by Ptolemy) and above Capricornus's tail-fin, the next day over Aquarius's right hip, and finally, on October 29–31, over his water-jar and water (fig. 10.31).

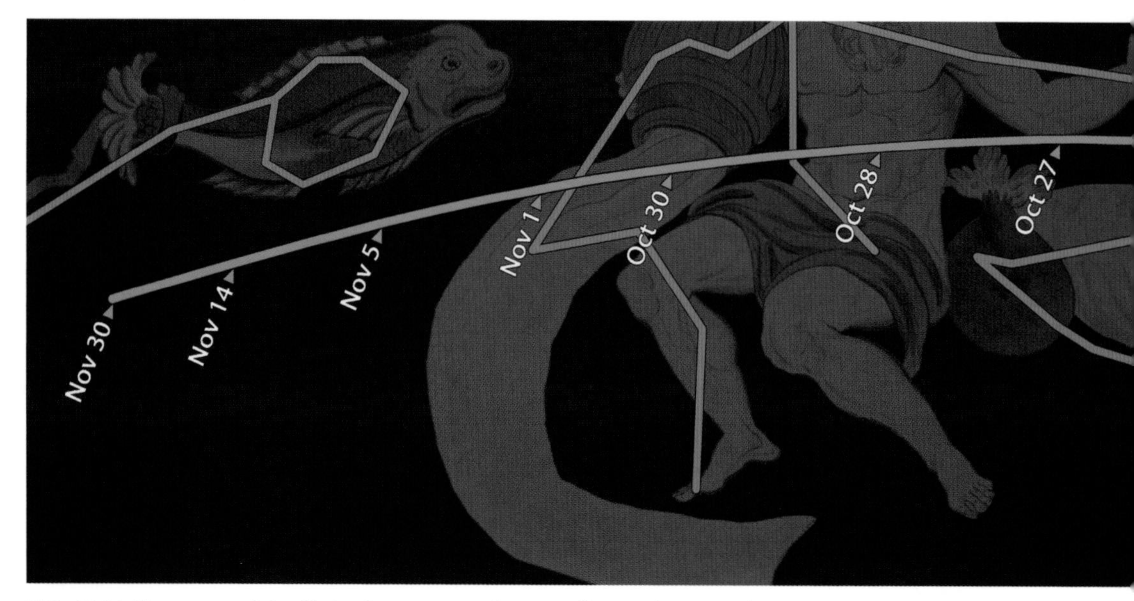

FIG. 10.31 The course of the Christ Comet across the constellations from October 23 to November 30, 6 BC. Image credit: Sirscha Nicholl, using constellation illustrations from Bloxam, *Urania's Mirror* (images credit: old bookart.com).

that fish's tail and the ribbon connected to it, where it was when the comet stood over the house where Jesus was, one night between November 23/24 and November 30/December 1, 6 BC (see fig. 10.31). Since it was in that vicinity of the sky that the comet had first appeared, this might well have seemed a very suitable climax, especially since Saturn and Jupiter were still nearby.

This description of the comet's movements after it switched to the western sky makes what the comet did at that time sound very ordinary. In truth it was anything but. For the whole time leading up to the point when the comet stood over one particular house in Bethlehem, it sported a long tail that was shortening. On October 23 the tail stretched across the entire dome of the sky. From that point on, it got smaller, initially in steep drops (by October 29 it had halved in length) and then more slowly. The length of the tail was actually a general measure of the progress that the Magi were making on their journey—the closer they got to Judea, the shorter the tail became. This may well have been a great encouragement to the Magi as they traveled westward. By November 23–30 the comet would have been about 33–38 degrees in length, considerably shorter than at its peaks, but still an impressive sight by any normal standard.

The comet's tail may have played important roles in the cometary apparition up to this point. It had possibly been the water being poured from Aquarius's water jug. It had probably formed the shaft and fletching of the Archer's Scorpion-slaying arrow. It may have looked like the tube of a magnificent trumpet announcing the Feast of Trumpets. And it had transformed the comet into the Messiah's long iron scepter to mark the Messiah's birth. Now, it turned the long-tailed comet into a majestic usher, guiding the Magi all the way to Judea, and more particularly to Bethlehem, and then directing them to their goal—the baby whose part the comet coma had played within Virgo.[83]

[83] On October 29 and 30 the large coma was located where, according to Greek and some Babylonian traditions, the water-jar of Aquarius (the Great One) was. It may have prompted the Magi to recall the celestial scene in the first week of August. If the coma was the water-jar and the tail was the water, the scene now was extraordinary—a great river was flowing from the water-jar!

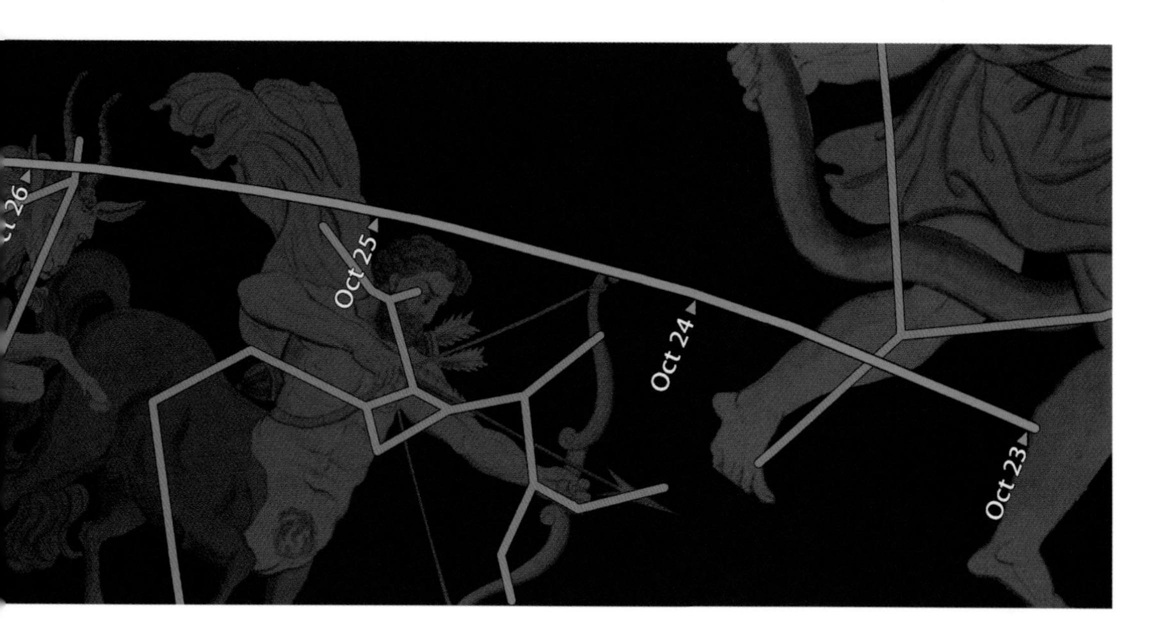

Having shifted to the evening sky, where was the comet's coma in the wake of sunset? For the first couple of weeks, the comet was migrating from the west to the southeast. On October 23 the Star was in the west-southwest. By October 25 it was in the southwest. By October 27 it was seen in the south. By the 29th the comet was in the south-southeast at sunset. From that point it steadily moved toward the southeast, getting there around November 9. Thereafter its movement within the celestial dome slowed to such an extent that it was less than the daily rotation of the sky, and so the comet appeared to edge slightly back the way it had come.[84] From November 9 to 23, even as it was advancing across the belly of the western fish of Pisces, it gave up about 5½ degrees of the progress it had made with respect to its compass position (its "azimuth").[85] By November 30 it had given up another 5½ degrees. Nevertheless, when the Magi were on the last stage of their journey to the Messiah's home in Bethlehem, one night between November 23 and 30, the comet was appearing in the south-southeast at sunset.

Of course, when the astrologers looked to their left as they traveled westward from their homeland toward Judea, they were looking southward. During the early days of their journey, when the comet (i.e., the coma) was quickly moving to the south, the Magi may have felt like they were catching up with it. When the comet then started off its nightly journeys from a position alongside or slightly behind them, so to speak, which it did for most of their journey, it was naturally a great encouragement to them that they had already made good progress and should keep going on their extraordinary pilgrimage.

How did the comet's altitude change in the weeks after it shifted to the western evening sky? At sunset on October 23, viewed from Babylon, the altitude of the comet nucleus had been 12 degrees at sunset. Every sunset

[84] That the comet ceased moving relative to the fixed stars at this time would have been duly noted by the Babylonian astronomers (see Stephenson, "Ancient History of Halley's Comet," 244).

[85] Azimuth is the distance in degrees, measured clockwise, along the horizon from due north to the point where a vertical line downward from a given celestial object intersects the horizon.

up until October 28 saw the comet move to a higher altitude, climaxing at just over 40 degrees, halfway up to the zenith. Thereafter it descended very slightly, bottoming out at around 38½ degrees altitude on November 4. After that, it again began ascending to a higher altitude until the end of November and beyond. On November 23–25 and 29–30 (calculating from Jerusalem) the nucleus would have been about 44–45 degrees and 46–46½ degrees respectively in altitude at sunset.

How did the comet tail's orientation change over the days and weeks? To use a crude analogy, in the weeks after October 22 the comet's orientation in the aftermath of each sunset was like a giant left-side (from the driver's perspective) windshield wiper sweeping from right to left that snapped off its pivot and went hurtling up into the air to the left and into the distance, even as it sought to complete the arc of its sweep. This was due to the fact that the comet in outer space was passing Earth and moving away from and indeed, so to speak, "behind" it.

Obviously, the higher up (altitude) and further south/southeast (azimuth) the comet was at sunset, the longer it took for the comet to set. On October 23 the comet (nucleus) started to set about 45 minutes later than it had on the 22nd. Each night thereafter until the night of the 26th/27th the comet set roughly an hour later. Over the following nights the gap between sunset and the comet's setting continued to increase, although the increase steadily became less impressive. The gap peaked on November 9/10, decreasing thereafter, albeit to an insignificant extent.

What did the comet look like as it moved across the heavens each night?

To employ an imperfect analogy, the comet was like a javelin. At sunset on October 23 it was as though the javelin were being thrust from close range into the ground in the west in front of the Magi. Over the following days it was as though the javelin were being thrown from ever greater distances to the left, each time landing ahead of them. After the first week, when the comet was in the southern-southeastern sky, it would have seemed that the javelin was being hurled upwards from a position left of the Magi and moving across the sky in an arc until it "landed" pointing downwards in front of them. On October 23 the javelin hit the ground at a 60-degree angle. The following evening it struck the earth at an approximately 55-degree angle. Thereafter the angle increased until, on November 23/24 and November 30/December 1, it was landing at an 80-degree angle.

To appreciate what the comet was doing each evening and night from about a week into their trip, when the comet was appearing at sunset in the southern sky, it may perhaps help to imagine a massive, transparent, hollow, rainbow-shaped arch. The bases of the arch remain on the ground while the arch is being hoisted up by its "keystone" until it is upright (the arch represents the ecliptic, the path of the Sun across the sky). Now imagine that, each night, a streamer (representing the tailed comet) is being launched inside the transparent arched tube while the arch is being raised. The streamer is launched from about halfway along the arch when the arch has been raised up to a 45-degree angle (this represents the comet at sunset). Then the streamer steadily makes its way to the end of the arch, reaching it when the arch is almost upright (this represents the setting of the comet). You are standing between the bases of the arch. As the streamer moves along the rising arch, its whole orientation seems to you to change from being almost horizontal to being almost vertical. This is just like the comet along the ecliptic. Since the comet was basically parallel to and right beside the ecliptic, its orientation was basically the same as that of the ecliptic. The ecliptic arc was steadily and increasingly rising up to a near-vertical position as the hours rolled by each night. As it did so, the stars and constellations—including, of course, those along the ecliptic,

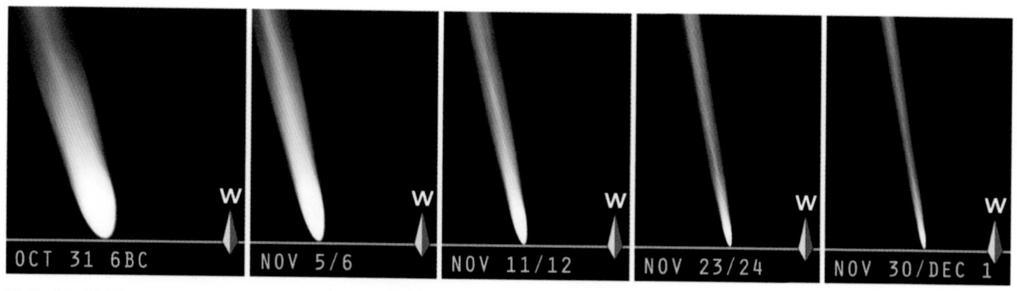

FIG. 10.32 The orientation and location of the Christ Comet as it set from October 31 to November 30/December 1, 6 BC. Image credit: Sirscha Nicholl.

together with the comet—followed their daily westward course. The result was that the straight, long-tailed comet's whole orientation appeared to change dramatically over the course of each night from the point of its appearance after sunset until its setting.

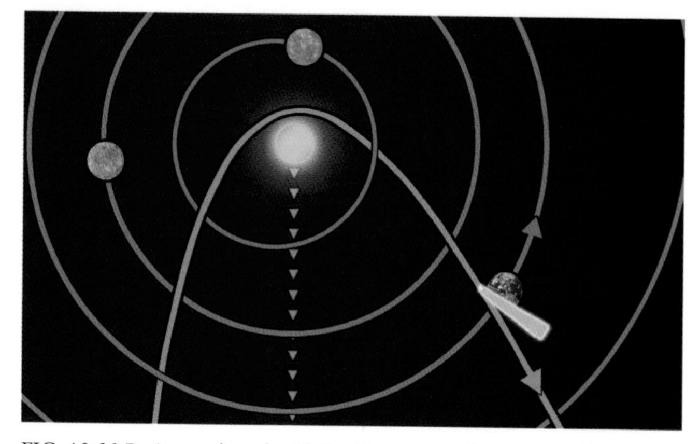

FIG. 10.33 Perigee: when the Christ Comet was closest to Earth (October 24/25, 6 BC). Image credit: Sirscha Nicholl.

What did the comet look like when it set from October 24 to November 30/December 1? During that period the comet was shrinking. Indeed, if the Magi had drawn images of the comet just before it set every night and flicked through an ordered collection of them quickly, they would have got a very clear vision of what was happening in outer space (fig. 10.32)—the comet was passing Earth and moving away into the distance. After all, the comet was passing by Earth (its closest approach to Earth [perigee] was October 24/25) and then getting farther and farther away from it (fig. 10.33).

From October 24 onward the comet's apparent magnitude also weakened. On October 24 the comet's apparent magnitude was at least -12.1,[86] the condensed part of its coma visible in the daytime. By November 23 the magnitude was down to (at least) -4.8; by November 30 it was down to (at least) -4.0.[87] Moreover, the brightness boost due to the phenomenon of "forward-scattering" quickly subsided as the comet's position relative to Earth changed radically on October 24–28. However, the effect of the weakening magnitude was counterbalanced by the comet's shrinkage, because the brightness that the comet did have was being distributed over a smaller surface area, concentrating it. In the early days of the Magi's journey the coma would have been massive, but from October 24/25 it steadily decreased in size. By November 23 and 30 it may have been about 1.75 to 2 degrees long and 0.7 to 0.8 degrees wide.

Throughout this period the comet functioned as a giant celestial guide, urging the Magi to journey westward on their way to

[86] Assuming n=4. This does not take into account the brightness boost that would have resulted from the "forward-scattering" effect—since the comet's nucleus as it set on October 24, 6 BC, had a phase angle of 143 degrees, the sunward side of the coma would have been brighter by about 2.5 magnitudes than our apparent magnitude values based on absolute magnitude and slope values would suggest (see Marcus, "Background and Model," 57 fig. 15).
[87] Assuming n=4.

Judea. At each nightly setting the Star moved over the horizon as though it were traveling in that direction. The advance of the long-tailed comet over the horizon must have given the Magi the distinct impression that it was moving on ahead of them. Just as, when you watch someone walking over a hill, the person's body gradually disappears, first the feet and then upwards from there, so the comet disappeared slowly but surely from the coma upwards. From very shortly after they set out on their journey, right through until their arrival in Jerusalem, the ever-shortening comet would have been last seen in the west, in front of them.[88]

One can well imagine that, as the Magi approached Jerusalem, the Star seemed to set right over the city, appearing to signal that this was their final destination.

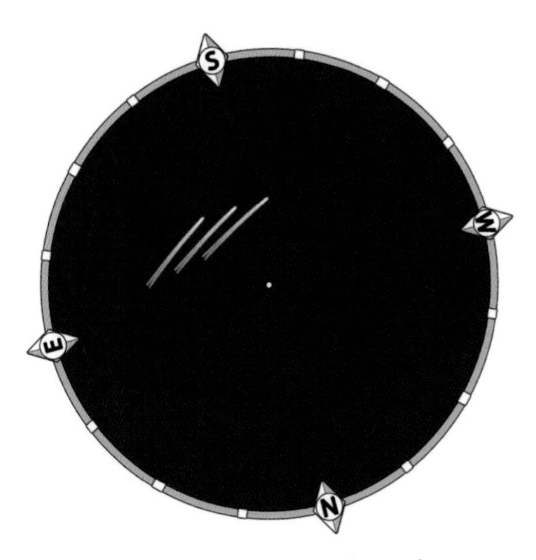

FIG. 10.34 The Christ Comet over the two hours starting at sunset on the evening when the Magi made their way from Jerusalem to Bethlehem. Image credit: Sirscha Nicholl.

THE STAR AT THE CLIMAX OF THE MAGI'S JOURNEY

The big question, of course, is: What did the comet do on the day when the Magi traveled from Herod's palace in Jerusalem to Bethlehem, and as they looked for the house where the baby Messiah was—between November 23/24 and November 30/December 1, 6 BC? The analogies used above are helpful for grasping what the comet did on each of those nights. As the Magi set out on their way, around sunset, the cometary coma appeared just under the ribbon connected to the tail of the western fish of Pisces[89] in the south-southeast, the comet stretching back approximately 33–38 degrees into Aries in the east. Over the couple of hours from the time when the comet did appear, as the Magi walked from Jerusalem to Bethlehem, it advanced, along with the fixed stars and constellations, to the south-southwest, precisely the direction of Bethlehem from Jerusalem.

We shall consider November 23 and 30 in turn.

With respect to the 23rd, when the approximately 38-degree-long comet became visible around sunset, it would have been seen at an altitude of 44 degrees in the south-southeast, with its tail sweeping back toward the east. As it progressed on its nightly course through the sky it would have increasingly moved toward the Magi as they traveled to Bethlehem. 1½ hours after sunset, the nucleus was at an altitude of 49 degrees in the south (the comet's highest point, or culmination). Thereafter, over the next 30 minutes it dropped in altitude very slowly and imperceptibly, down a fraction of a degree (to 48½ degrees). At this point it was in the south to south-southwest. If the 23rd was the night on which the Magi traveled from Jerusalem to Bethlehem, the Magi probably arrived at the town of their destination around this time, for the coma was then high in the sky in the direction of Bethlehem, and hence would have been di-

[88] Initially the Sun rose when part of the comet was still setting in the west. From near the start of November, the comet would have been sufficiently short that its end could be seen disappearing below the western horizon prior to sunrise.

[89] Could it be that the popular fish (*ichthus*) symbol in early Christianity was ultimately attributable to the Christ Comet climaxing (and conceivably also beginning) its apparition in Pisces the fishes? That seems more plausible than the proposal of Carl G. Jung (*On Christianity* [Princeton, NJ: Princeton University Press, 1999], 214) that it originated in "astrology."

rectly in front of the Magi, with its long tail streaking back to their left. Accordingly, the Sun must have set and the Star appeared toward the beginning of their evening journey from Jerusalem to Bethlehem.

As for the 30th, the situation was very similar: the now 33-degree-long comet appeared at sunset in the south-southeast, with the tail sweeping toward the east. Its altitude at that time was about 46½ degrees. Over the next 70 minutes or so it ascended to about 50 degrees, at which point it was at its culmination in the south. After that it began to descend; over the next 50 minutes it dropped almost 2 degrees. At the end of that time it was in the south-southwest, the tail sweeping high into the southeast. Again, in this scenario the Magi would have arrived in Bethlehem around that time. Accordingly, the Sun must have set and the Star appeared toward the start of their short journey.

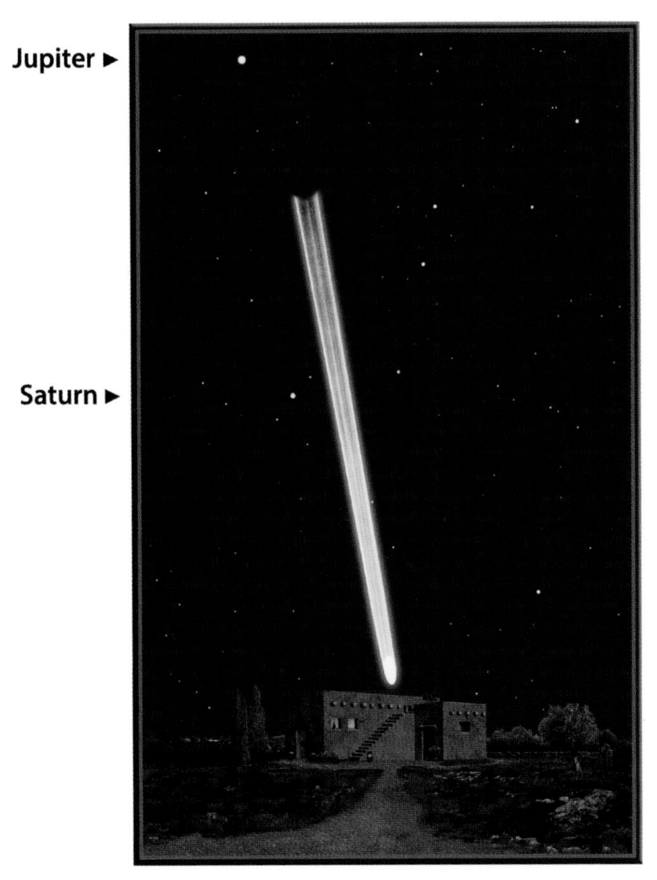

FIG. 10.35 The Christ Comet as it set on the night the Magi arrived in Bethlehem. Image credit: Sirscha Nicholl.

Either way, as Matthew reports, it really would have seemed that the Star was going ahead of them to Bethlehem (Matt. 2:9b). See fig. 10.34.

In both scenarios, assuming that the Magi arrived in Bethlehem 2 hours after sunset, it would have been something like 5 hours later that night that the comet set in the west. As the comet advanced westward with the rotating sky, it was descending in altitude, moving toward the Magi, and adjusting its apparent angle (along with the ecliptic) until it set at a slight tilt—leaning about 10 degrees off vertical. This behavior is what Matthew had in mind when he stated that it "came" and then "stood" over the house (v. 9b).[90] From the vantage point of the Magi, the comet as it set seemed to be pinpointing one particular house on the west-southwestern horizon of Bethlehem. It seems that the Magi were located to the east-northeast of the house, directly opposite the comet. The sunward side of the coma appeared to be located just above the house, from their perspective, and the tail was streaking up alongside the ecliptic toward the roof of the sky (fig. 10.35).[91]

[90] It is unclear whether, during the time that the Star was descending in altitude, the Magi anticipated that it was about to pinpoint the very house where the messianic baby was staying. Certainly, however, when they saw the Star standing over the house, they believed that its movements that night had been intended to usher them first to Bethlehem and then to the Messiah.

[91] If the narrowly inclined comet had a visible spike antitail at this point, this could have contributed to the impression that the comet was pinpointing very precisely one particular house.

Magnitude Slope (value of n)	Apparent Magnitude on the evening of November 23, 6 BC, if first observed on November 21–28, 8 BC	Apparent Magnitude on the evening of November 23, 6 BC, if first observed on February 5, 7 BC	Apparent Magnitude on the evening of November 23, 6 BC, if first observed on May 29, 7 BC	Apparent Magnitude on the evening of November 23, 6 BC, if first observed on August 17, 7 BC	Apparent Magnitude on the evening of November 23, 6 BC, if first observed on September 30, 7 BC	Apparent Magnitude on the evening of November 23, 6 BC, if first observed on December 10–17, 7 BC
3	-6.6	-6.4	-5.4	-4.3	-4.0	-3.7
4	-8.4	-8.2	-7.0	-5.8	-5.4	-4.8
5	-10.2	-9.9	-8.5	-7.2	-6.7	-6.0

TABLE 10.8 The Christ Comet's apparent magnitude on the evening of November 23, 6 BC. The values are essentially the same for the comet's setting later that night.

Magnitude Slope (value of n)	Apparent Magnitude on the evening of November 30, 6 BC, if first observed on November 21–28, 8 BC	Apparent Magnitude on the evening of November 30, 6 BC, if first observed on February 5, 7 BC	Apparent Magnitude on the evening of November 30, 6 BC, if first observed on May 29, 7 BC	Apparent Magnitude on the evening of November 30, 6 BC, if first observed on August 17, 7 BC	Apparent Magnitude on the evening of November 30, 6 BC, if first observed on September 30, 7 BC	Apparent Magnitude on the evening of November 30, 6 BC, if first observed on December 10–17, 7 BC
3	-5.8	-5.6	-4.6	-3.5	-3.2	-2.9
4	-7.6	-7.4	-6.2	-5.0	-4.6	-4.0
5	-9.3	-9.0	-7.6	-6.3	-5.8	-5.1

TABLE 10.9 The Christ Comet's apparent magnitude on the evening of November 30, 6 BC. The values are essentially the same for the comet's setting later that night.

By November 23/24–November 30/December 1 the comet would have decreased considerably in apparent magnitude from what it had been at the start of their journey. At the same time, however, the comet was shrinking in size, intensifying the brightness that it did have. The comet's brightness was evidently sufficient for the Magi to see it shortly after sunset as they walked from Jerusalem to Bethlehem. Gary W. Kronk comments, "The comet would have been located far from twilight and certainly would have been a striking object in the evening sky, being visible from about 6 p.m."[92] Indeed the comet would have been a stunning sight. Its magnitude was at least -4.8 as it set on November 23/24, 6 BC.[93] On the night of November 30/December 1 it was at least -4.0, comparable to the magnitude of Venus.[94] (See tables 10.8–9.)

The 13-day waxing gibbous and 19-day waning gibbous Moon were in the sky for the comet's setting on the 23rd/24th and the 30th/1st, respectively.[95] On the 23rd/24th the comet in outer space was 1.09 AU from Earth and 1.54 AU from the Sun. On November 30/December 1 it was 1.33 AU from Earth and 1.66 AU from the Sun. The comet was just

[92] Personal email message to the author, September 26, 2012.

[93] Gary W. Kronk, personal email message to the author, September 26, 2012, reckons that the magnitude was at least -4.

[94] Venus, which would have set just over an hour after the Sun, on November 23–24 and 30 had an apparent magnitude of around -3.9.

[95] The presence of the 13- to 15-day-old Moon or 18- to 19-day-old waning gibbous Moon nearby on November 23/24–25/26 and November 29/30–November 30/December 1 respectively may have dimmed the view slightly, but the sight would nevertheless have been magnificent. As regards the earlier part of the night, if they were traveling to Bethlehem on November 23–25, 7 BC, the light of the Moon and the brightness of the comet together would have guided the Magi there. If they were traveling from Jerusalem to Bethlehem on the 29th or 30th, the Moon would not have risen at the point that they were on the road. Regardless of whether they traveled on November 23–25 or 29–30, the Moon would have been present when the Magi were looking for the child in Bethlehem (on the 29th to 30th the Moon would have risen within a couple of hours of their arrival in Bethlehem).

beyond the orbit of Mars.[96] See figs. 10.36–37.

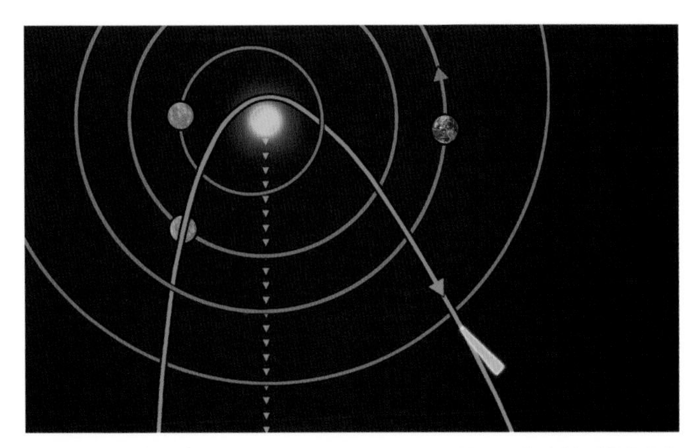

FIG. 10.36 The Christ Comet in outer space on November 23, 6 BC. It is just beyond the orbit of Mars. Image credit: Sirscha Nicholl.

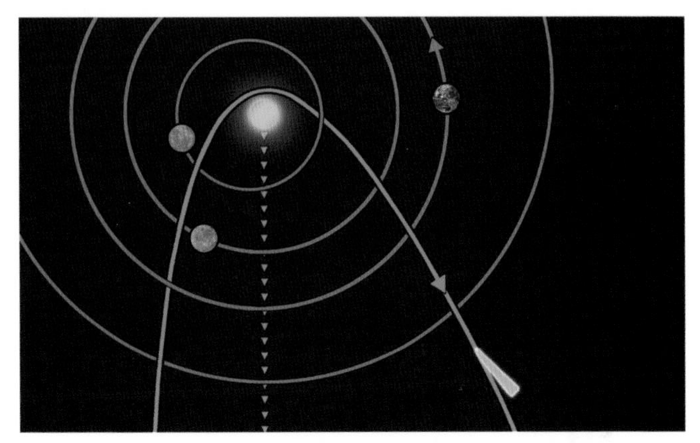

FIG. 10.37 The Christ Comet in outer space on November 30, 6 BC. It is beyond the orbit of Mars. Image credit: Sirscha Nicholl.

As the comet set between November 23/24 and November 30/December 1, Jupiter was above the coma in the constellation Aries, 42½–45 degrees off the horizon, while Saturn, in the constellation Pisces, was even closer to the coma, hovering just 20½–22½ degrees above the horizon, approximately halfway between the coma and Jupiter. The comet's impressively long tail would have extended upwards from the horizon right past Saturn toward Jupiter. To someone surveying the scene as a whole from the Magi's perspective, around the time of the comet's setting, the tailed comet would have seemed like a massive pointer directed downwards from Jupiter, past Saturn, to the Messiah's house. It may have looked as if Jupiter was pinpointing the location of the Messiah, the one destined to reign on behalf of Israel, represented by Saturn. This was an awesome climax to the Magi's journey, and indeed to the entire cometary apparition.

Of course, the comet had first appeared during a year dominated by the three Jupiter-Saturn conjunctions of 7 BC (climaxing on May 27–31, September 26–October 5, and December 1–8). Indeed it is very possible that the Bethlehem Star was first observed in connection with one of those conjunctions. In light of this, the fact that Jupiter and Saturn were present for the climax of the comet's apparition may well have been deemed especially significant. The Magi may already have come to the settled conviction that Jupiter had played the part of the Most High God, Saturn the role of Israel, and the "star" that of the one destined to exercise sovereignty over the earth on behalf of Israel. Needless to say, if the comet did first appear during the Jupiter-Saturn triple conjunction, this climactic scene in Bethlehem would have meant that a beautiful celestial inclusio bracketed the whole cometary apparition.

The comet, the coma of which had played the role of the messianic baby in Virgo in the great celestial drama in the eastern sky just over a month before, had now led the Magi to the very building on the earth where the baby was.

[96]It should be noted, however, that the more one moves into December, the more the comet would struggle to guide a traveler in the right direction from Jerusalem to Bethlehem (since it was beginning each night farther to the west).

11

"Brightest and Best of the Sons of the Morning"

The Greatest Comet in History

What the Bethlehem Star comet did in 7–6 BC was astonishing and obviously unique. It was clearly a magnificent comet, but how does it compare with the other great historical comets?

THE MARKS OF GREAT COMETS

In an earlier chapter we set out the marks of great comets. They must have some, if not all, of the following characteristics. They should (1) make a close pass by the Sun, no more than about 1 AU from it; (2) make a close pass by Earth; (3) provide good viewing opportunities for human observers; (4) be very productive; (5) have a clearly visible tail, at least 10 degrees in length; (6) have a large nucleus and notable coma; and (7) be at least as bright as the sky's more distinguished stars. Some great comets may have deficiencies (e.g., they may be farther from the Sun and/or Earth), but their other traits will make up for these.

The Christ Comet has all the marks of a truly great comet. Indeed its qualifications for greatness supersede those of any other comet in recorded history. It is quite simply the greatest of the great historical comets.

A CLOSE PASS BY THE SUN

The Bethlehem Star Comet came as close as 0.119 AU, or 17.8 million km, from the Sun. In the league of the Great Comets, as established by Don Yeomans in April 2007,[1] it would be 8th out of the 66 in this regard.[2] It is superseded only by sungrazers—the Great Comet of 1680 (0.006 AU)[3] and established members of the Kreutz sungrazing family, specifically the Great March Comet of 1843 (0.0055 AU), the Great September Comet of 1882 (0.00775 AU), and Ikeya-Seki in 1965 (0.0078 AU)—and two sunskirters, the Great Southern Comet of 1865 (0.026 AU) and the Great Comet of 1668 (0.07 AU), and by one NI comet, the Great Comet of 1665 (0.106 AU). The 6 BC Comet's perihelion distance

[1] http://ssd.jpl.nasa.gov/?great_comets (posted April 2007).

[2] We are counting the 65 great comets concerning which Yeomans denominates a perihelion distance.

[3] Comet ISON of 2013 had a perihelion distance of 0.0124 AU, although it disintegrated catastrophically before it reached perihelion on American Thanksgiving Day of that year.

is slightly closer to the Sun than Messier's Comet of 1769 (0.123 AU), which had a tail of over 90 degrees,[4] and the Great January Comet of 1910 (0.13 AU), which sported a 50-degree tail. It is worth remembering that the Great Comet of 1811 achieved greatness even though it was 1.04 AU from the Sun at perihelion, and the magnificent Hale-Bopp made it only to 0.91 AU from the Sun.

A CLOSE PASS BY EARTH

As regards making a close pass by Earth (perigee distance), the Christ Comet came as close as 0.1045 AU, or 15.63 million km, on October 24/25, 6 BC. In that regard it would rank 12th in Yeomans's list of great comets. The great comets ahead of it in perigee distance are Halley's Comet in 837 (0.03 AU), followed by the Great Comet of 1132 (0.04 AU), the Great Comet of 1472 (0.07 AU), the Great Comet of 1556 (0.08 AU), the Great Comet of AD 400 (0.08 AU), Halley's Comet in AD 374 and 607 (0.09 AU), the Comet of AD 568 (0.09 AU), Hyakutake in 1996 (0.1018 AU), the Comet of AD 390 (0.1037 AU), and Halley's Comet in 1066 (0.104 AU).[5] It should be noted that 4 of these 11 are different apparitions of the same comet, 1P/Halley. When one considers how magnificent the great comets of 1807 and 1811 were, when they came no closer than 1.15 AU and 1.22 AU respectively from Earth, and how stunning Hale-Bopp was, even though it never came closer to Earth than 1.32 AU, one begins to appreciate the significance of the Christ Comet's high perigee ranking. Moreover, it should be remembered that the Christ Comet actually made two close passes by Earth—on August 14, 6 BC, on its way toward perihelion, it came as close as 0.334 AU to Earth.

A COMBINATION OF CLOSE PASSES BY THE SUN AND BY EARTH

When one considers the perihelion and perigee distances together, the Christ Comet emerges at the top of the list.

Of the 7 great comets with perihelion distances closer than that of the Christ Comet, the one that came nearest to Earth was the Great Comet of 1680 (Kirch's Comet)—it was 0.42 AU away. The next closest was the Great Comet of 1665, which came to about 0.57 AU from Earth.

Of those great comets that came as close to Earth as the Christ Comet, or closer, the ones that came nearest the Sun were the Great Comet of AD 400, which was only 0.21 AU from the Sun, Hyakutake (fig. 11.1), whose perihelion distance was 0.23 AU, and the comets of 1471 and 1556, both of which came as close as 0.49 AU to the Sun.

When one considers the perihelion and perigee distances of all of Yeomans's great comets, it is the AD 400 comet that holds the record—0.21 AU from the Sun and 0.08 AU from Earth. After it comes Hyakutake—0.23 AU from the Sun and 0.1 AU from Earth—and then Great Comet of 905—0.2 AU from the Sun and 0.21 AU from Earth.

In its combination of perihelion and perigee distances—0.119 AU and 0.1045 AU respectively—the Christ Comet surpasses even the Great Comet of AD 400. It is worth discovering a little more about that comet.

Like the Christ Comet, the Comet of AD 400 made its closest approach to Earth after perihelion. Philostorgius[6] referred to this comet as a "star which appeared in the form of a sword," and Socrates Scholasticus[7] mentioned that it was a huge comet of extraordi-

[4] David Seargent, *The Greatest Comets in History: Broom Stars and Celestial Scimitars* (Berlin: Springer, 2009), 122.
[5] Of course, many other comets that made close passes by Earth are not regarded as "great," such as comets Lexell in 1770 (perigee: 0.0151 AU) and Tempel-Tuttle in 1366 (perigee: 0.0229 AU). For lists of close approaches, see the Near Earth Object Program's "Historic Comet Close Approaches," http://neo.jpl.nasa.gov/ca/historic_comets.html (last modified October 27, 2009); Martin Mobberley, *Hunting and Imaging Comets* (Berlin: Springer, 2011), 36–37; and Gary W. Kronk, "The Closest Approaches of Comets to Earth," http://cometography.com/nearcomet.html (last modified September 30, 2006).
[6] *Ecclesiastical History* 11.7; translation by Edward Walford, *Epitome of the Ecclesiastical History of Philostorgius* (London: Henry G. Bohn, 1855), http://www.tertullian.org/fathers/philostorgius.htm (accessed March 26, 2014).
[7] *Ecclesiastical History* 6.

nary magnitude that stretched from the sky to the ground, something unprecedented for those alive at the time.[8] Hermias Sozomen[9] described it as having "extraordinary magnitude" and being "larger . . . than any that had previously been seen."[10] David Seargent points out that it "holds the record for a double bill of small perihelion and small perigee."[11] He also suggests that its coma was "several degrees in diameter," that its tail stretched "like a broadening beam of light across tens of degrees of sky,"[12] and that it would have cast such strong shadows that people would have been able to read by its light.[13] Gary Kronk estimates that this comet had an absolute magnitude of only +6,[14] with a maximum apparent magnitude of around 0. However, the fact that this comet was described as simultaneously massive and of extraordinary brightness suggests that its apparent magnitude peak must have been much greater than this. It is possible that the comet had a temporary but dramatic flare-up in the aftermath of its closest approach to the Sun, and that this fizzled out within a couple of weeks of the comet passing Earth. At the same time, Seargent points out that an unimpressive absolute magnitude might have been partly compensated for by the fact that the comet's angle with respect to the ecliptic was becoming ever narrower throughout the apparition.[15]

FIG. 11.1 Comet Hyakutake on April 17, 1996. Image credit: Bojan Dintinjana and Herman Mikuz, Črni Vrh Observatory, Slovenia, http://www.observatorij.org.

As great as the Great Comet of AD 400 undoubtedly was, it pales in comparison to the Great Comet of 6 BC, that is, the Great Christ Comet. The comet of 6 BC was intrinsically and apparently brighter and was larger in coma and tail size than the Comet of AD 400.

GOOD VIEWING OPPORTUNITIES

With regard to viewing opportunities, the general rule is that cometary apparitions should occur in the evening and in the northern hemisphere to be regarded as great. It should be qualified that, in the ancient world, rather more people would be expected to be awake in the hour or two before dawn than in the modern world. Nevertheless, as now, so then, a comet would have been more widely observed in the evening hours. The ideal time for a comet to be on show is the early part of a night when the Moon is not very bright. At the same time, comets are best seen against the backdrop of a dark sky, meaning earlier mornings or later evenings.

The Bethlehem Star was visible for a long time, but, like most comets, it saved its most magnificent display for the weeks before and after perihelion.

Prior to perihelion it put on a stunning

[8] Gary W. Kronk, *Cometography: A Catalog of Comets*, 6 vols. (Cambridge: Cambridge University Press, 1999–), 1:71.
[9] *Ecclesiastical History* 8.4.
[10] Translation by Edward Walford, *The Ecclesiastical History of Sozomen* (London: Henry G. Bohn, 1885), 367–368. See Kronk, *Cometography*, 1:71–72.
[11] Seargent, *Greatest Comets*, 80.
[12] Ibid.
[13] Ibid., 82.
[14] Kronk, *Cometography*, 1:72.
[15] Seargent, *Greatest Comets*, 80. The Comet of AD 400 had an inclination of 32 degrees.

show as it made its way across the sky from Pisces to Virgo. Much of that glorious procession occurred in the evening and early-night sky when most people were out and about and liable to notice anything unusual in the skies. Among other things, we have suggested that they might have perceived the comet to be Aquarius's water-jar and the water coming out of it, to be an arrow fired from the Archer's bow flying through the sky to slay the Scorpion in August, and to be a trumpet and bright "evening star" in September. For the latter part of its procession it may well have been visible during the daytime.

The comet as a whole, taking the form of a spectacular scepter, began to rise heliacally around the time of perihelion. Two or three days later, at the point when the coma heliacally rose in the eastern sky, it beamed with the brightness of the full Moon, or greater, looking like a baby in Virgo's belly. It was an extraordinarily "bright morning star." At that time the comet would also have been a dramatic daytime object. Over the following days the comet moved to a higher altitude before the Sun rose, so that it could be seen in a darker sky, although the growth in the comet's size would have diluted the intensity of its brightness. The comet climaxed its apparition in the eastern sky on October 20. At that time the coma looked like a newborn baby and the comet as a whole like a massive iron scepter stretching across the whole sky and resting on the ground in the west. Over the following days it would have appeared very large, long, and bright as it made its close pass by Earth.

Then the comet shifted to the west of Earth and hence the focus of the majestic show moved from the morning sky to the evening and night sky, and from the eastern sky to the western sky. The comet's long-tailed appearances dominated the skies for at least another 30–40 days. At that time, as on its procession to the Sun, the celestial visitor to the inner solar system made its presence abundantly clear to everyone.

EXTRAORDINARY PRODUCTIVITY

Regarding productivity, the Christ Comet was very active, beginning its degassing farther from the Sun than any other comet on record, including Hale-Bopp. Moreover, it remained in view from the point at which it was first observed, at least 9½ months before perihelion,[16] until 2 or more months after it.[17] As mentioned above, the hyperactivity of the comet is reminiscent of Hale-Bopp, which pumped out dust and gas like a locomotive, becoming visible to the naked eye almost 10½ months before rounding the Sun. Hale-Bopp maintained naked-eye visibility for 18 months. That was double the time that the previous record-holder, the Great Comet of 1811, had remained observable to the naked eye (9 months, becoming visible 5 months before perihelion), and that comet had been a long way in front of the next in line.[18] We do not know how long in total the Christ Comet remained visible to the naked eye, but it is likely that it at least rivaled Hale-Bopp, and, quite possibly, surpassed it. Hale-Bopp was first observed by the naked eye at magnitude +6.7, because Terry Lovejoy knew exactly where to look, thanks to published data and telescopic and binocular aid.[19] Needless to say, the Christ Comet lacked this advantage and would have been first observed only when considerably brighter (approximately +3.4). Of course, the Bethlehem Star Comet's productivity is why its coma and tail were so large and bright immediately before and after perihelion.

[16] The period of naked-eye visibility obviously includes times when the comet was not observable due to its proximity to the Sun.
[17] Only 7 of Yeomans's select group of 73 great cometary apparitions in history up to 1996 (and hence before the peak of Hale-Bopp's performance) (for the list, see appendix B in Fred Schaaf, *Comet of the Century* [New York: Springer, 1997], 335–338) were visible to the naked eye for 100+ days, and only 2 were visible to the naked eye for 120+ days—the Great September Comet of 1882 (135 days) and the Great Comet of 1811 (260 days) (ibid., 202–203).
[18] It is worth recalling great ancient comets: for example, Josephus mentions a comet that remained visible for a year, apparently in AD 65–66 (Josephus, *J.W.* 6.5.3 [§289]), and Pliny mentions a comet that had lasted for 180 days (*Natural History* 2.22).
[19] See Gary W. Kronk, "C/1995 O1 [Hale-Bopp]," http://cometography.com/lcomets/1995o1.html (last modified October 3, 2006).

A LONG TAIL

With respect to the criterion that a great comet must have a tail at least 10 degrees in length, the Christ Comet excels.[20] We have seen that calculations suggest that the comet tail was majestically long, longer than most of Yeomans's great comets, in the western sky prior to perihelion. In addition, of course, after perihelion, on October 20, 6 BC, it seems to have sported an impressively long tail that caused the comet as a whole to look like an iron scepter that stretched from the eastern horizon to the western.[21] The tail would have continued to stretch across the entire dome of the sky for a few days after this. Moreover, its long tail streamed across much of the evening/night sky for the 28–37 days that the Magi were journeying from Babylon to Jerusalem, seeming to go before them to the west. Then one night, between November 23/24 and November 30/December 1, after ushering the Magi to Bethlehem in the south-southwest, the comet stood over the house where Jesus was, about 33–38 degrees in length.

The longest tail lengths among Yeomans's great comets are Tebbutt's Comet of 1861, which grew to 120 degrees, and the Great comet of 1618 (C/1618 W1), which was 104 degrees long. In regards to tail length, the Great Christ Comet seems to have surpassed even these impressive comets.

The Christ Comet's tail is reminiscent of the Great March Comet of 1843. On March 1, 1843, that comet was 30 degrees long, by March 4 it was between 69 and 90 degrees, on March 21 about 64 degrees, and on March 30 approximately 38 degrees.[22] Such a sustained period of tail length is most impressive. However, it is outdone by the Christ Comet, for its tail was longer and remained so for a considerably greater time. Nevertheless, if one wants to get a good idea of what the Christ Comet looked like as it processed from Pisces to Virgo or as the Magi journeyed westward to Judea and then found the house in Bethlehem, one can do little better than look at images of the Great March Comet of 1843.[23] See figs. 5.27–28; 6.6, 10; 10.16.

A LARGE COMA

As for the size of the Christ Comet's coma: in view of the very early first observation of the comet and its close perihelion and perigee distances, the coma, not surprisingly, grew very large. Revelation 12:1–5 may imply that the coma appeared to be the size of a newborn baby at the point of the baby's birth. Our calculations suggest that it may have been about 11 degrees long (and something like 4.4 degrees wide). It should also be appreciated that, had the comet not also been extremely bright, observers would not have been able to see the large coma.

In the recorded history of comets, comets Lexell in 1770, Hyakutake in 1996, and IRAS-Araki-Alcock in 1983[24] stand out as having had especially large comas when they made a close approach to Earth—2–4 degrees in diameter. The size of these comas was due to the proximity of the comet to Earth, for each of these comets' nuclei was relatively small. Only one of the three was a great comet—Hyakutake. Comet Holmes grew as large as 3 degrees by March 2008 due to a major outburst, but it is not classified as a great comet. Another coma worthy of note is that of the Great Comet of 1811. Even though this comet never came closer to the Sun than 1.04 AU or closer to Earth than 1.22 AU, its coma nevertheless at one point was reckoned

[20] The Chinese recorded comets with tail lengths of 200 and even 300 degrees, but none are regarded by Yeomans as among the great comets (http://ssd.jpl.nasa.gov/?great_comets [posted April 2007]).

[21] The "iron" color may suggest that the comet tail was not merely dusty but also gassy, like Hale-Bopp.

[22] Gary W. Kronk, *Comets: A Descriptive Catalog* (Hillside, NJ: Enslow, 1984), 36.

[23] This comet's tail in outer space was longer than the distance from the Sun to Mars (Patrick Moore and Robin Rees, *Patrick Moore's Data Book of Astronomy* [Cambridge: Cambridge University Press, 2011], 257).

[24] IRAS-Araki-Alcock grew to be a maximum of 2.0–2.5 degrees in size (Storm Dunlop and Will Tirion, *Collins Night Sky and Starfinder* [London: Collins, 2011], 123; Andreas Kammerer, personal email correspondence, October 30, 2012).

to be larger than the Sun, approximately 2 million km in diameter in space.[25]

As we have already commented, had Hale-Bopp come as close as 0.1 AU from the Sun and Earth, its coma would probably have been some 4 or perhaps even 5 degrees in diameter.[26] With respect to coma size, therefore, the Christ Comet was a much greater comet even than Hale-Bopp. In space, Hale-Bopp's coma extended up to 3 million km in diameter. The Christ Comet's coma may have been something like 5.75 million km long and 2.3 million km wide (equivalent to a round coma with a diameter of 3.6 million km—about the same overall size as the coma of Comet Holmes in late November or early December of 2007). As such it was larger than all known comet comas in history except for that of Comet Holmes in 2007/2008. It would unquestionably have been the largest object in the inner and outer solar system at the time and indeed one of the largest in recent millennia. The Christ Comet's great coma size was due to the larger nucleus size and greater productivity.

A LARGE NUCLEUS

For the coma to have grown so big, the nucleus must have been very large as well.[27] Small nuclei can produce relatively large comas if they are especially active. But the remarkable size and sustained brightness of the coma of the Christ Comet would seem to require both that it was very productive and that it had a large nucleus.

Estimating the precise size of a nucleus is an uncertain business, based on the comet's magnitude, activity, and coma diameter. The asteroidal comet Chiron is reckoned to be 166–233 km in diameter. Sarabat's Comet of 1729 was at least 100 km (comparable in diameter to metropolitan New York City), possibly as large as 300 km. Neither is classified as a "great comet," because they have never come close to the Sun—at its nearest, Comet Sarabat was still more than 4 AU from the Sun.[28] Had Sarabat come close to the Sun and Earth, it would have been a truly spectacular comet. Certainly, the Bethlehem Star Comet nucleus seems to have been larger than all recorded historically great comets, including Hale-Bopp, the diameter of which was 40 km (think London) to 70 km (think Boston), and the Great Comet of 1882, which was about 50 km in diameter (think Seattle).[29] The Christ Comet was probably 100 km or more in diameter, comparable to the more modest estimates of Comet Sarabat's size.

UNPRECEDENTED ABSOLUTE MAGNITUDE

In view of how long the Magi had been tracking the Christ Comet prior to perihelion, it is obvious that it had unprecedented intrinsic brightness.

In table 11.1, I set out the absolute magnitude if n=3, 4, or 5 for different points within the time period when the Star was first observed. The absolute magnitude values for the earliest and latest dates of the first sighting by the Magi are included. I also set out the absolute magnitude values assuming a first observation on February 5, 7 BC, when it would have been in conjunction with Jupiter. May 29 and September 30, 7 BC, are also included, because that is when the first two of the three conjunctions of Jupiter and Saturn in Pisces occurred. If the comet was first spotted during the peak of the third conjunction (i.e., on December 1–8, 7 BC), its absolute magnitude

[25] Patrick Moore, *Comets: An Illustrated Introduction* (New York: Scribner, 1973), 84; Mobberley, *Hunting and Imaging Comets*, 46.
[26] Andreas Kammerer, personal email message to the author, October 30, 2012.
[27] As Richard Schmude, *Comets and How to Observe Them* (New York: Springer, 2010), 35, has noted, generally speaking, the larger the nucleus, the larger the coma and the brighter the comet.
[28] See Kronk, *Cometography*, 1:396.
[29] Zdenek Sekanina, "Statistical Investigation and Modeling of Sungrazing Comets Discovered with the Solar and Heliospheric Observatory," *Astrophysical Journal* 566.1 (2002): 582.

Magnitude Slope (value of n)	Absolute Magnitude if first observed on November 21–28, 8 BC	Absolute Magnitude if first observed on February 5, 7 BC	Absolute Magnitude if first observed on May 29, 7 BC	Absolute Magnitude if first observed on August 17, 7 BC	Absolute Magnitude if first observed on Septem-ber 30, 7 BC	Absolute Magnitude if first observed on December 10–17, 7 BC
3	-8.1	-7.9	-6.9	-5.8	-5.5	-5.2
4	-10.4	-10.2	-9.0	-7.8	-7.4	-6.8
5	-12.7	-12.4	-11.0	-9.7	-9.2	-8.5

TABLE 11.1 The Christ Comet's range of possible absolute magnitude values.

values would have been essentially the same as on December 10–17, 7 BC.[30] I also include the values of absolute magnitude for August 17, 7 BC.[31]

Assuming n=4, the absolute magnitude was between -6.8 (if first spotted on December 17, 7 BC) and -10.4 (if first spotted on November 21, 8 BC). Kronk, having done brightness calculations on the Christ Comet, commented that it would have been "the biggest comet ever observed."[32]

Sarabat's Comet (1729) had an absolute magnitude of between -3 and -6. Hale-Bopp, in the early stages of its apparition, had an intrinsic brightness as great as absolute magnitude -2.7, although this changed to about -0.8. The Great Comet of AD 418 is plausibly reckoned by Seargent to have been -2 or -2.5.[33] The absolute magnitude of Tycho's Comet of 1577 was -1.8 and that of the Great Comet of 1746 was -0.5. After that comes the Great Comet of 1811, the absolute magnitude of which was 0. These comets are intrinsically the brightest comets on record. However, the Christ Comet's extraordinary absolute magnitude means that it surpasses even these to qualify as the intrinsically brightest comet in recorded history.[34]

NOTABLE APPARENT MAGNITUDE

When one considers that a comet with such an extraordinary absolute magnitude made close passes by both the Sun and Earth and was clearly visible when extremely large in Virgo, one realizes that its apparent magnitude must have been extraordinary. Indeed the Christ Comet may have been the brightest comet in human history. It is certainly in the super-league of cometary brightness.

The historical comets with the most notable values of maximum apparent magnitude are the Great Comet of 1680 (-18), the Great September Comet of 1882 (-17 or -15 to -20), Ikeya-Seki in 1965 (-15), the Comet of 1577 (-8), and the Great Southern Comet of 1865 (-8). Kronk observes that the Christ Comet would have been "a spectacular object that was visible in daylight for a time" and that its "maximum brightness is similar to what was observed for the Great September Comet of 1882 and Ikeya-Seki in 1965."[35]

The climax of the Christ Comet's apparent

[30] For December 1–3 and 4–8 the values for n=4 are lower than the stated value for December 10–17, 7 BC, by 0.2 and 0.1 respectively.

[31] That is approximately when John the Baptist was conceived.

[32] Personal email message to the author, September 26, 2012. Kronk's comments were based on my orbit and the fact that the comet was first spotted between one and two Jewish years before Herod's slaughter of the innocents.

[33] Seargent, *Greatest Comets*, 82.

[34] David Seargent stated back in 1982 that "we have no reason to suspect, from examination of the appearances and durations of historical comets, that any comet for which an orbit has not been calculated was of noticeably higher absolute magnitude [i.e., greater intrinsic brightness] than these [the comets of 1577, 1811 and 1882, among others]" (*Comets: Vagabonds of Space* [Garden City, NY: Doubleday, 1982], 110–111). The year-long comet mentioned by Josephus may be an exception to this, as may be the progenitor of the sungrazer system of comets (reckoned to have had a nucleus of 120 km in diameter and an absolute magnitude of -5—see Schaaf, *Comet of the Century*, 73–74; cf. Brian G. Marsden, "The Sungrazing Comet Group," *Astronomical Journal* 72 [1967]: 1170–1183; Peter Jenniskens, *Meteor Showers and Their Parent Comets* [Cambridge: Cambridge University Press, 2006], 424), and certainly the Christ Comet was an exception.

[35] Personal email message to the author, January 9, 2013. Kronk's comments were based on his calculations of the comet's brightness, assuming my orbit and the fact that the comet was first sighted at least one Jewish year before the slaughter of the innocents.

magnitude would have occurred at perihelion on September 27, 6 BC.[36] However, while part of the tail would have been observable at that point, the coma was not. The coma may have become visible on September 29, 6 BC, when its apparent magnitude was between -14.8 and -18.4.[37] On September 30, 6 BC, when it was almost certainly observable, its apparent magnitude was between -14.1 and -17.7.[38] In view of the possibility that the brightness slope may have been higher than 4, and in view of the unpredictable nature of comets, it is conceivable that the comet was even brighter than these estimates, particularly in the period after perihelion. Moreover, throughout the month of October in 6 BC, the comet benefited from a brightness boost as it was in the zone between Earth and the Sun (at that time the nucleus had a phase angle from 90 to 180 degrees); it peaked on October 22 at 174.75 degrees, which would have meant a boost in brightness of some 7.5 magnitudes in the sunward side of the coma due to forward-scattering! Joseph Marcus has pointed out that this special geometry is a factor that endows comets, particularly those of medium or large size, with greatness.[39] At the same time, the massive size of the Christ Comet's coma at some stages of the apparition would have meant that the brightness was being distributed over a large surface area, diluting some of the effect of the impressive apparent magnitude values.

LONG SPELL OF BETTER-THAN-ZERO APPARENT MAGNITUDE

With regard to maintaining a minimum apparent magnitude of 0 for the longest period, Hale-Bopp is the record-holder, having maintained this level of brightness for 8 weeks. It is followed by de Chéseaux's Comet of 1744 and Tycho's Comet of 1577, both of which maintained better-than-zero magnitude for approximately 6 weeks, and by the Great Comet of 1882, which managed it for about 5 weeks. The Bethlehem Star Comet, however, easily trumps all of these great comets. Our calculations of its overall brightness suggest that it was zero or greater in apparent magnitude for more than 8 months (May 13, 6 BC, to January 30, 5 BC),[40] although the Bible gives us no information about its behavior after its climactic performance on the night when the Magi found the baby Messiah (between November 23/24 and November 30/December 1, 6 BC), and the rate at which comets fade is famously variable![41]

CONCLUSION

In conclusion, the Christ Comet satisfies the criteria for cometary greatness. In fact, all things considered, it is undeniably the single greatest comet in recorded history. Even if one considers only the combination of its perihelion and perigee distances, its status as an elite comet is clear. When one adds to this the great viewing opportunities, the long tail length, the extremely large coma, the remarkable productivity, the long duration of the apparition, and the extraordinary brightness, it is seen to be in a league of its own. Truly the Bethlehem Star comet is sitting at the top of the tree of cometary greatness! The Christ Comet is the only very large and intrinsically very bright comet we know of

[36] Had the comet been visible on September 27, 6 BC, the apparent magnitude might have been as dramatic as -15.8 to -19.4 (if n=4).
[37] Assuming n=4.
[38] Assuming n=4.
[39] Joseph N. Marcus, "Forward-Scattering Enhancement of Comet Brightness. I. Background and Model," *International Comet Quarterly* 29 (2007): 61–62; idem, "Forward-Scattering Enhancement of Comet Brightness. II. The Light Curve of C/2006 P1 (McNaught)," *International Comet Quarterly* 29 (2007): 119. Marcus points out that, of the thirteen comets counted "great" by Bortle in his 1997 study of great comets (John E. Bortle, "Great Comets in History," *Sky and Telescope* 93.1 [1997]: 44–50), nine passed between Earth and the Sun, producing good forward-scattering geometry, and eight had phase angles peaking at between 155 and 180 degrees, with six of them having phase angles peaking at between 166 and 180 degrees ("Background and Model," 62).
[40] Assuming n=4 and the latest possible date of first observation. If n=5 and the comet was spotted at the latest possible date, the comet would have had zero magnitude or greater from May 1, 6 BC, to February 14, 5 BC, or over 9 months.
[41] See Seargent, *Comets: Vagabonds of Space*, 51, on the Great Comet of 1843.

that made very close passes by both the Sun and Earth.

David Seargent, in his book *Comets: Vagabonds of Space*, commented that "People sometimes ask 'What was the best comet ever seen?' To this there is no positive answer. No single object can definitely be said to have wiped the rest off the field."[42] "No single object stands out from the rest as obviously *the* greatest comet."[43] However, in light of what we have discovered about the Great Comet of 6 BC, this statement is in need of revising. The Christ Comet stands out from the rest as obviously *the* greatest comet of recent millennia.

Ignatius's description (in chapters 18–19 of his letter *To the Ephesians*) fits excellently with what we have discovered concerning the Christ Comet, and it is well worth citing as we close this chapter.[44] Most scholars believe that Ignatius is, in 19:2–3a,[45] quoting from very early Christian tradition (many believe it was a hymn[46]) that goes back well into the first century AD.[47] It certainly appears to be independent of Matthew, which makes it a very important source concerning the Star.[48] Strikingly, Ignatius makes it clear that the Star was a mystery hidden in the Hebrew Prophets.[49] In particular, his reference to "Mary's virginal conception and her giving birth" suggests that the fulfillment of Isaiah 7:14 was on his mind:[50]

> Where is the wise man? Where is the debater? Where is the boasting of those who are deemed intelligent? Our God, Jesus the Messiah, was conceived by Mary according to God's plan, both from the seed of David and of the Holy Spirit. . . . Now Mary's virginal conception and her giving birth [to the Messiah] had been kept hidden from the Prince of this Age (likewise also the death of the Lord—these are three mysteries to be loudly proclaimed, which were done in the quietness of God).[51] How, then, was [the virgin birth][52] revealed to the Aeons?[53]
>
> A star shone in heaven
> with a brightness beyond
> all the stars;

[42] Ibid., 109.
[43] Ibid., 115.
[44] This was one of the most frequently cited extracanonical texts in the history of the early church (Harry O. Maier, "Ignatius *Ephesians* 19.1–3," in *Prayer from Alexander: A Critical Anthology*, ed. Mark Christopher Kiley [London: Routledge, 1997], 267; J. B. Lightfoot, *Apostolic Fathers, Pt. II. S. Ignatius. S. Polycarp. Revised texts, with Introductions, Notes, Dissertations, and Translations* [London: Macmillan, 1885], 76).
[45] If vv. 2–3a are part of a hymn, v. 3b ("That which had been prepared by God began to come into effect. Therefore all things were perturbed, because the abolishing of Death was being worked out") is best regarded as Ignatius's own theological comment regarding the cited tradition.
[46] For example, H. F. Stander, "The Starhymn in the Epistle of Ignatius to the Ephesians (19:2–3)," *Vigiliae Christianae* 43 (1989): 209–214; Maier, "Ignatius *Ephesians* 19.1–3," 267–269; Matthew E. Gordley, *Teaching through Song in Antiquity* (Tübingen: Mohr Siebeck, 2011), 353–354. The style and hymnic/poetic vocabulary are regarded as evidence that 19:2–3a were originally a hymn. Schoedel, however, maintains that Ignatius himself composed vv. 2–3, albeit heavily dependent on preexisting tradition (William R. Schoedel, *Ignatius of Antioch*, Hermeneia [Philadelphia: Fortress, 1985], 87–88).
[47] Schoedel, *Ignatius of Antioch*, 92.
[48] William R. Schoedel, "Ignatius and the Reception of the Gospel of Matthew in Antioch," in *Social History of the Matthean Community: Cross-Disciplinary Approaches*, ed. David L. Balch (Minneapolis: Fortress, 1991), 156.
[49] Schoedel, *Ignatius of Antioch*, 89.
[50] So ibid., 90n18.
[51] Lightfoot, *Apostolic Fathers*, 80, nicely summarized Ignatius's thought: "These mysteries . . . were foreordained and prepared in silence by God, that they might be proclaimed aloud to a startled world."
[52] Although Michael W. Holmes, *The Apostolic Fathers: Greek Texts and English Translations*, 3rd ed. (Grand Rapids, MI: Baker, 2007), 199, renders the text, "How, then, were *they* revealed to the ages?" (italics his), implying that Ignatius had in mind the conception, birth, and death of Jesus, this is most unlikely. After all, the death of Jesus is introduced with "likewise also," underlining its parenthetical nature, and vv. 2–3 are exclusively focused on the birth of Jesus, as Schoedel, *Ignatius of Antioch*, 90, points out. The apostolic father is referring to a single (note the singular: "How then was *it* revealed to the Aeons?") complex mystery consisting of Mary's virginal conception and her delivery of Jesus—hence our rendering "the virgin birth." A surprising number of translations render the subject of the first verb in v. 2 "he," namely Christ (e.g., Bart Ehrman, *The New Testament and Other Early Christian Writings: A Reader*, 2nd ed. [Oxford: Oxford University Press, 2004], 328; Schoedel, *Ignatius of Antioch*, 87). However, the contrast between v. 1 and v. 2 is between what was "hidden from the prince of this Age," first and foremost Mary's virginal status when she conceived and gave birth to Jesus (note 18:2 also), and the revealing of this to the Aeons (cosmic powers).
[53] Schoedel, *Ignatius of Antioch*, 91.

its light was indescribable,
and its newness caused astonishment.

And all the rest of the stars,
together with the Sun and the Moon,
formed a chorus to the star,
yet its light far exceeded them all.
And there was perplexity regarding from
where this new entity came,
so unlike anything else [in the
heavens] was it.

Consequently, all Magianism[54]
began to be destroyed,[55]
every bond of wickedness began to
disappear,
ignorance began to be removed,
and the old kingdom began to be
destroyed,
when God was manifest as man
to bring the newness of
eternal life.

That which had been prepared by God began to come into effect. Therefore all things were perturbed, because the abolishing of Death was being worked out.

[54]The theology, rites, and ceremonies of the Magi are probably in view. The word had also come to refer more generally to magic, but in this context, where the historical Star of the Magi is in view, it most likely retains at least something of its original sense.
[55]We are translating the imperfects in v. 3a as inceptive (see ibid., 94). As the end of v. 3 makes clear, the victory that God won over the forces of evil began with the Star but will not be fully worked out until the eschaton.

12

"The Light Everlasting That Fades Not Away"

The Ongoing Story

In the course of this book we have made mention of many participants in the Christmas narrative—especially the Magi, the king of Judea, Mary and Joseph, Jesus, and the Star. It now remains to update the story. What became of these main characters, in particular Jesus and the Star?

THE MAGI

The Magi went back home to Babylon, no doubt excitedly telling others about their amazing adventure with the comet and the one whom it represented and to whom it led them. Ignatius (*Eph.* 19:3) may imply that the Magi who followed the Star turned away from the theology of the Magians (*mageia*) and the practices of sorcery. Likewise, Justin Martyr (*Dial.* 78–79) claims that the Magi escaped the demonic dominion of evil when they came to worship Jesus as Messiah.[1]

HEROD THE GREAT

Herod—Hydra in human form—failed in his horrendous quest to assassinate the messianic child. As Matthew tells the story, he was, quite simply, outmaneuvered by God. A dream warned the Magi of Herod's malicious scheming against the Messiah, prompting them to avoid Jerusalem and Herod on their way back home to Babylon. Thus deprived of the information necessary to mount a targeted strike at the Messiah (Plan A), Herod sent in his troops to massacre the baby boys of Bethlehem (Plan B). However, Joseph was warned in a dream about Herod's assassination attempt and so fled with Mary and Jesus to Egypt and remained there until Herod died.

The hard-hearted king of Judea got his comeuppance. On March 12/13 in 4 BC there was a striking omen in the heavens that seemed to proclaim in terms the sick king could well understand that he would soon die: the dreaded partial lunar eclipse.[2]

When Herod awoke, he found that his illness had taken a dramatic turn for the worse. As soon as the report of the location of the lunar eclipse in the sky reached him, the thought must have crossed his brain that

[1] Cf. Clement of Alexandria, *Excerpta ex Theodoto* 69–75; Origen, *Contra Celsum* 1.60.
[2] Josephus, *Ant.* 17.6.4 (§§164–167).

God had issued the death penalty against him for his attack on the baby Messiah. For, incredibly, the lunar eclipse was over the womb of Virgo, right where the Sun had been on September 15, 6 BC, and where the cometary coma had been on September 29/30 and over the following days (fig. 12.2). Instead of determining to worship the Messiah upon hearing of the great sign in the eastern sky that had brought the Magi to Judea, Herod had sought to assassinate him. Indeed he had even tried to outwit God, using information gleaned from the Magi's records of the comet (in particular, when it first appeared) and from Scripture to assist him in this enterprise. Herod, the chief agent of Hydra, had schemed to murder the one whose birth had been announced by a great wonder focused on Virgo's womb. Ironically, it was in this same womb that the divine realm seemed to announce Herod's death sentence. Within the next few weeks, despite his desperate efforts to resist his fate, he died a horrendous death.[3]

FIG. 12.1 An imagined portrait of Herod the Great. Image credit: Wikimedia Commons.

MARY AND JOSEPH

Joseph and Mary remained in Egypt until Herod the Great died. At that point, "an angel of the Lord appeared in a dream to Joseph in Egypt, saying, 'Rise, take the child and his mother and go to the land of Israel'" (Matt. 2:19b–20a). Straightaway, that is, in the immediate aftermath of Herod's death, Joseph "rose," "took the child and his mother," and "went to the land of Israel" (v. 21). Joseph and the holy family entered the territory of Greater Israel and prepared to go into Judea, obviously assuming that the angel meant Judea when he said "Israel." However, when the announcement that Archelaus had been made ruler (ethnarch) of Judea, Samaria, and Idumea in Herod's place reached Joseph's ears, he was surprised and filled with fear. That Joseph had to be warned in another dream not to enter Judea implies that his fear concerning Archelaus was entirely justified—the ruler *did* constitute a serious threat to Jesus. In response to the dream, Joseph took Mary and Jesus to another part of Greater Israel, namely lower Galilee, and in particular Nazareth, where they could live in safety. This was familiar territory to Joseph and Mary (Luke 2:1–5).

Joseph evidently died before Jesus began his ministry. Mary, however, remained alive throughout Jesus's ministry and suffered the trauma of watching her precious son being executed in the most brutal way imaginable—by crucifixion. It was almost certainly she who was the primary source of the birth narratives preserved in the Gospels.

JESUS

Jesus grew up in lower Galilee, in the town of Nazareth, with his mother and father and, in due course, brothers and sisters. He seems

[3] Herod's main symptoms in the run-up to his death were fever, intense whole-body itching, severe intestinal inflammation and pain, voracious hunger, foul breath, edema of the feet and lower abdomen, painful and ulcerated bowels, genital gangrene (Fournier's Disease), the production of worms, asthma, and convulsions. See Josephus, *Ant.* 17.6.5 §§168–173.

to have labored as a carpenter (Mark 6:3). When he was in his 30s, Jesus began a ministry of teaching and healing from a base in the Galilean village of Capernaum. Although Jesus did travel outside of lower Galilee, for example visiting Judea and Tyre and Sidon, he spent the lion's share of his time in rural Galilee, the very area which Tiglath-pileser had crushed so devastatingly in 733/732 BC. According to the Gospels, Jesus taught and preached and did miracles throughout the region. It was there that he preached the Sermon on the Mount and did countless wonders. His impact in Galilee was seismic, with multitudes from all across Galilee, as well as surrounding areas like Judea and the Decapolis, flocking to him (Matt. 4:23–25).

FIG. 12.2 The partially eclipsed Moon (enlarged) in Virgo's womb on the night of March 12/13, 4 BC. Image credit: Sirscha Nicholl; credit for the eclipsed Moon: Starry Night® Pro 6.4.3.

Yet, astonishingly, he was treated with contempt by many who should have known better. His own hometown of Nazareth rejected him and tried to kill him (Luke 4:29–30). Early in his Galilean ministry his own unbelieving siblings tried to seize him against his will (Mark 3:21). And, in the end, one of his own disciples betrayed him for 30 pieces of silver (14:10–22, 43–46). The religious leaders in Galilee and especially Jerusalem hated him and were eager to discredit and kill him. They denounced him as mad or demon-possessed. And eventually the Jewish Sanhedrin had him arrested and handed him over to the Roman authorities to be tried for fomenting revolutionary sentiment within the nation, forbidding the giving of tribute to the Roman emperor, and claiming to be the King of the Jews (Luke 23:2–3). By order of Pontius Pilate, the Roman procurator, Jesus was flogged and condemned to death, in fulfillment of his own repeated predictions (Mark 8:31; 9:31; 10:33–34, 45). Pilate sentenced him to be crucified, which he was, along with two criminals. And so he died and was buried. . . .

In light of what has been presented in the preceding chapters, any consideration of Jesus's life must answer the question, Did he live up to his celestial billing? Did he fulfill what the Hebrew Scriptures and in particular Isaiah had prophesied concerning the Messiah?

Isaiah 9:1–2 foretold that the Messiah would be metaphorically what the comet signaling his birth would be literally: a great light shining in the deep darkness. The prophet also claimed that the Messiah would glorify Galilee in particular by his presence and work. According to him, the Messiah would be divine in nature ("God with us" [7:14] and "Mighty God" [9:6]). At the same time, Isaiah made it clear that the execution of the divine plan would have a hidden aspect and not only would escape the notice of the people of Israel as a whole but also would offend them (8:14–15; cf. 6:9–10). Indeed Isaiah 53 disclosed that the Messiah would be rejected by his own and suffer and die as an atonement for sins. Nevertheless, he would ultimately reign over the earth (e.g., 11:4, 9–10).

Matthew explicitly asserts that Jesus fulfilled Isaiah 9:1–2 by his ministry in Galilee (Matt. 4:12–16).

According to the Gospel of John, Jesus himself expressly claimed to be "the light of the world" shining in the darkness in fulfillment of Isaiah's oracle (John 8:12; cf. 12:35).

However, the Gospels also indicate that Jesus regularly made efforts to restrict the disclosure of his identity as the Messiah. The reason for Jesus's reticence to go public regarding it is not difficult to explain. Most Jews in the first century AD had a strongly political conception of the Messiah—they expected him to be a human ruler who would overthrow the Roman government and establish a new world kingdom centered in Jerusalem—and the Roman authorities were well aware of this. Careless disclosure of the messianic claim would inevitably have stirred up misguided revolutionary hopes centered on Jesus, increasing crowd densities and provoking governmental intervention, and thus preventing Jesus from fulfilling his mission. Nevertheless, those who listened carefully to what he said and watched what he did were able to work out the fundamental claim Jesus was making about himself.

According to Mark's Gospel, when Jesus's chief disciple Peter declared to him, "You are the Messiah" (Mark 8:29), Jesus endorsed his judgment (v. 30) and privately took him, along with James and John, to the top of a mountain to give them a glimpse of his future messianic glory (9:1–8; cf. 2 Pet. 1:16–18).

Jesus declared that the Hebrew Bible portrayed the Messiah as divine in nature. He strongly challenged the contemporary understanding of the Messiah as a mere mortal—since David, in Psalm 110:1, referred to the Messiah as distinct from God and yet as being "my Lord," there could be no doubt that the Messiah was a distinct person of the Godhead (Mark 12:35–37).

According to the Gospels, Jesus did great deeds that implied that he had authority over nature—for example, he calmed a storm, walked on water, and fed multitudes; over death—for example, he raised from the dead the daughter of a synagogue ruler and a young man from Nain; and over disease—for example, he restored the limbs of paralytics, sight to the blind, hearing to the deaf, and speech to the dumb, and healed those with leprosy, epilepsy, dropsy, and hemorrhages.

Moreover, Jesus had an extraordinary divine self-consciousness. According to the Synoptic Gospels (Matthew, Mark, and Luke), he maintained that he possessed "all authority in heaven and on earth" (Matt. 28:18–20a) and that he was Immanuel ("God with us") (v. 20b). Furthermore, to substantiate his assertion that he shared the divine prerogative to forgive sins, he publicly healed a paralytic (Mark 2:1–12). Elsewhere Jesus strongly intimated that he was the unique Son of God (12:6–12) with a wholly unique relationship to God the Father (Matt. 11:27). John 8:56–58 reports that Jesus stated to unbelieving Jewish opponents that "before Abraham was, I am," plainly claiming preexistence and indeed the divine name (Ex. 3:14) for himself. According to John, Jesus also proclaimed that he and the Father were "one," prompting the Jews to try to stone him for blasphemy (John 10:30–33), and he accepted Thomas's remarkable appellation, "My Lord and my God!" (20:28–29).

At his trial before the Sanhedrin, Jesus was asked straightforwardly whether he was "the Messiah, the Son of God," and he responded, "I am," and proceeded to declare that he would soon be seated at the right hand of God the Father and would eventually return to the earth on clouds (Mark 14:60–62; cf. Matt. 26:63–65; Luke 22:67–70). Jesus's statement was rightly interpreted by the Jewish high priest as a claim to deity, and on this basis, without any evaluation of its truthfulness, the high priest called on the Sanhedrin to sentence him to death for blasphemy (Mark 14:63–64).

Furthermore, when Jesus was executed, his death was far from ordinary. The Roman centurion overseeing the execution was so deeply moved by the circumstances of his death, particularly Jesus's loud triumphant cry just before he breathed his last, that he exclaimed, "Truly this man was the Son of God!" (Mark 15:39).

When Jesus was buried, his tomb was sealed and guarded by Roman soldiers (Matt. 27:57–66) and yet, two days later, this very tomb was found to be empty (Matt. 28:1–8; Luke 24:1–12, 22–24). Over the following weeks, many hundreds of people testified that they saw him in individual and/or group encounters (1 Cor. 15:3–7). At the end of this period, his closest disciples insisted that they had seen him ascend into heaven in a cloud, and that angels in attendance had pledged that he would return in the same manner (Acts 1:9–11).

If we believe the testimony of the Gospels, therefore, Jesus certainly did live up to his celestial billing. In particular, what Isaiah prophesied concerning the Messiah came to fulfillment in and through him. Jesus was indeed Immanuel; he was indeed the metaphorical great light that shone in the deep darkness. The comet not only had taken Jesus's part in a celestial drama, but had also by its extraordinary brightness revealed his glory.

THE COMET

With respect to what happened to the comet after the Magi's arrival at their destination, we simply do not know for sure. If it maintained the same brightness slope as it had around the time of its U-turn around the Sun, it would have remained potentially visible for many months thereafter (although it would have been below the horizon for a few weeks in March/April).[4] According to our orbital elements, the comet would have gone on to spend the latter part of the spring and most of the summer in Aries the Ram[5] before slowly returning to Pisces. But the Bible gives us no explicit indication of what happened with respect to the comet from the end of November of 6 BC onward, and comets are so variable in their rates of fading that it would be unwise to make any dogmatic pronouncements. Besides, whatever happened in the rest of the cometary apparition after the comet had stood over the house in Bethlehem is basically irrelevant. After all, the most magnificent comet in human history had announced that the Messiah was born, who he was, what he was destined to do, and where he was.[6]

Will the comet soon return to the inner solar system? The slight ambiguity in the orbit, particularly its shape (or eccentricity), means that we cannot at this point be sure when, if ever, the comet will be back. If the comet has an eccentricity of 1.0, it is now 900 times farther away from the Sun than Earth is.[7] However, the orbit may be less elongated (i.e., have an eccentricity of less than 1.0), in which case the comet might be closer to our part of the solar system than that. It is even possible that the comet is already well on its way back toward the Sun and may soon make a reappearance. If it does return, it will no doubt put on another majestic show for all humans, a display that will inevitably push

[4] Of course, since December–March tends to be rainy in Israel and Babylon, it was presumably not always easy to see the comet. Assuming that the comet remained intact, when the clearer skies associated with spring came, it may well have been easily missed by an untrained eye.

[5] As we saw above, the constellation Aries was sometimes associated with Israel (see Ptolemy, *Tetrabiblos* 2.3).

[6] In his letter to the Galatians, the apostle Paul declares that "When the fullness of time came, God sent forth his son, born of a woman" (Gal. 4:4, my translation). On what ground is Paul able to claim that "the fullness of time" had come when Jesus was born? Ethelbert Stauffer comments, "Perhaps Paul, too, is thinking of the appearance of this star [i.e., the star seen by the Magi] in Gal. 4.3f." (Ethelbert Stauffer, *Jesus and His Story* [New York: Knopf, 1960], 36. Stauffer goes on to write, "At any rate, Ignatius of Antioch [*Eph.* 19:2–3] understood [Paul] to mean this when he combined the themes of Matt. 2 and Gal. 4 quite naturally in an apocalyptic advent hymn to the star of Bethlehem"). Although Stauffer's proposal regarding Gal. 4:3–4 has been largely ignored by subsequent scholarship, it merits attention. After all, Paul must have had some objective basis for his striking claim that Jesus was born at the divinely ordained moment for the initiation of the plan of salvation.

[7] 900 AU is 135 billion km, 0.014 light years, or 1/307 of the distance to the closest star system, Alpha Centauri. The approximately 1,000-km-diameter trans-Neptunian object 90377 Sedna, the largest solar system object discovered since Pluto, has an eccentricity of 0.84 and an aphelion distance of about 900 AU.

every human spectator to reflect on the religious significance of what it did in 6 BC.

What the Great Christ Comet did in 7–6 BC was extraordinary and merits wide telling. People of all disciplines—astronomers, theologians, historians, artists, etc.—must come to grips with its story. In an era when science is often viewed as the enemy of religion, the Christ Comet suggests that science may be the best friend of religion. In a period when the claims of Christ are commonly disregarded, the Star calls upon all to give his claims a fresh reappraisal. At a time when humans are so preoccupied with "the now," this Comet of comets exhorts us to reflect on the past and the future, giving us perspective for the present. In a world where Creation's wonders are often assigned to nothing other than random chance, the Magi's Star prompts us to consider whether there is not in truth a Sovereign Creator and Governor of it all.

Could there be a clearer example of God's mastery over the Cosmos than the celestial events that marked the birth of Jesus? The comet's size, shape, and chemical composition were all tailor-made for this occasion. Its orbit had its programmer's fingerprints all over it. In fact, the plan for the messianic sign was already in motion at the point that the solar system came into existence, and the precise moment of the Messiah's birth was firmly established then, guaranteed by the laws of physics. After all, the comet, the planets, and the Moon all had to be in their appointed locations within their orbits, as did Earth, for only then could the Sun, the stars, the constellations, and the meteoroids be in their proper positions. Moreover, Earth's revolution had to be in perfect sync with the comet's for this wonder to unfold. In addition, Earth's atmosphere and weather had to cooperate fully. A different comet had to have a fragmentation event at one particular historical moment so that its meteoroids could be in place, at their ascending node, to give rise to the Hydrid meteor storm on the eve of Christ's birth. The precession of the equinoxes (the slight wobble of Earth on its axis) played a key role in determining where within Virgo the cometary baby was located and where the radiant of the meteor storm was. Further, Earth's historical rate of rotation had to be such that the meteor storm occurred over the Middle East.

And it should be remembered that all of this was orchestrated to put on a dramatic celestial show tailored particularly for one small group of people, even when they were on the move. It disclosed to them detailed information regarding the Messiah and his birth, prompting them to leave their homeland on a pilgrimage to worship him. It accompanied and encouraged them as they traveled. After ushering the Magi to Bethlehem, it pinpointed the Messiah's precise location by setting near-vertically on the other side of the house from where they were. Having journeyed with the Magi all the way from Babylon to Bethlehem, we can appreciate just how awestruck they must have been as they bowed down and worshiped baby Jesus. The comet likewise beckons all humans to fall to their knees, doing so with greater awe than the Magi, reflecting a deeper appreciation of what God did to authenticate Jesus as the Messiah on the occasion of his birth—and how he did it.

In the words of Psalm 19:1–4a,

> The heavens declare the glory of God,
> and the sky proclaims the work of his hands.
> Day after day they pour out speech,
> night after night they communicate knowledge.
> There is no literal speech, nor are there literal words,
> no sound is literally heard from them.
> Yet their voice goes out into all the earth,
> their words to the ends of the world.[8]

[8] My translation.

Appendix 1

The Chinese Comet Records

We have made a case that the Star of Bethlehem was a comet. Why, then, is there no mention of it in the extant Chinese astronomical records?

As great a boon as the Chinese astronomical records are to students of ancient astronomy, there can be no question but that only a small percentage of records from the first century BC and the first century AD have survived.

Many comets were observed by the ancients that are not present in the extant Chinese records: for example, those in 480 BC (Greece only—a horn-shaped comet), 426 BC (Greece only), 373–372 BC (Greece only—this was a spectacular comet described by Aristotle, *Meteorologica* 343b as "great" and by Ephorus [Seneca, *Natural Questions* 7.16.2] as splitting in two), 345–344 BC (Italy only), 341–340 BC (Greece only), 210 BC (Babylonia only), 164 BC (Babylonia only), 163 BC (Babylonia only),[1] 43 BC (Italy only), 42 BC (Italy/Greece only[2]), 31 BC (Italy/Greece only), 30 BC (Italy/Greece only), 17 BC (Italy/Greece only), AD 9 (Germany/Italy only), AD 14 (Italy only), AD 59 (Korea only), AD 79 (Korea and Italy only, or Korea only and Italy only[3]), and AD 85,[4] 128, 153, 154 and 158 (all Korea only).[5]

Indeed we may add other comets to the list, for example, the Great Comet of 44 BC. It appeared in the aftermath of Julius Caesar's death during the Games of Venus Genetrix at around the eleventh hour of the day[6] and hence was a daytime comet that justified Plutarch's description of "a great comet which shone brilliantly for seven nights after Caesar's murder."[7] The Chinese record a comet, whether the same one as Caesar's Comet or a different one, in May–June of that year, but

[1] The 87 BC apparition of Halley's Comet may also be preserved only by the Babylonians, since the Chinese record does not prove a natural fit. To get the Chinese record to agree with Halley's Comet, one must assume that the Chinese made a mistake in the month or direction (see Tao Kiang, "The Past Orbit of Halley's Comet," *Memoirs of the Royal Astronomical Society* 76 [1972]: 56).

[2] Virgil (*Georgic* 1.488), a contemporary writing in 36–29 BC, speaks of frequent fearsome cometary apparitions at the time of the battle of Philippi (as does Manilius, *Astronomica* 1.907–908). The use of the Latin *cometae* and the context make it clear that comets are in view. Cassius Dio 47.40.2 speaks of the "Sun" shining at night.

[3] Cassius Dio 66.17.2, and Suetonius, *Vespasian* 23.4, refer to a "long-haired" comet that portended Vespasian's death. We simply cannot be sure that it was the same comet recorded in April by the Koreans.

[4] However, this may not have been a comet.

[5] Donald K. Yeomans, *Comets: A Chronological History of Observation, Science, Myth, and Folklore* (New York: John Wiley, 1991), 361–424; and A. A. Barrett, "Observations of Comets in Greek and Roman Sources before A.D. 410," *Journal of the Royal Astronomical Society of Canada* 72 (1978): 81–106. It should be noted, however, that the reliability of Korean records up until about AD 400 is questionable (see Ho Peng-Yoke, "Ancient and Mediaeval Observations of Comets and Novae in Chinese Sources," *Vistas in Astronomy* 5 [1962]: 149). Thomas John York, "The Reliability of Early East Asian Astronomical Records" (PhD thesis, Durham University, 2003), 12, comments that the Korean source of these early comet reports contains few records and that most of them are simply copies of Chinese records (York's thesis is available online at http://etheses.dur.ac.uk/3080/).

[6] Seneca, *Natural Questions* 7.17.2.

[7] Plutarch, *Caesar* 69.3 (my translation).

there is no surviving Chinese record of the extraordinary daytime cometary phenomenon that occurred toward the end of July.[8]

Octavia mentions a comet of "brilliant radiance" in the constellation Bootes in AD 62 that is not present in the extant Chinese records.[9]

In addition, Pliny[10] mentions a comet that shone almost continually and had a terrible glare in AD 64[11] that some[12] have identified as the Chinese "guest star" in May–July, but, as Gary Kronk notes,[13] Tacitus made it clear that Pliny's comet appeared "at the end of the year."[14] There is therefore no surviving Chinese record of Pliny's comet. We must remember that comets do sometimes come in clusters. For example, the 1530s saw notable comets in 1531, 1532, 1533, 1538, and 1539,[15] and in the latter half of 1618 three magnificent comets graced the skies,[16] followed by a 100-degree comet early in 1619.[17] The years 1880–1882 featured four great comets.[18]

Furthermore, Josephus[19] mentions a comet that remained visible for one year (apparently in AD 65–66), which may or may not at one point have resembled a sword, of which there is no surviving Chinese record.

With respect to the first century BC, the surviving Chinese records mention only 13 (or 14)[20] comets for the whole of the first century BC and indeed a paltry 10 (or 11) comets for the period 50 BC to AD 50:[21] 49 BC, 47 BC, May–June 44 BC, 32 BC, 12 BC, 5 BC (possibly two), 4 BC, AD 13, AD 22, AD 39.[22]

These Chinese records are derived from the

[8] With respect to the chronology and the question of the relationship between the late-July comet and the Chinese reports of a comet in May–June of 44 BC, John T. Ramsey and A. Lewis Licht (*The Comet of 44 B.C. and Caesar's Funeral Games* [Oxford: Oxford University Press, 1997]) claim that the same comet is in view. On this basis they seek to use the two reported positions of the comet to "arrive at a relatively narrow range of orbital parameters that fit our evidence" (12). The Roman apparition is taken to be a sudden cometary flare-up due to "change in the internal structure of the comet's nucleus" or nucleus splitting "almost two months after the likely date of perihelion" (119–124). Assigning May 30 to the Chinese observation and July 23 to the Roman sightings, they develop hypothetical orbits of the comet (125–132). Ramsey and Licht may be correct in suggesting that the same comet was described at different stages of its apparition (their development of a single orbit that holds together the two apparitions is impressive), but their hypothesis requires that there was a massive outburst well after perihelion. It seems equally, if not more, likely that a different comet was being reported (cf. Alexandre Guy Pingré, *Cométographie ou Traité Historique et Théoretique des Comètes*, 2 vols. [Paris: Imprimerie Royale, 1783–1784], 277–279), a very bright one around perihelion time.

[9] *Octavia* 231–232.

[10] Pliny the Elder, *Natural History* 2.23.

[11] Also Tacitus, *Annals* 15.47; Suetonius, *Nero* 36.

[12] E.g., Yeomans, *Comets*, 368.

[13] Gary W. Kronk, *Cometography: A Catalog of Comets*, 6 vols. (Cambridge: Cambridge University Press, 1999–), 1:33.

[14] Tacitus, *Ann.* 15.47 (my translation).

[15] Kronk, *Cometography*, 1:298–308.

[16] C/1618 Q1, V1, and W1.

[17] See Kronk, *Cometography*, 1:342.

[18] C/1880 C1 (Great Southern Comet), C/1881 K1 (Great Comet), C/1882 F1 (Wells), and C/1882 R1 (Great September Comet). We could also include other naked-eye comets like C/1880 S1 (Hartwig) and C/1881 N1 (Schaeberle). Peter Grego, *Blazing a Ghostly Trail: ISON and Great Comets of the Past and Future* (New York: Springer, 2014), 105, points out that six of the nineteenth century's eight great comets appeared within a 40-year period.

[19] Josephus, *J.W.* 6.5.3 (§289; cf. Cassius Dio 64.8.1).

[20] Fourteen comets if we include a rather peculiar record in January–February of 5 BC: "a white vapor emerged in the southwest, reaching from the ground up to the sky. It emerged beneath Shen and penetrated Tiance, as wide as a bolt of cloth and over 10 *zhang* [100 degrees] long. It lasted more than 10 days before departing" (David W. Pankenier, Zhentao Xu, and Yaotiao Jiang, *Archaeoastronomy in East Asia* [Amherst, NY: Cambria, 2008], 23–24). Curiously, I find that this phenomenon coincided with the birth of future Emperor Guangwu of Han, who ruled from AD 25 to 57. When I asked David Pankenier how sure he was that this was a comet, he replied (email correspondence, October 6, 2012):

> The record would be dubious, were it not for the fact that it says that the vapor was extremely long and persisted for more than 10 days, with the stellar location explicitly noted. With that it met our selection criteria. We can rule out an aurora, a bolide trail, or an atmospheric phenomenon, and the record is consistent with the appearance of a comet whose coma was invisible below the horizon but whose tail stretched up into the sky at the terrestrial place and time in question.

[21] Pankenier et al., *Archeoastronomy in East Asia*, 21–25.

[22] Christopher Cullen, "Halley's Comet and the 'Ghost' Event of 10 BC," *Quarterly Journal of the Royal Astronomical Society* 32 (1991): 113–119, discusses a Chinese record in the *Thung Chien Kang Mu* (1189) that has been interpreted by some as referring to a comet in 10 BC. He argues that it is actually referring to Halley's Comet in 12 BC. Of course, if the 10 BC comet recorded in the *Thung Chien Kang Mu* is not a wrongly dated record of Halley's Comet in 12 BC, then we would have to add it to our count of Chinese comets from 50 BC to AD 50, making 11 (or 12). It is also sometimes claimed (dubiously, in my view) that the 4 BC cometary record is misdated and originally referred to the comet in the spring of 5 BC (see Kronk, *Cometography*, 1:26–27).

Han shu, that is, *The History of the Former Han Dynasty*, which, as noted, was composed in the late first and early second centuries AD, and completed in AD 111. From 50 BC to AD 50, the *Han shu* preserves less than half of all the surviving comet records from the period and about 11–13% of the total number of comets that would generally be expected to be visible to the naked eye over such a period. The large gaps between recorded comets—43–33 BC, 31–13 BC, 11–6 BC, and 3 BC–AD 12—reinforce the conclusion that many cometary apparitions are missing. Indeed there is a notable lull in *Han shu* records of portents generally between 10 and 6 BC. In comparison, 20–11 BC and 5–1 BC are "peak" periods. To some extent the peaks and troughs may reflect political developments within China.[23] For example, the resurgence of records in 5–1 BC coincides with the end of the honeymoon period of Emperor Ai's reign and the upswelling of widespread disillusionment regarding his rule.[24]

Even if we judge the extant Chinese records in these centuries by the standard of later centuries—for example, the third or fourth century AD—the patchy nature of the surviving Chinese records is clearly seen. And we have good reason to believe that major comets are missing from the surviving Chinese records in the fourth century. According to Zdenek Sekanina and P. W. Chodas, the intrinsically very bright giant parent of the Kreutz Sungrazing Family of comets arrived at perihelion in AD 356, but we lack any mention of it in the Chinese records.[25] If the progenitor of the Kreutz family, unquestionably one of the most spectacular comets in history, did indeed arrive in that year, the absence of a Chinese record of it might say a lot about the state of the surviving Chinese records even from that period.

What confirms that there were other comets in our period is A. A. Barrett's catalog of cometary observations collected from scattered references in Greco-Roman literature:[26] the southern Europeans happen to mention comets during this very period which are not to be found among the extant Chinese records. Barrett points out that comets are mentioned in Greco-Roman literature for the following years: 49 BC, 48 BC, (July) 44 BC, 43 BC, 42 BC (multiple comets), 31 BC, 30 BC, 17 BC, 12 BC, AD 9 (multiple comets), and AD 14.[27] The historical reliability of some of these Greco-Roman cometary references may be questioned by some, but a good number of them have excellent historical credentials and cannot legitimately be discounted.

Moreover, Korean records make reference to yet other comets unpreserved in the extant

[23] Wolfram Eberhard, "The Political Function of Astronomy and Astronomers in Han China," in *Chinese Thought and Institutions*, ed. John K. Fairbank (Chicago: University of Chicago Press, 1957), 57–58. Michael Loewe, *Divination, Mythology and Monarchy in Han China* (Cambridge: Cambridge University Press, 2008), 67, 71n25, 75–76, points out the close association between cometary records and military campaigns. On pp. 81–83 he demonstrates that the 12 BC Chinese comet was preserved because it was regarded as having successfully augured events of great political significance. Note the AD 22 records of a "fuzzy star" that coincided with the beginning of Liu Yan's rebellion against Wang Mang (who had usurped the throne in AD 9).

[24] See Eberhard, "Political Function," 57–58. In 7–6 BC, the honeymoon period of Emperor Ai's reign, the officials and people were united in believing that, after the incompetent rule of Emperor Yuan and the extravagant reign of Emperor Cheng, they now had a proficient emperor (Wikipedia, s.v. "Emperor Ai of Han," http://en.wikipedia.org/wiki/Emperor_Ai_of_Han [last modified March 22, 2013]). Any comet occurring at that time was liable to have been regarded as an auspicious omen and hence may have been of less interest to the court astrologers and the historian. Note how Homer H. Dubs, *The History of the Former Han Dynasty*, 3 vols. (Baltimore: Waverly, 1938–1955), 1:261, explains the lack of eclipses and comets recorded during Emperor Wen's reign in the second century BC: "it looks as though the recorders of phenomena deliberately refused to record eclipses or comets, for the good reign of Emperor Wen made them think that Heaven was sending no admonitions, hence they concluded that there were no 'visitations.'" Notably the historian introduces inauspicious comets, along with an increased number of other portents, in the period when the positive sentiment toward the emperor dissipated, namely in 5 BC and 4 BC. At that very time tensions within the palace were growing and Ai began to be perceived as harsh, indecisive, easily offended, and always sick, and became involved in a homosexual relationship with Tung Hsien. These negative omens were no doubt included by the historian writing the *Han shu* in order to make the point that doom was now destined for Ai and the Former Han dynasty.

[25] Zdenek Sekanina and P. W. Chodas—see, for example, "Fragmentation Hierarchy of Bright Sungrazing Comets and the Birth and Orbital Evolution of the Kreutz System. I. Two-Superfragment Model," *Astrophysical Journal* 607 (2004): 624.

[26] Barrett, "Observations," 81–106.

[27] Ibid., 94–98: catalog numbers 32–42.

Chinese records in the relevant period, one in March 44 BC and another in AD 46–47.[28]

Hunger, Stephenson, Walker, and Yau emphasize what a tiny proportion of astronomical records from the Former Han period (206 BC–AD 9) has survived. They point out that during the entire Former Han period we never find more than 3 observations per year, and they sharply contrast this with the surviving nightly reports in the Babylonian Diaries and with the extant Chinese records from subsequent centuries.[29]

Lest one think that the surviving Chinese records would have included all of the *major* cometary phenomena in the latter part of the first century BC and first half of the first century AD, we should remember that they lack any reference to the following: the July 44 BC daylight comet; the multiple cometary apparitions in 42 BC;[30] the 17 BC comet as bright as the full Moon that stretched across the whole sky, north to south;[31] the 12 BC comet that experienced a fragmentation event;[32] and "several comets [that] appeared at the same time" in AD 9.[33] Notable comets are therefore missing from the extant Chinese records in the very period during which the Bethlehem Star appeared.

It is clear, then, that the extant Chinese records are very patchy indeed. Indeed so paltry is the number of cometary apparitions in the extant Chinese records in the relevant period that it is preferable to ask not why the surviving Chinese records lack certain comets, but rather why the historians of the *Han shu* preserved the particular ones that they did. Records of most comets, including a number of spectacular cometary phenomena, did not make it into the *Han shu*, and these would have been lost to history without the writings and records of other countries. In the period from 50 BC to AD 50, Greco-Roman literature and Korean astronomical records together contribute more than half of all extant references to cometary apparitions, and make it abundantly clear that the extant Chinese records lack great cometary apparitions.

What factors determined whether the writers of the *Han shu* included or excluded cometary records?

One important factor was that only a limited number of records were available to the historians of the *Han shu*. Hunger et al. state that it is certain that many astronomical records were lost prior to the composition of the *Han shu*.[34] It should be remembered that the efficiency of astronomical record-keeping and record preservation during the Former Han dynasty fluctuated greatly.[35] In explaining the absence of a record of the Halley's Comet 164 BC apparition in the Chinese records, and the lack of a sure reference to the 87 BC apparition, Stephenson comments: "In order to offer an explanation for this deficiency we must examine the statistics of astronomical observations throughout the Former Han dy-

[28] If Cullen, "Halley's Comet and the 'Ghost' Event of 10 BC," is wrong and the 10 BC comet is one that went unrecorded in the *Han shu*, it would be further evidence that those responsible for the *Han shu* were selective regarding which cometary records of the imperial astronomers they included in their work.

[29] Ramsey and Licht, *Comet of 44 B.C.*, 47. David Pankenier, a respected authority on Chinese astronomical records, concurred with this assessment (personal email message to the author, October 6, 2012).

[30] Virgil, *Georgic* 1.488–489; Manilius, *Astronomica* 1.907–908; cf. Cassius Dio 47.40.2.

[31] Cassius Dio 54.19.7; Julius Obsequens, *Liber de prodigiis* 71.

[32] Cassius Dio 54.29.8.

[33] Ibid., 56.24.4; cf. Manilius, *Astronomica* 1.899–900.

[34] Hermann Hunger, F. Richard Stephenson, C. B. F. Walker, and K. K. C. Yau, *Halley's Comet in History* (London: British Museum, 1985), 45. Ramsey and Licht, *Comet of 44 B.C.*, 109n48, point out that astronomical records in the imperial archives "no doubt perished in the turmoil that followed the overthrow of the usurper Wang Mang" in AD 23 or 25.

[35] David W. Pankenier, "On the Reliability of Han Dynasty Solar Eclipse Records," *Journal of Astronomical History and Heritage* 15.3 (2012): 211, states that, at times during the Former Han dynasty, the standard of record-keeping declined on account of cronyism, negligence, civil unrest, etc. Speaking more generally, York ("Reliability," 121), in his doctoral study of the reliability of East Asian astronomical records, concludes that the fluctuating numbers of surviving comet records are explicable with reference to changes in the proficiency of astronomical observations, record-making, and record-keeping. He points out that, in their observations of conjunctions, the efficiency of Chinese astronomers ranged from 0% to 20%, and he suggests that this low rate probably applies to other astronomical phenomena (123).

nasty. . . . Clearly much data must have been missing by the time the *Han-shu* was compiled . . . , a conclusion that is supported by the highly irregular form of the histogram."[36]

Second, as Hunger et al. also highlight, the content of a history is determined by what its author/editor sees fit to include, and it is therefore likely that a lot of data that would have been of great interest and significance to modern astronomers was deliberately excluded by the ancient historians and therefore has not survived.[37] Pankenier emphasizes that the editors of the *Han shu*, working on behalf of the ruling emperor, made their determinations regarding which astronomical records to include in their history based largely on whether the particular reports showed the working out of gan-ying theory, that every human action is met with a cosmic reaction.[38] They were more likely to include a particular astronomical report if the observation played a key role in political or military history or seemed to augur the end of the dynasty. Halley's Comet in 12 BC made a deep impression on the Chinese at the time because of the astrological messages it seemed to convey against the backdrop of the constellations. According to the *Han shu*, one contemporary astrologer (Gu Yong) stated, "This is an omen of extreme disorder such as has been rarely seen since high antiquity. If we examine the rapid movement [of this object], the variations in the length of its flaming rays, and the [constellations] on which it has trespassed successively, [it clearly signifies] harm to the women of the rear palace within, and the disaster of rebellion in the realm without."[39] The *Han shu* goes on to record that another official warned that this was a terrible omen for the dynasty.[40] The extraordinary detail given concerning this comet and its interpretation in the *Han shu* is present because it fit so perfectly with the historians' agenda: the heavens had spelled doom for the Former Han dynasty.

Could a reason for the patchy nature of the cometary records be traceable back to the process of reporting observations? The astronomers were civil servants tasked with maintaining a daily watch of the sky, both day and night, on the lookout for celestial omens.[41] They probably spotted most astronomical abnormalities, but did their observations get entered into the Register of the T'ai Shih Ling, the court official whose records were the source of the *Han shu*'s astronomical references?

Hans Bielenstein has made the case that, for a portent to get logged in the Register, high court officials had to be concerned enough about its astrological implications for the imperial order to make it the subject of a memorial to the emperor.[42] However, Martin Kern has challenged this interpretation, arguing instead that all unusual events were made subjects of memorials to the emperor.[43] If a portent was reported that caused alarm, it required action by the emperor to restore

[36] F. Richard Stephenson, "The Ancient History of Halley's Comet," in *Standing on the Shoulders of Giants*, ed. Norman Thrower (Berkeley: University of California Press, 1990), 238. The histogram he mentions is found on page 239 of Stephenson's essay (fig. 13.3).
[37] Hunger et al., *Halley's Comet in History*, 45.
[38] David Pankenier, *Popular Astrology and Border Affairs in Early Imperial China*, Sino-Platonic Papers (Philadelphia: University of Pennsylvania, 2000), 4–5.
[39] See Cullen, "Halley's Comet and the 'Ghost' Event of 10 BC," 117, who emphasizes that a number of the Chinese constellations through which the comet traversed were of great astrological importance and that it is this fact that explains the "apocalyptic note" in the commentary. See also Loewe, *Divination*, 82.
[40] David W. Pankenier, "Notes on Translations of the East Asian Records Relating to the Supernova of AD 1054," *Journal of Astronomical History and Heritage* 9.1 (2006): 80, which is a modification of a translation in Cullen, "Halley's Comet and the 'Ghost' Event of 10 BC," 117.
[41] Hunger et al., *Halley's Comet in History*, 47.
[42] Hans Bielenstein, "An Interpretation of the Portents in the Ts'ien Han Shu," *Bulletin of the Museum of Far Eastern Antiquities* 22 (1950): 127–143; idem, "Han Prognostications and Portents," *Bulletin of the Museum of Far Eastern Antiquities* 56 (1984): 97–112. Cf. R. de Crespigny, *Portents of Protest in the Later Han Dynasty: The Memorials of Hsiang-k'ai to Emperor Huan* (Canberra: Australian National University Press, 1976), 45n15.
[43] Martin Kern, "Religious Anxiety and Political Interest in Western Han Omen Interpretation: The Case of the Han Wudi Period (141–87 B.C.)," *Chūgoku shigaku* 10 (2000): 1–31.

cosmic order by making edicts (as, for example, in 44 BC).

A few decades ago it was common for scholars of the Han dynasty to claim that astronomy and astrology had an essentially political function within China, being used as a tool by officials to manipulate the emperors.[44] It was and sometimes still is believed that the astronomers would even stoop to falsifying celestial phenomena to achieve their ends.[45] Scholars such as Dubs, however, have argued that this notion is inaccurate.[46] For one thing, phenomena such as comets were difficult and dangerous to fabricate or try to hide.[47] It is likely that court officials did seek to manipulate the emperor by means of comets, but by the interpretation they assigned to them, not by free invention of them.

In conclusion, the surviving Chinese comet records are manifestly very incomplete. It would therefore be most unwise to assume that, if the Bethlehem Star was a comet, the apparition would have been preserved in the extant Chinese astronomical records.

[44] So, for example, Eberhard, "Political Function," 70; cf. 66.

[45] Ibid., 60–62, lists the 15 people whose reported portents were included in *Han shu*, series b, noting their political allegiances and biases and their tendencies to submit omens that furthered their private political agenda. Eberhard comments that "most of them had outspoken political loyalties and utilized their 'science' for the realization of their political aims" (62). With respect to fabrication, Zhang Lan and Zhao Gang, "The Identification of Comets in Chinese Historical Records," *Science China—Physics, Mechanics and Astronomy* 54 (2011): 150–151, caution those working with the Chinese records to appreciate that "some records" may have been "fabricated for political or other reasons." See also Eberhard, "Political Function," 50–51, 56, 59–60, and Loewe, *Divination*, 69, including footnote 18. Loewe states that because of, among other things, the possibility of fabrications in the Han astronomical records, "it is difficult to count the number of different [cometary] appearances that were recorded" (69).

[46] Dubs, *History*, 1:151; cf. Pankenier, "Reliability," 200–212.

[47] Pankenier, "Reliability," 211 (cf. Dubs, *History*, 3:559), highlights that deceiving the emperor by fabricating astronomical phenomena was a capital crime and indeed maintains that it was "suicidal" because such inventions could be "easily detected."

Appendix 2

The Meteor Storm of 6 BC

In our treatment of Revelation 12:1–5, because our focus was on the cometary apparition, we passed quickly over verses 3–4a. However, these verses merit closer inspection. They read, "And another sign appeared in heaven: behold, a great fire-colored[1] dragon, with seven heads and ten horns, and on his heads seven crowns.[2] His tail dragged/swept[3] a third of the stars of heaven and cast them to the earth."

Verses 3–4a should not be divorced from their subsequent context, which extends through to the end of the chapter and indeed all the way to the end of chapter 13. As we have already noted, these verses are clearly set apart from verses 1–2 not only by their content but also by the new introduction ("And another sign . . ."; v. 3). The events of verses 3ff. belong chronologically after those of verses 1–2 and, unlike them, are framed in terms of a great conflict of sovereignty. The celestial sign described in verses 3–4 belongs to the time after Virgo has begun "crying out in birth pains and the agony of giving birth" (v. 2) but before she has brought forth her child (v. 5).

Verse 4b locates the action of verses 3–4 in the latter stage of fetal expulsion, on the eve of the child's birth. One must remember that this celestial play was unfolding during brief windows of time in advance of the Sun's rising each morning. When therefore the dragon is said to stand before Virgo when she was "about to bring forth the child," it almost certainly implies that the child was born on the following film frame, seen the next morning before dawn. If the birth occurred on October 20, 6 BC, the meteor storm would have taken place on October 19.

That what is described in verses 3–4 occurred while the celestial woman, playing the part of Israel, was giving birth is theologically significant. In the Hebrew Bible, Israel is envisioned as enduring labor pains in connection with the birth of the Messiah in Bethlehem (Mic. 5:2–3[4]) and in connection with the establishment of the messianic kingdom on the earth at the end of the age (Hos. 13:13; Isa. 26:17–18; Mic. 4:10). Therefore the fact that verses 3–4 of Revelation 12 occur while the woman is giving birth provides a natural bridge between the events of 6 BC and the great tribulation immediately before the coming of the kingdom of God. As we shall see, verse 3 strongly alludes to Hebrew tradition regarding the world tyrant who persecutes the people of God dur-

[1] Again, see Johannes P. Louw and Eugene A. Nida, *Greek-English Lexicon of the New Testament: Based on Semantic Domains* (New York: United Bible Societies, 1988), §79.31.

[2] With KJV, NIV, and NLT ("crowns"); also CEB and ISV ("royal crowns").

[3] With NIV ("swept") (cf. NASB, NET, and HCSB ["swept away"]), and KJV and RV ("drew").

[4] However, this text seems to have in mind both the Messiah's birth and the transition into the messianic age.

ing the great tribulation and is ultimately conquered by "one like a son of man," namely the Messiah. We have already seen that Hydra (or the Serpent, in Babylon) is in view here. This massive constellation is located alongside the zodiacal constellations of Virgo and Leo, to their south. It is no surprise that Hydra is presented as a serpentine dragon. However, it is unexpected that the celestial dragon is fire-colored and has seven heads, seven crowns, and ten horns (v. 3). In addition, we are astonished that Hydra with its tail throws to the earth one third of the stars (v. 4).

Since a sign is, of course, something that is seen with the eyes, it is important to ask what caused the constellation figure of Hydra to turn the color of fire, to look like it had seven heads and seven crowns, and ten horns on its heads, and to appear to throw countless stars toward the earth.

That a third of the stars seemed to be hurled to the earth can only refer to a great meteor storm.[5]

In the words of Pete Bias,

> A meteor storm is one of the most amazing natural spectacles seen on Earth. An intense storm of meteors may fill all sections of the sky with fireworks. Hundreds of meteors per minute can sometimes be seen, often accompanied by beautifully bright flashes and bursting meteors that are so remarkable that people are frightened or awestruck. . . .
>
> Of course, meteor storms are a very rare phenomenon.[6]

Indeed they are. In the entire twentieth century there was a grand total of four meteor storms—two Draconid storms—1933 and 1946—and two Leonid storms—1966 and 1999.[7] It is therefore an immeasurable privilege to see one.

In Revelation 12 the emphasis is on the extent and source of the meteor storm. One third of the stars seem to be thrown from Hydra's tail. Modern meteor enthusiasts make records of the tracks of the meteors that they see during a given meteor shower. By extending all of the lines back and seeing where they converge, they can identify the radiant. During meteor storms, however, it is very easy to tell where the radiant is. In the wake of the Leonid meteor storm of November 13, 1833, many observers reported that the meteors were radiating from close to the star γ (Gamma) Leonis.[8] Indeed it was the 1833 Leonid storm that prompted astronomers to realize that all meteor showers have radiants.[9] With respect to the meteor storm that is described in Rev. 12:3–4 and that may be dated to 6 BC, attributing to Hydra's tail the dragging and throwing of such a huge swath of stars strongly suggests that the meteors seemed to radiate out from it.

Since the lowest part of the tail (π [Pi] Hydrae) rose above the horizon, so that Hydra could be regarded as standing, only at the

[5] By definition, meteor storms have at least 1,000 meteors an hour (or, more technically, would have a rate of 1,000 meteors per hour if the radiant were at the zenith) (Mark Littmann, *The Heavens on Fire: The Great Leonid Meteor Storms* [Cambridge: Cambridge University Press, 1999], 276 note c).

[6] Peter V. Bias, *Meteors and Meteor Showers: An Amateur's Guide to Meteors* (Cincinnati, OH: Miracle Publishing, 2005), 11. I am grateful to the author for helping me secure a copy of this outstanding introduction to meteors.

[7] Ibid., 12. Although it is common to refer to "the 1998 Leonid Fireball Storm," it was technically only a meteor shower.

[8] Littmann, *Heavens on Fire*, 16.

[9] Ibid., 17–18. To observers it seemed that the meteors, regardless of where in the sky they started, were radiating out from a particular point in the constellation Leo (ibid., 1–2). At the radiant, the meteors were little more than brightening dots (heading straight for the observer!); near the radiant they were short streaks; and farther out from the radiant they consisted of longer streaks (ibid., 253). Although meteoroids travel in parallel paths, striking Earth's atmosphere at one point, they seem to radiate out as they come closer, like parallel railway tracks or like snowflakes sweeping toward a speeding car's windscreen during a blizzard. Littmann (253) gives a wonderful description of what a meteor shower/storm looks like if you gaze directly at the radiant (italics his):

> Out of the corners of your eyes, you will catch meteors streaking past, creating the impression that you are flying through space . . . *which of course you are*. Your spaceship Earth is racing around the Sun at 18.5 miles per second (29.8 kilometers per second), but almost never can you sense this motion. The one exception is during a meteor storm, as the Earth dashes through a stream of particles. Then and only then can you truly sense the Earth in motion, in high-speed flight, a little like in *Star Trek* when the *Enterprise* travels at warp speed.

FIG. 14.1 The meteor storm of 1866 over Greenwich (Collection of Gerald H. Morris, London). Anonymous chromolithograph. From Agnes Giberne, *Sun, Moon, and Stars: Astronomy for Beginners* (New York: American Tract Society, 1880), opposite page 218.

conclusion of or after the meteor storm, the meteors probably radiated from the upper part of Hydra's tail, that is, from between γ (Gamma) Hydrae and the section of Hydra on which Corvus/the Raven perched. With one-third of the visible stars seeming to be thrown from the sky to the earth,[10] observers could have quickly detected that they all seemed to issue from Hydra's tail (see fig. 14.2).

Meteor showers and storms are due to meteoroid streams, consisting of debris that has been expelled from comets (or asteroids), which cross Earth's orbital path when Earth is present.

In recent history there have been two meteoroid streams that have had a propensity to create meteor storms on Earth: the Draconids, related to the comet 21P/Giacobini-Zinner; and the Leonids, related to 55P/Tempel-Tuttle. Many impressive meteor storms of the past millennium were due to the Leonid meteoroid stream. Leonid meteor storms or heavy meteor showers occur in a cycle of 33.25 years, which is how long their parent comet takes to complete one revolution.

Mark Littmann helpfully pointed out in 1998 that every Leonid meteor storm from 902 to 1966 took place when the comet that had parented the meteoroid stream, Tempel-Tuttle, was no farther than 3 years from Earth. He suggested that the Leonid meteor storms were caused by a concentration of meteoroids within a one-eighth section of the meteoroid stream, orbiting in sync with the

[10] One witness of the 1799 Leonid meteor storm stated that "there was not a space on the firmament equal in extent to three diameters of the moon which was not filled every instant with bolides and falling stars" (Alexander von Humboldt and Aimé Bonpland, *Personal Narrative of Travels to the Equinoctial Regions of the New Continent during the Years 1799–1804*, vol. 1, trans. T. Ross [London: George Bell & Sons, 1907], chapter 1.10).

FIG. 14.2 An artistic representation of the meteor storm radiating from Hydra's tail in 6 BC. Image credit: Sirscha Nicholl.

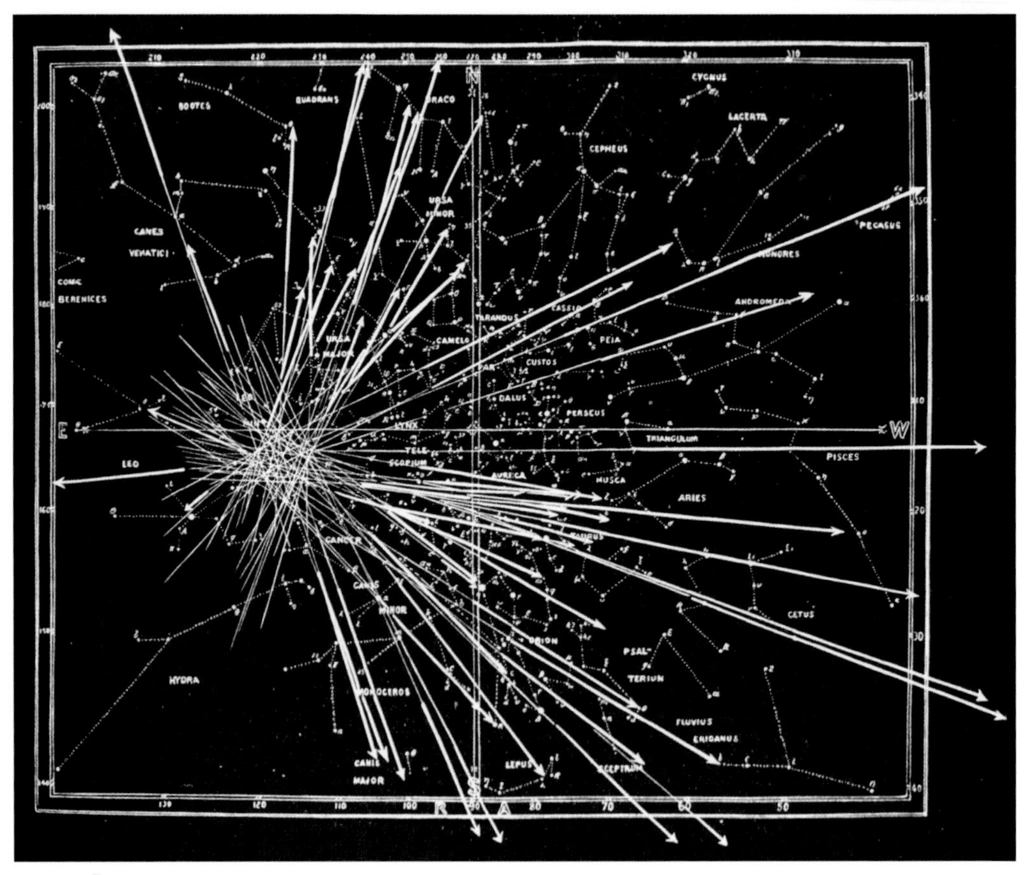

FIG. 14.3 The radiant of the Leonid meteors seen at Greenwich on November 13, 1866. From George F. Chambers, *The Story of the Comets* (Oxford: Clarendon, 1909), 195 fig. 103.

FIG. 14.4 The Leonid Meteor Storm on November 13, 1833. From *Bible Readings for the Home Circle* (Toronto: Review & Herald, 1889). Image source: Wikimedia Commons.

FIG. 14.5 The Leonid Meteor Storm of November 12, 1799, as witnessed by Andrew Ellicott during a full Moon from Floridian waters. From Edwin Dunkin, *The Midnight Sky* (London: Religious Tract Society, 1869), 293.

parent comet, with most of those meteoroids dragging behind it.[11]

Therefore when we read of a major meteor storm in Revelation 12:4, we should give a thought to where the parent comet is.

Sometimes meteor storms may be due to the parent comet fragmenting in whole or in part—in the case of the Andromedid meteor storms of 1872 and 1885, Comet 3D/Biela's complete disintegration over a decade beginning in 1842–1843.[12]

Meteor storms are not only rare, but they also tend to be short-lived—they endure for at most a few hours, sometimes no more than a quarter of an hour. They "spawn from very narrow streams of densely packed particles, perhaps only 25,000 to 75,000 km wide—or only a few to several times the diameter of Earth."[13]

And then we must consider that only a narrow slice of the world's population is in a position to observe the phenomenon, and only if weather conditions are favorable.[14]

[11] Littmann, *Heavens on Fire*, 273. David Asher and Robert McNaught famously developed a model explaining how the occurrences of the Leonids can be matched against particular past perihelion passages of the parent comet, enabling remarkably precise predictions of meteor outbursts. See R. H. McNaught and D. J. Asher, "Leonid Dust Trails and Meteor Storms," *WGN, Journal of the International Meteor Organization* 27 (April 1999): 85–102; D. J. Asher, "Leonid Dust Trail Theories," in *Proceedings of the International Meteor Conference, Frasso Sabino, Italy, 23–26 September 1999*, ed. R. Arlt (Potsdam: IMO, 2000), 5–21; and especially the section on the Armagh Observatory's website entitled "Leonid dust trails": http://www.arm.ac.uk/leonid/dustexpl.html (last modified September 8, 2010).

[12] See Peter Jenniskens and Jeremie Vaubaillon, "3D/Biela and the Andromedids: Fragmenting versus Sublimating Comets," *Astronomical Journal* 134 (2007): 1037–1045.

[13] Bias, *Meteors and Meteor Showers*, 12.

[14] Ibid.

Modern meteor storm hunters, even with all of their technological sophistication, would be the first to emphasize that seeing one is remarkably fortuitous. That one radiated from Hydra's tail, right next to Virgo, on the eve of the birth of her cometary baby is astonishing. Although observers would have had no idea what caused meteor storms, they would have been aware of how rare they were, and they must have been deeply impressed that one occurred right at that very time.

As we shall see, the meteoroid stream responsible for the 6 BC meteor storm was direct or prograde (i.e., it orbited in a counterclockwise direction), unlike the Christ Comet, which was retrograde (orbiting in a clockwise direction). The meteoroid stream responsible was passing from the south to the north of the ecliptic plane (i.e., it was crossing its ascending node) at the time when the meteor storm occurred.

FIG. 14.6 The Leonid Meteor Storm of November 12–13, 1833 (from E. Weiß, *Bilderatlas der Sternenwelt* [Esslingen: J. F. Schreiber, 1888]). Image credit: Wikimedia Commons.

Astronomically speaking, the sign in Revelation 12:3–4a is entirely unrelated to the Christ Comet. Yet the cometary apparition and meteor storm worked together in harmony as two distinct actors with different main parts in a single, great unfolding celestial play marking the nativity of the Messiah.

Is there any way to match the meteor storm described in Revelation with a modern meteor shower or meteoroid stream or with a particular comet? It is not necessarily a hopeless venture. It is theoretically possible that the meteoroid stream that gave rise to the meteor storm still exists and that it still delivers, or will in the future deliver, meteors on an annual or sporadic basis. We know approximately when this meteor storm of 6 BC occurred and roughly where it radiated from, and so we are in a position to determine the range of orbital possibilities for the meteoroid stream responsible. By comparing the possible orbits of the Hydrid meteoroid stream to the known orbits of meteoroid streams and comets (and asteroids), we might conceivably be able to find a match. Needless to say, the implications of two millennia of gravitational effects on the evolution of orbits would have to be taken into account. We would also need to allow for the impact of precession of the equinoxes, the 26,000-year cycle due to the wobble of Earth on its axis. It would, of course, be wonderful if we could identify the meteoroid stream or comet responsible for the meteor storm of 6 BC. However, we do well to remember what Pete Bias has written:

> Most meteor showers are short-lived and many are unpredictable. The same gravitational processes that occasionally pull meteor streams into the Earth's orbit also pull them away. Moreover, because of the continual dispersion of meteoroids away from a parent comet's orbit (by a combination of planetary perturbations, unique initial ejection

FIG. 14.7 The Chelyabinsk Meteor (superbolide) on February 15, 2013, as captured by a dashcam at dawn in Kamensk-Uralsky. Image credit: Aleksandr Ivanov. Image source: YouTube/Wikimedia Commons.

> conditions for each meteoroid, Poynting-Robertson effects, and others), even old faithful streams like the Perseids eventually disperse unless new meteoroids are continually added. Meteor showers must, perforce, be finite.[15]

It is worth recalling what exactly a meteor is. It is essentially a small "pebble" of space dust that collides with Earth's atmosphere. The collision begins some 130–150 km up in the atmosphere. The friction the meteoroid experiences causes it to become superheated and bright and to lose some of its mass. This dispelled material forms a vapor cloud around the remaining core of the meteoroid, and it too begins to disintegrate as it impacts the atmosphere. Thus there is "an ever expanding, ever disintegrating nebulous mass of dust and superheated vapor that is thrown back and away from the main particle as it is continually, violently slowed by the collisions with the atmosphere."[16] Most meteoroids have fully disintegrated by the time they are about 75 km above the surface of Earth.[17] To human observers, what happens high in the atmosphere manifests as a bright light streaking across the heavens for a brief moment.

Especially bright meteors, caused by larger pieces of comet debris striking Earth's atmosphere, are called fireballs. They are brighter than all the stars and at least as bright as the planets Jupiter or Venus (approximately magnitude -2 to -5 respectively). Sometimes they are as bright as, or even brighter than, the full Moon. Fireballs that attain to a magnitude of -14 or greater are called bolides; those that reach -17 or greater, like the Chelyabinsk meteor of February 15, 2013 (see fig. 14.7), are called superbolides.[18]

Meteoroids responsible for fireballs that are very bright (over -8 in apparent magnitude)

[15] Ibid., 196–197.
[16] Ibid., 2.
[17] Cf. ibid., 2–3.
[18] Mario Di Martino and Alberto Cellino, "Physical Properties of Comets and Asteroids Inferred from Fireball Observations," in *Mitigation of Hazardous Comets and Asteroids*, ed. M. J. S. Belton (Cambridge: Cambridge University Press, 2004), 156.

sometimes survive the passage through the atmosphere and make it to the ground as meteorites (Acts 19:35 mentions a "sacred stone that fell from the sky"). Often fireballs remain visible for several seconds. Some very bright fireballs, bolides, and superbolides are visible for many seconds. In their wake they may leave trails that can last for quite a few minutes or, in rare cases such as the Chelyabinsk meteor, for up to 9 hours.

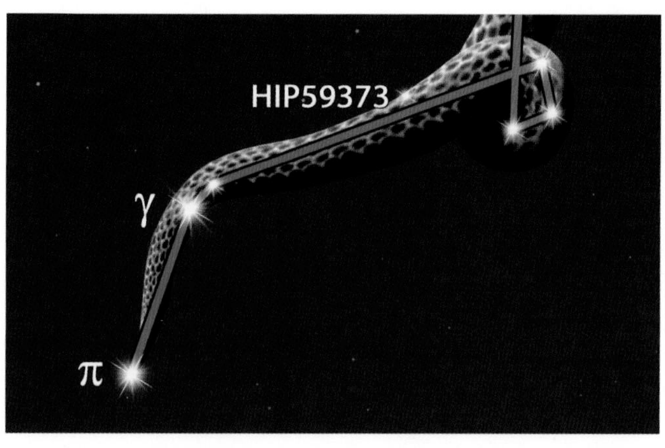

FIG. 14.8 The tail of Hydra, the Serpent. The tip was π (Pi) and the tail extended up approximately to where the very faint star HIP59373 is. Image credit: Sirscha Nicholl.

Concentrated numbers of fireballs are associated with particular heavy meteor showers or storms. On November 17, 1998, Earth passed through a broad section of a Leonid meteoroid stream, with the result that there was "a glorious rain of fireballs" across the world.[19] The reason for the dense concentration of larger meteoroids that gives rise to such "fireball showers" is almost certainly that the parent comet underwent some kind of fragmentation or splitting event at the time that it deposited these meteoroids, whether recently or hundreds or even thousands of years earlier. In the case of the 1998 Leonid "storm," the meteoroids were deposited by Comet Tempel-Tuttle during its 1333 return.[20] Bias reports that for 2½ hours he "was dazzled by the most beautiful shower of fireballs that I'll ever hope to see. Fireballs of -3 magnitude or brighter were being seen almost every other minute! All were terrifically bright and impressive. . . . Several lit up the entire backyard. One evidently was so bright that it lit up the southern horizon like a reddish false dawn despite the fact that the meteor was below my horizon."[21]

Jenniskens comments that "The biggest Leonid fireballs associated with the 1998 Filament had a magnitude of [about] -14.5 [magnitude],"[22] which is brighter than the full Moon.

The meteor storm described by Revelation 12 gave the impression that a third of the visible stars in the dome of the sky at the time were pulled and thrown toward Earth by Hydra's tail. It seems that the tail of Hydra has always tended to be understood as ending at π (Pi) Hydrae. As for how far up Hydra's body the tail extends, we gather from Pseudo-Eratosthenes (*Catasterismi* 41) that Corvus/the Raven was regarded as being perched on the tail. π (Pi) Hydrae and γ (Gamma) Hydrae together form a distinctive last part of the tail, which is essentially parallel to the ecliptic.[23]

[19] Peter Jenniskens, *Meteor Showers and Their Parent Comets* (Cambridge: Cambridge University Press, 2006), 208.
[20] This was discovered by David J. Asher. See David J. Asher, Mark E. Bailey, and V. V. Emel'yanenko, "The Resonant Leonid Trail from 1333," *Irish Astronomical Journal* 26.2 (1999): 91–93; David J. Asher, Mark E. Bailey, and V. V. Emel'yanenko, "Resonant Meteoroids from Comet Tempel-Tuttle in 1333: The Cause of the Unexpected Leonid Outburst in 1998," *Monthly Notices of the Royal Astronomical Society* 304.4 (April 16, 1999): L53–L56.
[21] Bias, *Meteors and Meteor Showers*, 67.
[22] Jenniskens, *Meteor Showers*, 238. Technically, a -14.5-magnitude fireball would classify as a bolide.
[23] This is how most artists have tended to portray it. Ptolemy denominated γ (Gamma) Hydrae as "the star after Corvus, in the section by the tail (*prope caudam*)" (*Ptolemy's Almagest*, trans. Toomer, 392)—the Latin *prope caudam* means "beside the tail." Perhaps Ptolemy is envisioning the tail as curving around a couple of degrees to the side of γ Hydrae. However, Ptolemy's predecessors Pseudo-Eratosthenes (*Catasterismi* 41) and Hyginus (*Poetica Astronomica* 2.40) certainly seem to have regarded γ Hydrae as part of Hydra's body, not as being beside it (see Theony Condos, *Star Myths of the Greeks and Romans: A Sourcebook* [Grand Rapids, MI: Phanes, 1997], 120 and 122). Moreover, although we lack the star catalog of Hipparchus, we gather from his extant *Commentary*

From γ (Gamma) Hydrae upwards, the tail curves sharply and heads toward the south. It was on this stretch, between γ (Gamma) Hydrae and the coil associated with β, ο, and ξ Hydrae, that Corvus/the Raven rested and the tail commenced (fig. 14.8).

The star γ (Gamma) Hydrae rose approximately 54 minutes before π (Pi) Hydrae. If γ (Gamma) Hydrae was the radiant of the meteors, then, the meteor storm would have had to occur during that 54-minute period. If the radiant was at, say, HIP59373 (the part of Hydra on which Corvus/the Raven perched), then the meteor storm would have had to have taken place within the 1 hour and 44 minutes between its rising and π (Pi) Hydrae's rising. It was a moonless sky.

The meteor display in Hydra is probably responsible for the peculiar appearance of Hydra, as described in Revelation 12:3—the constellation figure's fiery appearance, its seven crowns, and its ten horns. Notably the author of Revelation counts this description of Hydra's appearance as an integral part of the second extraordinary sign in heaven (vv. 3–4). At the time of the meteor storm, Hydra would have stretched upwards just over a third of the way from the eastern (ESE/SE) horizon to the western (W/WSW). That is striking, because Revelation 12:3 states that one-third of the stars in the dome of the sky seemed to be thrown to the earth. That would seem to mean that no meteor streak commenced beyond that point, although many meteors presumably extended beyond it.

To discover the orbit of the meteoroid stream responsible for the meteor storm radiating from a point between γ (Gamma) Hydrae and HIP59373, in the relevant window of time, from the Near East, I approached David Asher of the Armagh Observatory.

Nailing down an orbit for an ancient meteoroid stream is no easy business—calculations must take into account the rate of Earth's rotation, precession of the equinoxes, and many other factors. Plus, it is unclear from Revelation 12:3–4 what the velocity of the meteors was.

Meteor storms are typically related to short-period comets, either Jupiter-family comets (orbital period: 3–20 years) or Halley-type comets (orbital period: 20–200 years). However, long-period comets with orbital periods up to 10,000 years can give rise to meteor outbursts, and cometary asteroids might also conceivably give rise to meteor storms.

David Asher worked out a range of possible orbits for the meteoroid stream that caused the 6 BC Hydrid meteor storm.[24] Any of the orbits in table 14.1 could theoretically have caused a meteor storm at the relevant time, radiating from γ (Gamma) Hydrae.

It is also possible that the meteors radiated from higher up on the tail, which extended approximately as far as the star HIP59373. A meteoroid stream with any of the sets of orbital elements shown in table 14.2 could theoretically have resulted in a meteor storm radiating from the uppermost part of the tail, where the feet of the Babylonian Raven (Greek, Corvus the Crow) rested.

We suggest that the meteors probably radiated from somewhere between γ (Gamma) Hydrae and HIP59373. Therefore these tables

on the Phenomena of Aratus and Eudoxus that he regarded the tail as ending at π (Pi) Hydrae and as encompassing HIP65835, which is just 2½ degrees below γ (Gamma) Hydrae. This suggests that he imagined the tail as passing through γ Hydrae (see the index to constellations in the forthcoming English translation of Hipparchus's Commentary by Roger MacFarlane and Paul Mills).

[24] Peter Jenniskens of NASA's SETI Institute (personal email messages to the author, October 15 and November 27, 2012) was the first to work on the orbit. Then David Asher (personal email messages to the author, December 31, 2012, January 4, 2013, and August 6–7, 2013) recalculated the orbit in a J2000 ecliptic frame, using positions along the upper section of Hydra's tail in this same reference frame, taking the vectors from NASA's Horizons website. He assumed a viewing on or around 1719522.708333333 (October 19, 6 BC, 05:00 Coordinate Time) from Babylon (a viewing from Jerusalem produces virtually identical results, because it is on essentially the same latitude as Babylon). David took account of zenith attraction, which is an important factor, since the meteor storm is radiating from relatively close to the horizon. I am most grateful to Dr. Asher, who developed a meteor orbit calculation program to determine the possible orbits of the meteoroid stream giving rise to the Hydrid meteor storm of 6 BC.

Vinf	Vg	Z_t	*lambda*	*beta*	a	q	e	i	Node	ω	f	P
Long-period												
45.00	43.61	85.01	208.62	-14.31	49.323	0.172	0.997	37.5	50.9	229.4	130.9	434.1
Halley-type												
44.00	42.58	85.08	208.70	-14.32	11.195	0.177	0.984	36.0	50.9	229.5	130.9	36.79
Jupiter-family												
43.00	41.54	85.17	208.78	-14.33	6.410	0.182	0.972	34.6	50.9	229.5	130.9	16.57
42.00	40.51	85.25	208.87	-14.35	4.540	0.188	0.958	33.2	50.9	229.6	130.8	9.47
41.00	39.47	85.35	208.96	-14.37	3.544	0.195	0.945	31.8	50.9	229.7	130.7	6.68
40.00	38.43	85.45	209.07	-14.39	2.927	0.202	0.931	30.4	50.9	229.8	130.6	5.01
39.00	37.39	85.56	209.18	-14.41	2.509	0.209	0.917	29.1	50.8	230.0	130.4	4.00
38.00	36.34	85.68	209.30	-14.43	2.206	0.217	0.902	27.9	50.8	230.2	130.2	3.30
Other												
37.00	35.30	85.81	209.44	-14.45	1.979	0.225	0.886	26.6	50.8	230.5	130.0	2.77
36.00	34.25	85.95	209.58	-14.48	1.802	0.234	0.870	25.4	50.8	230.7	129.8	2.41
35.00	33.20	86.11	209.74	-14.51	1.661	0.244	0.853	24.2	50.7	231.0	129.5	2.14
34.00	32.14	86.28	209.91	-14.54	1.546	0.254	0.836	23.1	50.7	231.4	129.1	1.93
33.00	31.08	86.47	210.10	-14.57	1.451	0.265	0.817	22.0	50.7	231.8	128.8	1.74
32.00	30.02	86.68	210.31	-14.61	1.373	0.277	0.798	20.9	50.7	232.2	128.4	1.61
31.00	28.95	86.90	210.54	-14.65	1.306	0.290	0.778	19.8	50.6	232.7	127.9	1.49
30.00	27.87	87.16	210.80	-14.70	1.250	0.304	0.757	18.8	50.6	233.3	127.4	1.39
29.00	26.79	87.44	211.09	-14.75	1.203	0.318	0.735	17.8	50.6	233.9	126.8	1.31
28.00	25.71	87.76	211.42	-14.80	1.163	0.334	0.713	16.8	50.5	234.5	126.2	1.26
27.00	24.62	88.12	211.78	-14.87	1.128	0.351	0.689	15.9	50.5	235.3	125.4	1.20
26.00	23.51	88.53	212.20	-14.94	1.099	0.370	0.664	14.9	50.4	236.2	124.6	1.16
25.00	22.40	89.00	212.68	-15.02	1.075	0.390	0.638	14.0	50.4	237.2	123.7	1.12
24.00	21.28	89.53	213.22	-15.11	1.056	0.412	0.610	13.2	50.3	238.3	122.6	1.09
23.00	20.15	90.15	213.85	-15.21	1.040	0.436	0.581	12.3	50.3	239.6	121.4	1.06
22.00	19.00	90.87	214.59	-15.33	1.029	0.462	0.551	11.5	50.2	241.1	119.9	1.04
21.00	17.83	91.73	215.47	-15.46	1.021	0.491	0.519	10.6	50.1	243.0	118.2	1.03
20.00	16.64	92.75	216.51	-15.62	1.017	0.523	0.486	9.8	50.0	245.2	116.1	1.03

TABLE 14.1 The orbital possibilities for the meteoroid stream that caused the Hydrid meteor storm if it was observed from Babylon to radiate from γ (Gamma) Hydrae 1 hour 6 minutes before sunrise* on October 19, 6 BC (as calculated by Dr. David Asher of the Armagh Observatory). "Vinf" is the pre-atmospheric velocity (km/sec); "Vg" the geocentric velocity (km/sec); "Z_t" the distance of the radiant from the zenith (taking into account the analysis of P. S. Gural, "Fully Correcting for the Spread in Meteor Radiant Positions Due to Gravitational Attraction," *WGN, Journal of the International Meteor Organization* 29.4 [2000]: 134–138); "*lambda*" and "*beta*" the radiant as it would have been observed (before correction for zenith attraction); "a" the semi-major axis of the meteoroid stream (in AU); "q" the perihelion distance (in AU); "e" the eccentricity of the orbit; "i" the inclination of the orbital plane; "Node" the ascending node; "ω" the argument of perihelion; "f" the true anomaly; and "P" the orbital period (in years).

NOTE: *Absolute precision regarding the moment when the meteor storm occurred is not required for our purpose, that is, getting a sense of the orbital characteristics of the meteoroid stream responsible.

contain the approximate outer limits of the meteoroid stream's actual orbital elements.[25]

A cometary asteroid, Jupiter-type comet, Halley-type comet, and long-period comet all remain on the table as possible parents of the meteoroid stream that caused the meteor storm of 6 BC (see fig. 14.9). If it was a long-period meteoroid stream, the meteoroids would have had a velocity of about 45 km/second. If the meteoroids hailed from a Halley-type or Jupiter-family comet, they would have had a medium-to-fast velocity (for a

[25]The higher the radiant of a meteor shower is relative to the horizon, the more of its meteors are visible (for more on this, see Bias, *Meteors and Meteor Showers*, 30–35). It is estimated that when a radiant is about 15 degrees above the horizon, you will see approximately one quarter of the meteors that would have been visible if the radiant had been at the zenith (ibid., 33 fig. 2.2). If the 6 BC Hydrid meteor storm radiated from HIP59373 and was seen 66 minutes before sunrise, the radiant would have been approximately 15 degrees above the horizon. The 1799 Leonid meteor storm was observed by Alexander von Humboldt when the

Vinf	Vg	Z_t	*lambda*	*beta*	a	q	e	i	Node	ω	f	P
Halley-type												
51.00	49.78	75.33	195.10	-25.33	15.376	0.259	0.983	80.0	51.2	241.0	119.3	59.47
Jupiter-type												
50.00	48.75	75.38	195.15	-25.34	7.228	0.254	0.965	78.3	51.2	239.3	120.9	19.55
49.00	47.73	75.42	195.20	-25.35	4.777	0.249	0.948	76.5	51.2	237.7	122.5	10.48
48.00	46.70	75.47	195.25	-25.37	3.597	0.245	0.932	74.7	51.2	236.2	124.0	6.84
47.00	45.67	75.52	195.31	-25.38	2.905	0.241	0.917	72.8	51.1	234.7	125.5	4.95
46.00	44.64	75.58	195.37	-25.40	2.450	0.237	0.903	70.9	51.1	233.3	126.9	3.82
45.00	43.61	75.64	195.43	-25.42	2.129	0.234	0.890	68.9	51.1	231.9	128.3	3.1
Other												
44.00	42.58	75.71	195.50	-25.44	1.891	0.232	0.877	66.9	51.1	230.6	129.6	2.59
43.00	41.54	75.77	195.57	-25.46	1.707	0.230	0.866	64.8	51.1	229.4	130.9	2.25
42.00	40.51	75.85	195.65	-25.48	1.562	0.228	0.854	62.7	51.1	228.2	132.0	1.95
41.00	39.47	75.93	195.74	-25.51	1.445	0.227	0.843	60.6	51.1	227.1	133.1	1.74
40.00	38.43	76.02	195.83	-25.53	1.348	0.227	0.832	58.5	51.1	226.1	134.1	1.57
39.00	37.39	76.11	195.93	-25.56	1.267	0.228	0.820	56.3	51.1	225.2	135.0	1.43
38.00	36.34	76.21	196.03	-25.59	1.199	0.229	0.809	54.1	51.1	224.4	135.9	1.31
37.00	35.30	76.32	196.15	-25.63	1.141	0.231	0.798	51.9	51.0	223.6	136.6	1.22
36.00	34.25	76.44	196.28	-25.66	1.092	0.233	0.786	49.7	51.0	222.9	137.3	1.14
35.00	33.20	76.58	196.42	-25.70	1.049	0.237	0.774	47.6	51.0	222.4	137.9	1.07
34.00	32.14	76.72	196.57	-25.74	1.012	0.241	0.762	45.4	51.0	221.8	138.4	1.02
33.00	31.08	76.88	196.74	-25.79	0.980	0.246	0.749	43.2	51.0	221.4	138.9	0.97
32.00	30.02	77.06	196.93	-25.84	0.952	0.252	0.735	41.1	51.0	221.1	139.2	0.93
31.00	28.95	77.25	197.14	-25.90	0.928	0.260	0.720	39.0	51.0	220.8	139.5	0.89
30.00	27.87	77.47	197.37	-25.96	0.908	0.268	0.705	36.9	50.9	220.7	139.7	0.87
29.00	26.79	77.71	197.62	-26.03	0.890	0.277	0.688	34.9	50.9	220.6	139.8	0.84
28.00	25.71	77.98	197.91	-26.11	0.875	0.288	0.671	32.9	50.9	220.6	139.8	0.82
27.00	24.62	78.29	198.24	-26.20	0.863	0.301	0.652	30.9	50.9	220.7	139.7	0.80
26.00	23.51	78.64	198.61	-26.29	0.853	0.314	0.631	29.0	50.8	220.9	139.5	0.78
25.00	22.40	79.03	199.04	-26.40	0.846	0.330	0.610	27.1	50.8	221.2	139.3	0.78
24.00	21.28	79.49	199.53	-26.53	0.841	0.348	0.586	25.3	50.8	221.6	138.9	0.77
23.00	20.15	80.01	200.09	-26.67	0.838	0.367	0.561	23.5	50.7	222.2	138.3	0.76
22.00	19.00	80.63	200.76	-26.84	0.837	0.390	0.534	21.7	50.7	223.0	137.6	0.77
21.00	17.83	81.36	201.54	-27.03	0.839	0.415	0.505	20.0	50.6	223.9	136.7	0.77
20.00	16.64	82.22	202.49	-27.25	0.843	0.444	0.474	18.4	50.6	225.2	135.5	0.78
19.00	15.43	83.27	203.63	-27.51	0.851	0.476	0.440	16.7	50.5	226.8	133.9	0.78

TABLE 14.2 The orbital possibilities for the meteoroid stream that caused the Hydrid meteor storm if it was observed from Babylon to radiate from HIP59373 1 hour 6 minutes before sunrise on October 19, 6 BC (as calculated by Dr. David Asher of the Armagh Observatory). For abbreviations, see Table 14.1.

Halley-type stream, 43–51 km/second; for a Jupiter-family stream, 38–50 km/second).[26] If a cometary asteroid gave rise to the meteoroid stream (3200 Phaethon is the parent of the Geminid meteor shower), the meteors would have had a slow or medium velocity.

Could the meteor storm of 6 BC be related to any current meteor shower? That is difficult to answer, since there is at this point no catalog detailing orbital elements of meteoroid streams or radiants of meteor showers that occurred two millennia ago. Gravitational

radiant was low on the eastern horizon from Cumana, Venezuela. During that window of time he reported seeing thousands and thousands of meteors and fireballs radiating out across the sky (von Humboldt and Bonpland, *Personal Narrative*, chapter 1.10). Similarly, the 1766 meteor storm was seen from Cumana, as well as from Quito, Ecuador, when the radiant was very low on the horizon (ibid.). Leonid meteor storms have been recorded with meteor hourly rates of several hundreds of thousands. If the 6 BC Hydrid storm was anything like the 1766, 1799, or 1833 Leonid storms in intensity, tens of thousands may have been visible per hour radiating from the Serpent's tail.

[26] Medium-velocity meteoroids tend to make for brighter meteors than low-velocity meteoroids. Moreover, Gary Kronk notes that we usually see brighter meteors (and more meteors generally) in the period before dawn ("What Is a Meteor Shower?," http://meteor

factors mean that orbits evolve and hence the orbital elements of the meteoroid stream *then* might no longer resemble what they are *now* (particularly in their argument of perihelion [ω] and longitude of the ascending node [Ω] values). However, it is still interesting to observe that a few meteoroid streams have orbits that are at least superficially similar to that of the meteoroid stream responsible for the meteor storm in 6 BC.

If the meteors radiated from high on the tail and the meteoroid stream orbit had a semi-major axis of 1.04–1.1 AU, it would be reminiscent of the μ and κ Hydrid meteor showers (and the related January Hydrids and Iota Sculptorids) not only in perihelion distance (0.233–0.237 AU compared with 0.215 AU and 0.249 AU respectively for the κ and μ Hydrids) but also in eccentricity (0.77–0.79, compared with 0.79 and 0.77), velocity (33–34 km/second compared with 37.6 and 39.1 km/second), and, to some extent, inclination (48–50 degrees compared with 66.5 and 71.8 degrees).[27]

In addition, it should be noted that within the large population of near-Earth asteroids are an unknown number that are cometary in nature (like Apollo asteroid 4015 Wilson-Harrington = Comet 107P/Wilson-Harrington) or are remnants of comets. It is therefore perhaps worth pointing out that some objects classified as Apollo asteroids have orbits that are similar to possible orbits of the meteoroid stream that caused the Hydrid meteor storm of 6 BC. Of course, we must remember that 2,000 years of orbital evolution may mean that the orbit of the Hydrid meteoroid stream is no longer recognizably similar to the orbits of these asteroids.

The orbit of Apollo asteroid 2009 HU58 (magnitude 19) is reminiscent of the orbit of the meteoroid stream responsible for the 6 BC meteor storm if the latter's meteors radiated from one-third of the way from γ Hydrae to HIP59373 (table 14.3).

The orbit of the Apollo asteroid 2000 UR16 (magnitude 23) is reminiscent of the orbit of the meteoroid stream responsible for the 6 BC meteor storm if the latter's meteors radiated from two-thirds of the way from γ Hydrae to HIP59373 (table 14.4).

The orbit of Apollo asteroid 2004 WK1 (magnitude 21) is reminiscent of the orbit of the meteoroid stream responsible for the 6 BC meteor storm if the latter's meteors radiated from HIP59373 (table 14.5).

If the meteoroid stream had a Jupiter-family orbit and its radiant was one-third of the way from γ (Gamma) Hydrae to HIP59373, it would have had a relatively small perihelion distance (q=0.173 AU), although larger than 96P/Machholz 1 (q=0.124 AU). It is interesting to compare the Hydrid meteoroid stream's orbit with this comet's (table 14.6).

Could it have been a Halley-type meteoroid stream? Since Halley-type meteoroids tend to peak just before dawn, whereas Jupiter-family meteoroids tend to peak just after midnight,[28] and since the velocity and inclination of the meteoroid stream responsible for the Hydrid meteor storm are on or over the upper threshold of typical Jupiter-family meteoroid streams (approximately 11–35 km/second[29] and 0–30 degrees respectively),[30] a good case for the parent being a Halley-type comet, like C/1917 F1 (Mellish), can be made. Actually, the orbit of the Hydrid meteoroid stream, assuming that the radiant of

showersonline.com/what_is.html [accessed March 26, 2014]). In addition, meteoroid deposits resulting from comet fragmentation events may consist of a higher proportion of larger meteoroids, which make for brighter (and larger) meteors. The fact that some fireballs seem to have occurred during the 6 BC meteor storm (Rev. 12:4; see below) suggests that it was indeed characterized by a preponderance of bright meteors.

[27] On these meteor showers, see P. Brown, D. K. Wong, R. J. Weryk, P. Wiegert, "A Meteoroid Stream Survey Using the Canadian Meteor Orbit Radar. II: Identification of Minor Showers Using a 3D Wavelet Transform," *Icarus* 207.1 (2010): 78 table 4.

[28] Jenniskens, *Meteor Showers*, 110.

[29] Ibid.

[30] If the meteor storm radiated from higher up the tail, i.e., closer to HIP59373 than γ (Gamma) Hydrae, the velocity and inclination would be above what is normal for Jupiter-family orbits.

Name	q	e	i	ω	Node	Period
2009 HU58	0.187	0.91	35.77	285.35	62.90	2.97 years
Hydrids (Vinf=40)	0.19	0.908	39.3	226.1	51.0	2.97 years

TABLE 14.3 A comparison of the orbit of asteroid 2009 HU58 to the orbit of the meteoroid stream responsible for the 6 BC Hydrids, assuming that the meteors radiated from one-third of the way from γ Hydrae to HIP59373 and that Vinf=40.

Name	q	e	i	ω	Node	Period
2000 UR16	0.507	0.4388	11.74	228.78	33.85	313.64 days
Hydrids (Vinf=19)	0.501	0.444	13.8	233.7	50.4	312.42 days

TABLE 14.4 A comparison of the orbit of asteroid 2000 UR16 to the orbit of the meteoroid stream responsible for the 6 BC Hydrids, assuming that the meteors radiated from two-thirds of the way from γ Hydrae to HIP59373 and that Vinf=19.

Name	q	e	i	ω	Node	Period
2004 WK1	0.293	0.73	34.5	223.1	51.85	413.14 days
Hydrids (Vinf=30)	0.274	0.704	36.0	222.0	50.9	325.3 days

TABLE 14.5 A comparison of the orbit of asteroid 2004 WK1 to the orbit of the meteoroid stream responsible for the 6 BC Hydrids, assuming that the meteors radiated from HIP59373 126 minutes before sunrise and that Vinf=30.

Name	q	e	i	ω	Node	Period
96P/Machholz 1	0.124	0.959	58.30	14.756	94.32	5.24 years
Hydrids (Vinf=44)	0.173	0.958	46.1	227.0	51	8.36 years

TABLE 14.6 A comparison of the orbit of Comet 96P/Machholz 1 to the orbit of the meteoroid stream responsible for the 6 BC Hydrids, assuming that the meteors radiated from one-third of the way from γ (Gamma) Hydrae to HIP59373 and that Vinf=44.

Name	q	e	i	ω	Node	Period
C/1917 F1 (Mellish)	0.190	0.993	32.68	121.32	88.67	145 years
Hydrids A (Vinf=49)	0.196	0.993	66.3	51.1	232.8	148.16 years
Hydrids B (Vinf=44)	0.177	0.984	36.0	50.9	229.5	36.79 years

TABLE 14.7 A comparison of the orbit of Comet C/1917 F1 (Mellish) to the orbit of the meteoroid stream responsible for the 6 BC Hydrids, assuming that the meteors radiated from two-thirds of the way from Gamma Hydrae to HIP59373 and that Vinf=49 (Hydrids A) or from γ (Gamma) and that Vinf=44 (Hydrids B).

the meteors was two-thirds of the way from γ (Gamma) Hydrae to HIP59373 or from γ (Gamma) Hydrae itself, is reminiscent of Comet Mellish in perihelion distance and eccentricity (table 14.7). David Asher, perceiving the similarity, backtracked the orbit of this comet to see if it matched, but found that Mellish's argument of perihelion and ascending node did not evolve in a way that realistically permitted it itself to be the parent of the meteoroid stream.[31]

Whether the meteoroid stream or its parent comet (or cometary asteroid) has already been recorded, will be recorded in the near future,

[31] Personal email message to the author, January 5, 2013. Comet Mellish has been linked to the December Monocerotid and November Orionid meteor showers—see Peter Veres, Leonard Kornos, and Juraj Toth, "Meteor Showers of Comet C/1917 F1 Mellish," *Monthly Notices of the Royal Astronomical Society* 412 (2011): 511–521.

or no longer exists, we do not know. If the original stream remains intact and crosses Earth's orbital path, and if the parent body, or even a part of it, still survives after two millennia, by means of orbital backtracking we might well be able to identify or associate them. I must leave this task to specialists in solar system dynamics.

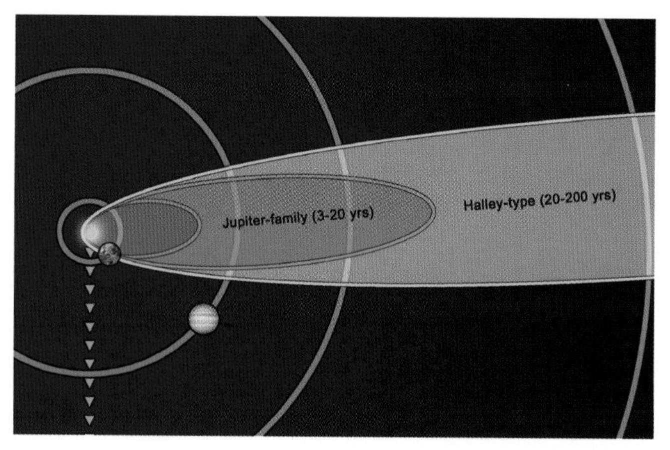

FIG. 14.9 Possible orbits of the meteoroid stream responsible for the meteor storm of October 19, 6 BC. Apollo asteroid-type orbits fall within the pink zone, Jupiter-family orbits within the green zone, and Halley-type orbits within the yellow zone. The more elongated the orbit, the more steeply inclined it is. The outermost planetary orbit is that of Uranus. Image credit: Sirscha Nicholl.

Assuming that there was one horn for each of the seven heads of the sea-dragon, we are presumably to envision that the 8th, 9th, and 10th heads have been cut off, but that the three horns nevertheless remain. Each of the seven heads has a crown, but there are no crowns to go with the three headless horns. The scene is strongly reminiscent of Daniel 7's fourth beast, which has ten horns, three of which are plucked up by the roots, leaving only seven (Dan. 7:7–8), although Daniel's 10-3=7 horns have become Revelation's 10-3=7 heads. This concurs with the fact that the dragon's throwing of many stars to the earth in Revelation 12:4a is reminiscent of Daniel 8:10, where the little horn, representing the latter-day tyrant, threw down to the ground "some of the host and some of the stars . . . and trampled on them." In other words, in Revelation 12:3–4a Hydra is introduced in such a way as to identify him with the eschatological rebellion of humanity against Yahweh, which is led by a blasphemous world ruler.[32]

How might the meteor activity of Revelation 12:4 elucidate the description of verse 3?

First, the fire color of Hydra is explicable astronomically with reference to the intense meteor activity in that part of the sky. The high frequency of meteors, fireballs, and bolides would have caused the constellation to look like it was on fire from its heads to its tail. A Macon, Georgia, newspaper described what it was like during the 1833 Leonid meteor storm: "We do not jest when we say that stubborn hearts were bent and flinty hearts melted into deep contrition at the alarming prospect of 'the heavens on fire.'"[33] So bright were the many thousands of meteors that copious witnesses spoke of the scene as one in which everything seemed to be on fire.[34]

[32]The eschatological dimension of this portrayal of Hydra is consistent with the presentation, in verse 1, of Virgo in terms that allude to Israel's eschatological exaltation and sovereignty. What we have in Revelation 12 is a war between the penultimate world empire, represented by Hydra, and the kingdom destined to conquer and replace it, represented by Virgo. The opening volley in this conflict occurred in connection with the birth of the Messiah, the one destined to vanquish Satan and his latter-day henchman and to exercise sovereignty on behalf of Israel over the nations.

[33]*Macon Georgia Messenger* (November 14, 1833), as cited by Littmann, *Heavens on Fire*, 6. Observers frequently employed fiery language to describe what they saw, referring to a "shower of fire," a "storm of fire," "the heavens being streaked with liquid fire," "the atmosphere above and all around rolling up and kindling into innumerable balls of rolling fire," "balls of livid fire, like burning rockets shooting toward the earth, and emitting numerous sparks," and "the heavens apparently on fire—millions of stars seeming to fall from their spheres, and the elements, as if about to melt with fervent heat" ("The Meteoric Shower," *The New-England Magazine* 6 [1834]: 47, 52; Denison Olmsted, "Observations on the Meteors of November 13th, 1833," *American Journal of Science* 25.2 (1834): 366, 368, 372, 382). In Greenland, the meteor storm of 1799 was described as having "the semblance of the heavens on fire above; for glowing points and masses, thick as hail, filled the firmament, as if some vast magazine of combustible materials had exploded in the far-off depths of space" ("Celestial Fireworks," *The National Magazine* 11 [1857]: 17).

[34]Littmann, *Heavens on Fire*, 4–7.

FIG. 14.10 The Leonid Meteor Storm, as illustrated in *Mechanics' Magazine* (November 1833).

One observer of the storm from Bowling Green, Missouri, wrote,

> Forcibly we were reminded of that remarkable passage in Revelations [*sic*] which speaks of the great red dragon, as drawing the third part of the stars of heaven and casting them to the earth; and if it be a figurative expression, that figure appeared to be fully painted on the broad canopy of the sky,—spread over with sheets of light, and thick with streams of rolling fire. There was scarcely a space in the firmament which was not filled at every instant with these falling stars. . . .[35]

Many witnesses reported that most of the meteors were about half the size of Jupiter, with some being larger and some smaller,[36] and a minority being larger than the full Moon.[37] A significant number of people

FIG. 14.11 The Leonid Meteor Storm of November 13, 1833. From W. A. Spicer, *Our Day in the Light of Prophecy* (Nashville: Southern Publishing Association, 1917), 92.

[35] From the *Salt River Journal* (November 20, 1833), as cited by Olmsted, "Observations," 382.

[36] Olmsted, "Observations," 368, 382, 383; idem, "Observations on the Meteors of November 13th, 1833," *American Journal of Science* 26.1 (1834): 138.

[37] Olmsted, "Observations," *American Journal of Science* 26.1 (1834): 155.

FIG. 14.12 The Great Meteor of August 18, 1783, by Henry Robinson. The bolide initially appeared as a single fiery ball, but then fragmented into a number of smaller balls. Note that each ball of light has a horn-like trail. Image credit: Dr. Arnaud Mignan, Tricottet Collection Image Archive, http://www.thetricottetcollection.com. The original has been slightly modified by Sirscha Nicholl with the kind permission of Dr. Mignan.

believed that the stars were actually abandoning their places: "The sky presented the appearance of a shower of stars, which many thought were real stars, and omens of dreadful events."[38] According to a minister in Annapolis, "Their appearance was so incessant during some part of the phenomenon that all the stars of the firmament seemed to be darting from their places."[39] The intense brightness of a great meteor storm is sufficient for people to read newspapers[40] and sufficient to awaken the sleeping, convincing them that their residences are on fire.[41]

Besides this, some meteors have an orange or red hue and some are yellow.[42] The color is a reflection of the physical constituency of the meteoroid, its velocity, and its brightness.[43] Silicate meteors tend to be red, while sodium-rich meteors tend to be orange and yellow, and iron-rich meteors may appear yellow.[44] At the same time, it is widely thought that slow-to-medium meteors tend to be more red, orange, and yellow, and that faster meteors (like the Leonids) tend to have a green or blue hue. In the case of the Hydrid meteor storm, the richness in reds, oranges, and yellows (= fire-colored) is probably due to both the constituency of the meteors and their medium velocity.[45]

[38] Ibid., 138.
[39] Olmsted, "Observations," *American Journal of Science* 25.2 (1834): 372, citing "Rev. Dr. Humphreys."
[40] According to a witness from Bowling Green cited by Olmsted, "Observations," 382.
[41] Ibid., 372, citing Humphreys.
[42] Bias, *Meteors and Meteor Showers*, 43.
[43] Ibid.
[44] Ibid.; Wikipedia, s.v. "Meteoroid," http://en.wikipedia.org/wiki/Meteoroid#Color (last modified April 27, 2013).
[45] Peter Jenniskens (personal email message to the author, October 16, 2012) suggested that "fire-colored" points to a predominance of red and yellow meteors (I would add orange). The medium-velocity Geminids (parented by cometary asteroid 3200 Phaethon),

It is likely that the meteors of the meteor storm of 6 BC were considerably fierier in color than those of the 1833 Leonid meteor storm.

Second, the seven heads, on which were crowns and horns, may be explained with reference to fireballs in the area associated with Hydra's head.[46] Since the scene is transpiring during a meteor storm, we can safely assume that the heads were not caused by ordinary meteors but rather by extraordinarily bright fireballs. Like the 1998 Leonid meteor display, this was a fireball-rich meteor outburst.

Fireballs may take various forms. Sometimes they consist simply of short or long bright streaks of light, but often they have extraordinarily bright heads at the front of the streaks. In shape, these heads may be round, oval, or (most commonly) pear-shaped. As anyone who has seen footage from the dashboard cameras that captured the astonishing fireball (technically, superbolide) over Chelyabinsk on February 15, 2013 (the brightness of which exceeded that of the Sun!) can testify, the streaks may look remarkably like horns, and the heads are well named because they are capable of looking very like creaturely heads. The conical region at the rear of a pear-shaped fireball head could readily pass for a stunning tall, or tiara-style, crown.[47]

FIG. 14.13 A fireball seen from a camera at NASA's Marshall Space Flight Center in Huntsville, Alabama, on September 30, 2011. Image credit: NASA/Meteoroid Environment Office/Bill Cooke.

The fireballs, together with the meteor storm, would have caused Hydra to come to life, indeed in 3D, with the dragon's heads appearing to move up and outward toward the observer, with their horns sticking out the back of their heads.

As regards the three horns without heads or crowns, we may presume that they consisted of very bright fireballs in the head region of Hydra that looked like horns, but that failed to develop notable heads.

In summary, on the eve of the celestial birth scene, a great meteor storm occurred, radiating from the tail of Hydra, the serpentine dragon. To observers convinced that the cometary phenomenon that was happening in the neighboring constellation in those days was the announcement of the Messiah's birth, the meteor storm would have seemed significant. It suggested that a great spiritual conflict was brewing between the forces of Order and the forces of Chaos, a conflict focused on the Messiah and his birth.

Hydra, the celestial representative of the forces of Evil and Disorder, appeared to have seven heads of power and crowns of

travel at 35 km/second and tend to be rich in yellows and oranges. Slower and higher velocity meteors can give rise to these same colors. The slow Taurids also have a high ratio of yellows. The 51 km/second Upsilon Pegasids, 59 km/second Perseids, and 66 km/second (high-velocity) Eta-Aquarids (parented by Halley's Comet) all tend to be yellowish, while the 57 km/second Epsilon-Eridanids are yellow-orange (Jenniskens, *Meteor Showers*, 311). Further, Jenniskens (246) writes of his observations of the 2001 Leonid meteor storm (71 km/second high-velocity meteors): "I was amazed by the bright red and orange colors of many Leonids, green on occasion." Red light may be emitted by air atoms and molecules as the meteoroids pass through the atmosphere (NASA's Leonid meteor shower page "Leonid Shower," http://leonid.arc.nasa.gov/meteor.html [last modified July 6, 2008]).

[46] The stars ζ, ε, δ, σ, and η in Hydra.

[47] It is possible that a single large meteoroid split into pieces as it collided with Earth's atmosphere, giving rise to the ten meteoric horns and seven heads of Rev. 12:3.

FIG. 14.14 A Leonid fireball during the November 2002 meteor shower—still images taken from a short movie. Images credit: NASA/MSFC/MEO/Bill Cooke. Image framing: Sirscha Nicholl.

FIG. 14.15 A fireball photographed by a fish-eye camera of the Czech Republic station of the European Fireball Network on January 21, 1999. Image credit and copyright: Pavel Spurný, Astronomical Institute, Academy of Sciences of the Czech Republic, Ondrejov.

sovereignty, arrogantly displaying his great royal authority. As the seven heads and ten horns streaked up into the upper half of Hydra, it must have looked like the serpent was rearing itself up self-assertively and aggressively, just as the highly venomous Black Mamba and King Cobra rear themselves up when they feel threatened and are about to attack a potential victim. At that time the dragon seemed to use its tail to hurl to the earth one-third of the stars of heaven. The celestial developments that night climaxed with the unforgettable image of the woman in the advanced stages of childbirth and the dragon beside her, looking like they were both standing on the western horizon. As the last predawn scene before the cometary baby's birth, this one set the stage for the climax of the celestial nativity drama, the birth scene, the following day.

What did it mean? Hydra was representing the forces of Chaos, and in particular the ultimate orchestrator of the worldwide rebellion of humanity at the end of the age and the authority and power behind the latter-day world tyrant, Satan (Rev. 12:17ff.). Hydra had come alive on the eve of the Messiah's birth in order to reveal that the forces of Disorder were intent on mounting a preemptive strike against the one who was destined to overcome them, who would put down the rebellion of humanity against God at the end of the age, and in particular who would vanquish the Devil and his eschatological henchman, the Antichrist. The evil empire felt gravely threatened by the Messiah's appearance on the earthly stage and was determined to kill him. Eager to thwart the divine plan to establish the rule of God on the earth, the ultimate possessor of the Antichrist's royal authority was dead set on destroying the Messiah as soon as he had fully emerged from his mother's belly. What was at stake was the ultimate fate of the world.

The dramatic events on October 19 climaxed with Hydra standing, as π (Pi) Hydrae rose sufficiently so that it was level with the eastern horizon. It is possible that the meteor storm had died down by this point, because Earth had already completed its pass through the dense section of the meteoroid stream. If not, the meteor storm would soon have fallen victim to the bleaching effect of the rising Sun.

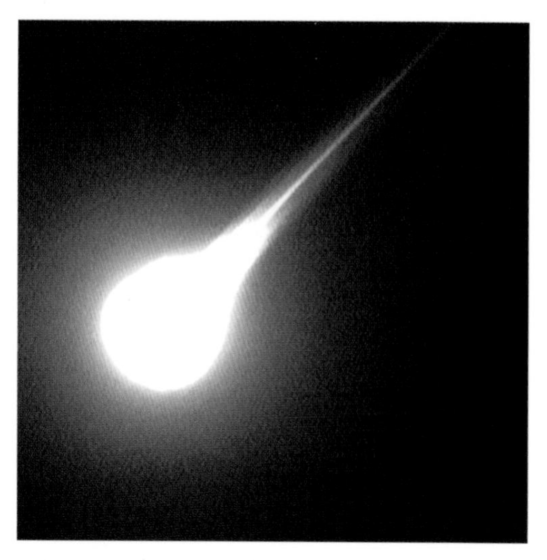

FIG. 14.16 This Leonid fireball in November 2002 left a train that lasted for more than 4 minutes. Image credit and copyright: George Varros, New Market, Maryland, www.gvarros.com.

Naturally, at sunrise most meteors become invisible because of the intensity of the sunlight. However, during the Leonid storms such was the preponderance of bright meteors that they could still be observed streaking across the sky after sunrise ("the beautiful shower of fire" in 1833 continued "till after daylight"[48]).

Undoubtedly the meteor storm ratcheted up the tension. Indeed the tension could hardly have been greater as the final night of observing before the birth came to a close. The touching scene of Virgo pregnant and giving birth to the cometary baby had been transformed into a taut thriller.

[48] Eyewitness Samuel Strickland of Ontario, *Twenty-Seven Years in Canada West: Or, The Experience of an Early Settler*, 2 vols. (London: Richard Bentley, 1853), 2:208–209. According to another (Henry R. Schoolcraft), they "continued to be visible until day light" (Olmsted, "Observations," *American Journal of Science* 26.1 [1834]: 139). Various witnesses speak of fireballs being observed after dawn (idem, "Observations," *American Journal of Science* 25.2 [1834]: 324, 381). Similarly, von Humboldt and Bonpland, *Personal Narrative*, chapter 1.10, recalled that bright Leonid meteors had been visible 15 minutes after sunrise early on November 12, 1799.

Glossary of Astronomical Terms

Absolute magnitude. A measurement of the intrinsic brightness of a celestial object: the brightness it would have if it were precisely 1 AU from both Earth and the Sun. On magnitude values, see Magnitude.

Acronychal rising. A celestial body's rising in the east as the Sun is setting in the west.

Almanac. A collection of astronomical predictions, especially relating to the positions of the planets, for an upcoming year.

Altitude. The apparent height (in degrees) of a celestial body relative to the horizon.

Antitail. A thin cometary "mini-tail" that points toward the Sun.

Aphelion. The point in a celestial body's orbit where it is farthest from the Sun.

Apparent magnitude. The brightness of a celestial object as it appears from Earth. On magnitude values, see Magnitude.

Apparition. The time during which a comet is visible in the sky.

Arcminute. An astronomical angular measurement equal to one sixtieth of one degree.

Arcsecond. An astronomical angular measurement equal to one sixtieth of one arcminute.

Argument of perihelion (ω). The angular distance (in degrees) from the longitude of the ascending node (on which, see Longitude of the ascending node) to the perihelion point, measured in the celestial body's orbital plane and in the direction of the body's motion.

Ascending node. The point at which a celestial body's orbit crosses the plane of the ecliptic as the orbit moves from the south to the north.

Asterism. A pattern of stars in the sky.

Asteroid. A planet-like rocky or metallic body that orbits the Sun.

Asteroid belt. The region of the solar system between Mars and Jupiter where most asteroids are found.

Asteroidal comet. A comet that is no longer active and therefore is mistakable for an asteroid.

Astronomical diaries. Babylonian records of daily astronomical observations; now preserved in the British Museum.

Astronomical twilight. The period before sunrise or after sunset that starts or ends when the Sun is 18 degrees below the horizon.

Astronomical unit (AU). A unit for measuring astronomical distances: 1 AU is the average distance between Earth and the Sun during the year.

Azimuth. The distance (in degrees, measured clockwise) from due north to the point where a vertical line downward from a given celestial object intersects the horizon.

Backscattering. A phenomenon that boosts the brightness of a comet when it is on the other side of Earth from the perspective of the Sun, or the other side of the Sun from the perspective of Earth. The sunlight is reflected back off the comet's larger dust particles.

Binary system. A system of two stars orbiting around a common center of mass.

Bolide. A fireball that attains to an apparent magnitude of between -14 and -17.

Brightness slope (n). The pattern of development of a comet's brightness, expressed as the value of "n."

Centaur. A comet- or asteroid-like "minor planet" that orbits between Jupiter and Neptune.

Circumpolar. Celestial bodies near a celestial pole that do not rise or set during a 24-hour day, because they do not drop below the horizon.

Coma. The comet's head, consisting of a gas and dust cloud enveloping a nucleus.

Comet. A celestial body consisting of dust and ice that, when it is close to the Sun, produces gas and dust, which form a head (coma) and tail(s) that point away from the Sun.

Cometary. Of or relating to comets.

Conjunction. The coming together of two celestial objects in the sky so that they seem to observers to be in the same location or very close to one another; or, the occasion when the celestial longitude of two astronomical bodies is the same; or, the occasion when a celestial body is too close to the Sun to be visible.

Constellation. A grouping of stars that seem to observers to form a pattern.

Culmination. The arrival of a celestial body at the meridian and therefore its highest daily altitude (relative to the horizon).

Daytime comet. A comet so bright that it is visible, for a while, during the daytime.

Degree. A unit for measuring angles. 180 degrees of the sky are visible above an unobstructed horizon. From the zenith to the unobstructed horizon there are 90 degrees; 1 degree is equal to 60 arcminutes or 3,600 arcseconds.

Delta Effect. The theory that, as a comet makes a close approach to Earth (within about 0.4 AU), the outer edges of the growing coma may go undetected by the human eye.

Descending node. The point at which an orbit crosses the plane of the ecliptic, as the comet moves from the north of it to the south of it.

Dust. A variety of, among other things, magnesium-rich silicates, sulfides, and carbon expelled from a comet.

Dust tail. The tail of a comet, consisting of dust particles lagging behind the comet's head (or coma). The dust is illuminated by the Sun and may become bright enough to be visible to Earth-dwellers.

Earth-Sun line. The imaginary straight line connecting Earth and the Sun.

Eccentricity (e). The extent to which a celestial body's orbit deviates from perfect circularity—the eccentricity of a circle is 0; the more stretched the oval (ellipsis) is, the higher the eccentricity is, up to 1; an eccentricity of greater than 1 means that the celestial body is incapable of completing a orbital revolution.

Ecliptic. The plane on which Earth orbits the Sun.

Elliptical. Oval.

Encke-type comets. Short-period comets that, like Comet Encke, have orbits so small that the comets do not have close encounters with Jupiter.

Fireball. A meteor at least as bright as Jupiter or Venus.

First Point of Aries. The location of the vernal equinox, a base line for astronomical measurements.

Forward-scattering. As an active comet moves closer to the imaginary line between Earth and the Sun, the coma and the dust tail are subject to an increasingly large spike in brightness, because the Sun's light hits the small dust particles and is scattered forward.

Full Moon. The lunar phase during which the whole of the Moon is illuminated, which occurs when it is in opposition to the Sun.

Galaxy. A gravitationally bound system of stars and associated matter.

Gas tail. The usually bluish, straight tail formed when electrically charged gas particles that exploded from the comet nucleus are pushed by the solar wind straight back behind the nucleus to point away from the Sun.

Gravitational effect. The effect of the gravitational pull of planets (such as Jupiter and Saturn) on a solar system object's orbit.

Great comet. A comet judged to be exceptional by virtue of its brightness and/or size.

Halley-type comet. A comet that, like Halley's Comet, has an orbital period of between 20 and 200 years.

Heliacal rising. The first visible rising of a celestial body over the (usually eastern) horizon after being invisible because of proximity to the Sun.

Heliacal setting. The final visible setting of a celestial body in the run-up to becoming invisible because of proximity to the Sun.

Hui-hsing. A Chinese expression for a broom-star comet, that is, a comet with a tail.

Hyperbolic orbit. A solar system object's orbit when it has greater than escape velocity (that is, an eccentricity value of greater than 1.0).

Inclination (i). The angle between the plane of a solar system object's orbit and the ecliptic (in degrees).

Inferior planets. Mercury and Venus.

Inner planets. Planets inside the asteroid belt, namely Mercury, Venus, Earth, and Mars.

Inner solar system. The region of the solar system up to and including the asteroid belt, incorporating the Sun, Mercury, Venus, Earth, Mars, the asteroid belt, and Ceres.

Intercalary month. A leap lunar month inserted into a lunar calendar to calibrate it with the solar calendar.

Julian calendar. A calendar that assumes that a full year consists of precisely 365.25 days and that, unlike the Gregorian calendar, does not allow for the fact that a full year is actually 11 minutes less than 365.25 days. From 45/44 BC to 9/8 BC a leap day was added every three years instead of four, with the result that three leap days too many had been intercalated. The Emperor Augustus remedied this by suspending leap years until AD 4 or 8.

Jupiter-family comets. Comets that complete one revolution around the Sun in less than 20 years.

Kracht Group. A group of comets that come as close as 0.047 AU to the Sun and have a relatively low inclination (roughly 13.4 degrees).

Kreutz Family. A large group of bright sungrazing comets, including the great comets of 1843 and 1882 and Ikeya-Seki in 1965, which have high-inclination and 600- to 1,100-year orbits.

Longitude of the ascending node (Ω). The angle from the First Point of Aries (the vernal equinox) to a celestial object's ascending node.

Longitude of perihelion. The angle between the First Point of Aries (the vernal equinox) and the point of perihelion.

Long-period comet. A comet that takes more than 200 years to complete one revolution around the Sun.

Lunar eclipse. When the Moon, or a part of it, moves into Earth's shadow, directly behind Earth.

Magnitude. A measurement of the relative brightness of celestial bodies. The brighter the object, the lower the magnitude value. A star of magnitude +1 is 2.5 times brighter than a star of magnitude +2. A star five magnitudes brighter than another star is 100 times brighter.

Major axis. The longest diameter of an ellipse (oval), measuring from the widest points through the foci and the center.

Massing. A grouping of three or more planets in the same area of sky.

Meridian. The great imaginary circle that passes through the observer's zenith and nadir, the celestial north and south poles, and the observer's due north and due south.

Meteor. The streak of light, or "shooting star," that appears when a body of matter from outer space (a meteoroid) is heated until incandescent due to friction as it encounters resistance from Earth's atmosphere.

Meteor shower. When a number of meteors seem to radiate from one particular point in the sky at one particular time of the year.

Meteor storm. When more than 1,000 meteors per hour radiate out of one point in the sky (or, more technically, when more than 1,000 meteors would occur per hour if the radiant were at the zenith).

Meteorite. A piece of space debris that survives its encounter with Earth's atmosphere and surface.

Meteoroid. A piece of rocky or metallic debris in space, commonly from comets or asteroids.

Meteoroid stream. A stream of particles ejected from a comet or asteroid.

Minor axis. The shortest diameter of an ellipse (oval), measuring from the narrowest points through the center.

Minor planet. A solar system object that is not classified as a comet or a planet. Included in the minor planets are asteroids, centaurs, and trans-Neptunian objects.

Morning star. A bright celestial entity, particularly Venus, that is present in the eastern sky in the predawn period.

Nebulous star. A cluster of hazy stars, or a star in a haze.

New Moon. The initial phase of the Moon's cycle, when the Moon first becomes visible after being in conjunction with the Sun.

Nongravitational effects. The acceleration or deceleration of a comet due to the recoil effect of its degassing.

Nova. A cataclysmic nuclear explosion on the surface of a white dwarf that causes a sudden brightening that lasts for weeks or months.

Nucleus. The icy and dusty core of a comet, which becomes active when near the Sun.

Occultation. When one celestial body is hidden by another body that moves between it and the observer.

Orbital elements. The six pieces of data that fully describe a solar system body's orbit.

Orbital period (P). The time a solar system body takes to complete one revolution around the Sun.

Orbital plane. The plane on which a body in the solar system orbits the Sun.

Outburst. The sudden, unexpected explosive release of dust by a comet, causing significant brightening.

Outer planets. The planets in the solar system that are beyond the asteroid belt, consisting of Jupiter, Saturn, Uranus, and Neptune.

Outer solar system. The region of the solar system beyond the asteroid belt, as far as Neptune.

Outgassing. The release of jets of gas from the comet nucleus when it is close to the Sun.

Parabolic orbit. A solar system object's orbit when it has an eccentricity value of 1.0; the boundary between a capture (elliptical) orbit and an escape (hyperbolic) orbit.

Partial lunar eclipse. When a portion of the Moon passes behind Earth through Earth's shadow.

Perigee. The point in a celestial body's orbit when it is closest to Earth.

Perihelion. The point in a celestial body's orbit when it is closest to the Sun.

Perihelion distance (q). The distance from a solar system body to the Sun when it is closest to it.

Perihelion time (T). The time in a celestial body's orbit when it is closest to the Sun.

Periodic comet. A comet that completes a single revolution around the Sun in less than 200 years. (Also called a "short-period comet.")

Phase angle. The angle formed where a line extending to a solar system body from the Sun and one extending to the same body from Earth converge.

Planetarium software. Software that simulates the sky at different points in history.

Planetary perturbations. A solar system body's deviations from its calculated orbit due to the gravitational forces exerted by other solar system bodies, especially planets such as Jupiter and Saturn.

Precession of the equinoxes. The slow westward shift of the equinoxes along the ecliptic, caused by the spinning-top-like wobble of Earth's axis. One cycle lasts about 26,000 years.

Progenitor comet. A parent or ancestor comet.

Prograde. Orbital movement that is in the same direction as most solar system bodies. When viewed from the vantage point of Earth's North Pole, it is counterclockwise motion.

Pseudonucleus. A cometary coma's star-like zone of concentrated brightness in the vicinity of the nucleus.

Radiant. The point from which the meteors in a given meteor shower seem to emanate.

Retrograde. Orbital movement in a direction opposite to that of most solar system bodies. When viewed from the vantage point of Earth's North Pole, it is clockwise motion.

Return. A comet's reappearance in connection with a particular perihelion pass.

Semi-major axis. Half of the longest diameter of an ellipse, measuring from the widest points through the foci and the center.

Short-period comet. A comet that completes one revolution around the Sun in less than 200 years. (Also called a "periodic comet.")

Solar wind. The constant stream of charged particles emanating from the Sun.

Spin axis. The imaginary line around which an object rotates.

Sporadic meteors. Meteors that have not been associated with a particular meteor shower.

Sungrazer. Any of the group of comets that make very close passes by the Sun, coming within 0.01 AU (or 0.05 AU) of it.

Sunskirter. Any of the group of comets that have a perihelion distance that is greater than 0.01 AU (or 0.05 AU) but less than 0.1 AU.

Superbolide. A bolide (meteor) that attains to an apparent magnitude of -17 or brighter.

Superior planets. Mars, Jupiter, Saturn (and, in recent times, Uranus and Neptune).

Supernova. The enormous, catastrophic nuclear explosion of a star that greatly intensifies its brightness for a period up to about three years. A supernova may be due to the cataclysmic ignition of nuclear fusion in a white dwarf's core, or due to a massive star's running out of fuel and then catastrophic collapse under its own gravity.

Surface brightness. The brightness of a celestial body's brightness per unit area, expressed in magnitudes per square arcsecond/arcminute.

Tail. Streams of dust and gas expelled from and pushed behind a coma that may become bright enough to be seen on Earth.

Trans-Neptunian object. Any solar system object with an orbit farther away from the Sun than Neptune (for example, Pluto).

Vernal equinox. The moment, in the Spring, when the Sun crosses the plane of Earth's equator, with the result that night and day are of equal length. Also, the celestial location of the Sun at the moment of the vernal equinox—known as the First Point of Aries.

Waning gibbous Moon. The phase of the lunar cycle after a full Moon when the Moon is less than fully, but more than half, illuminated.

Waxing gibbous Moon. The phase of the lunar cycle before a full Moon when the Moon is more than half, but less than fully, illuminated.

White dwarf. A still-white-hot, old, dying star that has used up all of its nuclear fuel and has cast off its outer material.

Zenith. The imaginary point directly above an observer.

Zenith attraction. An effect in which meteor radiants appear to observers to be closer to the zenith than they really are.

Zodiac (zodiacal signs). The 12 divisions, each 30 degrees long, into which the zodiacal band is divided in astrology. The divisions are named after the constellations that occupy the zodiacal band.

Zodiacal band. The band of sky either side of the ecliptic, along which the Sun, Moon, and major planets appear to traverse the heavens.

Zodiacal constellations. The constellations located along the zodiacal band.

Bibliography

Adair, Aaron. *The Star of Bethlehem: A Skeptical View*. Fareham, England: Onus, 2013.

Albright, W. F., and C. S. Mann. *Matthew*. Anchor Bible. Garden City, NY: Doubleday, 1971.

"Al Hillah: Full Year Climatology." http://www.nrlmry.navy.mil/CLIMATOLOGY/pages/arabian_sea_region/iraq/pdf/alhillah_iraq.pdf. Last modified April 24, 2013.

Allen, Richard Hinckley. *Star Names: Their Lore and Meaning*. New York: G. E. Stechert, 1899. http://penelope.uchicago.edu/Thayer/E/Gazetteer/Topics/astronomy/_Texts/secondary/ALLSTA.

Allen, W. C. *The Gospel according to St. Matthew*. 2nd ed. International Critical Commentary. New York: Scribner, 1907.

Allison, Dale C. "What Was the Star That Guided the Magi?" *Bible Review* 9.6 (1993): 20–24. Also in *The First Christmas: The Story of Jesus' Birth in History and Tradition*. Edited by Sara Murphy. 25–31. Washington, DC: Biblical Archaeology Society, 2009.

Asher, D. J. "Leonid Dust Trail Theories." In *Proceedings of the International Meteor Conference, Frasso Sabino, Italy, 23–26 September 1999*. Edited by R. Arlt. 5–21. Potsdam: IMO, 2000.

Asher, D. J., and S. V. M. Clube. "An Extraterrestrial Influence during the Current Glacial-Interglacial." *Quarterly Journal of the Royal Astronomical Society* 34 (1993): 481–511. http://adsabs.harvard.edu/abs/1993QJRAS..34..481A.

Asher, D. J., Mark E. Bailey, and V. V. Emel'yanenko. "The Resonant Leonid Trail from 1333." *Irish Astronomical Journal* 26.2 (1999): 91–93.

———. "Resonant Meteoroids from Comet Tempel-Tuttle in 1333: The Cause of the Unexpected Leonid Outburst in 1998." *Monthly Notices of the Royal Astronomical Society* 304.4 (April 16, 1999): L53–L56.

Ashley, Timothy R. *The Book of Numbers*. New International Commentary on the Old Testament. Grand Rapids, MI: Eerdmans, 1993.

Asimov, Isaac. *Asimov's Guide to Halley's Comet: The Awesome Story of Comets*. New York: Walker, 1985.

Aune, David E. "Greco-Roman Biography." In *Greco-Roman Literature and the New Testament*, 107–126. Atlanta: Scholars, 1988.

Aune, David E. "The Gospel as Hellenistic Biography." *Mosaic* 20 (1987): 1–10.

———. *The New Testament in Its Literary Environment*. Cambridge: James Clarke, 1988.

———. *Revelation*. 3 vols. Nashville: Thomas Nelson, 1997–1998.

"The Average Weather in September for Iraq." http://weatherspark.com/averages/31352/9/Iraq-Babil. Accessed May 13, 2014.

Baddeley, Alan D. *Human Memory: Theory and Practice*. Rev. ed. Hove, England: Psychology Press, 1997.

Bailey, Mark E., V. V. Emel'yanenko, G. Hahn, N. W. Harris, K. A. Hughes, K. Muininen, and J. V. Scotti. "Orbital Evolution of Comet 1995 O1 Hale-Bopp." *Monthly Notices of the Royal Astronomical Society* 281 (1996): 916–924. http://adsabs.harvard.edu/full/1996MNRAS.281..916B.

Bailey, Mark E., S. V. M. Clube, G. Hahn, W. M. Napier, and G. B. Valsecchi. "Hazards Due to Giant Comets: Climate and Short-Term Catastrophism." In *Hazards Due to Comets and Asteroids*. Edited by T. Gehrels. 479–533. Tucson: University of Arizona Press, 1995.

Bailey, Mark E., S. V. M. Clube, and William M. Napier. *The Origin of Comets*. Oxford: Pergamon, 1990.

Ball, David Mark. *"I Am" in John's Gospel: Literary Function, Background and Theological Implications*. Sheffield: Sheffield Academic Press, 1996.

Barnes, Timothy D. "The Date of Herod's Death." *Journal of Theological Studies* 19 (1968): 204–209. DOI: 10.1093/jts/XIX.1.204.

———. "The Triumphs of Augustus." *Journal of Roman Studies* 64 (1974): 21–26.

Barrett, A. A. "Observations of Comets in Greek and Roman Sources before A.D. 410." *Journal of the Royal Astronomical Society of Canada* 72 (1978): 81–106. http://adsabs.harvard.edu/full /1978JRASC..72 . . . 81B.

———. "The Star of Bethlehem: A Postscript." *Journal of the Royal Astronomical Society of Canada* 78 (1984): L23. http://adsabs.harvard .edu/full/1984JRASC..78L..23B.

Barton, Tamsyn. *Ancient Astrology*. London: Routledge, 1994.

Bauckham, Richard. *Jesus and the Eyewitnesses: The Gospels as Eyewitness Testimony*. Grand Rapids, MI: Eerdmans, 2007.

———. "Luke's Infancy Narrative as Oral History in Scriptural Form." In *The Gospels: History and Christology: The Search of Joseph Ratzinger–Benedict XVI*. Edited by Bernardo Estrada, Ermenegildo Manicardi and Armand Puig i Tàrrech. Vol. 1, 399–417. Vatican City: Libreria Editrice Vaticana, 2013.

———. "The Qumran Community and the Gospel of John." In *The Dead Sea Scrolls Fifty Years After Their Discovery: Proceedings of the Jerusalem Congress, July 20–25, 1997*. Edited by L. H. Schiffman, E. Tov and J. C. Vanderkam. 105–115. Jerusalem: Israel Exploration Society, 2000. http://cojs.org/cojswiki/The_Qumran _Community_and_the_Gospel_of_John ,_Richard_Bauckham.

Bauer, W., W. F. Arndt, F. W. Gingrich, and F. W. Danker. *A Greek-English Lexicon of the New Testament and Other Early Christian Literature*. 3rd ed. Chicago: University of Chicago Press, 2000.

Beale, G. K. *The Book of Revelation: A Commentary on the Greek Text*. New International Greek Testament Commentary. Grand Rapids, MI: Eerdmans, 1999.

Beare, Francis Wright. *The Gospel according to Matthew: A Commentary*. Oxford: Blackwell, 1981.

Beasley-Murray, George R. *Revelation*. New Century Bible Commentary. Grand Rapids, MI: Eerdmans, 1981.

Beaton, Richard. "Isaiah in Matthew's Gospel." In *Isaiah in the New Testament*. Edited by S. Moyise and M. J. J. Menken. 63–78. London: Continuum, 2005.

Beckwith, Isbon T. *The Apocalypse of John*. New York: Macmillan, 1919.

Beitzel, Barry. "Travel and Communication (OT World)." In *The Anchor Bible Dictionary*. Edited by D. N. Freedman. 6 vols., 6:644–648. New York: Doubleday, 1992.

Benko, Stephen. *The Virgin Goddess: Studies in the Pagan and Christian Roots of Mariology*. Leiden: Brill, 1993.

Benn, David. "Surface Brightness Calculator." http:// www.users.on.net/~dbenn/ECMAScript/surface _brightness.html. Last modified June 16, 2005.

"Bethlehem's Star." http://www.unmuseum.org /bstar.htm. Accessed May 15, 2014.

Bias, Peter V. *Meteors and Meteor Showers: An Amateur's Guide to Meteors*. Cincinnati, OH: Miracle Publishing, 2005.

Bielenstein, Hans. "Han Prognostications and Portents." *Bulletin of the Museum of Far Eastern Antiquities* 56 (1984): 97–112.

———. "An Interpretation of the Portents in the Ts'ien Han Shu." *Bulletin of the Museum of Far Eastern Antiquities* 22 (1950): 127–143.

Bishop, Eric F. F. "Some Reflections on Justin Martyr and the Nativity Narratives." *Evangelical Quarterly* 39 (1967): 30–39. http://www .biblicalstudies.org.uk/pdf/eq/1967-1_030.pdf.

Blass, Friedrich, Albert Debrunner, and R. W. Funk. *A Greek Grammar of the New Testament and Other Early Christian Literature*. Chicago: University of Chicago Press, 1961.

Blomberg, Craig. *The Historical Reliability of the Gospels*. Downers Grove, IL: InterVarsity Press, 1987.

———. "Matthew." In *Commentary on the New Testament Use of the Old Testament*. Edited by G. K. Beale and D. A. Carson. 1–110. Grand Rapids, MI: Baker Academic, 2007.

———. *Matthew*. New American Commentary. Nashville: Broadman, 1992.

Blount, Brian K. *Revelation: A Commentary*. New Testament Library. Louisville, KY: Westminster John Knox, 2009.

Bloxam, Richard Rouse. *Urania's Mirror*. London: Samuel Leigh, 1825.

Blunt, Anne. *A Pilgrimage to Nejd, The Cradle of the Arab Race*. 2nd ed. 2 vols. London: John Murray, 1881.

Boa, Kenneth D. "The Star of Bethlehem." ThM thesis, Dallas Theological Seminary, 1972. http://www.kenboa.org/downloads/pdf/TheStarofBethlehem.pdf. Accessed March 12, 2013.

Boa, Kenneth, and William Proctor. *The Return of the Star of Bethlehem: Comet, Stellar Explosion, or Signal from Above?* New York: Doubleday, 1980.

Bock, Darrell L. *Jesus according to Scripture: Restoring the Portrait from the Gospels*. Grand Rapids, MI: Baker, 2002.

———. *Luke 1:1–9:50*. Baker Exegetical Commentary on the New Testament. Grand Rapids, MI: Baker, 1994.

———. *Proclamation from Prophecy and Pattern: Lucan Old Testament Christology*. Sheffield: JSOT, 1987.

Boiy, T. *Late Achaemenid and Hellenistic Babylon*. Leiden: Brill, 2005.

Boll, Franz. *Aus der Offenbarung Johannis: hellenistische Studien zum Weltbild der Apokalypse*. Leipzig and Berlin: Teubner, 1914.

———. "Der Stern der Weisen." *Zeitschrift für die neutestamentliche Wissenschaft und die Kunde des Urchristentums* 18 (1917/1918): 40–48.

———. *Sphaera*. Leipzig: Teubner, 1903; Hildesheim: Georg Olms, 1967.

Boll, Franz, ed. and trans. *Catalogus Codicum Astrologorum Graecorum*. Vol. 7. Brussels: Lamertin, 1908.

Borg, Marcus J., and John Dominic Crossan, *The First Christmas: What the Gospels Really Teach About Jesus's Birth*. New York: HarperCollins, 2007.

Bornkamm, Gunther. "The Risen Lord and the Earthly Jesus: Matthew 28:16–20." In *The Future of Our Religious Past*. Edited by J. Robinson. 203–229. New York: Harper & Row, 1971.

Bortle, John E. "The Bright-Comet Chronicles." *International Comet Quarterly* (1998). http://www.icq.eps.harvard.edu/bortle.html. Accessed March 26, 2014.

———. "Great Comets in History." *Sky and Telescope* 93.1 (1997): 44–50.

Bousset, Wilhelm. *Die Offenbarung Johannis*. Göttingen: Vandenhoeck und Ruprecht, 1904.

Bovon, François. *Das Evangelium nach Lukas*. Evangelisch-Katholischer Kommentar zum Neuen Testament. 3 vols. Zurich: Benziger, 1989–2001.

Brandt, John C., and Robert D. Chapman. *Introduction to Comets*. 2nd ed. Cambridge: Cambridge University Press, 2004.

———. *Rendezvous in Space: The Science of Comets*. New York: W. H. Freeman, 1992.

Brewer, W. F. "What Is Recollective Memory?" In *Remembering Our Past: Studies in Autobiographical Memory*. Edited by D. C. Rubin. 19–66. Cambridge: Cambridge University Press, 1996.

"Brightest Comets Seen since 1935." *International Comet Quarterly*. http://www.icq.eps.harvard.edu/brightest.html. Accessed March 26, 2014.

Brown, P., D. K. Wong, R. J. Weryk, P. Wiegert. "A Meteoroid Stream Survey Using the Canadian Meteor Orbit Radar. II: Identification of Minor Showers Using a 3D Wavelet Transform." *Icarus* 207.1 (2010): 66–81. DOI: 10.1016/j.icarus.2009.11.0.

Brown, Raymond. *The Birth of the Messiah: A Commentary on the Infancy Narratives in the Gospels of Matthew and Luke*. 2nd ed. New York: Doubleday, 1993.

Brown, Robert. *Researches into the Origin of the Primitive Constellations of the Greeks, Phoenicians, and Babylonians*. 2 vols. Oxford: Williams & Norgate, 1899.

Bruner, Frederick Dale. *The Gospel of John: A Commentary*. Grand Rapids: Eerdmans, 2012.

Bruns, J. Edgar. "The Magi Episode in Matthew 2." *Catholic Biblical Quarterly* 23 (1961): 51–54.

Buber, Martin. *Der Glaube der Propheten*. Zürich: Conzett & Huber, 1950.

Bulmer-Thomas, Ivor. "The Star of Bethlehem—A New Explanation—Stationary Point of a Planet." *Quarterly Journal of the Royal Astronomical Society* 33 (1992): 363–374. http://adsabs.harvard.edu/full/1992QJRAS..33..363B.

Burkert, Walter. *Structure and History in Greek Mythology and Ritual*. Berkeley: University of California Press, 1979.

Burnham, Robert. *Great Comets*. Cambridge: Cambridge University Press, 2000.

Burridge, Richard A. "About People, by People, for People." In *The Gospels for All Christians: Rethinking the Gospel Audiences*. Edited by Richard Bauckham. 113–145. Grand Rapids, MI: Eerdmans, 1998.

———. *What Are the Gospels? A Comparison with Graeco-Roman Biography*. 2nd ed. Grand Rapids, MI: Eerdmans, 2004.

Byrskog, Samuel. *Story as History—History as Story*. Leiden: Brill, 2002.

Caird, George Bradford. *The Revelation of Saint John*. Peabody, MA: Hendrickson, 1966.

Calvin, John. *Calvin's Bible Commentaries: Matthew, Mark and Luke, Part 1*. Translated by William Pringle. Edinburgh: Calvin Translation Society, 1845.

———. *Commentary on the Book of the Prophet Isaiah—Volume 1*. Translated by William Pringle. Edinburgh: Calvin Translation Society, 1850.

Campion, Nicholas. *A History of Western Astrology: Volume 1, The Ancient World*. London and New York: Continuum, 2008.

Carroll, Susan S. "The Star of Bethlehem: An Astronomical and Historical Perspective." http://www.tccsa.tc/articles/star_susan_carroll.pdf. Last modified February 22, 2010.

Carson, Donald A. "Matthew." In *Expositor's Bible Commentary*. Edited by Tremper Longman III and David E. Garland. Vol. 9, 23–670. Rev. ed. Grand Rapids, MI: Zondervan, 2010.

Carter, Warren. "Matthew 1–2 and Roman Political Power." In *New Perspectives on the Nativity*. Edited by Jeremy Corley. 77–90. Edinburgh: T. & T. Clark, 2009.

Cassuto, Umberto. *A Commentary on the Book of Genesis (Part 1)*. Jerusalem: Magnes, 1989.

Cathcart, Kevin J. "Numbers 24:17 in Ancient Translations and Interpretations." In *The Interpretation of the Bible: The International Symposium in Slovenia*. Edited by J. Krašovec. 511–519. Sheffield: Sheffield Academic Press, 1998.

Cats, Jacob. *Aenmerckinghe op te tegenwoordige steert-sterre*. Middelburg, The Netherlands: n. p., 1619.

"Celestial Fireworks." *The National Magazine* 11 (1857): 15–19. http://books.google.co.uk/books?id=-ZsmAQAAIAAJ. Accessed April 8, 2013.

Chadwick, Henry. *Origen: Contra Celsum*. Cambridge: Cambridge University Press, 1965.

Chambers, George F. *The Story of the Comets*. London: Clarendon, 1909.

Charles, R. H. *The Assumption of Moses*. London: A. & C. Black, 1897.

———. *A Critical and Exegetical Commentary on the Revelation of St. John*. 2 vols. International Critical Commentary. Edinburgh: T. & T. Clark, 1920.

Childs, Brevard S. *Isaiah: A Commentary*. Old Testament Library. Louisville, KY: Westminster John Knox, 2000.

Chilton, Bruce. *The Isaiah Targum*. Wilmington, DE: Michael Glazier, 1987.

Clark, David H., John H. Parkinson, and F. Richard Stephenson. "An Astronomical Re-Appraisal of the Star of Bethlehem—A Nova in 5 BC." *Quarterly Journal of the Royal Astronomical Society* 18 (1977): 443–449. http://adsabs.harvard.edu/full/1977QJRAS..18..443C.

Clark, Roger Nelson. *Visual Astronomy of the Deep Sky*. Cambridge: Cambridge University Press, 1990.

Clube, S. V. M., F. Hoyle, W. M. Napier, and N. C. Wickramasinghe. "Giant Comets, Evolution, and Civilization." *Astrophysics and Space Science* 245 (1996): 43–60. http://articles.adsabs.harvard.edu/full/1996Ap%26SS.245 . . . 43C.

Clube, Victor, and Bill Napier. *Cosmic Winter*. Oxford: Blackwell, 1990.

Clube, Victor, and Bill Napier. *The Cosmic Serpent: A Catastrophist View of Earth History*. New York: Universe, 1982.

Coates, Richard. "A Linguist's Angle on the Star of Bethlehem." *Astronomy and Geophysics* 49.5 (October 2008): 27–32. http://www.staff.science.uu.nl/~gent0113/hovo/downloads/text1_10a.pdf.

Cohen, Gillian. *Memory in the Real World*. Hillsdale, NJ: Erlbaum, 1989.

Collins, Adela Yarbro. *The Combat Myth in the Book of Revelation*. Missoula, MT: Scholars Press, 1976.

Collins, David. *The Star of Bethlehem*. Stroud, England: Amberley, 2012.

Collins, J. J. *The Scepter and the Star: The Messiahs of the Dead Sea Scrolls and Other Ancient Literature*. 2nd ed. Grand Rapids, MI: Eerdmans, 2010.

———. "Sibylline Oracles." In *Old Testament Pseudepigrapha*. Edited by James H. Charlesworth. 2 vols., 1:317–472. New York: Doubleday, 1983.

Collins, Ken. "The Star of Bethlehem." http://www.kencollins.com/explanations/why-01.htm. Last modified March 23, 2013.

"Comet Hale-Bopp: The Great Comet of 1997." http://stardust.jpl.nasa.gov/science/hb.html. Last modified November 26, 2003.

"Comet Hale-Bopp—Still Enormous!" http://www.eso.org/public/news/eso9933. Last modified June 29, 1999.

"Comet in Major Outburst Now Visible in Evening Sky." http://star.arm.ac.uk/press/2007/cometholmes. Last modified October 10, 2012.

Condos, Theony. *Star Myths of the Greeks and Romans: A Sourcebook*. Grand Rapids, MI: Phanes, 1997.

Consolmagno, Guy. "Looking for the Star, or Coming to Adore?" *Thinking Faith*. http://www.thinkingfaith.org/articles/20101231_1.htm. Last modified December 31, 2010.

Cramer, Frederick H. *Astrology in Roman Law and Politics*. Philadelphia: American Philosophical Society, 1954.

Cranfield, C. E. B. *On Romans and Other New Testament Essays*. Edinburgh: T. & T. Clark, 1998.

Crovisier, Jacques, and Thérèse Encrenaz. *Comet Science: The Study of Remnants from the Birth of the Solar System*. Translated by Stephen Lyle. Cambridge: Cambridge University Press, 2000.

Cullen, Christopher. "Can We Find the Star of Bethlehem in Far Eastern Records?" *Quarterly Journal of the Royal Astronomical Society* 20 (1979): 153–159. http://adsabs.harvard.edu/full/1979QJRAS..20..153C.

———. "Halley's Comet and the 'Ghost' Event of 10 BC." *Quarterly Journal of the Royal Astronomical Society* 32 (1991): 113–119. http://adsabs.harvard.edu/full/1991QJRAS..32..113C.

Daniélou, Jean. *Primitive Christian Symbols*. Translated by Donald Attwater. London: Burns & Oates, 1964.

Davies, Graham. "The Significance of Deuteronomy 1:2 for the Location of Mount Horeb." *Palestine Exploration Quarterly* 111 (1979): 87–101.

Davies, W. D., and Dale C. Allison. *A Critical and Exegetical Commentary on the Gospel according to Saint Matthew*. 3 vols. International Critical Commentary. Edinburgh: T. & T. Clark, 1988–1997.

Dawes, Gregory W. "Why Historicity Still Matters: Raymond Brown and the Infancy Narratives." *Pacifica* 19 (2006): 156–176. DOI: 10.1177/1030570X0601900203.

Delling, G. "*magos*." In *Theological Dictionary of the New Testament*. Edited by Gerhard Kittel and Gerhard Friedrich. 10 vols., 4:356–359. Grand Rapids, MI: Eerdmans, 1964–1976.

de Crespigny, R. *Portents of Protest in the Later Han Dynasty: The Memorials of Hsiang K'ai to Emperor Huan*. Canberra: Australian National University Press, 1976.

de Sousa, Rodriga F. *Eschatology and Messianism in LXX Isaiah 1–12*. New York: T. & T. Clark, 2010.

de Strycker, Emile. *La forme la plus ancienne du Protevangile de Jacques*. Brussels: Société des Bollandistes, 1961.

de Vaux, Roland. *Ancient Israel: Its Life and Institutions*. Grand Rapids, MI: Eerdmans; Livonia, MI: Dove, 1997.

de Veragine, Jacobus. *The Golden Legend*. Edited by W. Caxton and F. S. Ellis. London: Wyman & Sons, 1878.

Depuydt, Leo. "Why Greek Lunar Months Began a Day Later than Egyptian Lunar Months." In *Living the Lunar Calendar*. Edited by J. Ben-Dov, W. Horowitz, and J. M. Steele. 153–164. Oxford: Oxbow, 2012.

Di Martino, Mario, and Alberto Cellino. "Physical Properties of Comets and Asteroids Inferred from Fireball Observations." In *Mitigation of Hazardous Comets and Asteroids*. Edited by M. J. S. Belton. 153–166. Cambridge: Cambridge University Press, 2004.

Dio's Rome, Volume 3. Translated by Herbert Baldwin Foster. New York: Pafraets, 1906.

Dio's Rome, Volume 4. Translated by Herbert Baldwin Foster. New York: Pafraets, 1905.

Dio's Rome, Volume 5. Translated by Herbert Baldwin Foster. New York: Pafraets, 1906.

Downey, Roma, and Mark Burnett, *The Bible*, television miniseries, produced by Roma Downey and Mark Burnett, aired on the History Channel, March 3–31, 2013 (Beverly Hills, CA: Lightworkers Media, 2013), DVD.

Dubs, Homer H. *The History of the Former Han Dynasty*. 3 vols. Baltimore: Waverly, 1938–1955.

Dunkin, Edwin. *The Midnight Sky*. London: Religious Tract Society, 1869.

Dunlop, Storm, and Will Tirion. *Collins Night Sky and Starfinder*. London: Collins, 2011.

Eberhard, Wolfram. "The Political Function of Astronomy and Astronomers in Han China." In *Chinese Thought and Institutions*. Edited by John K. Fairbank. 33–70. Chicago: University of Chicago Press, 1957.

Edberg, Stephen J., and David H. Levy. *Observing Comets, Asteroids, Meteors, and the Zodiacal Light*. Cambridge: Cambridge University Press, 1994.

Edwards, Ormond. "Herodian Chronology." *Palestine Exploration Quarterly* 114 (1982): 29–42. DOI: http://dx.doi.org/10.1179/peq.1982.114.1.29.

Ehrman, Bart. *The New Testament and Other Early Christian Writings: A Reader*. 2nd ed. Oxford: Oxford University Press, 2004.

Ephraem Syrus. *The Book of the Cave of Treasures*. Translated by E. A. Wallis Budge. London: Religious Tract Society, 1927.

———. *Opera Syriaca*. Rome: Vatican, 1740.

Evans, Craig A. *Jesus and His Contemporaries: Comparative Studies*. Leiden: Brill, 1995.

———. *Matthew*. New Cambridge Bible Commentary. Cambridge: Cambridge University Press, 2012.

Evans, James. *The History and Practice of Ancient Astronomy*. Oxford: Oxford University Press, 1998.

Evans, James, and J. Lennart Berggren. *Geminos's* Introduction to the Phenomena*: A Translation and Study of a Hellenistic Survey of Astronomy*. Princeton, NJ: Princeton University Press, 2006.

Farrar, F. W. *The Herods*. London: James Nisbet, 1899.

Farrer, A. A. *The Revelation of St. John the Divine*. Oxford: Clarendon, 1964.

Fernández, Yanga R. "The Nucleus of Comet Hale-Bopp (C/1995 O1): Size and Activity." *Earth, Moon, and Planets* 89 (2002): 3–25. http://www.pha.jhu.edu/~weaver/science_paper_28-mar-1997.pdf.

Ferrari-D'Occhieppo, Konradin. "The Star of the Magi and Babylonian Astronomy." In *Chronos, Kairos, Christos*. Edited by Jerry Vardaman and E. M. Yamauchi. 41–53. Winona Lake, IN: Eisenbrauns, 1989.

Filmer, W. E. "Chronology of the Reign of Herod the Great." *Journal of Theological Studies* 17 (1966): 283–298. DOI: 10.1093/jts/os-XXXVI.141.22.

Finegan, Jack. *Handbook of Biblical Chronology*. Peabody, MA: Hendrickson, 1998.

Fitzmyer, Joseph A. *The One Who Is to Come*. Grand Rapids: Eerdmans, 2007.

Ford, J. Massyngberde. *Revelation: Introduction, Translation, and Commentary*. Anchor Bible. Garden City, NY: Doubleday, 1975.

Fotheringham, J. K. "The New Star of Hipparchus and the Date of the Birth and Accession of Mithridates." *Monthly Notices of the Royal Astronomical Society* 89 (1919): 162–167. http://adsabs.harvard.edu/full/1919MNRAS..79..162F.

France, R. T. "The Authenticity of the Sayings of Jesus." In *History, Criticism, and Faith*. Edited by Colin Brown. 101–143. Downers Grove, IL: InterVarsity Press, 1976.

———. *The Gospel of Matthew*. New International Commentary on the New Testament. Grand Rapids, MI: Eerdmans, 2007.

———. "Herod and the Children of Bethlehem." *Novum Testamentum* 21 (1979): 98–120. http://www.jstor.org/discover/10.2307/1560717.

———. *Jesus and the Old Testament*. Grand Rapids, MI: Baker, 1982.

Frontinus, Sextus Julius. *Strategemata 2.1.17*. Edited by Gotthold Gundermann. Leipzig: Teubner, 1888.

Gaechter, Paul. *Das Matthäus-Evangelium*. Innsbruck: Tyrolia, 1963.

Galactic Photography. "Planetary Photo Techniques." http://www.galacticphotography.com/astro_Planetary_technique_3.html. Last modified September 9, 2012.

Gautschy, Rita. "Last and First Sightings of the Lunar Crescent." http://www.gautschy.ch/~rita/archast/mond/mondeng.html. Last modified February 15, 2013.

Geminos. *Introduction aux phénomènes*. Edited and translated by Germaine Aujac. Paris: Les Belles Lettres, 1975.

Gemser, Berend. "Der Stern aus Jacob (Num. 24.17)." *Zeitschrift für die Alttestamentliche Wissenschaft* 43 (1925): 301–302.

Genuth, Sara Schechner. *Comets, Popular Culture, and the Birth of Modern Cosmology*. Princeton, NJ: Princeton University Press, 1997.

Giberne, Agnes. *Sun, Moon, and Stars: Astronomy for Beginners*. New York: American Tract Society, 1880.

Gibson, J. C. L. *Canaanite Myths and Legends*. 2nd ed. Edinburgh: T. & T. Clark, 1978.

Gingerich, Owen. "Review Symposium: The Star of Bethlehem." *Journal of Biblical Literature* 33 (2002): 391–394. http://adsabs.harvard.edu/full/2002JHA. . . . 33..386G.

Gnilka, Joachim. *Das Matthäusevangelium*. Herders theologischer Kommentar zum Neuen Testament. 2 vols. Freiburg: Herder, 1986–1992.

Goetz, S. C., and C. L. Blomberg. "The Burden of Proof." *Journal for the Study of the New Testament* 11 (1981): 39–63. DOI: 10.1177/0142064X8100401103.

Gordley, Matthew E. *Teaching through Song in Antiquity*. Tübingen: Mohr Siebeck, 2011.

Goulder, Michael D. *Midrash and Lection in Matthew*. London: SPCK, 1974.

Grant, Mary Amelia. *The Myths of Hyginus*. Lawrence: University of Kansas, 1960. http://www.theoi.com/Text/HyginusAstronomica2.html#25.

Greeven, H. "*proskuneō*." In *Theological Dictionary of the New Testament*. Edited by Gerhard Kittel and Gerhard Friedrich. 10 vols., 6:758–766. Grand Rapids, MI: Eerdmans, 1964–1976.

Grego, Peter. *Blazing a Ghostly Trail: ISON and Great Comets of the Past and Future*. New York: Springer, 2014.

Grossfield, B. *The Targum Onqelos to Leviticus and Numbers*. The Aramaic Bible. Wilmington, DE: Michael Glazier, 1988.

Guide 9.0. Project Pluto. 168 Ridge Road, Bowdoinham, ME 04008. http://www.projectpluto.com.

Guillemin, Amédée. *Le Ciel, notion élémentaire d'Astronomie physique*. Paris: Librairie Hachette et Cie, 1877.

———. *The World of Comets*. Edited and translated by James Glaisher. London: Sampson Low, Marston, Searle, & Rivington, 1877.

Gundel, Wilhelm. "Kometen." In *Paulys Realencyclopädie der Classischen Altertumswissenschaft* 11.1, 1143–1193. Stuttgart: Druckmüller, 1921.

———. "Parthenos." In *Paulys Realencyclopädie der Classischen Altertumswissenschaft* 18.4, 1936–1957. Stuttgart: Druckmüller, 1949.

Gundry, Robert H. *Matthew: A Commentary on His Handbook for a Mixed Church under Persecution*. 2nd ed. Grand Rapids, MI: Eerdmans, 1994.

Gural, P. S. "Fully Correcting for the Spread in Meteor Radiant Positions Due to Gravitational Attraction," *WGN, Journal of the International Meteor Organization* 29.4 (2000): 134–138.

Hadorn, W. *Die Offenbarung des Johannes*. Leipzig: Deichert, 1928.

Hagner, Donald A. "Interpreting the Gospels: The Landscape and the Quest." *Journal of the Evangelical Theological Society* 24 (1981): 23–37. http://www.etsjets.org/files/JETS-PDFs/24/24-1/24-1-pp023-037_JETS.pdf.

———. *Matthew*. Word Biblical Commentary. 2 vols. Dallas: Word, 1993–1995.

Hanson, Jeanne K. *The Star of Bethlehem: The History, Mystery, and Beauty of the Christmas Star*. New York: Hearst, 1994.

Hard, Robin. *The Routledge Handbook of Greek Mythology*. London: Routledge, 2004.

Hare, Douglas. *Matthew*. Interpretation. Louisville, KY: Westminster John Knox, 1993.

Harris, Murray J. *The Second Epistle to the Corinthians*. New International Commentary on the New Testament. Grand Rapids, MI: Eerdmans, 2005.

Hauerwas, Stanley. *Matthew*. Brazos Theological Commentary on the Bible. Grand Rapids, MI: Brazos, 2006.

Hayes, John H., and Stuart Irvine. *Isaiah the Eighth-Century Prophet: His Times and His Preaching*. Nashville: Abingdon, 1987.

Hayman, A. P., ed. *The Old Testament in Syriac, According to the Peshitta Version: Numbers*. Leiden: Brill, 1991.

Hazzard, R. A. *Imagination of a Monarchy: Studies in Ptolemaic Propaganda*. Toronto: University of Toronto Press, 2000.

———. "Theos Epiphanes: Crisis and Response." *Harvard Theological Review* 88 (1995): 415–436. DOI: http://dx.doi.org/10.1017/S0017816000031692.

Hedrick, W. K. "The Sources and Use of the Imagery in Apocalypse 12." Unpublished ThD dissertation. Graduate Theological Union, Berkeley, CA, 1971.

Hegedus, Tim. *Early Christianity and Ancient Astrology*. New York: Peter Lang, 2007.

———. "The Magi and the Star in the Gospel of Matthew and Early Christian Tradition." *Laval théologique et philosophique* 59 (2003): 81–95. http://www.valentino-salvato.com/Astrology/pdf/Magi_and_Star.pdf.

———. "Some Astrological Motifs in the Book of Revelation." In *Religious Rivalries and the Struggle for Success in Sardis and Smyrna*. Edited by Richard S. Ascough. 67–85. Waterloo, ON: Wilfred Laurier University Press, 2005.

Hengel, Martin. *Acts and the History of Earliest Christianity*. London: SCM, 1979.

Hephaistio of Thebes. *Apotelesmatics, Book I*. Translated by Robert Schmidt. Edited by Robert Hand. Berkeley Springs, WV: Golden Hind Press, 1994.

Hevelius, Johannes. *Cometographia*. Danzig: Simon Reiniger, 1668.

Heyob, Sharon Kelly. *The Cult of Isis among Women in the Graeco-Roman World*. Leiden: Brill, 1975.

Ho, Peng-Yoke. "Ancient and Mediaeval Observations of Comets and Novae in Chinese Sources." *Vistas in Astronomy* 5 (1962): 127–225. DOI: 10.1016/0083-6656(62)90007-7.

Hoehner, Harold W. *Herod Antipas*. Cambridge: Cambridge University Press, 1972.

Hoffmeier, James K. *Ancient Israel in Sinai: The Evidence for the Authenticity of the Wilderness Tradition*. Oxford: Oxford University Press, 2005.

Holden, James H., ed. and trans. *Rhetorius the Egyptian*. Tempe, AZ: American Federation of Astrologers, 2009.

Holmes, Michael W. *The Apostolic Fathers: Greek Texts and English Translations*. 3rd ed. Grand Rapids, MI: Baker, 2007.

Holzmeister, U. "La stella dei Magi." *Civiltà Cattolica* 93 (1942): 9–22.

Horsley, R. A. *The Liberation of Christmas: The Infancy Narratives in Social Context*. New York: Crossroad, 1989.

Hübner, Wolfgang. "Teukros im Spätmittelalter." *International Journal of the Classical Tradition* 1.2 (1994–1995): 45–57. DOI: 10.1007/BF02678994.

Hughes, David W. "Apian's Woodcut and Halley's Comet." *International Halleywatch Newsletter* 5 (1984): 24–25.

———. "Early Long-Period Comets: Their Discovery and Flux." *Monthly Notices of the Royal Astronomical Society* 339 (2003): 1103–1110. http://articles.adsabs.harvard.edu/full/2003MNRAS.339.1103H.

———. "The Magnitude Distribution, Perihelion Distribution, and Flux of Long-Period Comets." *Monthly Notices of the Royal Astronomical Society* 326 (2001): 515–523. DOI: 10.1046/j.1365-8711.2001.04544.x.

———. *The Star of Bethlehem Mystery*. London: J. M. Dent, 1979.

Hughes, David W., Kevin K. C. Yau, and F. Richard Stephenson. "Giotto's Comet—Was It the Comet of 1304 and Not Comet Halley?" *Quarterly Journal of the Royal Astronomical Society* 34 (1993): 21–32. http://adsabs.harvard.edu/full/1993QJRAS..34 . . . 21H.

Hughes, David W., N. McBride, J. Boswell, and P. Jalowiczor. "On the Variation of Cometary Coma Brightness with Comet-Earth Distance (the Delta Effect)." *Monthly Notices of the Royal Astronomical Society* 263 (1993): 247–255. http://articles.adsabs.harvard.edu/full/1993MNRAS.263..247H.

Humphreys, Colin J. "The Star of Bethlehem." In *Science and Christian Belief* 5 (1995): 83–101. http://www.asa3.org/ASA/topics/Astronomy-Cosmology/S&CB%2010-93Humphreys.html.

———. "The Star of Bethlehem, A Comet in 5 B.C., and the Date of the Christ's Birth." *Tyndale Bulletin* 43 (1992): 31–56. http://www.tyndalehouse.com/tynbul/library/TynBull_1992_43_1_02_Humphreys_StarBethlehem.pdf.

———. "The Star of Bethlehem—A Comet in 5 B.C.—And the Date of the Birth of Christ." *Quarterly Journal of the Royal Astronomical Society* 32 (1991): 389–407. http://adsabs.harvard.edu/full/1991QJRAS..32..389H.

Hunger, Hermann, and David Edwin Pingree, eds. *MUL.APIN: An Astronomical Compendium in Cuneiform*. Horn, Austria: Ferdinand Berger, 1989.

Hunger, Hermann, F. Richard Stephenson, C. B. F. Walker, and K. K. C. Yau. *Halley's Comet in History*. London: British Museum, 1985.

Ipatov, Sergei I. "Cavities as a Source of Outbursts from Comets." In *Comets: Characteristics, Composition, and Orbits*. Edited by Peter G. Melark. 101–112. Hauppauge, NY: Nova Science, 2011. http://arxiv.org/ftp/arxiv/papers/1103/1103.0330.pdf.

Irvine, Stuart A. *Isaiah, Ahaz, and the Syro-Ephraimite Crisis*. Atlanta: Scholars Press, 1990.

Isbouts, Jean-Pierre. *Young Jesus: Restoring the "Lost Years" of a Social Activist and Religious Dissident*. New York: Sterling, 2008.

James, N. D. "Comet C/1996 B2 (Hyakutake): The Great Comet of 1996." *Journal of the British Astronomical Association* 108 (1998): 157–171. http://adsabs.harvard.edu/full/1998JBAA..108..157J.

James, Nick, and Gerald North. *Observing Comets*. London: Springer, 2003.

Jenkins, R. M. "The Star of Bethlehem and the Comet of AD 66." *Journal of the British Astronomical Association* 114 (2004): 336–343. http://www.bristolastrosoc.org.uk/uploaded/BAAJournalJenkins.pdf.

Jenniskens, Peter. *Meteor Showers and Their Parent Comets*. Cambridge: Cambridge University Press, 2006.

Jenniskens, Peter, and Jeremie Vaubaillon. "3D/Biela and the Andromedids: Fragmenting versus Sublimating Comets." *Astronomical Journal* 134 (2007): 1037–1045. DOI: 10.1086/519074.

Jeremias, Alfred. *Handbuch der altorientalischen Geisteskultur*. Leipzig: J. C. Hinrichs, 1913.

Jeremias, Joachim. *New Testament Theology*. Translated by John Bowden. New York: Scribner, 1971.

John of Damascus. *Exposition of the Orthodox Faith*. Translated by S. D. F. Salmond. A Select Library of Nicene and Post-Nicene Fathers of the Christian Church. Second Series. Vol. 9. Oxford: J. Parker, 1899.

Johnson, Luke Timothy. *The Gospel of Luke*. Sacra Pagina. Collegeville, MN: Liturgical Press, 1991.

Jordan, Tony. *The Nativity*, television miniseries, directed by Coky Giedroyc, produced by Ruth Kenley-Letts, aired on the BBC, December 2010 (Ampthill, Bedford: Red Planet Pictures, 2011), DVD.

Jung, Carl G. *On Christianity*. Princeton, NJ: Princeton University Press, 1999.

Justinus. *Epitome of the Philippic History of Pompeius Trogus*. Translated by J. C. Yardley. New York: Oxford University Press, 1994.

Kaiser, Walter C. "The Promise of Isaiah 7:14 and the Single-Meaning Hermeneutic." *Evangelical Journal* 6 (1988): 55–70. http://www.jashow.org/wiki/index.php/Isaiah_7:14%E2%80%94Would_the_Messiah_be_%22Virgin_Born%22%3F.

Kaler, Jim. "PI HYA (Pi Hydrae)." http://stars.astro.illinois.edu/sow/pihya.html. Last modified April 24, 2011.

Kamel, L. "The Comet Light Curve Catalogue/Atlas." *Astronomy and Astrophysics Supplement Series* 92 (1992): 85–149. http://adsabs.harvard.edu/full/1990acm..proc..363K.

———. "The Delta-Effect in the Light Curves of 13 Periodic comets." *Icarus* 128 (1997): 145–159. DOI: http://dx.doi.org/10.1006/icar.1997.5732.

Kamesar, Adam. "The Virgin of Isaiah 7:14: The Philological Argument from the Second to the Fifth Century." *Journal of Theological Studies* 41 (1990): 51–75. DOI: 10.1093/jts/41.1.51.

Kammerer, Andreas. "Analysis of Past Comet Apparitions: C/1995 O1 (Hale-Bopp)." http://kometen.fg-vds.de/koj_1997/c1995o1/95o1eaus.htm. Last modified June 26, 2007.

Kee, H. C. "Testaments of the Twelve Patriarchs." In *The Old Testament Pseudepigrapha*. Edited by J. H. Charlesworth. 2 vols., 1:782–828. New York: Doubleday, 1983.

Keener, Craig S. *1–2 Corinthians*. New Cambridge Bible Commentary. Cambridge: Cambridge University Press, 2005.

———. *The Gospel of Matthew: A Socio-Rhetorical Commentary*. Grand Rapids, MI: Eerdmans, 2009.

———. *The Historical Jesus of the Gospels*. Grand Rapids, MI: Eerdmans, 2009.

———. *Revelation*. Downers Grove, IL: InterVarsity Press, 2009.

Keller, Werner. *The Bible as History*. Rev. ed. London: Hodder & Stoughton, 1980.

Kelley, David H., and Eugene F. Milone. *Exploring Ancient Skies*. 2nd ed. New York: Springer, 2011.

Kern, Martin. "Religious Anxiety and Political Interest in Western Han Omen Interpretation: The Case of the Han Wudi Period (141–87 B.C.)." *Chūgoku shigaku* 10 (2000): 1–31. http://www.princeton.edu/~mkern/Religious%20Anxiety.pdf.

Kiang, Tao. "The Past Orbit of Halley's Comet." *Memoirs of the Royal Astronomical Society* 76 (1972): 27–66. http://adsabs.harvard.edu/full/1972MmRAS..76 . . . 27K.

Kidger, Mark R. *The Star of Bethlehem: An Astronomer's View*. Princeton, NJ: Princeton University Press, 1999.

King, Henry C. *The Christmas Star*. Toronto: Royal Ontario Museum, 1970.

Kingsbury, J. D. *Matthew as Story*. 2nd ed. Philadelphia: Fortress, 1986.

Knoblet, Jerry. *Herod the Great*. Lanham, MD: University Press of America, 2005.

Koch, J. Neue *Untersuchungen zur Topographie des babylonischen Fixsternhimmels*. Wiesbaden: Otto Harrassowitz, 1989.

Koch-Westenholz, Ulla. *Mesopotamian Astrology: An Introduction to Babylonian and Assyrian Celestial Divination*. Copenhagen: Museum Tusculanum Press, 1995.

Kokkinos, Nikos. "Crucifixion in A.D. 36." In *Chronos, Kairos, Christos*. Edited by Jerry Vardaman and E. M. Yamauchi. 133–163. Winona Lake, IN: Eisenbrauns, 1989.

Kraft, H. *Die Offenbarung des Johannes*. Tübingen: Mohr, 1974.

Kronk, Gary W. "C/1861 J1 (Great Comet of 1861)." http://cometography.com/lcomets/1861j1.html. Last modified September 30, 2006.

———. "C/1882 R1 (Great September Comet)." http://cometography.com/lcomets/1882r1.html. Last modified October 3, 2006.

———. "C/1975 V1 (West)." http://cometography.com/lcomets/1975v1.html. Last modified October 3, 2006.

———. "C/1995 O1 (Hale-Bopp)." http://cometography.com/lcomets/1995o1.html. Last modified October 3, 2006.

———. "C/1996 B2 (Hyakutake)." http://cometography.com/lcomets/1996b2.html. Last modified September 30, 2006.

———. "The Closest Approaches of Comets to Earth." http://cometography.com/nearcomet.html. Last modified September 30, 2006.

———. *Cometography: A Catalog of Comets*. 6 vols. Cambridge: Cambridge University Press, 1999–.

———. *Comets: A Descriptive Catalog*. Hillside, NJ: Enslow, 1984.

———. *Meteor Showers: An Annotated Catalog*. 2nd ed. New York: Springer, 2014.

———. "What Is a Meteor Shower?" http://meteorshowersonline.com/what_is.html. Accessed March 26, 2014.

Kugler, F. X. *Die Babylonische Mondrechnung. Zwei Systeme der Chaldäer über den Lauf des Mondes und der Sonne*. Freiburg: Herder, 1900.

Kümmel, W. G. *Heilsgeschehen und Geschichte*. Vol. 2. Marburg: N. G. Elwert Verlag, 1978.

Lan, Zhang, and Zhao Gang. "The Identification of Comets in Chinese Historical Records." *Science China—Physics, Mechanics and Astronomy* 54 (2011): 150–155. http://link.springer.com/article/10.1007%2Fs11433-010-4135-6.

Lange, Rainer, and Noel M. Swerdlow. *Planetary, Lunar, and Stellar Visibility* software. Version 3.1.0. November 20, 2006. http://www.alcyone.de.

Larson, Frederick A. *The Star of Bethlehem*. DVD. Directed by Stephen Vidano. Santa Monica, CA: Mpower Pictures, 2006.

———. "To Stop a Star." http://www.bethlehemstar.net/starry-dance/to-stop-a-star. Accessed March 26, 2014.

———. "Westward Leading." http://www.bethlehemstar.net/starry-dance/westward-leading. Accessed March 26, 2014.

Le Boeuffle, André. *Les Noms Latins d'astres et de constellations*. Paris: Les Belles Lettres, 1977.

"Leonid Shower." http://leonid.arc.nasa.gov/meteor.html. Last modified July 6, 2008.

Levine, Baruch A. *Numbers 21–36*. Anchor Yale Bible. New Haven, CT: Yale University Press, 2000.

Levison, H. F., A. Morbidelli, L. Domes, R. Jedicke, P. A. Wiegert, and W. F. Bottke, Jr. "The Mass Disruption of Oort Cloud Comets." *Science* 296 (2002): 2212–2215. https://www.sciencemag.org/content/296/5576/2212.

Levy, David. *Comets: Creators and Destroyers*. New York: Simon & Schuster, 1998.

———. *David Levy's Guide to Observing and Discovering Comets*. Cambridge: Cambridge University Press, 2003.

"Le Zodiaque de Dendéra." http://cartelen.louvre.fr/cartelen/visite?srv=car_not_frame&idNotice=19044. Accessed April 4, 2014.

Licht, A. Lewis. "The Rate of Naked-Eye Comets from 101 BC to 1970 AD." *Icarus* 137 (1999): 355–356. http://www.sciencedirect.com/science/article/pii/S0019103598960481.

Liddell, Henry George, and Robert Scott. *A Greek-English Lexicon*. Edited by Henry Stuart Jones. Oxford: Clarendon Press, 1940.

Lightfoot, J. B. *Apostolic Fathers, Pt. II. S. Ignatius. S. Polycarp. Revised Texts, with Introductions, Notes, Dissertations, and Translations*. London: Macmillan, 1885.

Lightfoot, J. L. *The Sibylline Oracles*. Oxford: Oxford University Press, 2007.

Littmann, Mark. *The Heavens on Fire: The Great Leonid Meteor Storms*. Cambridge: Cambridge University Press, 1999.

Littmann, Mark, and Donald K. Yeomans. *Comet Halley: Once in a Lifetime*. Washington, DC: American Chemical Society, 1985.

Loewe, Michael. *Divination, Mythology and Monarchy in Han China*. Cambridge: Cambridge University Press, 2008.

Louw, Johannes P., and Eugene A. Nida. *Greek-English Lexicon of the New Testament: Based on Semantic Domains*. New York: United Bible Societies, 1988.

Luciuk, Mike. "Astronomical Magnitudes: Why Can We See the Moon and Planets in Daylight?" http://www.asterism.org/tutorials/tut35%20Magnitudes.pdf. Last modified April 25, 2013.

Luczycki, Rebecca. "Starring in the Night," *Alaska* magazine. http://www.alaskamagazine.com/article/77/09/starring_in_the_night. Accessed May 3, 2014.

Lunsford, Robert. *Meteors and How to Observe Them*. New York: Springer, 2008.

Luz, Ulrich. *Matthew 1–7: A Continental Commentary*. Translated by Wilhelm C. Linss. Minneapolis: Augsburg Fortress, 1989.

Lyytinen, Esko, and Peter Jenniskens. "Meteor Outbursts from Long-Period Comet Dust Trails." *Icarus* 162 (2003): 443–452. http://www.sciencedirect.com/science/article/pii/S0019103502000714.

Maier, Harry O. "Ignatius *Ephesians* 19.1–3." In *Prayer from Alexander: A Critical Anthology*. Edited by Mark Christopher Kiley. 267–272. London: Routledge, 1997.

Malan, S. C., trans. *The Book of Adam and Eve, also Called the Conflict of Adam and Eve with Satan*. London: Charles Carrington, 1882.

Malina, Bruce J. *On the Genre and Message of Revelation: Star Visions and Sky Journeys*. Peabody, MA: Hendrickson, 1995.

Malina, Bruce J., and John J. Pilch. *Social-Science Commentary on the Book of Revelation*. Minneapolis: Augsburg Fortress, 2000.

Manilius. *Astronomica*. Translated by G. P. Goold. Cambridge, MA: Harvard University Press, 1989.

Manitius, C., ed. *Hipparchi in Arati et Endoxi Phaenomena Commentariorum Libri Tres*. Leipzig: Teubner, 1894.

Marcus, Joseph N. "Forward-Scattering Enhancement of Comet Brightness. I. Background and Model." *International Comet Quarterly* 29 (2007): 39–66. http://www.icq.eps.harvard.edu/marcus_icq29_39.pdf.

———. "Forward-Scattering Enhancement of Comet Brightness. II. The Light Curve of C/2006 P1 (McNaught)." *International Comet Quarterly* 29 (2007): 119–130. http://www.icq.eps.harvard.edu/marcus_icq29_119.pdf.

———. "The Need for Cometary Photometry." *International Amateur-Professional Photoelectric Photometry Communication* 8 (1982): 26–30. http://adsabs.harvard.edu/full/1982IAPPP . . . 8 . . . 26M.

Marks, Richard G. *The Image of Bar Kokhba in Traditional Jewish Literature: False Messiah and National Hero*. University Park: Pennsylvania State University Press, 1994.

Marsden, Brian G. "The Sungrazing Comet Group." *Astronomical Journal* 72 (1967): 1170–1183. http://adsabs.harvard.edu/full/1967AJ. . . . .72.1170M.

Marsden, Brian G., and Zdenek Sekanina. "Comets and Nongravitational Forces. VI. Periodic Comet Encke 1786–1971." *Astronomical Journal* 79 (1974): 413–419. http://adsabs.harvard.edu/abs/1974AJ. . . . .79..413M.

Marshall, I. Howard. *The Gospel of Luke: A Commentary on the Greek Text*. New International Greek Testament Commentary. Grand Rapids, MI: Eerdmans, 1984.

———. *I Believe in the Historical Jesus*. Grand Rapids, MI: Eerdmans, 1977.

Martin, Ernest L. *The Birth of Christ Recalculated*. 2nd ed. Pasadena, CA: Foundation for Biblical Research, 1980.

———. *The Star of Bethlehem: The Star That Astonished the World*. 2nd ed. Portland, OR: Associates for Scriptural Knowledge, 1996. http://www.askelm.com/star.

Matthews, Victor. "Perfumes and Spices." In *The Anchor Bible Dictionary*. Edited by D. N. Freedman. 6 vols., 5:226–228. New York: Doubleday, 1992.

Maunder, E. W. *The Astronomy of the Bible: An Elementary Commentary on the Astronomical References of Holy Scripture*. New York: Mitchell Kennerley, 1908.

———. "Star of the Magi." In *International Standard Bible Encyclopedia*. Edited by James Orr. 5 vols., 5:2848–2849. Grand Rapids, MI: Eerdmans, 1939. http://www.internationalstandardbible.com/S/star-of-the-magi.html.

Maximus the Confessor. *The Philokalia: The Complete Text*. Vol. 2. Translated and edited by G. E. H. Palmer, P. Sherrard, and K. Ware. London: Faber & Faber, 1981.

Mayer, Adrienne. *The Poison King: The Life and Legend of Mithridates, Rome's Deadliest Enemy*. Princeton, NJ: Princeton University Press, 2010.

McCarter Jr., P. Kyle. "The Balaam Texts from Deir 'Alla: The First Combination." *Bulletin of the American Schools of Oriental Research* 239 (1980): 49–60. http://www.jstor.org/discover/10.2307/1356759.

McDaniel, Thomas F. "Problems in the Balaam Tradition." http://tmcdaniel.palmerseminary.edu/Balaam.pdf. Accessed August 1, 2014.

McGrath, James F., trans. *Mandaean Book of John*. http://rogueleaf.com/book-of-john/2011/06/04/18-portents-of-the-birth-of-john-the-baptist. Posted June 4, 2011.

McIvor, R. S. *Star of Bethlehem, Star of Messiah*. Toronto: Overland, 1998.

McNamara, Martin. "Early Exegesis in the Palestinian Targum (Neofiti 1) Numbers Chapter 24." *Proceedings of the Irish Biblical Association* 16 (1993): 57–79.

———. *Targum Neofiti 1: Numbers*. The Aramaic Bible. Collegeville, MN: Liturgical Press, 1995.

McNaught, R. H., and D. J. Asher. "Leonid Dust Trails and Meteor Storms." *WGN, Journal of the International Meteor Organization* 27 (April 1999): 85–102.

McNeile, A. H. *The Gospel according to St. Matthew: Greek Text with Introduction, Notes, and Indices*. London: Macmillan, 1915.

Meier, John P. *Matthew*. New Testament Message. Dublin: Veritas, 1980.

"The Meteoric Shower." *The New-England Magazine* 6 (1834): 47–54. http://books.google.co.uk/books?id=d8NSAAAAcAAJ.

Metzger, Bruce M. "Ancient Astrological Geography and Acts 2:9–11." In *Apostolic History and the Gospel: Biblical and Historical Essays Presented to F. F. Bruce*. Edited by W. Ward Gasque and Ralph P. Martin. 123–133. Exeter: Paternoster, 1970.

Milgrom, Jacob. *Numbers*. The JPS Torah Commentary. New York: Jewish Publication Society, 1992.

Mobberley, Martin. *Hunting and Imaging Comets*. Berlin: Springer, 2011.

Molnar, Michael R. *The Star of Bethlehem: The Legacy of the Magi*. New Brunswick, NJ: Rutgers University Press, 1999.

Montefiore, H. W. "Josephus and the New Testament." *Novum Testamentum* 4 (1960): 139–160. http://www.jstor.org/discover/10.2307/1560122.

Moore, Patrick. *Comets: An Illustrated Introduction*. New York: Scribner, 1973.

———. *The Star of Bethlehem*. Bath, England: Canopus, 2001.

Moore, Patrick, and Robin Rees. *Patrick Moore's Data Book of Astronomy*. Cambridge: Cambridge University Press, 2011.

Morehouse, A. J. "The Christmas Star as a Supernova in Aquila." *Journal of the Royal Astronomical Society of Canada* 72 (1978): 65–68. http://adsabs.harvard.edu/full/1978JRASC..72...65M.

Morris, Leon. *The Gospel According to Matthew*. Downers Grove, IL: InterVarsity Press, 1992.

Mowinckel, Sigmund. *He That Cometh: The Messiah Concept in the Old Testament and Later Judaism*. New York: Abingdon, 1954.

Moyise, Steve. *Was the Birth of Jesus according to Scripture?* Eugene, OR: Wipf & Stock, 2013.

Mueller, Walter. "A Virgin Shall Conceive." *Evangelical Quarterly* 32 (1960): 203–207.

Mullaney, James. "The Star of Bethlehem." *Science Digest* 80 (December 1976): 61–65.

Müller, W. W. "Frankincense." In *The Anchor Bible Dictionary*. Edited by D. N. Freedman. 6 vols., 2:854. New York: Doubleday, 1992.

Murphy, Frederick J. *Fallen Is Babylon*. Harrisburg, PA: Trinity Press International, 1998.

Napier, W. M. "Evidence for Cometary Bombardment Episodes." *Monthly Notices of the Royal Astronomical Society* 366 (2006): 977–982.

NASA JPL HORIZONS System. http://ssd.jpl.nasa.gov/?horizons.

NASA JPL Small-Body Database Browser. http://ssd.jpl.nasa.gov/sbdb.cgi.

The Natural History of Pliny. Vol. 1. Translated by John Bostock and Henry Thomas Riley. London: Henry G. Bohn, 1893.

Near Earth Object Program. "Historic Comet Close Approaches." http://neo.jpl.nasa.gov/ca/historic_comets.html. Last modified October 27, 2009.

Neugebauer, Otto. *The Exact Sciences in Antiquity*. Mineola, NY: Dover, 1969.

———. *A History of Ancient Mathematical Astronomy*. 3 vols. Berlin: Springer, 1975.

Nolland, John. *The Gospel of Matthew*. New International Greek Testament Commentary. Grand Rapids, MI: Eerdmans, 2005.

North, J. D. *Horoscopes and History*. London: The Warburg Institute, University of London, 1986.

Olmsted, Denison. "Observations on the Meteors of November 13th, 1833." *American Journal of Science* 25.2 (1834): 363–411; 26.1 (1834): 132–174.

Olson, Roberta J. M. *Fire and Ice: A History of Comets in Art*. New York: Walker, 1985.

Olson, Roberta J. M., and Jay M. Pasachoff. *Fire in the Sky: Comets and Meteors, the Decisive Centuries, in British Art and Science*. Cambridge: Cambridge University Press, 1998.

———. "New Information on Comet Halley as Depicted by Giotto Di Bondone and Other Western Artists." In *20th ESLAB Symposium on the Exploration of Halley's Comet: Proceedings of the International Symposium, Heidelberg, Germany, 27–31 October*. Vol. 3, C201–C213. Noordwijk, Netherlands: European Space Agency, 1986.

Öpik, Ernst Julius. "Sun-Grazing Comets and Tidal Disruption." *Irish Astronomical Journal* 7 (March 1966): 141–161. http://adsabs.harvard.edu/full/1966IrAJ. . . . 7..141O.

Oriti, R. A. "The Star of Bethlehem." *The Griffith Observer* 39.12 (December 1975): 9–14.

Osborne, Grant R. *Revelation*. Baker Exegetical Commentary on the New Testament. Grand Rapids, MI: Baker, 2002.

Oswalt, John N. *The Book of Isaiah*. 2 vols. New International Commentary on the Old Testament. Grand Rapids, MI: Eerdmans, 1986–1998.

———. "The Significance of the 'Almah Prophecy in the Context of Isaiah 7–12." *Criswell Theological Review* 6.2 (1993): 223–235. http://faculty.gordon.edu/hu/bi/ted_hildebrandt/otesources/23-isaiah/text/isa-articles/oswalt-almahisa-ctr.pdf.

Paffenroth, Kim. "The Star of Bethlehem Casts Light on Its Modern Interpreters." *Quarterly Journal of the Royal Astronomical Society* 34 (1993): 449–460. http://adsabs.harvard.edu/full/1993QJRAS..34..449P.

Palmer Lake Historical Society. "Star Light, Star Bright." http://palmerdividehistory.org/startale.html. Last modified May 7, 2011.

Pane, John F. "Comet 17P/Holmes." http://www.cs.cmu.edu/~pane/holmes. Accessed July 1, 2014.

Pankenier, David W. "Notes on Translations of the East Asian Records Relating to the Supernova of AD 1054." *Journal of Astronomical History and Heritage* 9.1 (2006): 77–82. http://articles.adsabs.harvard.edu/full/2006JAHH. . . . 9 . . . 77P.

———. "On the Reliability of Han Dynasty Solar Eclipse Records." *Journal of Astronomical History and Heritage* 15.3 (2012): 200–212. http://articles.adsabs.harvard.edu/full/2012JAHH . . . 15..200P.

———. *Popular Astrology and Border Affairs in Early Imperial China*. Sino-Platonic Papers. Philadelphia: University of Pennsylvania, 2000.

Pankenier, David W., Zhentao Xu, and Yaotiao Jiang. *Archaeoastronomy in East Asia*. Amherst, NY: Cambria, 2008.

Parker, Richard A., and Waldo H. Dubberstein. *Babylonian Chronology 626 B.C.–A.D. 75*. Providence, RI: Brown University Press, 1956.

Parpola, Simo. "The Magi and the Star: Babylonian Astronomy Dates Jesus' Birth." In *The First Christmas: The Story of Jesus' Birth in History and Tradition*. Edited by Sara Murphy. 13–24. Washington, DC: Biblical Archaeology Society, 2009.

Pauli, C. W. H., trans. *The Chaldee Paraphrase on the Prophet Isaiah*. London: London Society's House, 1871.

Phipps, William. "The Magi and Halley's Comet." *Theology Today* 43 (1986–1987): 88–92. DOI: 10.1177/004057368604300109.

Pinch, Geraldine. *Egyptian Myth: A Very Short Introduction*. Oxford: Oxford University Press, 2004.

———. *Egyptian Mythology: A Guide to the Gods, Goddesses, and Traditions of Ancient Egypt*. Oxford: Oxford University Press, 2004.

Pingré, Alexandre Guy. *Cométographie ou Traité Historique et Théoretique des Comètes*. 2 vols. Paris: Imprimerie Royale, 1783–1784.

Pingree, David. "Astronomy and Astrology in India and Iran." *Isis* 54 (1963): 229–246. http://www.iranicaonline.org/articles/astrology-and-astronomy-in-iran.

———. "Mesopotamian Astronomy and Astral Omens in Other Civilizations." In *Mesopotamien und seine Nachbarn: Politische und kulturelle Wechselbeziehungen im alten Vorderasien vom 4. bis 1. Jahrtausend v. Chr.*. Edited by Hans Jorg Nissen. 613–631. Berlin: Reimer, 1982.

Plummer, Alfred. *An Exegetical Commentary on the Gospel according to S. Matthew*. London: Elliot Stock, 1909.

Prabhu, George M. Soares. *The Formula Quotations in the Infancy Narratives of Matthew: An Enquiry into the Tradition History of Mt 1–2*. Rome: Pontifical Biblical Institute, 1976.

Pritchard, James B. *Ancient Near East in Pictures Relating to the Old Testament with Supplement*. 2nd ed. Princeton, NJ: Princeton University Press, 1969.

———. *Ancient Near Eastern Texts Relating to the Old Testament with Supplement*. 3rd ed. Princeton, NJ: Princeton University Press, 1969.

Pseudo-Hegesippus. *On the Ruin of the City of Jerusalem*. Translated by Wade Blocker. http://www.tertullian.org/fathers/hegesippus_05_book5.htm. Last modified November 25, 2005.

Ptolemy. *Tetrabiblos: Or Quadripartite*. Translated by Frank Egleston Robbins. Loeb Classical Library. Cambridge, MA: Harvard University Press, 1940.

Puech, Emile. "Palestinian Funerary Inscriptions." In *The Anchor Bible Dictionary*. Edited by D. N. Freedman. 6 vols., 5:126–135. New York: Doubleday, 1992.

Quarles, Charles L. *Midrash Criticism: Introduction and Appraisal*. Lanham, MD: University Press of America, 1998.

Rahe, Jürgen, Bertram Donn, and Karl Wurm. *Atlas of Cometary Forms: Structures Near the Nucleus*. Washington, DC: NASA, 1969.

Ramirez, Frank. *The Christmas Star*. Lima, OH: CSS, 2002.

Ramsay, William. *Was Christ Born at Bethlehem? A Study on the Credibility of St. Luke*. New York: G. P. Putnam's Sons, 1898.

Ramsey, John T. "Mithridates, the Banner of Ch'ih-Yu, and the Comet Coin." *Harvard Studies in Classical Philology* 99 (1999): 197–253. http://www.jstor.org/discover/10.2307/311482.

Ramsey, John T., and A. Lewis Licht. *The Comet of 44 B.C. and Caesar's Funeral Games*. With a foreword by Brian G. Marsden. Oxford: Oxford University Press, 1997.

Rao, Joe. "The Greatest Comets of All Time." SPACE.com, January 19, 2007. http://www.space.com/3366-greatest-comets-time.html. Accessed March 26, 2014.

Redshift 7. United Soft Media Verlag GmbH., Thomas-Wimmer-Ring 11, D-80539 Munich, Germany. http://www.redshift-live.com.

Reznikov, A. I. "La comète de Halley: une démystification de la légende de Noël?" *Recherches d'astronomie historique* 18 (1986): 65–80.

Rhys, Steffan. "Star of Bethlehem Comet Theory." http://www.walesonline.co.uk/news/wales-news/2008/12/22/star-of-bethlehem-comet-theory-91466-22528488. Accessed March 26, 2014.

Rich, Mike. *The Nativity Story*, directed by Catherine Hardwicke, produced by Wyck Godfrey, cinematic release December 1, 2006 (Los Angeles: New Line Cinema, 2007), DVD.

Richardson, Peter. *Herod: King of the Jews and Friend of the Romans*. Columbia, SC: University of South Carolina Press, 1996.

Richardson, Robert S. "The Star of Bethlehem—Fact or Myth?" *The Griffith Observer* 22 (December 1958): 162–164.

Riddle, M. B. "Arabic Gospel of the Infancy of the Saviour." In *The Ante-Nicene Fathers*. Edited by Alexander Roberts and James Donaldson. 10 vols., 8:405–415. Grand Rapids, MI: Eerdmans, 1979.

Ritchie, David. *Comets: Swords of Heaven*. New York: New American Library, 1985.

Roberts, Alexander, and James Donaldson, eds. *The Ante-Nicene Christian Library: The Writings of the Fathers Down to A.D. 325*. Vol. 2. Edinburgh: T. & T. Clark, 1868.

Roberts, Alexander, James Donaldson, and A. Cleveland Coxe, eds. *The Ante-Nicene Fathers. Vol.*

IV. Translations of the Writings of the Fathers Down to A.D. 325. New York: Scribner, 1926.

Roberts, Courtney. *The Star of the Magi: The Mystery That Heralded the Coming of Christ*. Franklin, NJ: Career Press, 2007.

Robinson, Bernard P. "Matthew's Nativity Stories: Historical and Theological Questions for Today's Readers." In *New Perspectives on the Nativity*. Edited by Jeremy Corley. 110–131. Edinburgh: T. & T. Clark, 2009.

Rochberg, Francesca. *Babylonian Horoscopes*. Philadelphia: American Philosophical Society, 1998.

Rodríguez Arribas, Josefina. "The Terminology of Historical Astrology according to Abraham Bar Hiyya and Abraham Ibn Ezra." *Aleph: Historical Studies in Science and Judaism* 11 (2011): 11–54. http://www.jstor.org/discover/10.2979/aleph.2011.11.1.10.

Rolek, Barbara. "Polish Christmas Traditions." http://easteuropeanfood.about.com/od/christmaseve/a/Polishxmas.htm. Last modified May 3, 2013.

Roloff, Jürgen. *The Revelation of John: A Continental Commentary*. Translated by John F. Alsup. Minneapolis: Augsburg Fortress, 1993.

Rösel, Martin. "Die Jungfrauengeburt des endzeitlichen Immanuel. Jesaja 7 in der Übersetzung der Septuaginta." *Jahrbuch für Biblische Theologie* 6 (1991): 135–151.

Rosenberg, Roy A. "The Star of the Messiah: Reconsidered." *Biblica* 53 (1972): 105–109. http://www.jstor.org/discover/10.2307/42609680.

Ross, Albert. *Mary, Mother of Jesus*, television film, directed by Kevin Connor, produced by Eunice Kennedy Shriver, aired on NBC TV, November 14, 1999 (Universal City, CA: Universal Studios, 2010), DVD.

Roughton, N. A., J. M. Steele, and C. B. F. Walker. "A Late Babylonian Normal and *Ziqpu* Star Text." *Archives of the History of the Exact Sciences* 58 (2004): 537–572. DOI: 10.1007%2Fs00407-004-0083-8.

Ruggles, Clive L. N. *Ancient Astronomy: An Encyclopedia of Cosmologies and Myth*. Santa Barbara, CA: ABC-CLIO, 2005.

Sachs, A. J. "A Late Babylonian Star Catalog." *Journal of Cuneiform Studies* 6 (1952): 146–150. http://www.jstor.org/discover/10.2307/1359538.

Sachs, A. J., and C. B. F. Walker. "Kepler's View of the Star of Bethlehem and the Babylonian Almanac for 7/6 B.C." *Iraq* 46 (1984): 43–51. http://www.jstor.org/discover/10.2307/4200210.

Sagan, Carl, and Ann Druyan. *Comet*. New York: Pocket Books, 1986.

Sanders, E. P. *The Historical Figure of Jesus*. London: Penguin, 1993.

Schaaf, Fred. *Comet of the Century*. New York: Springer, 1997.

———. *Wonders of the Sky: Observing Rainbows, Comets, Eclipses, the Stars, and Other Phenomena*. Mineola, NY: Dover, 1983.

Schaaf, Fred, and Guy Ottewell. *Mankind's Comet: Halley's Comet in the Past, the Future, and Especially the Present*. Greenville, SC: Furman University, 1985.

Schiffman, Lawrence H. "From Observation to Calculation: The Development of the Rabbinic Lunar Calendar." In *Living the Lunar Calendar*. Edited by J. Ben-Dov, W. Horowitz, and J. M. Steele. 231–243. Oxford: Oxbow, 2012.

Schlyter, Paul. "Radio and Photometry in Astronomy." http://stjarnhimlen.se/comp/radfaq.html. Last modified April 13, 2010.

Schmude, Richard. *Comets and How to Observe Them*. New York: Springer, 2010.

Schnackenburg, Rudolf. *The Gospel of Matthew*. Grand Rapids, MI: Eerdmans, 2002.

Schoedel, William R. "Ignatius and the Reception of the Gospel of Matthew in Antioch." In *Social History of the Matthean Community: Cross-Disciplinary Approaches*. Edited by David L. Balch. 129–177. Minneapolis: Fortress, 1991.

———. *Ignatius of Antioch*. Hermeneia. Philadelphia: Fortress, 1985.

Schrader, Eberhard. *Die Keilinschriften und das Alte Testament*. 3rd ed. Edited by H. Zimmern and H. Winkler. Berlin: Reuther und Reichard, 1903.

Schwyzer, Hans-Rudolf. *Chairemon*. Leipzig: G. Harrassowitz, 1932.

Scott, James M. *Geography in Early Judaism and Christianity: The Book of Jubilees*. Cambridge: Cambridge University Press, 2005.

Seargent, David. *Comets: Vagabonds of Space*. Garden City, NY: Doubleday, 1982.

———. *The Greatest Comets in History: Broom Stars and Celestial Scimitars*. Berlin: Springer, 2009.

———. *Sungrazing Comets: Snowballs in the Furnace*. Kindle Digital book, Amazon Media, 2012.

———. *Weird Astronomy*. New York: Springer, 2011.

Sekanina, Zdenek. "Activity of Comet Hale-Bopp (1995 O1) beyond 6 AU from the Sun." *Astronomy and Astrophysics* 314 (1996): 957–965. http://adsabs.harvard.edu/full/1996A%26A . . . 314..957S.

———. "Statistical Investigation and Modeling of Sungrazing Comets Discovered with the Solar and Heliospheric Observatory." *Astrophysical Journal* 566.1 (2002): 577–598. http://iopscience.iop.org/0004-637X/566/1/577.

Sekanina, Zdenek, and P. W. Chodas. "Fragmentation Hierarchy of Bright Sungrazing Comets and the Birth and Orbital Evolution of the Kreutz System. I. Two-Superfragment Model." *Astrophysical Journal* 607 (2004): 620–639. http://trs-new.jpl.nasa.gov/dspace/bitstream/2014/39288/1/03-3526.pdf.

———. "Fragmentation Hierarchy of Bright Sungrazing Comets and the Birth and Orbital Evolution of the Kreutz System. II. The Case for Cascading Fragmentation." *Astrophysical Journal* 663 (2007): 657–676. http://trs-new.jpl.nasa.gov/dspace/bitstream/2014/40925/1/04-3531.pdf.

Sela, Shlomo. *Abraham Ibn Ezra and the Rise of Medieval Hebrew Science*. Leiden: Brill, 2003.

Seymour, John, and Michael W. Seymour. "The Historicity of the Gospels and Astronomical Events concerning the Birth of Christ." *Quarterly Journal of the Royal Astronomical Society* 19 (1978): 194–197. http://adsabs.harvard.edu/full/1978QJRAS..19..194S.

Seymour, P. A. H. *The Birth of Christ: Exploding the Myth*. London: Virgin, 1998.

"The Shocking Size of Comet McNaught." http://www.ras.org.uk/news-and-press/157-news2010/1782-the-shocking-size-of-comet-mcnaught. Last modified April 13, 2010.

"The Star of Bethlehem: Historical and Astronomical Perspectives—A Multi-Disciplinary Discussion." http://www.astro.rug.nl/~khan/bethlehem/scientific-rationale.php. Accessed July 5, 2014.

Swerdlow, Noel M. *The Babylonian Theory of the Planets*. Princeton, NJ: Princeton University Press, 1998.

Smith, Fiona Veitch. "Did the Star of Bethlehem Really Exist?" http://www.veitchsmith.com/2009/12/10/did-the-star-of-Bethlehem-really-exist. Posted December 10, 2009.

Smith, M. S. "The Near Eastern Background of Solar Language for Yahweh." *Journal of Biblical Literature* 109 (1990): 29–39. http://www.jstor.org/discover/10.2307/3267327.

Warner, Brian. *Charles Piazzi Smyth: Astronomer-Artist: His Cape Years, 1835–1845*. Cape Town: A. A. Balkema, 1983.

Spicer, W. A. *Our Day in the Light of Prophecy*. Nashville: Southern Publishing Association, 1917.

Staerk, W. *Die jüdische Gemeinde des Neuen Bundes in Damaskus*. Gütersloh: C. Bertelsmann, 1922.

Stander, H. F. "The Starhymn in the Epistle of Ignatius to the Ephesians (19:2–3)." *Vigiliae Christianae* 43 (1989): 209–214. http://www.jstor.org/discover/10.2307/1584061.

Stanton, Graham. *Jesus and Gospel*. Cambridge: Cambridge University Press, 2004.

"Star of Bethlehem: Behind the Myth." BBC2 Television documentary. Aired December 24, 2008. London: Atlantic Productions, 2008.

Starry Night® Pro 6.4.3. Simulation Curriculum Corporation. 11900 Wayzata Blvd, Suite 126, Minnetonka, MN 55305. http://astronomy.starrynight.com.

Stauffer, Ethelbert. *Jesus and His Story*. New York: Knopf, 1960.

Steel, Duncan. *Eclipse*. London: Headline, 1999.

———. *Marking Time: The Epic Quest to Invent the Perfect Calendar*. New York: John Wiley, 2000.

———. *Rogue Asteroids and Doomsday Comets*. London: John Wiley, 1997.

Steele, John M. "Living with a Lunar Calendar in Mesopotamia and China." In *Living the Lunar Calendar*. Edited by J. Ben-Dov, W. Horowitz, and J. M. Steele. 373–387. Oxford: Oxbow, 2012.

Stein, Robert H. "The 'Criteria' for Authenticity." In *Gospel Perspectives*, vol. 1. Edited by R. T. France and David Wenham. 225–253. Sheffield: JSOT Press, 1980.

———. *Jesus the Messiah: A Survey of the Life of Christ*. Downers Grove, IL: InterVarsity Press, 1996.

Steinmann, Andrew E. "When Did Herod the Great Reign?" *Novum Testamentum* 51 (2009): 1–29. DOI: 10.1163/156853608X245953.

Stenzel, Arthur. *Jesus Christus und sein Stern*. Hamburg: Verlag der Astronomischen Korrespondenz, 1913.

Stephenson, F. Richard. "The Ancient History of Halley's Comet." In *Standing on the Shoulders of Giants*. Edited by Norman Thrower. 231–253. Berkeley: University of California Press, 1990.

———. "SN 1006: The Brightest Supernova." *Astrophysics and Geophysics* 51.5 (2010): 27–32. DOI: 10.1111/j.1468-4004.2010.51527.x.

Stephenson, F. Richard, Kevin Yau, and Hermann Hunger. "Records of Halley's Comet on Babylonian Tablets." *Nature* 314 (April 18, 1985): 587–592. http://www.nature.com/nature/journal/v314/n6012/abs/314587a0.html.

Stevenson, Gregory M. "Conceptual Background to Golden Crown Imagery in the Apocalypse of John (4:4, 10; 14:14)." *Journal of Biblical Literature* 111.2 (1995): 257–272.

Stern, Sacha. "The Rabbinic New Moon Procedure: Context and Significance." In *Living the Lunar Calendar*. Edited by J. Ben-Dov, W. Horowitz, and J. M. Steele. 211–230. Oxford: Oxbow, 2012.

Stol, Marten. *Birth in Babylonia and the Bible: Its Mediterranean Setting*. Groningen: Styx, 2000.

Strange, James F. "Nazareth." In *The Anchor Bible Dictionary*. Edited by D. N. Freedman. 6 vols., 4:1050–1051. New York: Doubleday, 1992.

Strickland, Samuel. *Twenty-Seven Years in Canada West: or, The Experience of an Early Settler*. 2 vols. London: Richard Bentley, 1853. http://books.google.co.uk/books?id=WRSCASqkqmkC.

Strobel, A. "Weltenjahr, große Konjunktion und Messiasstern, Ein themageschichtlicher Überblick." *Aufstieg und Niedergang der Römischen Welt* 2.20.2 (1987): 988–1187.

Suetonius. *The Lives of the Twelve Caesars*. Edited and translated by J. C. Rolfe. Rev. ed. Vol. 2. Loeb Classical Library. Cambridge, MA: Harvard University Press, 1914.

Sukenik, Eleazar Lipa, and Steven Fine. *The Ancient Synagogue of Beth Alpha*. London: Oxford University Press, 1932.

Sumners, Carolyn, and Carlton Allen. *Cosmic Pinball: The Science of Comets, Meteors, and Asteroids*. New York: McGraw-Hill, 2000.

Sweeney, Marvin A. *Isaiah 1–39, with an Introduction to Prophetic Literature*. The Forms of the Old Testament Literature. Grand Rapids, MI: Eerdmans, 1996.

Syme, Ronald. *The Crisis of 2 B.C.* Munich: Verlag der Bayerischen Akademie der Wissenschaften, 1974.

Tacitus. *The Annals*. Translated by A. J. Woodman. Indianapolis: Hackett, 2004.

Talbert, Charles H. *What Is a Gospel? The Genre of the Canonical Gospels*. Philadelphia: Fortress, 1977.

Te Velde, Herman. *Seth, God of Confusion: A Study of His Role in Egyptian Mythology and Religion*. Leiden: Brill, 1977.

Thierens, A. E. *Astrology in Mesopotamian Culture: An Essay*. Leiden: Brill, 1935.

Thompson, M. E. W. "Israel's Ideal King." *Journal for the Study of the Old Testament* 7 (1982): 79–88. DOI: 10.1177/030908928200702405.

———. *Situation and Theology: Old Testament Interpretations of the Syro-Ephraimite War*. Sheffield: Almond Press, 1982.

Thompson, R. Campbell. *The Reports of the Magicians and Astrologers of Nineveh and Babylon in the British Museum*. Vol. 2. London: Luzac, 1900.

Thorley, John. "When Was Jesus Born?" *Greece and Rome*. Second Series 28.1 (April 1981): 81–89. DOI: http://dx.doi.org/10.1017/S0017383500033520.

Thureau-Dangin, F. *Tablettes d'Uruk à l'usage des prêtres du Temple d'Anu au temps des Séleucides*. Paris: Paul Geuthner, 1922.

Toomer, G. J. "Hipparchus and Babylonian Astronomy." In *A Scientific Humanist: Studies in Memory of Abraham Sachs*. Edited by Erle Leichty, Maria Ellis, and Pamel Gerardi. 353–362. Philadelphia: Occasional Publications of the Samuel Noah Kramer Fund, 1988.

Toomer, G. J., trans. *Ptolemy's Almagest*. Princeton, NJ: Princeton University Press, 1998.

Tromp, Johannes. *The Assumption of Moses: A Critical Edition with Commentary*. Leiden: Brill, 1993.

Trumbull, H. Clay. *Kadesh-Barnea: Its Importance and Probable Site*. Philadelphia: J. D. Wattles, 1895.

Tsevat, M. "bᵉthûlâh." In *Theological Dictionary of the Old Testament*. Edited by G. J. Botterweck, H. Ringgren, and H.-J. Fabry. Translated by J. T. Willis. Vol. 2, 338–343. Grand Rapids, MI: Eerdmans, 1977.

Turner, David L. *Matthew*. Baker Exegetical Commentary on the New Testament. Grand Rapids, MI: Baker Academic, 2008.

"Ulysses's Surprise Trip through Comet's Tail Puts Hyakutake in Record Books." http://www.ras.org.uk/news-and-press/70-news2000/377-pn00-07. Last modified May 8, 2012.

van der Horst, Pieter Willem, ed. *Chaeremon: Egyptian Priest and Stoic Philosopher. The Fragments Collected and Translated with Explanatory Notes*. Leiden: Brill, 1984.

van der Toorn, Karel. "Sun." In *The Anchor Bible Dictionary*. Edited by D. N. Freedman. 6 vols., 6:237–239. New York: Doubleday, 1992.

van der Waerden, H. *Science Awakening II*. Leyden, Netherlands: Noordhoff, 1974.

van Henten, Jan Willem. "Dragon Myth and Imperial Ideology in Revelation 12–13." In *The Reality of Apocalypse: Rhetoric and Politics in the Book of Revelation*. Edited by D. Barr. 181–203. Atlanta: Society of Biblical Literature, 2006.

van Unnik, W. C. "*Dominus Vobiscum*: The Background of a Liturgical Formula." In *New Testament Essays*. Edited by A. J. B. Higgins. 270–305. Manchester, England: Manchester University Press, 1959.

Vardaman, Jerry. "Jesus' Life: A New Chronology." In *Chronos, Kairos, Christos*. Edited by Jerry Vardaman and E. M. Yamauchi. 55–84. Winona Lake, IN: Eisenbrauns, 1989.

Veres, Peter, Leonard Kornos, and Juraj Toth. "Meteor Showers of Comet C/1917 F1 Mellish." *Monthly Notices of the Royal Astronomical Society* 412 (2011): 511–521.

Vermes, Géza. *The Nativity: History and Legend*. London: Penguin, 2006.

Vettius Valens. *Anthologies*. Translated by Mark Riley. http://www.csus.edu/indiv/r/rileymt/Vettius%20Valens%20entire.pdf. Last modified January 5, 2011.

von Humboldt, Alexander, and Aimé Bonpland. *Personal Narrative of Travels to the Equinoctial Regions of the New Continent during the Years 1799–1804*. Vol. 1. Translated by T. Ross. London: George Bell & Sons, 1907.

von Staufer, Maria Hubert. "Christmas in Poland." http://www.christmasarchives.com/wpoland.html. Last modified October 25, 2010.

Vsekhsvyatskii, S. K. *Physical Characteristics of Comets*. Jerusalem: Israel Program for Scientific Translations, 1964.

Walford, Edward, trans. *The Ecclesiastical History of Sozomen*. London: Henry G. Bohn, 1885. https://archive.org/details/ecclesiasticalh00walfgoog.

———. *Epitome of the Ecclesiastical History of Philostorgius*. London: Henry G. Bohn, 1855. http://www.tertullian.org/fathers/philostorgius.htm.

Wallace, Daniel. *Greek Grammar: Beyond the Basics*. Grand Rapids, MI: Zondervan, 1996.

Walton, John H. "Isaiah 7:14: What's in a Name?" *Journal of the Evangelical Theological Society* 30 (1987): 289–306. https://www.galaxie.com/article/jets30-3-04.

Warner, Brian. *Charles Piazzi Smyth, Astronomer-Artist: His Cape Years, 1835–1845*. Cape Town: A. A. Balkema, 1983.

Watson, F. G. *Between the Planets*. Rev. ed. Cambridge, MA: Harvard University Press, 1956.

Watson, James C. *A Popular Treatise on Comets*. Philadelphia: James Challen & Son, 1861.

Wegner, Paul. *An Examination of Kingship and Messianic Expectation in Isaiah 1–35*. Lewiston, NY: Edwin Mellen, 1992.

Weidner, Ernst F. "Eine Beschreibung des Sternenhimmels aus Assur." *Archiv für Orientforschung* 4 (1927): 73–85. http://www.jstor.org/discover/10.2307/41679031.

Weingarten, Judith. "The Magi and Christmas." http://judithweingarten.blogspot.com/2007/12/magi-and-christmas.html. Posted December 22, 2007.

Whipple, Fred L. *The Mystery of Comets*. Washington, DC: Smithsonian Institution Press, 1985.

Whipple, Fred L., and S. E. Hamid. "A Search for Encke's Comet in Ancient Chinese Records: a Progress Report." In *The Motion, Evolution of Orbits, and Origin of Comets*. Edited by Gleb Aleksandrovich Chebotarev, E. I. Kazimirchak-Polonskaia, and B. G. Marsden. 152–154. Dordrecht, Netherlands: Reidel, 1972.

White, Gavin. *Babylonian Star-Lore: An Illustrated Guide to the Star-Lore and Constellations of*

Ancient Babylonia. 2nd ed. London: Solaria, 2007.

Wijngaards, J. N. M. "The Episode of the Magi and Christian *Kerygma*." *Indian Journal of Theology* 16 (1967): 30–41. http://www.biblicalstudies.org.uk/pdf/ijt/16-1-2_030.pdf.

Wildberger, Hans. *Isaiah 1–12*. Translated by Thomas H. Trapp. Continental Commentary. Minneapolis: Augsburg Fortress, 1991.

Wilkins, Michael J. *Matthew*. Zondervan Illustrated Bible Backgrounds Commentary. Grand Rapids, MI: Zondervan, 2002.

Williams, John. *Observations of Comets, from B.C. 611 to A.D. 1640*. London: Strangeways & Walden, 1871.

Williams, Mary Francis. "The *Sidus Iulium*, the Divinity of Men, and the Golden Age in Virgil's *Aeneid*." *Leeds International Classical Studies* 2 (2003): 1–29. http://lics.leeds.ac.uk/2003/200301.pdf.

Wilson, R. Dick. "The Meaning of *'Almah* (A.V. 'Virgin') in Isaiah VII. 14." *Princeton Theological Review* 24 (1926): 308–316. http://journals.ptsem.edu/id/BR1926242/dmd006.

Witherington, Ben. *The Christology of Jesus*. Minneapolis: Augsburg Fortress, 1990.

Witts, R. E. *Isis in the Ancient World*. Baltimore: Johns Hopkins University Press, 1971.

Wolf, Herbert M. "A Solution to the Immanuel Prophecy in Isaiah 7:14–8:22." *Journal of Biblical Literature* 91 (1972): 449–456. http://www.jstor.org/discover/10.2307/3263678.

Yamauchi, Edwin M. "The Episode of the Magi." In *Chronos, Kairos, Christos*. Edited by Jerry Vardaman and E. M. Yamauchi. 15–39. Winona Lake, IN: Eisenbrauns, 1989.

———. *Persia and the Bible*. Grand Rapids, MI: Baker, 1990.

Yau, Kevin, Donald Yeomans, and Paul Weissman. "The Past and Future Motion of Comet P/Swift-Tuttle." *Monthly Notices of the Royal Astronomical Society* 266 (1994): 305–316. http://adsabs.harvard.edu/full/1994MNRAS.266..305Y.

Yeomans, Donald K. "Cometary Astronomy." In *History of Astronomy: An Encyclopedia*. Edited by John Lankford. 154–159. New York: Routledge, 1996.

———. *Comets: A Chronological History of Observation, Science, Myth, and Folklore*. New York: John Wiley, 1991.

———. "Great Comets in History." http://ssd.jpl.nasa.gov/?great_comets. Posted April 2007.

Yeomans, Donald K., and Paul W. Chodas, "Predicting Close Approaches of Asteroids and Comets to Earth." In *Hazards Due to Comets and Asteroids*. Edited by T. Gehrels. 241–258. Tucson: University of Arizona Press, 1995.

York, Thomas John. "The Reliability of Early East Asian Astronomical Records." PhD thesis, Durham University, 2003. http://etheses.dur.ac.uk/3080/.

Young, Edward J. *The Book of Isaiah: A Commentary*. 3 vols. Grand Rapids, MI: Eerdmans, 1965.

Zobel, Hans-Jürgen. "šēḇeṭ." In *Theological Dictionary of the Old Testament*. Edited by G. J. Botterweck, H. Ringgren, and H.-J. Fabry. Vol. 14, 302–311. Grand Rapids, MI: Eerdmans, 2004.

Zolli, I. "Il significato di '*shēbheṭ*' nel Salmo CXXV." In *Atti del XIX congress Internazionale degli Orientalisti*, 455–462. Rome: G. Bardi, 1938.

Sources for Carol Quotations

Chapter 1: "Star of Wonder"
John H. Hopkins, "We Three Kings."

Chapter 2: "We Beheld (It Is No Fable)"
George Ratcliffe Woodward, trans. "Shepherds, in the Field Abiding."

Chapter 3: "They Looked Up and Saw a Star"
William B. Sandys, "The First Noel."

Chapter 4: "What Star Is This?"
John Chandler, "What Star Is This, with Beams So Bright?"

Chapter 5: "What Sudden Radiance from Afar?"
Philipp Nicolai and Johann A. Schlegel. "How Brightly Beams the Morning Star."

Chapter 6: "A Stranger midst the Orbs of Light"
John Chandler, "What Star Is This, with Beams So Bright?"

Chapter 7: "Yon Virgin Mother and Child"
Joseph Mohr, "Silent Night, Holy Night."

Chapter 8: "With Royal Beauty Bright"
John H. Hopkins, "We Three Kings."

Chapter 9: "Lo, the Star Appeareth"
James A. Blaisdell, "Christians, Lo, the Star Appeareth."

Chapter 10: "Following Yonder Star"
John H. Hopkins, "We Three Kings."

Chapter 11: "Brightest and Best of the Sons of the Morning"
Reginald Heber, "Brightest and Best of the Sons of the Morning."

Chapter 12: "The Light Everlasting That Fades Not Away"
Eliza E. Hewitt, "Beautiful Star."

General Index

Scripture Index